Genome Organization
and Expression in Plants

NATO ADVANCED STUDY INSTITUTES SERIES

A series of edited volumes comprising multifaceted studies of contemporary scientific issues by some of the best scientific minds in the world, assembled in cooperation with NATO Scientific Affairs Division.

Series A: Life Sciences

Recent Volumes in this Series

Volume 21a – Chromatin Structure and Function
Molecular and Cytological Biophysical Methods
edited by Claudio A. Nicolini

Volume 21b – Chromatin Structure and Function
Levels of Organization and Cell Function
edited by Claudio A. Nicolini

Volume 22 – Plant Regulation and World Agriculture
edited by Tom K. Scott

Volume 23 – The Molecular Biology of Picornaviruses
edited by R. Pérez-Bercoff

Volume 24 – Humoral Immunity in Neurological Diseases
edited by D. Karcher, A. Lowenthal, and A. D. Strosberg

Volume 25 – Synchrotron Radiation Applied to
Biophysical and Biochemical Research
edited by A. Castellani and I. F. Quercia

Volume 26 – Nucleoside Analogues: Chemistry, Biology, and Medical Applications
edited by Richard T. Walker, Erik De Clercq, and Fritz Eckstein

Volume 27 – Developmental Neurobiology of Vision
edited by Ralph D. Freeman

Volume 28 – Animal Sonar Systems
edited by René-Guy Busnel and James F. Fish

Volume 29 – Genome Organization and Expression in Plants
edited by C. J. Leaver

Volume 30 – Human Physical Growth and Maturation
edited by Francis E. Johnston, Alex F. Roche, and Charles Susanne

The series is published by an international board of publishers in conjunction with NATO Scientific Affairs Division

A Life Sciences	Plenum Publishing Corporation
B Physics	New York and London
C Mathematical and Physical Sciences	D. Reidel Publishing Company Dordrecht and Boston
D Behavioral and Social Sciences	Sijthoff International Publishing Company Leiden
E Applied Sciences	Noordhoff International Publishing Leiden

Genome Organization and Expression in Plants

Edited by
C. J. Leaver
University of Edinburgh
Edinburgh, Scotland

PLENUM PRESS • NEW YORK AND LONDON
Published in cooperation with NATO Scientific Affairs Division

Library of Congress Cataloging in Publication Data

Nato Advanced Study Institute on Genome Organization and Expression in Plants, Edinburgh, 1979.
   Genome organization and expression in plants.

   (NATO advanced study institutes series: Series A, Life sciences; v. 29)
   Includes index.
   1. Plant genetics–Congresses. 2. Gene expression–Congresses. 3. Genetic regulation–Congresses. I. Leaver, C. J. II. North Atlantic Treaty Organization. Division of Scientific Affairs. III. Title. IV. Series.
QH433.N23 1979   581.1′5   79-28255
ISBN-13: 978-1-4613-3053-0   e-ISBN-13: 978-1-4613-3051-6
DOI: 10.1007/978-1-4613-3051-6

Proceedings of the NATO Advanced Study Institute on Genome Organization and Expression in Plants, held in Edinburgh, Scotland, United Kingdom, July 11–21, 1979.

© 1980 Plenum Press, New York
Softcover reprint of the hardcover 1st edition 1980
A Division of Plenum Publishing Corporation
227 West 17th Street, New York, N.Y. 10011

Preface

In the summer of 1976 a successful workshop on nucleic acids and protein synthesis in plant systems was organised in Strasbourg by Jacques Weil and Lawrence Bogorad. The participants in the workshop, were, without exception, excited both by the quality of the work discussed and by the rapid progress being made in several areas of genomic analysis and expression in plants. It also became apparent that there was a need for an international assembly of this sort at regular intervals. These workshops not only encourage stimulating discussion and constructive thinking but also result in increased collaboration and productive liaison between laboratories with common interests. Hence a ten-day advanced studies institute course was organised in Edinburgh from 11-21 July 1979, and in this volume we have published the contributions given by the invited speakers.

The subjects discussed covered most areas of plant molecular biology and the lecturers were asked to balance a review of their chosen subject with the results of their own recent research and likely future advances. Probably the most important technical advance since the previous meeting of this group in Strasbourg, was the application of restriction enzyme analysis and cloning techniques. This is illustrated in many of the published lectures and was the basis for many of the more informal discussion sessions.

The participation of a large number of highly qualified young investigators was particularly encouraging and suggests that in the near future we will see impressive progress in this important, but previously neglected area of molecular biology.

The organisation of this advanced course and the attendance of lecturers and participants from some twenty-eight countries would not have been possible without the generous financial support of the following organisations: the North Atlantic Treaty Organisation, the Federation of European Biochemical Societies, the International Union of Biological Sciences, Commission of the European Communities, the Underwood Fund of the Agricultural Research Council, the British Council, the United States Department of Agriculture, the National

Science Foundation, Imperial Chemical Industries, Monsanto Agricultural Products Co., Shell Research Ltd., Unilever Research Ltd., Tate and Lyle Ltd., the Radiochemical Centre and the University of Edinburgh.

I am particularly grateful to many colleagues and friends who gave their time, energy and experience in assisting with the organisation and running of the course. Finally I thank the speakers for their very considerable time and effort which made the task of organising the workshop and editing this volume a pleasure.

C. J. Leaver
Edinburgh

Contents

MACROMOLECULAR PROPERTIES, BIOSYNTHESIS AND GENETIC
REGULATION OF LEGUME SEED STORAGE PROTEINS

ORGANIZATION AND EXPRESSION OF THE
CHLOROPLAST GENOME

CONTRASTING PATTERNS OF DNA SEQUENCE ORGANIZATION IN PLANTS[*],[**]

W. F. Thompson,[†] M. G. Murray and R. E. Cuellar[†]

Department of Plant Biology
Carnegie Institution of Washington
Stanford, California 94306

One of the major outstanding questions in the recent history of molecular biology concerns the origin and significance of repetitive DNA sequences in eukaryotic genomes. Although higher plant species have not yet received much concentrated attention in this context, an analysis of what data is available shows that plants frequently have much more repetitive DNA than most animals (reviewed in Thompson and Murray, 1979). Therefore it seems logical to view plant genomes as particularly favorable material for studies on the origin and evolution of repeated sequences.

Variation in genome size is also more pronounced among plants than among animals, and there is very little correlation between DNA content and organismic complexity. For example, data summarized by Bennett and Smith (1976) indicate at least a 36-fold range in the Leguminosae alone, and up to about seven-fold variation within a single genus. Since there is probably a reasonably direct

_____

*This work was supported in part by grants from the National Science Foundation (PCM 7705656) and the Competitive Grants Office of the Science and Education Administration, U.S. Department of Agriculture (5901-0410-8-0009-0). R.E.C. received support from a Ford Foundation predoctoral fellowship.

**Publication No. _671_ from the Carnegie Institution of Washington, Department of Plant Biology.

†Also from the Department of Biological Sciences, Stanford University.

relationship between repetitive DNA content and changes in genome size, one good way to approach the problem of the origin and evolution of repetitive DNA is to carry out comparative studies on organisms with widely differing DNA contents. Such comparisons would be especially interesting if the species involved were closely related.

SEQUENCE REPETITION AND DIVERGENCE IN PEA AND MUNG BEAN DNA

The two species we have been studying offer just such a comparison. Both the garden pea (*Pisum sativum*) and the mung bean (*Vigna radiata*) are members of the same legume subfamily (Faboideae), but the haploid genome of the pea contains about 4.6-4.8 pg of DNA while that of the mung bean has only 0.48-0.53 pg (Murray et al., 1978, 1979).* Most eukaryote genome sizes fall between these two values, with a major peak in the vicinity of about 1 pg (Hinegardner, 1976). Figures 1 and 2 show that the pea genome also contains a larger fraction of repetitive DNA than does the mung bean. At a length of 300 nucleotides, about 85% of the pea DNA fragments but less than 50% of the mung bean fragments reassociate with repetitive kinetics. We will see later that the mung bean value is an overestimate, since it includes the ca. 11% chloroplast sequences present in the leaves from which the DNA was isolated (Palmer et al., 1979). After subtracting the chloroplast component, the fraction of nuclear repetitive DNA in mung beans is less than half of that in peas, and the total mass is about 20 times smaller.

It is also clear from a comparison of the data in Figures 1 and 2 that highly repetitive sequences make up a much larger fraction of the pea genome than they do in mung bean DNA. The dashed lines illustrate theoretical second order components which can be fitted to the data by least squares procedures (Pearson et al., 1977). While an adequate fit for mung bean DNA requires only one repetitive component with a copy number of about 300, the fit to the pea data requires a very large component at about 10,000 copies. Thus a very large portion of the pea genome is composed of a relatively small number of different kinds of sequences repeated many times. By way of illustration, we may calculate that if the high frequency component shown in Figure 1 were composed of sequence elements averaging 500 nucleotides in length (Murray et al., 1978) it would include only about 400 families of different sequences.

---

*In each case, the first value given is based on a careful analysis of reassociation kinetics using internal *E. coli* DNA rate standards, while the second was determined cytophotometrically by Dr. M. D. Bennett.

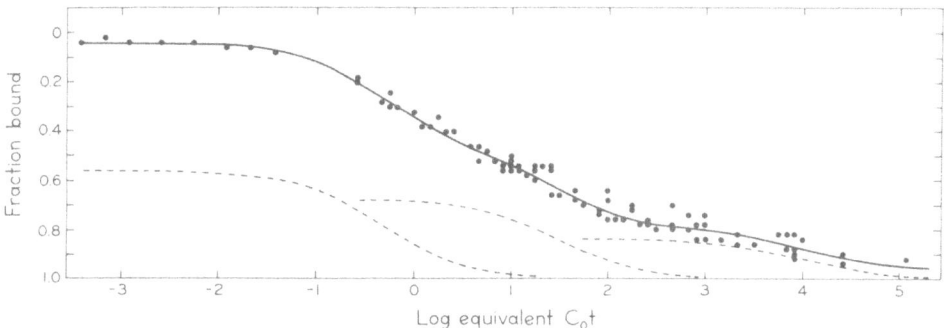

Fig. 1. Reassociation kinetics of short pea DNA fragments. DNA sheared to a modal single strand length of 300 nucleotides was reassociated and the fraction of fragments able to bind to hydroxyapatite (i.e., containing duplex regions) plotted as a function of C_0t. The dashed lines represent elements of the best three-component model describing the data (see text). From Murray et al. (1978).

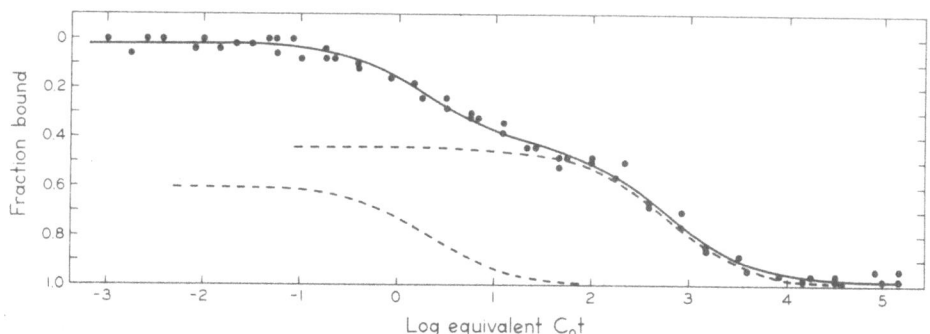

Fig. 2. Reassociation of short fragments of mung bean leaf DNA. Total cell DNA from mung bean leaves was sheared and reassociated as in Fig. 1. The dashed lines are elements of the simplest model solution to the data. From Murray et al., (1979).

In addition to the large mass of DNA derived from high frequency repeats, the pea genome also contains a much larger number of low frequency repetitive families than we find in mung bean DNA. If we again use the components in Figures 1 and 2 for purposes of illustration, we find that the 300-copy component of pea repeats would contain about 10,000 families of sequences 500 nucleotides long. Since this fraction of the mung bean genome contains less DNA and the average size of repetitive sequences is somewhat greater, (Murray et al., 1979) a similar calculation yields less than 1,000 different families of repeats in mung beans. In summary, we may conclude that pea DNA differs greatly from mung bean DNA both in maximum repetition frequency and in the numbers of different sequences which have been amplified to produce repetitive families.

It is important to remember that the theoretical components used in these calculations come from simplified mathematical descriptions of the data, and do not necessarily correspond to individual entities. In both pea and mung bean DNA, detailed experiments (Murray et al., 1978, 1979) reveal a more heterogeneous distribution of frequency classes than is implied by component analysis. In the case of pea DNA, it is quite reasonable to suppose that repetition frequencies vary more or less continuously over a range of at least 1,000 fold. A similarly continuous distribution has been reported for a sample of cloned repeats from sea urchin DNA (Klein et al, 1978).

There is also considerable heterogeneity with respect to the precision with which different sequences are repeated. Typically, reassociated repetitive sequences in most eukaryotes melt about 10° below the T_m of native DNA, indicating an average of about 10% base pair mismatch in the products of random reassociation. Figure 3 shows that the melting temperatures of different duplexes in the repetitive fraction are broadly distributed, with some duplexes melting very close to the reassociation temperature and some exhibiting nearly the same thermal stability as native DNA. Thus some of the repetitive sequences must reassociate almost perfectly while others cover the entire range of mispairing permitted by the reassociation criterion.

These complex melting profiles may be modeled in terms of gaussian components with different degrees of mismatch (Cuellar et al., 1978). As in the case of the kinetic components used to model reassociation data, there is no reason to suppose that these thermal components necessarily represent discrete individual entities. However, they may indicate that repetitive families tend to be arranged into a few major groups with varying amounts of sequence divergence. One way to explain such a situation would be to assume that different groups of families arose during periods of relatively frequent amplification events in the evolutionary history of a species. After the initial amplification, the accumulation of random mutations would lead to sequence divergence among the members of a given family, so that less homology would exist in the older families than in those created more recently. The fact that amplification events have occurred periodically during the evolution of plant genomes has been demonstrated directly by Flavell et al. (1977) and can often be inferred from indirect evidence (e.g., Stein et al., 1979).

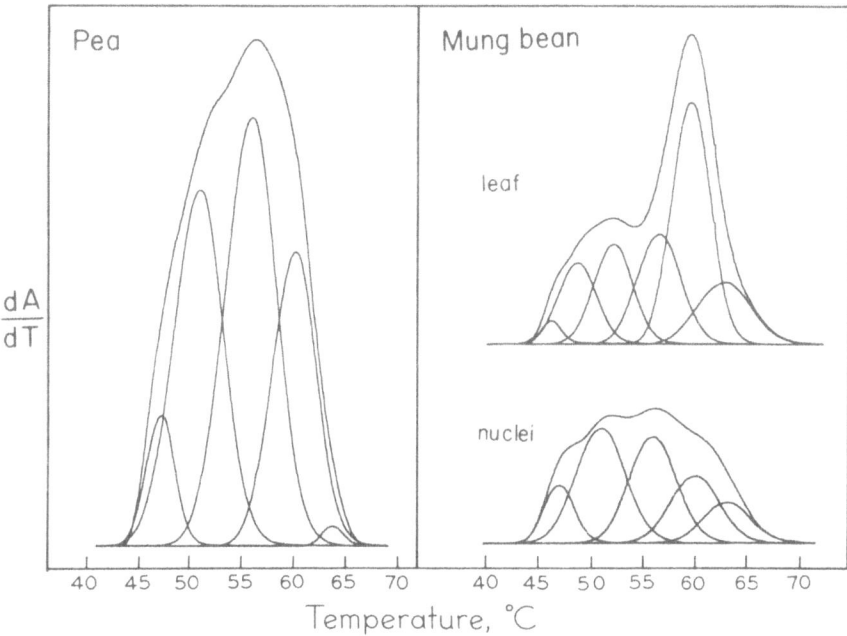

Fig. 3. First derivatives of optical thermal denaturation profiles
for reassociated repetitive sequences from the pea and
mung bean genomes. DNA fragments were reassociated and
melted in 2.4 M tetraethylammonium chloride ("TEACl") to
eliminate the effect of base composition on thermal denaturation and thus achieve a higher level of resolution than is
available in conventional buffers. In all cases, reassociation was carried out at 45°C. Results were analyzed by
higher derivative procedures to provide a basis for fitting
the theoretical Gaussian components as described by
Cuellar et al. (1978). The curves are scaled so that the
area under each is proportional to the fraction of nucleotides paired at the end of reassociation.

Figure 3 also shows thermal stability profiles for repetitive
duplexes from mung bean DNA. In this case, chloroplast DNA
sequences contribute about 25% of the total leaf "repetitive" fraction and reassociate to form essentially perfect duplexes (Preisler
and Thompson, 1978; Palmer et al., 1979). However, when nuclear
DNA is used the resulting profile is qualitatively similar to that
for pea DNA. Thus although the repetitive fractions of pea and
mung bean DNA differ greatly in size and repetition frequency, the
distribution of sequence divergence among the repeats is similar
in both genomes.

INTERSPERSION

Interest in the linear arrangement of repetitive and single copy DNA was greatly stimulated in 1969, when Britten and Davidson proposed that repetitive sequence elements interspersed at intervals among single copy sequences might function in gene regulation. Subsequent studies have demonstrated extensive interspersion of short (*ca*. 300 nucleotide) repetitive sequences at intervals of about 2000 base pairs or less in the DNA of a wide variety of unrelated organisms, including higher plants (e.g., Davidson et al., 1975; Thompson and Murray, 1979). It is sometimes assumed that the widespread occurrence of short period interspersion indicates that sequence arrangements necessary for gene regulation are conserved in evolution. However, a number of exceptions to this general pattern have been found, and a consideration of the range of variation in repetitive DNA content - especially among higher plants - suggests that such specific hypotheses are still premature (reviewed in Thompson and Murray, 1979).

From the data reviewed in the previous section it can be seen that the pea and mung bean genomes offer a unique opportunity to examine interspersion patterns in two closely related species differing greatly in both total and repetitive DNA content. Through such studies we can obtain a better indication of the limits within which interspersion patterns may vary, and, once the general patterns have been established, it will be possible to compare the organization of transcribed sequences with the overall organization in species whose gene regulatory mechanisms should be quite similar.

The ability of hydroxyapatite to bind DNA molecules which are only partially double stranded is a powerful tool in the study of repetitive sequence interspersion. Since the formation of a duplex in the repetitive portion of an otherwise single copy sequence fragment will cause the entire fragment to bind, it is possible to assess the fraction of DNA fragments of any given length which contain some portion of a repeated sequence. When the fraction of fragments bearing a repetitive element is measured as a function of fragment length, it is possible to estimate the length distribution for single copy sequences between repeats (Davidson et al., 1973).

In practice, tracers of different lengths are usually incubated with short unlabeled "driver" DNA to a C_0t value at which only repetitive sequences have reassociated in the driver DNA, and the fraction of tracer fragments binding to hydroxyapatite is taken as a measure of the fraction containing repetitive elements. Although conceptually simple, this approach can be subject to severe - and frequently unrecognized - complications. Most of these complications arise from the fact that long tracers reassociate more rapidly than short ones containing the same sequences (e.g., Hinnebusch et al., 1978). Since long single copy sequences, long repeats,

or clusters of short repeats of similar frequency will reassociate
more rapidly as the fragment length is increased, a C_0t value which
gives an adequate separation between repeat and single copy fragments 300 nucleotides long may give very poor resolution for 1500
or 3000 nucleotide tracers.

This point can be illustrated by the pea DNA data in Figure 4.
The inset demonstrates length effects for partially reassociated
E. coli DNA fragments. The lower curve for pea DNA was constructed
from binding measurements at C_0t 50, when about 85% of the repeats
in the driver have reacted. This condition is comparable to those
used in most published reports of such experiments, and the slight
slope of the curve beyond 1000 nucleotides would normally be interpreted as indicating (by extrapolation to the Y-axis) that as much
as 13% of the genome is composed of single copy regions longer than
2500 nucleotides. However, when the fraction of unreacted repeats
is large in comparison to the amount of single copy DNA - as it is
here and in several other plant genomes - much of the putative "long
single copy" fraction may actually be composed of repetitive DNA.
This interpretation is supported by the fact that when the same
experiment is carried out at C_0t 680 the binding of longer tracers
remains constant at about 94%, a value which is close to the expected
maximum (Davidson et al., 1973). Thus these binding experiments
provide no evidence for *any* significant fraction of long single
copy DNA.

Instead, it appears that the dominant length of single copy
sequences in the pea genome is more like 300-400 nucleotides, indicated by the sharp break in slope which occurs in both sets of data
in Figure 4. At least for fragments longer than 150 nucleotides,
this result cannot be explained by failure of short duplexes to bind
to hydroxyapatite. This conclusion is supported by model experiments
with *E. coli* DNA, as well as by the fact that extrapolation of the
data between 150 and 300 nucleotides to zero fragment length yields
an estimate of base pairing which is in excellent agreement with
independent measurements of optical hypochromicity. The absence of
a significant fraction of single copy sequences longer than 1000
nucleotides has also been confirmed by more sensitive experiments
in which we analyzed the complete reassociation kinetics of long
tracers (Murray et al., 1978).

So far, we have established that pea plants contain a tremen-
dous amount of repetitive DNA, and that this DNA is interspersed at
shorter intervals throughout a larger fraction of the genome than
has yet been reported for any other plant or animal. Initial
attempts to analyze sequence interspersion in mung bean DNA by means
of binding curves similar to those in Figure 4 produced results
which appeared to indicate that most of the DNA in this species might
also be organized in a relatively short period pattern. However,
further investigation showed that most of the apparent short period

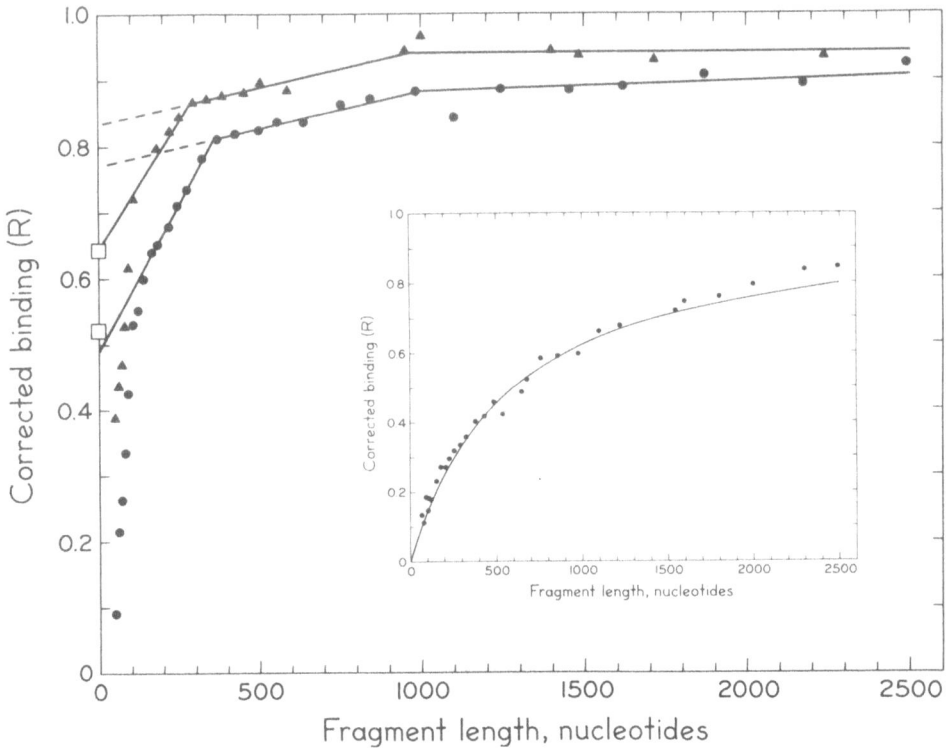

Fig. 4.   Hydroxyapatite binding of partially reassociated pea and
          E. coli DNA fragments of different lengths. Tracers were
          fractionated according to length in alkaline agarose gels
          and each fraction then reassociated with an excess of
          unlabeled 300 nucleotide driver fragments. The fraction
          bound to hydroxyapatite (after correcting for zero time
          binding) is plotted as a function of tracer length
          measured after reassociation. For pea DNA, experiments
          were conducted at both C_0t 50 (•) and C_0t 680 (▲). *E.
          coli* DNA was reassociated to C_0t 3.4 (inset). The fraction
          of base paired nucleotides (□) in pea DNA at each C_0t value
          was calculated from optical hyperchromicity data. From
          Murray et al. (1978).

interspersion in mung beans was actually attributable to the tracer
length effects mentioned above, and in fact it proved impossible to
resolve mung bean repetitive and single copy DNA well enough to
obtain meaningful binding curves. To analyze interspersion in mung
bean DNA we have therefore measured the complete reassociation
kinetics for tracers of several different lengths. Figure 5 shows
a series of experiments using this approach. About 65% of the

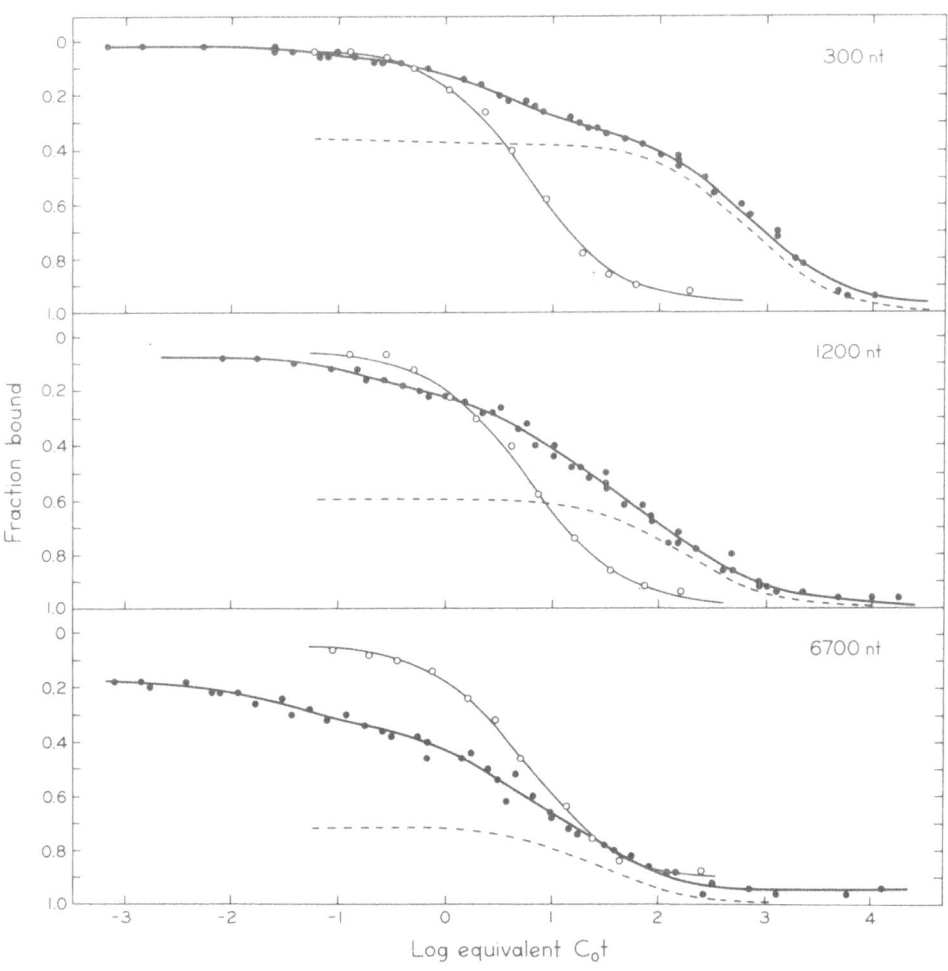

Fig. 5.   Reassociation kinetics of mung bean DNA tracers of different lengths.   Tracers of the indicated modal single strand
length were prepared from root DNA (in order to minimize
the contribution of chloroplast DNA sequences) and reassociated with an excess of unlabeled 300 nucleotide fragments.
Each C_0t curve was analyzed by fitting three second order
components, the rate constant of the slowest component
being fixed at the value expenced for single copy fragments of the indicated length (see text).   Dashed lines
illustrate the reassociation of this component. *E. coli*
DNA fragments (o) were included in each reaction to provide
an internal kinetic standard.   From Murray and Thompson
(1979).

fragments 300 nucleotides long reassociate with a rate constant of
1.6×10^{-3}, as expected for single copy sequences in a genome of
this size. Since the rate of tracer reassociation increases almost
in direct proportion to fragment length (e.g., Hinnebusch et al.,
1978), 1200 nucleotide single copy fragments will react with a rate
of about 6.4×10^{-3} and those 6700 nucleotides long with a rate of
3.6×10^{-2}. Least squares analysis of the data using single copy
components with these rate constants indicates that over 40% of
the 1200 nucleotide tracer fragments contain only single copy DNA,
while some 30% still show no repeats even at 6700 nucleotides.
Thus more than 60% (40%/65%) of the mung bean single copy sequences
are longer than 1200 nucleotides - a length at which we can detect
no single copy DNA at all in the pea genome - and nearly half
(30%/65%) have lengths in excess of 6700 nucleotides. No other
higher plant has yet been shown to contain such a large fraction of
its single copy DNA in a long period interspersion·pattern.

SEQUENCE ORGANIZATION AND GENOME EVOLUTION

        Figure 6 presents a summary comparison of the overall patterns
of repetition and interspersion in pea and mung bean DNA. It is
striking that the *4 billion* or so extra nucleotides in the pea
genome are all contained either in repetitive or short single copy
sequences. Since the majority of the repetitive sequences are also
short, it is possible to imagine that most of the single copy
sequences in pea DNA are actually members of ancient interspersed
repetitive families which have accumulated so many mutations that
we can no longer recognize them as repetitive. Such sequences
would be what we might call "fossil repeats." As noted above, the
distribution of sequence divergence among repetitive sequences is
quite broad, and it is perfectly reasonable to suppose that some of
the more diverged repeats would escape detection under standard conditions. Other authors have occasionally reported that the fractioi
of a genome behaving as repetitive sequences can be a sensitive
function of the reassociation criterion, especially in higher plant:
(e.g., Bendich and McCarthy, 1970; Bendich and Anderson, 1977;
Flavell and Smith, 1976). Although this fact seems not to be
generally recognized, it deserves to receive much more attention
in future studies.

        We have recently shown that even the "single copy" fraction of
pea DNA produces mismatched duplexes upon reassociation, and that
sequences showing single copy kinetics at the usual experimental
criterion of T_m-25° reassociate about 20 times faster when the
criterion is lowered to T_m-35°. Mung bean single copy tracers do
not show this effect to nearly the same extent, and a cDNA probe
complementary to polyadenylated messenger RNA of pea seedlings does
not show it at all (Murray and Thompson, in preparation; Peters et
al., 1979). We conclude that nearly all of the pea "single copy"

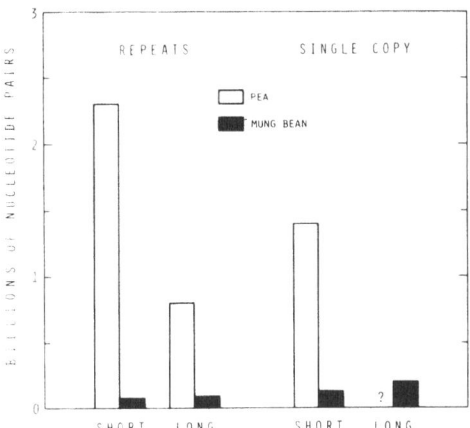

Fig. 6.   Comparison of the pea and mung bean genomes in terms of
          repetitive and single copy sequence lengths. The bars
          indicate the absolute amount of DNA in each genome which
          is composed of repetitive or single copy sequences shorter
          or longer than 1200 nucleotides, based on data presented
          in Murray et al. (1978, 1979). From Thompson et al. (1979).

DNA is actually composed of fossil repeats, and that coding sequences
transcribed into messenger RNA constitute so small a fraction as to be
undetectable in experiments with total DNA.

     The dramatic differences we see between pea and mung bean DNA
indicate to us that there is probably little or no direct relationship between sequence organization and organismic complexity or
developmental programming. Nor does it seem likely that any very
large fraction of the vast excess of highly interspersed DNA in peas
could be essential for processes such as chromosomal "housekeeping."
Especially in large genomes such as that of the pea plant, it is
likely that the majority of the DNA corresponds to what Hinegardner
(1976) has called "secondary DNA" - that is, DNA which has no direct
role in the immediate activities of the organism and is therefore
free to evolve rapidly.

     The evolution of secondary DNA must include large numbers of
amplification, translocation, mutation, and deletion events. In
some cases, random changes may be selected by virtue of a chance
association with other changes having a positive effect on the
phenotype (e.g., Schimke et al., 1978), while in other cases fixation might occur by genetic drift (e.g., Smith, 1976). In still
other cases there may be selection for increases or decreases in
DNA mass *per se*, regardless of its sequence (e.g., Bennett, 1972,

1977). Whatever the detailed mechanisms may be, it appears that
these essentially stochastic processes probably account for much of
the variation we see in the size and organization of the larger
eukaryotic genomes (reviewed by Thompson and Murray, 1979).

Figure 7 shows that a remarkably consistent relationship can be
demonstrated between the amount of single copy DNA (defined kinetic-
ally under standard criterion conditions of T_m-25°) and total genome
size in different plant and animal species. The apparent consistency
of this relationship is surprising, as is the clear difference
between plant and animal genomes which becomes apparent above about
10^9 nucleotide pairs of total DNA. Since there is good evidence

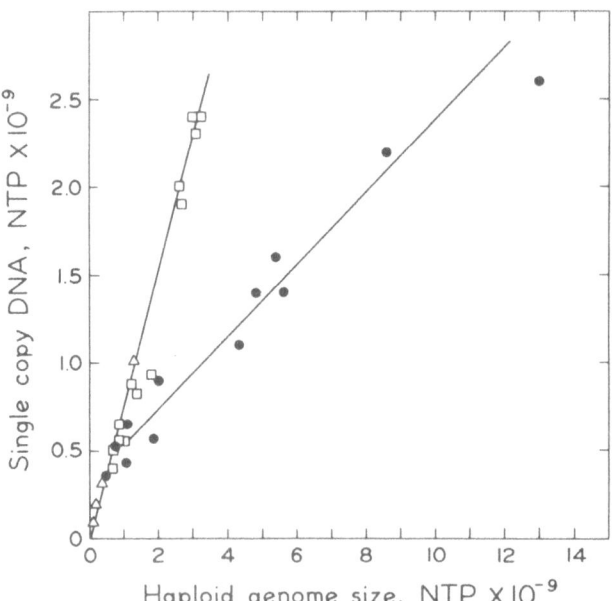

7. Single copy DNA content as a function of total genome size
    in plants (o) and animals (□, Δ). All estimates of single
    copy content are based on measurements using techniques
    which minimize the effect of different patterns of inter-
    spersion. From Thompson (1978).

that amplification events occur frequently in the course of evolution and that much of the single copy DNA in organisms such as the
pea is actually composed of ancient repetitive sequences, it is
possible to imagine that the amount of what we normally regard as
single copy DNA is determined by a balance between the rates of
amplification, mutation, and deletion in the past history of a
genome. According to this hypothesis, one way in which we might
explain the different slopes of the curves for plant and animal
genomes in Figure 7 is to assume that amplification events in plants
tend to be more frequent and/or to add more repetitive DNA per unit
of "initial" DNA. Although highly speculative, this notion might
help to explain the fact that plant genomes as a group tend to be
larger and more variable in size than those of animals.

If these ideas are correct, it follows that the organization
and function of higher plant genomes should be studied in a context
that includes evolutionary as well as developmental perspectives.
This does not mean that functions such as gene regulation might not
have evolved to take advantage of particular patterns of sequence
arrangement, or that changes in sequence arrangement may not sometimes have profound consequences for gene regulation and developmental programming. However, if our understanding of repetitive
sequences is to proceed very far beyond its current rather primitive
level we believe that many more careful comparative studies must be
conducted, and the overall problem viewed in a much more explicitly
evolutionary perspective. It seems likely that more intensive
studies of higher plant genomes will be particularly rewarding in
this context.

REFERENCES

Bendich, A., and Anderson, R. S., 1977. Characterization of families
        of repeated DNA sequences from four vascular plants, *Biochemis-
        try* 16:4655-4663.
Bendich, A. J., and McCarthy, B. J., 1970, DNA comparisons among
        barley, oats, rye, and wheat, *Genetics*, 65:545-565.
Bennett, M. D., 1972, Nuclear DNA and minimum generation time in
        herbaceous plants, *Proc. Royal Soc. Lond.* B 181:109-135.
Bennett, M. D., 1977, The time and duration of meiosis, *Phil. Trans.
        Royal Soc. Lond.* B 277:201-226.
Bennett, M. D.,and Smith, J. B., 1976, Nuclear DNA amounts in
        angiosperms, *Phil. Trans. Royal Soc. Lond.* B 274:227-273.
Britten, R. J., and Davidson, E. H., 1969, Gene regulation for
        higher cells: a theory, *Science* 165:349-357.
Cuellar, R. E., Ford, G. A., Briggs, W. R., and Thompson, W. F.,
        1978, Application of higher derivative techniques to analysis
        of high resolution thermal denaturation profiles of reassoci-
        ated repetitive DNA, *Proc. Nat. Acad. Sci. U.S.A*, 75:6026-6030.

Davidson, E. H., Hough, B. R., Amenson, C. S., and Britten, R. J.,
    1973, General interspersion of repetitive with non-repetitive
    sequence elements in the DNA of *Xenopus*, *J. Mol. Biol.* 77:1-23.
Davidson, E. H., Galau, G. A., Angerer, R. C. and Britten, R. J.,
    1975, Comparative aspects of DNA organization in metazoa,
    Chromosoma (Berl.) 51:253-259.
Flavell, R. B., and Smith, D. B., 1976, Nucleotide sequence organi-
    zation in the wheat genome, *Heredity* 37:231-252.
Flavell, R. B., Rimpau, J., and Smith, D. B., 1977, Repeated sequence
    DNA relationships in four cereal genomes, *Chromosoma* (Berl.)
    63:205-222.
Graham, D. E., Neufeld, B. R., Davidson, E. H., and Britten, R. J.,
    1974, Interspersion of repetitive and non-repetitive DNA
    sequences in the sea urchin genome, *Cell* 1:127-137.
Hinegardner, R., 1976. Evolution of genome size, *in*: "Molecular
    Evolution," F. J. Ayala, ed., Sinauer Associates, Inc.,
    Sunderland, Massachusetts.
Hinnebusch, A. G., Clark, V. E., and Klotz, L. C., 1978. Length
    dependence in reassociation kinetics of radioactive tracer
    DNA, *Biochemistry* 17:1521-1529.
Klein, W. H., Thomas, T. L., Lai, C., Scheller, R. H., Britten,
    R. J., and Davidson, E. H., 1978, Characteristics of indi-
    vidual repetitive sequence families in the sea urchin genome
    studied with cloned repeats, *Cell* 14:889-900.
Murray, M. G., and Thompson, W. F., 1979, DNA sequence organization
    in the mung bean genome, *in*: "Year Book 78," (in press),
    Carnegie Institution of Washington, Stanford, California.
Murray, M. G., Cuellar, R. E., and Thompson, W. F., 1978, DNA
    sequence organization in the pea genome, *Biochemistry* 17:
    5781-5790.
Murray, M. G., Cuellar, R. E., and Thompson, W. F., 1979, DNA
    sequence organization in the mung bean genome, submitted
    for publication.
Palmer, J. D., Murray, M. G., and Thompson, W. F., 1979, Studies
    on chloroplast DNA of mung bean and pea, *in*: "Year Book 78,"
    (in press), Carnegie Inst. of Washington, Stanford, California
Peters, D. L., Murray, M. G., and Thompson, W. F., 1979, The use
    of cDNA for studies of transcribed sequences, *in*: "Year Book
    78," (in press), Carnegie Inst. of Washington, Stanford,
    California.
Preisler, R. S., and Thompson, W. F., 1978, Distribution of nucleo-
    tide sequence divergence among families of repetitive sequences
    in mung bean DNA, *in*: "Year Book 77," Carnegie Inst. of
    Washington, Stanford, California.
Pearson, W. R., Davidson, E. H., and Britten, R. J., 1977, A program
    for least squares analysis of reassociation and hybridization
    data, *Nucleic Acids Research* 4:1727-1735.
Schimke, R. T., Kaufman, R. J., Alt, F. W., and Kellems, R. F.,
    1978, Gene amplification and drug resistance in cultured
    murine cells, *Science* 202:1051-1055.

Smith, G. P., 1976, Evolution of repeated DNA sequences by unequal crossover, *Science* 191:528-535.

Stein, D. B., Thompson, W. F., and Belford, H. S., 1979, Studies on DNA sequences in the Osmundaceae, *J. Mol. Evol.*, in press.

Thompson, W. F., 1978, Perspectives on the evolution of plant DNA, *in*: "Year Book 77," Carnegie Inst. of Washington, Stanford, California.

Thompson, W. F., and Murray, M. G., 1979. The nuclear genome: Structure and function, *in*: "The Biochemistry of Plants," vol. 6, A. Marcus, ed., Academic Press, N. Y. (in press).

Thompson, W. F., Murray, M. G., Belford, H. S., and Cuellar, R. E., 1979, Patterns of DNA sequence repetition and interspersion in higher plants, *in*: "Genetic Improvement of Crops," I. Rubenstein, R. L. Phillips, C. E. Green, and B. G. Gegenbach, eds., University of Minnesota Press, Minneapolis, Minnesota (in press).

ON THE EVOLUTION AND FUNCTIONAL SIGNIFICANCE OF DNA

SEQUENCE ORGANIZATION IN VASCULAR PLANTS

Arnold J. Bendich and Bernard L. Ward

Department of Botany
University of Washington
Seattle, Washington 98195

We have investigated properties of repeated DNA sequences in vascular plants for the purposes of assessing the way in which they may be useful to the organism and for probing the evolution of the nuclear genome. In pursuit of the first goal we have compared DNA sequence organization both in a diverse group of land plants and in a group of species within a single genus. We reasoned that if a particular arrangement of repeated sequences is important in basic metabolic events, such as the regulation of gene expression, then we might find a common pattern among closely related plants and perhaps among distantly related ones as well. Regardless of the role of repeated sequences in the genome, the processes by which the size of the genome has grown during the evolution of vascular plants is of intrinsic interest. Although little experimental evidence is currently available on the mechanisms for genome growth and rearrangement we suppose have occurred, we do have theory from which to work. The proposal that long blocks of tandemly repeating sequences (sometimes recognized as a satellite band in density gradients) could be periodically created and serve as precursor material for families of dispersed repeated sequences was first made by Britten and Kohne in 1968. We show here that sequences homologous to a satellite DNA component are present in longer DNA fragments at the density of the mainband component in muskmelon DNA and thus provide evidence for the proposal of genome growth by sequence amplification and translocation. The DNA at satellite density in muskmelon is also shown to contain ribosomal RNA genes (rDNA) and a surprising large amount of mitochondrial DNA (mtDNA).

17

On Sequence Organization and its Functional Significance

     DNA was incubated to a point at which only sequences that
appear to be repeated have reassociated. The single stranded regions
on duplex-containing structures were then removed with Sl nuclease
and the length distribution of the remaining duplexes determined by
exclusion chromatography and by electrophoresis on agarose gels
(Figure 1). The fractionation of duplex lengths by gel filtration
on an agarose A-50 column is poor since all duplexes greater than
about 1200 nucleotides (1.2 kb, either in single or double stranded
form) elute together in the excluded volume. Furthermore the resolution of shorter lengths in the included fractions is also poor.
Nevertheless these elution profiles are presented to facilitate comparison with previously published A-50 profiles. From the electrophoretic measurements it can be seen that the length of repeated
sequences varies from about 0.1 to greater than 10 kb but there
appears to be no obvious pattern of length distribution common to
all of the plants. There is a large amount of long (5-10 kb) Sl-
resistant DNA in both barley and watermelon. It is likely that a
minor fraction of these long repeats represent organelle DNA and/or
rDNA. We show below that the large satellite DNA in muskmelon is
composed of rDNA, a simple repeating sequence that is quite long
(Bendich and Taylor, 1977) and a large amount of mtDNA. The chloroplast DNA (ctDNA) content in this total watermelon DNA preparation
from etiolated hypocotyls is 1.2% of total DNA (Lamppa and Bendich,
1979). The Sl-resistant DNA shown here represented 44% of the total
and thus ctDNA might account for $1.2/0.44 = 2.7\%$ of the watermelon
DNA depicted in Figure 1. Whereas about 3% of watermelon DNA is
present in a dense satellite (Ingle et al., 1973; our unpublished
results), total barley DNA has no detectable dense satellite and
the preparation of barley DNA from fully-greened seedling shoots
used here contains 1.1% ctDNA (Lamppa & Bendich, 1979). We do not
know the mtDNA content of either barley DNA or any of the other
plants represented in Figure 1, but the ctDNA contents of all the
plant DNA preparations used here (except watermelon) is probably
about 1% of the total (Lamppa & Bendich, 1979) and would not contribute appreciably to these data.

     Previous work with animal DNA has shown a relationship between
the length of repeated DNA sequence elements and the type of sequence organization present in the genome (Britten et al., 1976).
The repeated sequences that appear in the Sl assay at 0.2-0.4 kb
are said to represent those repeated elements in a "short period"
interspersion pattern consisting of short repeated sequences
flanked by sequences that appear not to be found elsewhere in the
genome when the criterion of reassociation is 25° below the melting
point of native DNA ($T_m - 25°$). These are the so-called "single-copy"

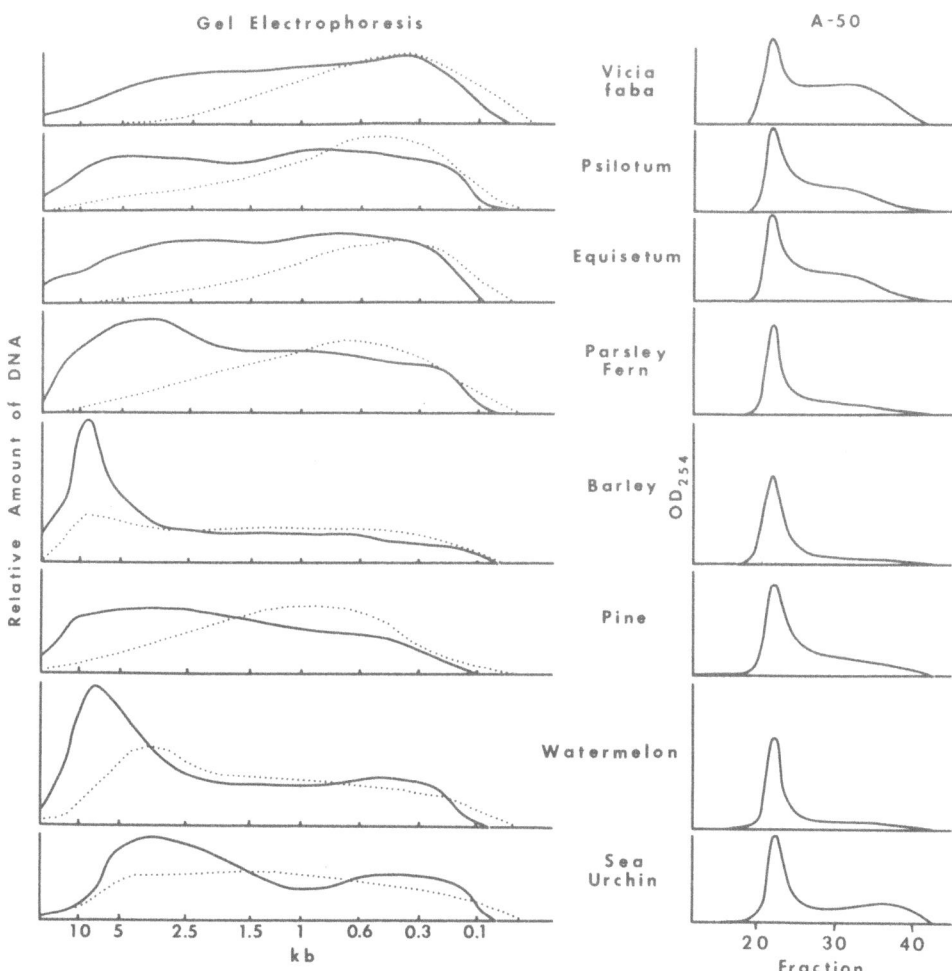

Fig. 1. Length distribution of repeated DNA sequences. Duplex
DNA fragments about 5 kb long were reassociated 3 hr to repetitive $C_o t$ values. Repeated sequences resistant to S1 nuclease
(digestion to maximum acid solubility) were analyzed by electrophoresis in 1% agarose gels at neutral (solid lines) and alkaline
(dotted lines) pH and by agarose A-50 chromatography. Since the
photographic response to the ethidium bromide (EB) stain is not
linear with increasing amounts of DNA, the densitometer tracings
have been adjusted using known amounts of sonicated DNA in other
lanes of the same gel to calibrate each gel. Lengths were determined using phage DNA restriction fragments. The A-50 column
eluant was monitored using a flow cell.

sequences[1]. However, as pointed out by R. Flavell at this conference (see his article in this volume), "compound" repeated sequences
that may be quite long in the genome may appear to be short after S1
digestion of reassociated repetitive DNA. This occurs when members
of repetitive sequence families are flanked by different sequences
in different locations in native genomic DNA. For example, the
sequence r_1 may be tandemly repeated in one location and alternate
with copies of r_2 in another. In addition, those repeated elements
flanked by other repeats in one location and by single-copy sequences
in another may be classified as short after reassociation and S1
digestion. The interpretation of the repeat length distribution of
S1-resistant repetitive sequences is therefore extremely difficult.
However we can say that dissimilar gel profiles for two plants do
indicate differences in the sequence organization within their
genomes and that the differences may be due to repeat:repeat and/or
repeat:single-copy sequence arrangements.

The length of about 0.3 kb is certainly represented in the
plants but is not particularly frequent and is not evident as a peak
in either the gel electrophoresis or A-50 profiles. Had plants been
so analyzed before animals, we would not have focused on a certain
length class of short repeats (the 300's) as particularly interesting. We would therefore either have missed a potentially important
aspect of repeated sequences or have determined that for plants the
short repeat length is not likely to be any more useful to genome
function than any other length class.

After finding no pattern in Figure 1 common to all the diverse
species, we turned to a group of species within a genus. Vicia was
chosen since the variation in genome size is large despite a haploid
chromosome set of 6 or 7 (and rarely 5). Thus we attempted to hold
the biological variable as close to constant as possible while
allowing the DNA content to vary. We reasoned that mechanisms used
for regulating genome expression and any other process fundamentally
important for viability would likely be the same among species in a
genus. The repeat sequence length distribution for the nine species
investigated is shown in Figure 2. It is interesting to note that
dasycarpa is considered in one taxonomic treatment to be a subspecies

---

[1]The portions of nuclear DNA that are classified as "repetitious"
and "single-copy" are strongly dependent upon the criterion of
reassociation (the number of degrees below the T_m of native DNA
used during measurement). This is an extremely important, though
usually ignored fact when considering possible functions for
repeated sequences. See Bendich and Anderson (1977) for a discussion of why almost all of the sequences regarded as chemically
similar by the investigator may not be functionally repetitious
for the cell.

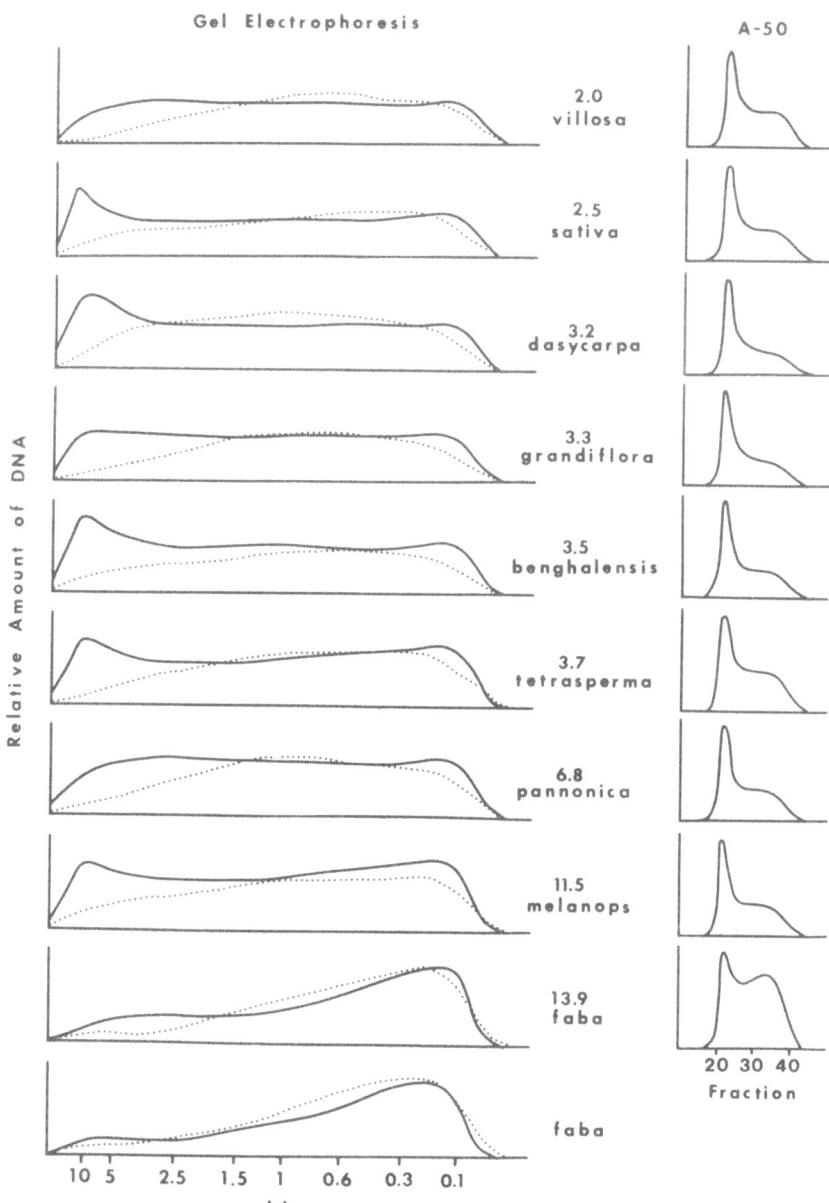

Fig. 2.   Length distribution of repeated DNA sequences for _Vicia_
species.   DNA fragments of 5 and 8 (bottom curve) kb (single
strand) for _faba_ and 7 to 13 kb for the other species were
treated as in Fig. 1.   The numbers show the haploid nuclear DN
content in pg (Bennett and Smith, 1976).

of villosa (Hermann, 1960) and in another is not delineated from
villosa at all (Kupicha, 1976). Donnelley and Clark (1962) were
able to produce some hybrid seedlings, but all lacked chlorophyll,
and thus the two might be considered as separate species on this
basis. Dasycarpa has a nuclear DNA content 1.6 times that of villosa,
and shows more long repeat sequence DNA than does villosa in Figure
2.

    There appears to be a strikingly different sequence organiza-
tion in V. faba as compared with all the other Vicia species examined. If we assume that all these Vicia species regulate the expression of genes in an essentially identical manner, then we must
conclude that the overall pattern sequence organization has little
or nothing to do with such a process. Of course there may be a
small fraction of sequences organized in the same fashion in all
these species including faba, and it is this minor fraction that is
responsible for crucial regulatory functions. However, in these
and all other sequence organization analyses using bulk repetitive
DNA (and this applies to essentially all such work with plants and
animals) the generalizations drawn and concepts developed must be
based on the available data. Thus if it is argued that models for
function of repetitive DNA are based on a minor fraction of DNA in
a pattern of organization other than that observed, then such models
are founded on conjecture rather than on experimental evidence.

    In conclusion it would appear that the pattern of sequence
organization observed in nuclear DNA may vary greatly without
affecting the proper regulation of gene expression or any other
process without which a plant cannot survive and reproduce. It
follows that although the further study of overall sequence organization may well enlighten us on the evolution of the genome, it is
not likely to be efficient in leading us to the function of repetitive DNA sequences.

On the Mechanism of Genome Growth and Rearrangement

    Muskmelon DNA contains a satellite in addition to a principal
or mainband component (Figure 5). The density-purified 1.706 g/cm^3
satellite contains three components: satellite I, satellite II and
rDNA. When this 1.706 DNA is cleaved with Eco RI, the rDNA is cut
into 5 major bands (one is a doublet) and one minor band, each of
which hybridizes with rRNA (Figure 3). Satellite I is not appreciably cut by Eco RI but is cleaved by Hind III yielding a prominent
band of 0.37 kb and minor bands at 2 and 3 times this length after
gel electrophoresis (Figure 10). The uncut satellite I was recovered from a gel, labeled by nick-translation and hybridized by
the blotting procedure to mainband DNA cleaved with Eco RI (Figure
4). A pattern of hybridization essentially identical to that shown
in Figure 4 was obtained when the 0.37 kb fragment produced by

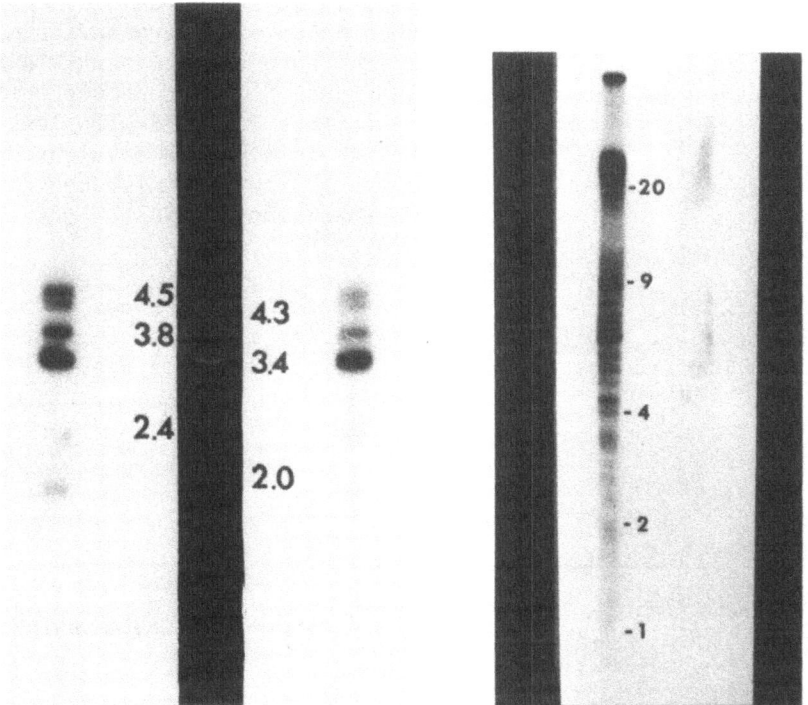

Fig. 3 (left). Blot-hybridization of ribosomal RNA with satellite DNA. The center lane shows EB stained products of muskmelon 1.706 satellite DNA digested with Eco RI; uncleaved satellite I is at the top of the lane. The outer lanes are autoradiograms of Eco RI digests of 1.706 DNA hybridized with kinase-labeled ^{32}P-rRNA (18S plus 25S) from zucchini squash. The numbers indicate the length in kb of the major and minor (3.8 kb) fragments.

Fig. 4 (right). Blot-hybridization of satellite I with mainband DNA. The probe is the uncleave DNA after digestion of 1.706 satellite DNA with Eco RI obtained from a gel similar to the center lane in Fig. 3. After removal from the gel, the uncleaved satellite I was labeled with ^{32}P by nick-translation. Muskmelon mainband DNA (left lane) and P22 phage DNA (right lane) were digested with Eco RI, electrophoresed and hybridized with the probe. Lengths in kb are indicated.

Hind III cleavage of 1.706 DNA was used as the nick-translated probe Thus satellite I shows the same pattern of blot-hybridization with mainband DNA whether cleaved or not cleaved by a restriction enzyme. The hybridization pattern in Figure 4 is not due to possible

contamination of our mainband fraction with 1.706 DNA because 1) the
hybridization of a satellite I probe with an Eco RI blot of 1.706
DNA looks not at all like the pattern in Figure 4, and 2) none of
the banding patterns of rDNA produced in an extensive series of Eco
RI digestions of 1.706 DNA ranging from none through partials to
complete digestion resemble that in Figure 4. We conclude that there
are short sequences homologous to satellite I linked with long molecules of mainband density. The linkage is likely a covalent one
since our earlier demonstration of short satellite I sequences in
long mainband molecules involved denaturation of DNA strands in an
aggregation assay (Bendich and Taylor, 1977).

We previously interpreted the presence of sequences in mainband DNA that are homologous with satellite I as representing a
stage in the dissemination of members of a family of dispersed and

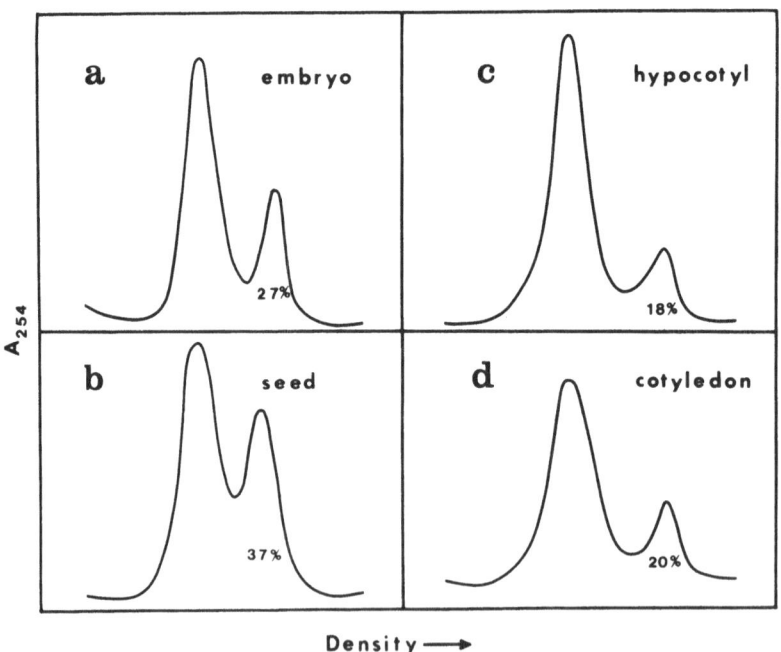

Fig. 5. CsCl gradients of DNA from muskmelon tissues. Dry seeds,
    embryos from dry seeds, and cotyledons and hypocotyls from 10
    day old etiolated seedlings were homogenized in sucrose buffer.
    The homogenate was made 1.5% in Triton X-100, centrifuged 10 min
    at 350 x g and DNA was extracted from the pellet.

intermediately repetitive DNA sequences from a source of tandemly repeating sequences (Bendich and Taylor, 1977). This amplification − translocation process is seen as a mechanism by which the genome grows. We plan to further investigate this evolutionary mechanism by examining the sequence environment of several mainband fragments that have presumably accepted a mobile satellite I sequence.

Satellite DNA in Different Tissues

The 1.706 satellite accounts for a different fraction of the DNA extracted from different muskmelon tissues (Figure 5). Pearson et al. (1974) previously noted such differences for melon DNA. We wanted to determine which component(s) of 1.706 DNA (satellites I, II and rDNA) were altered in amount in the tissues and used a melting analysis for this purpose. The upper part of Figure 6 shows

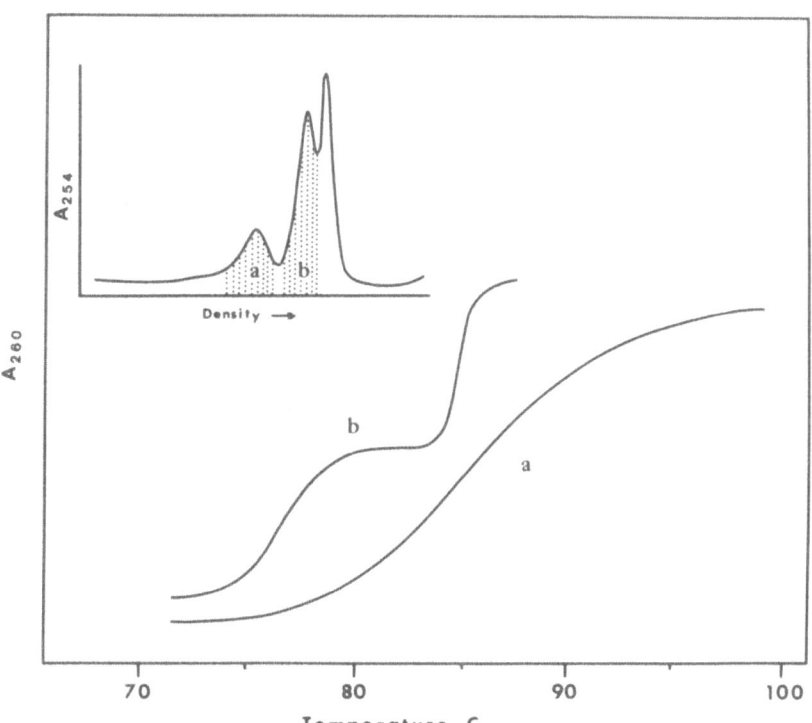

Fig. 6. Melting curves of components of muskmelon satellite DNA. The upper figure shows a Ag⁺−Cs₂SO₄ gradient of the 1.706 satellite DNA from etiolated hypocotyls. Hind III digestion of DNA from a similar gradient indicates that peak a is rDNA. Melting was in 26 mM Na phosphate, pH 6.8.

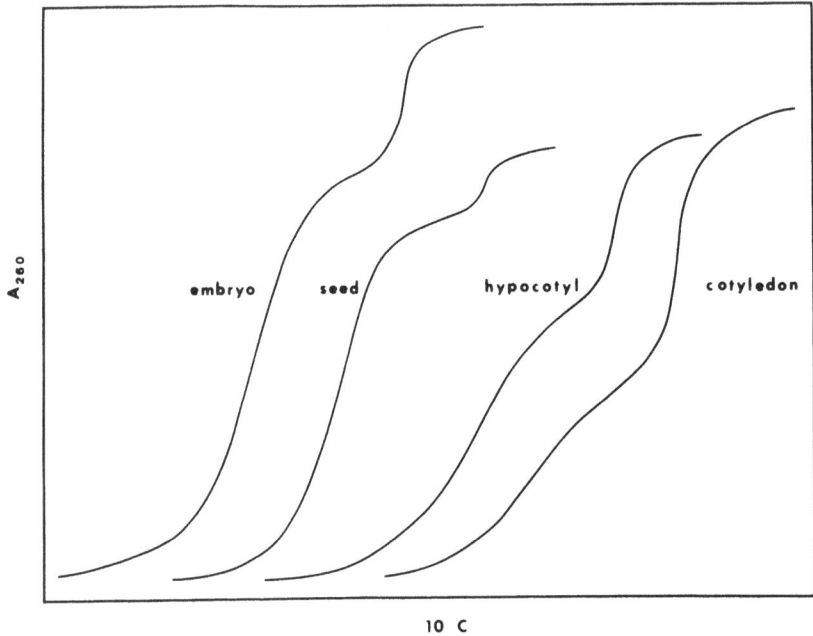

Fig. 7. Melting curves of muskmelon 1.706 satellite DNA from
    different tissues. The satellite peaks in Fig. 5 were melted in
    26 or 120 mM (for cotyledon) Na phosphate.

the 1.706 DNA fractionated on a $Ag^+-Cs_2SO_4$ gradient. Peak a contains rDNA and this DNA melts over a very broad temperature range.
Peak b contains satellites I and II; the very sharply melting
fraction is satellite I and the lower melting fraction is satellite
II. Taking the T_m values and breadths of each melting transition
as characteristic for each of the three components in 1.706 DNA, we
were able to resolve the melting profiles of 1.706 DNA from the
different tissues depicted in Figure 7. Multiplying the fraction of
the total DNA in the 1.706 satellite (Figure 5) by the fraction
each component represents in 1.706 DNA gives the percentage each
component contributes to the total DNA. In Figure 8 we see that
the rDNA content does not change among these four tissues. Ingle
and Sinclair (1972) using saturation hybridization as the assay
similarly found no evidence for rRNA gene amplification in somatic
cell DNA of maize and wheat. Germ cells have not yet been assayed.
The content of satellite I is seen to vary almost two-fold between
seed and cotyledon and this may represent a meaningful difference.
A most dramatic change is seen for satellite II in seeds compared
to the other tissues. Since the embryos we cut from dry seeds un-
avoidably contained contaminating seed cotyledon tissue, the
relatively high satellite II content observed may actually be due
to the satellite II-rich contaminant and not to embryo tissue.

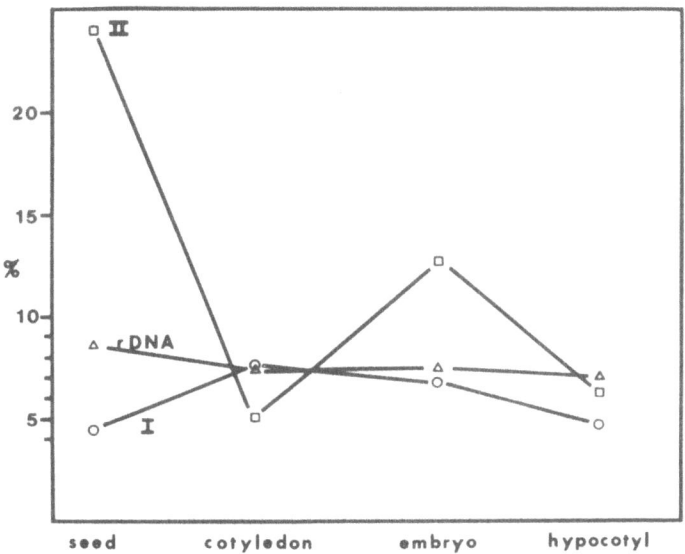

Fig. 8.   Satellite components as a percentage of total muskmelon
   DNA.   The curves in Fig. 7 were resolved into their components
   using the melting characteristics of each component from Fig. 6.
   The fraction each component represents in total DNA was calculated using the data in Fig. 5.

We next turn to the identity of satellite II.

Satellite II is Mitochondrial DNA

     Some time ago we reported that the slowly reassociating frac-
tion of 1.706 DNA had a kinetic complexity of 11×10^8 daltons
(Bendich and Anderson, 1974; Bendich and Taylor, 1977).   This was
a spectrophotometric measurement on the entire 1.706 DNA.   Sinclair
et al. (1975) subsequently reported a value of 18×10^8 for the
slowly reassociating component of melon satellite DNA, again with
the optical method after first enriching for this component.   Figure 9 shows the hydroxyapatite assay of reassociation kinetics for
satellite II purified by $Ag^+-Cs_2SO_4$ gradient centrifugation.   The
kinetic complexity is about 6×10^8.

     Mitochondrial DNA prepared from hypocotyls and from cotyledons
of 10-day old etiolated muskmelon seedlings showed a single, uni-
modal peak at 1.705 g/cm^3 by analytical ultracentrifuge analysis.
There was no peak whatsoever at 1.692 g/cm^3, the density of mainband (presumably nuclear) DNA.   The mtDNA and 1.706 satellite DNA
from hypocotyls were digested with Hind III and the resulting

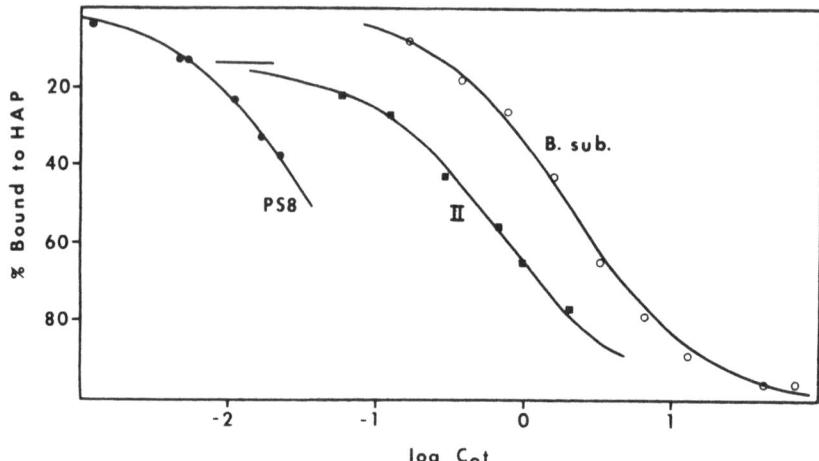

Fig. 9. Reassociation kinetics of muskmelon satellite II. The
   kinetics for ^{3}H-labeled DNA (about 1.1 kb fragments) were measured using hydroxyapatite at $T_m-25°$ in buffer with 1 M NaClO$_4$
   (II and B. subtilis) or 0.2M monovalent cation (PS8). Equivalent C_0t values in 0.2M cation are shown. The zero-time bound
   ^{3}H for II, indicated by the horizontal line, was 14% (4% for PS8).
   Ideal second order curves drawn are 100% component curves for PS8
   and B. subtilis, and 85% for II. The kinetic complexity of II is
   6.0×10^8 daltons based on 21×10^8 for B. subtilis or 5.2×10^8 based
   on 33×10^6 for PS8. Satellite II was taken as the dense peak
   from a Ag$^+$-Cs$_2$SO$_4$ gradient similar to that in Fig. 6 of Bendich
   and Anderson (1974).

fragments were separated on a gel (Figure 10). The 1.706 DNA digest
shows the satellite I bands at 0.37 and 0.74 kb and the rDNA band at
11 kb. In addition there are a series of less prominent bands that
are mirrored in the mtDNA digest. The precise correspondence of the
many bands and interband regions leaves little doubt that satellite
II is, in fact, mtDNA.

     From the results in Figure 8 we see that dry seeds (90% cotyle-
dons, 10% embryo by weight) are a rich source of mtDNA. It should be
noted that DNA represented in Figure 8 was prepared from a crude
"nuclear" pellet after the homogenate was brought to 1.5% Triton X-100
and centrifuged 10 minutes at 350 x g. Such a pellet is apparently
heavily contaminated with mtDNA. in previous work (Bendich and
Anderson, 1974; Bendich and Taylor, 1977; and numerous unpublished
observations) with DNA extracted from hypocotyl or entire seedling
tissue homogenates without first preparing a "nuclear" fraction,
satellite II represented some 10-15% of the total DNA. It is also of
interest that watermelon DNA contains a satellite at 1.707-1.708
g/cm^3 but this satellite represented only about 3% of the DNA

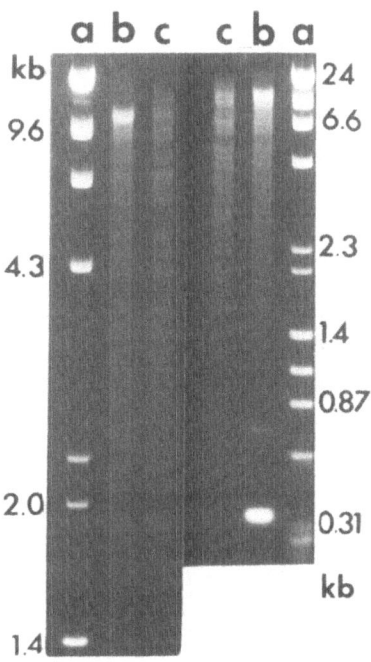

Fig. 10.   Restriction of 1.706
satellite and mtDNA from muskmelon hypocotyls.   The mtDNA
(lanes c) and satellite DNA
(lanes b) were digested with
Hind III.   Lanes a contain λ
DNA cut with Hind III and ϕX-
174 RF DNA cut with Hae III.
The 0.7% agarose gel was
stained and photographed
(right) and after continued
electrophoresis, was again
photographed (left).

extracted from leaf tissue by Ingle, et al. (1973) and from etiolated
hypocotyls (our unpublished results); and some of this is probably
rDNA.   Several other cucurbit species show from 0-6% of their DNA in
the 1.706 density region and for many other plants there is no detect
able band at this density (Ingle, et al., 1973; our unpublished results).   Since the density of mtDNA from higher plants, including
muskmelon and cucumber (Vedel, 1975) is about 1.705-1.706 g/cm^3 (F.
Quetier, personal communication; see his article in this volume) it
would appear that mtDNA can represent a widely different percentage
of bulk DNA extracted from different plants.

References

Bendich, A. J., and Anderson, R. S., 1974, Novel properties of satellite DNA from muskmelon, Proc. Nat. Acad. Sci. USA, 71:1511.

Bendich, A. J., and Anderson, R. S., 1977, Characterization of families of repeated DNA sequences from four vascular plants, Biochem., 16:4655.

Bendich, A. J., and Taylor, W. C., 1977, Sequence arrangement in satellite DNA from the muskmelon, Plant Physiol., 59:604.

Bennett, M. D., and Smith, J. B., 1976, Nuclear DNA amounts in angiosperms, Phil. Trans. Royal Soc. Lond., series B, 274:28.

Britten, R. J., Graham, D. E., Eden, F. C., Painchaud, D. M., and Davidson, E. H., 1976, Evolutionary divergence and length of repetitive sequences in sea urchin in DNA, J. Mol. Evol., 9:1

Britten, R. J., and Kohne, D. E., 1968, Repeated sequences in DNA. Science, 161:529.

Donnelly, E. D., and Clark, E. M., 1962, Hybridization in the genus Vicia, Crop Sci., 2:141.

Hermann, F. J., 1960, "Vetches in the United States - native, naturalized, and cultivated," U.S. Dept. Agric. Handbook No. 168.

Ingle, J., Pearson, G. G., and Sinclair, J., 1973, Species distribution of nuclear satellite DNA in higher plants, Nature New Biol., 242:193.

Ingle, J., and Sinclair, J., 1972, Ribosomal RNA genes and plant development, Nature, 235:30.

Kupicha, F. K., 1976, The infrageneric structure of Vicia, Notes Royal Bot. Gard., Edinb., 34:287.

Lamppa, G. K., and Bendich, A. J., 1979, Chloroplast DNA sequence homologies among vascular plants, Plant Physiol., 63:660.

Pearson, G. G., Timmis, J. N., and Ingle, J., 1974, The differential replication of DNA during plant development, Chromosoma, 45:281.

Sinclair, J., Wells, R., Deumling, B., and Ingle, J., 1975, The complexity of satellite deoxyribonucleic acid in a higher plant, Biochem. J., 149:31.

Vedel, F., 1975, Purification and quantitative changes of mitochondrial DNA in etiolated cucumber seedlings, Planta, 125:171.

PLANT DNA: LONG, PURE AND SIMPLE

Arnold J. Bendich, Robert S. Anderson and
Bernard L. Ward

Department of Botany, University of Washington
Seattle, Washington 98195, USA

The difficulties in obtaining undegraded plant nucleic acids free of contaminants are legendary. We describe a method for iso-lating DNA that is long and apparently free of polysaccharide con-taminants. The strategy for preserving DNA length is to initially prevent nuclease activity and to avoid shear forces. The strategy for purification is to pellet insoluble polysaccharides and float proteins as a first step and then separate the DNA from soluble polysacchardes using the buoyant effect of ethidium bromide (EB). Our purpose here is not to describe how to obtain DNA from one organelle free of DNA from any other organelle. For that objective either purified organelles are first prepared or individual DNA components in a mixture are resolved by physical methods such as density gradient centrifugation.

Crude nuclei are prepared by homogenizing soft tissue in the cold in buffers containing an osmoticum using a tissue homogenizer or mortar and pestle and centrifuging for 10 min at 350 x g (to discriminate against plastids and mitochondria) or at 10,000 x g (to include the smaller organelles and chromatin). Seeds can either be similarly homogenized or ground dry in an inexpensive coffee grinder or mill used to make flour from grain. Stored, fro-zen tissue can be ground to a powder at dry ice temperature by usin the frozen tissue and solid CO_2 in a chilled coffee grinder or by homogenizing in liquid nitrogen. The centrifuged pellet will con-tain nuclei and if 10,000 x g centrifugation is used, it will also contain chromatin and small organelles. If 1.5-4% Triton X-100 is added to the homogenate, the pellet after 350 x g centrifugation will contain DNA from the small organelles (likely bound to mem-branes) as well.

DNA is released from the pellet by resuspending in buffer with osmoticum, bringing to about 20 mM EDTA and 1% sarkosyl (1) at $0\text{-}2^{\circ}$ and rocking and rolling the vessel gently and intermittently for 5 min at $0\text{-}2^{\circ}$ until the lysate appears uniformly dispersed. Next add sufficient CsCl so that DNA will band properly upon subsequent addition of EB (2-4). Before adding the EB, centrifuge the lysate containing CsCl at 17,000 x g for 15-30 min at 4°. The pellet will contain starch (and will be very large with seeds) and a protein pellicle will float. If the lysate is centrifuged before adding CsCl, most of the DNA will be found in the pellet. Remove the solution using a wide-bore tube to a preparative ultracentrifuge tube and now adjust the density using refractive index measurement. Add EB from a 10 mg/ml stock to 300-500 µg/ml and centrifuge 15-20 hr at about 40,000 rpm at $20\text{-}25^{\circ}$. Using an ultraviolet light source locate the fluorescent band of DNA and remove it (5). The DNA is added to a tube of 60% saturated CsCl without additional EB and the centrifugation repeated. The 15-20 hr of centrifugation is insufficient to produce equilibrium banding, but adequate to separate DNA from the polysaccharide (6). EB may be removed from the DNA in CsCl by alcohol extraction or by use of a Dowex-50W column (7).

Notes

(1)   Sodium lauryl sulfate is not very soluble in CsCl in the cold, so use sodium lauryl sarkosinate: "sarkosyl".

(2)   Optical grade CsCl is not needed here. We use a relatively inexpensive technical grade from Kawecki Berylco (220 E. 42nd St., N.Y.; about $35/pound in 1979). Before dialyzing the DNA to its final solvent we dialyze against 1 mM EDTA to remove possible metal contaminants in the technical grade CsCl.

(3)   Use either solid CsCl or, preferably, 100% saturated CsCl so that the final concentration is about 60% saturated CsCl [e.g. 6 ml of completely (100%) saturated CsCl plus 4 ml of lysate and EB]. The final density is 1.57 g/cm^3. Do not rely on refractive index measurement for density estimation in this crude mixture.

(4)   We feel that it is important to get the lysate into CsCl quickly to minimize nuclease activity and do not use proteolytic enzymes, ethanol precipitation, chloroform, phenol or ribonuclease (RNA pellets in CsCl).

(5)   The fluorescent band can be removed by side-puncture of the tube. However, a better method is to use a syringe attached to a wide-bore needle (e.g. 15 gauge, 1.8 mm inside diameter) bent to a J shape. Insert this hooked needle through the fluorescent band in the gradient and withdraw the band from below.

(6) The polysaccharide is likely to be pectin according to Karl
     Jakob. To separate pectin from RNA see K.M. Jakob, J. Gressel,
     and I. Tal, Removal of pectins from methylated rRNA and pre-
     cursors by LiCl, Analyt. Biochem. 63:457 (1975).

(7) Use two extractions with n-butanol or a Dowex column as in R.
     Radloff, W. Bauer, and J. Vinograd, A dye-buoyant-density
     method for the detection and isolation of closed circular
     duplex DNA: the closed circular DNA in HeLa cells, Proc. Nat.
     Acad. Sci. USA, 57:1514 (1967).

(8) If the DNA accumulates a brown pigment (presumably due to
     polyphenoloxidase activity), we include 0.1 M diethyldithio-
     carbamate, disodium in the homogenization buffer. This agent
     competes with oxygen for the copper in the enzyme.

(9) Polysaccharides can be removed from previously "purified" DNA
     by the EB banding procedure. Total tissue DNA can be prepared
     from powdered frozen tissue by alternate methods (such as
     digesting detergent extracts with proteolytic enzymes) but we
     recommend the EB banding step in addition to any other steps.

(10) Since polysaccharides do not absorb strongly at 260 nm, a con-
     siderable level of contamination may go undetected. With many
     "purified" total tissue DNA preparations from many vascular
     plants not subjected to EB banding we never observed greater
     than a 37% hyperchromic effect upon melting ($A_{100}°/A_{60}°$ =
     1.37). With EB banded DNA we can obtain about 42% hyper-
     chromicity. To test the affect of contaminants on the ability
     of plant DNA to serve as a substrate for restriction enzymes
     a comparison can be made of the gel patterns upon restriction
     of "purified" DNA before and after EB banding.

(11) Most of the mass of muskmelon DNA prepared from a crude
     nuclear pellet by the EB banding method was in linear mole-
     cules greater than 150,000 base pairs as viewed in the electron
     microscope.

THE EVOLUTION OF PLANT GENOME STRUCTURE

R. Flavell, J. Rimpau, D.B. Smith, M. O'Dell and
J.R. Bedbrook

Dept. of Cytogenetics, Plant Breeding Institute
Cambridge CB2 2LQ

INTRODUCTION

In the past few years, data on the general sequence organisation in about a dozen or so higher plant genomes have been published. These studies have involved the classification of DNA into repeated or non-repeated sequences by renaturation kinetics and measurement of the lengths and arrangement of the sequences in the non repeated classes. Sequence organisation patterns have been described but the significance of these patterns is still a mystery. Our laboratory has studied the genomes of wheat, rye, barley and oats – four cereal species that have diverged from a common ancestor (Flavell, Rimpau and Smith, 1977). Our approach has frequently been an interspecies, comparative one because we believed that it would be instructive to understand not only the present day structures of these genomes but also the ways in which the genomes have evolved. A number of general concepts about genome structure and evolution have emerged from these studies. We therefore thought it would be useful at this workshop to outline these concepts and some of the evidence supporting them. In addition, we have drawn the ideas together in presenting a general scheme for the evolution of much of the DNA in these complex plant genomes. The conclusions drawn from our cereal genome studies, and the evolutionary scheme, should be applicable to the genomes of other higher plants. The paper starts with the description of the genome in terms of repeated and non repeated DNA, then considers the organisation and evolution of non repeated DNA and of families of repeated sequences. It concludes with the general scheme for plant genome evolution.

CLASSIFICATION OF CHROMOSOMAL SEQUENCES AS REPEATED OR NON-REPEATED

The analysis of complex genomes into repeated and non-repeated sequences is most easily carried out by studying the renaturation kinetics of small DNA fragments. What proportion of a genome is classed as repeated or non-repeated depends upon the stringency of the renaturation conditions. For example, when wheat DNA fragments are renatured at 50°C in 0.18 M Na$^+$ at least 80% of the genome renatures with repeated sequence DNA kinetics while at 60°C in 0.18 M Na$^+$ and 80°C in 0.18 M Na$^+$, 75% and 30-40% respectively renatures with repeated sequence kinetics; Flavell and Smith, 1976; Rimpau, Smith and Flavell, 1978). This is because repeated sequences accumulate mutations and diverge from other members of the same family. Under more stringent renaturation conditions more diverged repeats are unable to form stable duplexes and so behave as low copy or non-repeated sequences (hereafter called non-repeated sequences for simplicity). The reduced thermal stability of in vitro renatured repeated sequences is illustrated in Fig. 1. Some renatured repeated sequence duplexes are only just stable under the renaturation conditions indicating they are highly diverged, while others are as stable as native DNA. Two important conclusions emerge from the above results (1) More than 80% of the DNA of wheat (and of the other three cereal genomes) has arisen in amplification

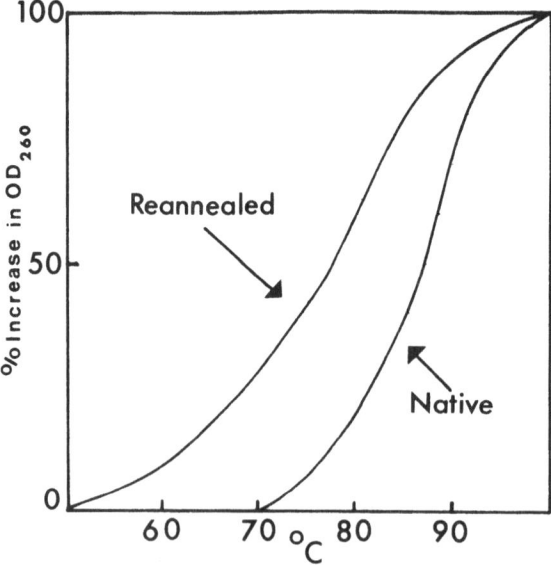

Fig. 1.   Thermal stability of native wheat DNA and repeated sequences renatured in 0.12 M phosphate buffer at 60°C. The duplex DNAs were melted in 0.12 M phosphate buffer and denaturation followed by the increase in OD_{260}

events of one kind or another and (2) Much of the DNA which behaves
as non-repeated under standard renaturation conditions (60°C 0.18 M
Na⁺) has evolved from families of repeated sequences.

NON REPETITIVE DNA; SIZE, ORGANISATION AND EVOLUTION

     The non-repeated DNA fraction contains not only diverged
repeated DNA but also structural gene sequences. However, these
genes are expected to occupy less than 1% of these cereal genomes.
The mean lengths of the non-repeated DNA sequences observed in the
three DNA renaturation   conditions described above have been
estimated (Flavell and Smith, 1976; Smith and Flavell, 1977;
Rimpau, Smith and Flavell, 1978, 1979).

     At the low and intermediate stringency conditions most of the
non-repeated DNA consists of sequences shorter than 1000 bp. This
short non-repeated DNA accounts for 10 (low stringency) to 15%
(intermediate stringency) by weight of these cereal genomes. In
wheat and rye some non-repeated sequences a few thousand bp may also
be present in the intermediate stringency conditions (Rimpau, Smith
and Flavell, 1978 but see also Murray, Cuellar and Thompson, 1978).

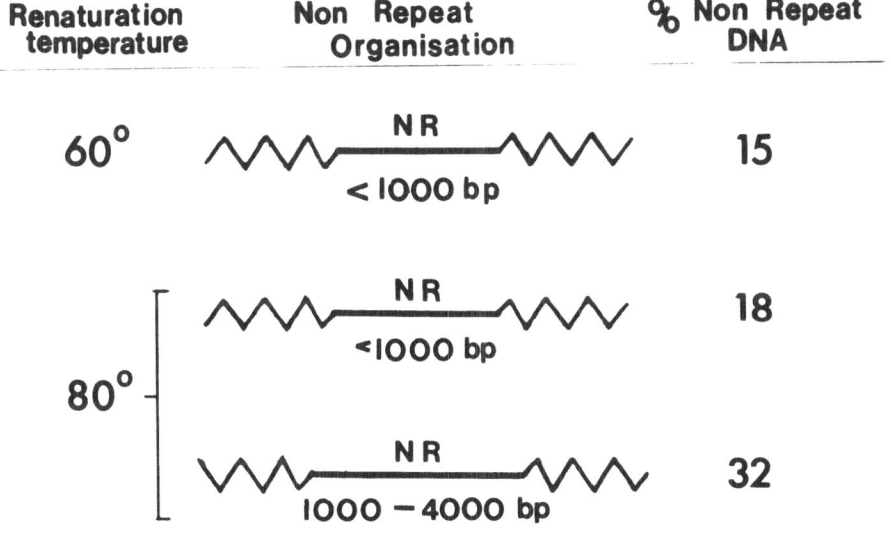

Renaturation temperature	Non Repeat Organisation	% Non Repeat DNA
60°	NR < 1000 bp	15
80°	NR < 1000 bp	18
	NR 1000 – 4000 bp	32

Fig. 2.   Organisation of wheat non-repeated DNA detected at two
          renaturation temperatures.〰〰 = repeated sequence DNA
          NR = non-repeated sequence DNA. The % non-repeat DNA
          refers to the percentage of the genome consisting of non-
          repeats of the given size class interspersed between
          repeated sequences.

The length and arrangement of the short non-repeated DNA assayed at 60°C in 0.18 M Na$^+$ is schematically represented in figure 2. The longer non-repeated DNA regions have been ignored because of lack of definitive evidence for their existence.

Under the high stringency renaturation conditions about 60% of the repeated sequences stable at 60°C in 0.18 M Na$^+$ cannot form stable duplexes and behave as non-repeated DNA. This can be inferred from the melting curve in figure 1. Some highly diverged repeated DNA may be clustered in long arrays but most of it occurs in short stretches between 2000 and 4000 bp long, interspersed with short lengths (<600 bp) of less diverged repeated DNA (Flavell and Smith, 1976; Smith and Flavell, 1977). The arrangement and proportion of DNA behaving as non-repeated at 80°C in 0.18 M Na$^+$ is also schematically presented in figure 2.

Comparison of the patterns of non-repeated DNA organisation at the two renaturation conditions emphasizes that the type of non-repeat interspersion pattern that is obtained i.e. whether only short or long and short non-repeated DNA regions are interspersed with repeated DNA regions, depends upon the experimental condition employed and the extent of repeated sequence DNA divergence in the genome.

Plant genome analyses are most commonly carried out using the renaturation conditions of 60°C 0.18 M Na$^+$. Under these conditions, as stated above, most of the non-repeated sequences appear shorter than 1000 bp in wheat, rye, barley and oats. The lengths of those repeated sequences interspersed with the non-repeated sequences have not been measured directly but most of them appear to be shorter than 1000 bp (Rimpau, Smith and Flavell, 1978, 1979). Using these values the proportions of the wheat, rye, barley and oats genomes consisting of stretches of short non-repeats interspersed with short repeats are between 25% and 40%. Most of the remaining DNA in the genomes consists of long regions of repeated sequences. There may be an additional few percent which includes longer stretches of non-repeated DNA (Rimpau, Smith and Flavell, 1978).

How do pieces of non-repeated DNA become interspersed within repeated DNA or vice versa? We have shown above that repeated sequence DNA diverges to behave as non-repeated DNA. Thus some short non-repeated regions interspersed within repeated DNA may reflect sequence divergence within arrays of repeats to a level which prevents short regions renaturing under given conditions. In addition, short pieces of repeated and non-repeated DNA may move around the genome to become inserted in new positions. Evidence will be presented later which implies that insertion of DNA pieces into new positions occurs during chromosome evolution.

EVOLUTION OF REPEATED SEQUENCE DNA

When sequences are amplified they are almost certainly tandemly arrayed initially. Amplification events occur sufficiently often during evolution to enable closely related species to be distinguished on the basis of their repeated sequence DNA complements (Flavell, Rimpau and Smith, 1977; Flavell, O'Dell and Smith, 1979). For example, 16, 23, 28 and 55% of the wheat, rye, barley and oats genomes respectively consist of repeated sequence DNA which is species-specific and not found in any of the other species. Similarly S. cereale contains several different tandem arrays of repeated sequences absent from S. silvestre (Bedbrook et al., 1980). The sizes of the pieces of DNA amplified in these sorts of events probably vary considerably. The repeating unit of one family common to barley, wheat and rye is only 12 bp (Dennis, Gerlach and Peacock, personal communication) while other families of rye which have been analysed in detail have repeating units between 100 and 2500 bp (see later). The average repeating unit length gained from whole genome studies also appears to be in the region of a few hundred bp (Rimpau, Smith and Flavell, 1978; Flavell and Smith, 1976).

What Pieces of DNA in a Genome Become Amplified?

Undoubtedly many kinds of sequences, especially coding sequences, are unlikely to survive as amplified arrays. However, because so little of these plant genomes consist of coding sequences, most of the DNA may be available as a substrate for amplification events whose products may survive. With this assumption, the knowledge of how a plant genome is organised and the evidence that the DNA pieces which are amplified are a few hundred bp on average, it is possible to make predictions about the expected structure of families of repeated sequences. First, because 75 to 80% of these genomes consist of repeated sequence DNA there is a high probability that already repeated sequence DNA will be reamplified. Second, because 25 to 40% of the genomes consist of interspersed short repeated and non-repeated DNA, there is a high probability that DNA pieces spanning a junction between different sequences will be amplified and the newly amplified repeating unit will be compound i.e. it will consist partly of previously amplified repeated DNA and partly of DNA that was previously non-repeated. This is illustrated in figure 3. There is also evidence that the parts of the genome that consist of long arrays of repeated sequences are permutations of different short repeats (Flavell and Smith, 1976; Rimpau, Smith and Flavell, 1978, 1979). If pieces of this DNA were amplified, the newly amplified repeating unit would also appear compound.

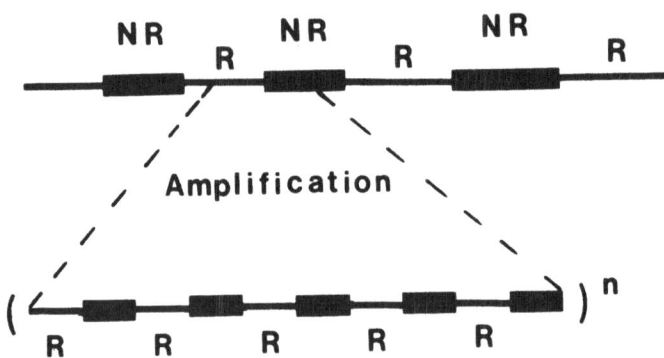

Fig. 3.   Amplification of a compound piece of DNA (repeat + non-
          repeat) to form a new family of repeated sequences.
          NR = non-repeat sequence DNA. R = repeated sequence DNA.

Evidence for Reamplification of Repeated DNA

     If a diverged repeated sequence family is present in two closely
related species and different members of the family become reamplified in the two species, then the related repeated sequences should
be more homogeneous within each species than between the two species.
The relationship between repeated sequences within and between
cereal species has been examined. Very closely related repeated
sequences were isolated from the wheat genome by carrying out
renaturation under very stringent conditions. These repeats were
then allowed to reanneal under less stringent conditions to wheat,
rye, barley and oats repeated sequences. The duplexes formed with
wheat DNA had a T_m of 80°C in 0.18 M Na$^+$ while those duplexes formed
with rye, barley and oats DNA had T_ms of 75°, 72° and 72° respectively (Flavell, Rimpau and Smith, 1977). This is therefore good
evidence for reamplification of repeated sequence variants during
genome evolution.

     Further evidence for reamplification has been gained recently by
the direct inspection of specific families of repeated sequences in
rye and wheat. The rye genome was cut with EcoRI or BamHI restriction
endonucleases fractionated on an agarose gel, transferred to a nitrocellulose filter and hybridised with a labelled cRNA probe made from
a cloned member of a rye family of repeats. Most copies of this
family of repeats are not cut by EcoRI, BglII, HindIII or BamHI but

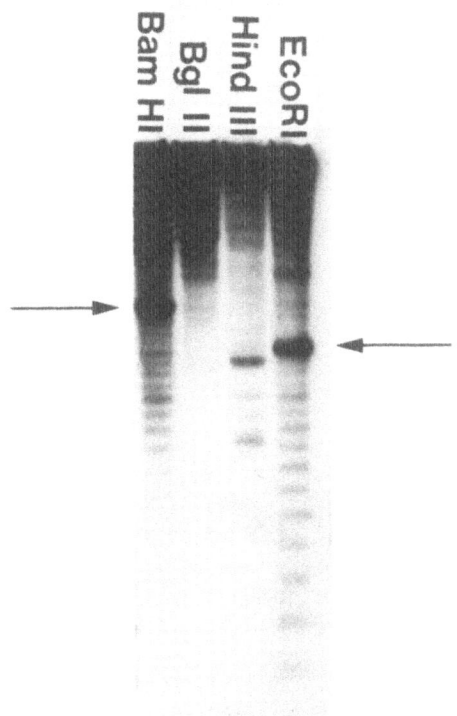

Fig. 4.   Reamplification of members of a family of repeats. The
          rye genome has been cut with BamHI, BglII, HindIII and
          EcoRI, fractionated on an agarose gel and sequences of a
          family detected by hybridisation to a cloned member of the
          family. Sequence heterogeneity within the family is
          indicated by the multimeric series (see text). Some of
          the reamplified variants in high copy number are indicated
          by the arrows.

some of the diverged members of the family have acquired sites for
these enzymes. Which members acquire a site would be expected to
be random. Consequently, sites in neighbouring members of the
tandem array causing the excision of the monomer would be expected
to be rarest. Dimers, trimers etc. would be expected to occur with
increasing frequency. This is what is found with some families.
However in this case some multimers are present in much higher
frequency than expected (see fig. 4) which is strong evidence for
reamplification of members of this particular family of repeats.

Evidence for Amplification of Compound Pieces of DNA

        Comparative studies on sequence organisation in the wheat, rye,
barley and oats genomes have revealed results most easily explained

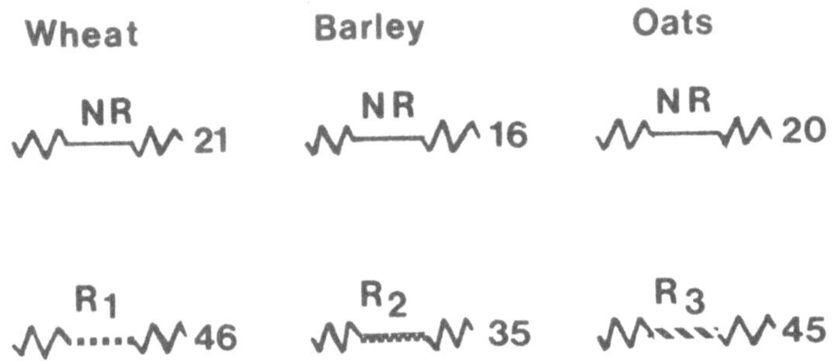

Fig. 5.   Organisation of "common" repeated DNA (〜) in the genomes
          of wheat, barley, and oats. NR = Non-repeated DNA R_1, R_2
          and R_3 = different sets of repeated sequences shorter than
          3500 bp found repeated only in wheat, barley and oats
          respectively. The numbers are the approximate percentages
          of each genome occupied by each sequence pattern.

by the amplification of compound pieces of DNA. Schematic comparative maps of substantial fractions of three of the genomes are shown
in figure 5 (see Rimpau, Smith and Flavell, 1978, 1979). The
fractions are defined as the portions of each genome in which
repeated sequences "common" to all four species are finely interspersed. These common repeated sequences are mostly short (<200 bp)
and are interspersed with non-repeated or other repeated sequences
shorter than 3000 bp. These latter repeated sequences are species
specific i.e. are not found in either of the two other species.
The pattern of the interspersed repeated sequences in all three
genomes is probably the consequence of amplification of compound
sequences. Each compound sequence would contain (1) repeated DNA,
homologous to repeated sequences first amplified in the ancestor
common to these cereal species i.e. "common" repeated DNA and (2)
non-repeated DNA. Thus the pattern of interspersed repeats in
fig. 5 visualised by interspecies DNA/DNA hybridisation is most
easily explained by the model already presented in fig. 3 with the
additional premise that the common repeats existed in ancestral
genomes interspersed with non-repeated DNA (fig. 5). This model
also provides an explanation for how related repeated sequences are
interspersed with different repeated sequences in different species.
If different compound sequences consisting of part common repeat and
part non-repeat become amplified during species divergence, then
arrays of common repeats interspersed with species-specific repeats
would result, as displayed in fig. 6 (Rimpau, Smith and Flavell,
1978, 1979).

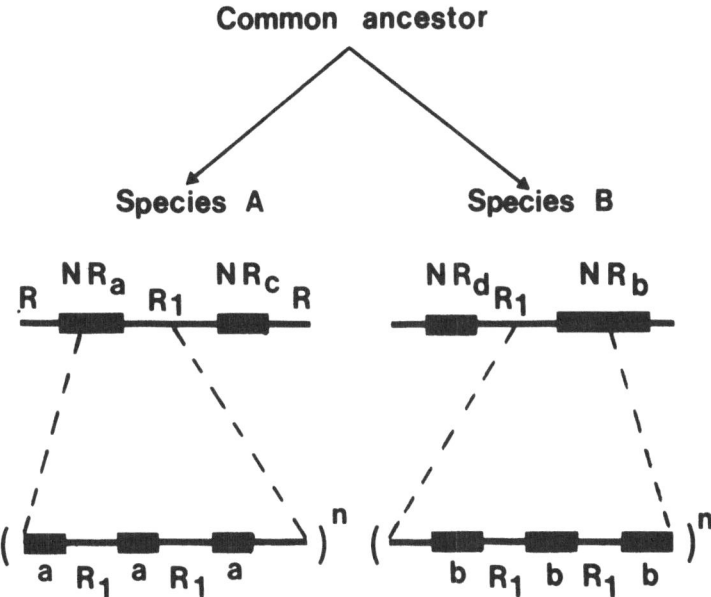

Fig. 6.   Amplification of compound (repeat + non-repeat) sequences
          in diverging species where the repeated sequences but not
          the non-repeated sequences are common to both species.
          NR = Non-repeated DNA.   R = repeated DNA.

     The sequence interspersion patterns in fig. 5 were derived from
renaturation experiments involving many families of repeated
sequences.   Recently, the repeating unit structures of six families
of rye repeated sequences have been elucidated.   They account for
approximately 10% of the rye genome.   Four of the families have a
compound repeating unit as shown in figure 7.   These units are
defined as compound because each contains sub-repeats and another
sequence unrelated to the sub-repeat.   The sub-repeats in families
3 and 4 are closely related to each other and to family 5 which
consists of tandem arrays of these sub-repeats.   Family 5 is found
in Secale cereale (rye) and in Secale silvestre while families 3 and
4 cannot be detected in Secale silvestre.   It therefore appears very
likely that families 3 and 4 arose from family 5 during divergence
of these Secale species by amplification of regions into which
another piece of DNA had been inserted as suggested in the model in
fig. 3.   These examples on specific families of rye repeated
sequences give strong support to the general conclusions derived
from the general genome organisation and renaturation studies
discussed above that amplification of compound sequences has occurred
frequently during cereal genome evolution.   The examples also provide
strong dircumstantial evidence that (1) pieces of DNA become inserted

R. FLAVELL ET AL.

44   R. FLAVELL ET AL.

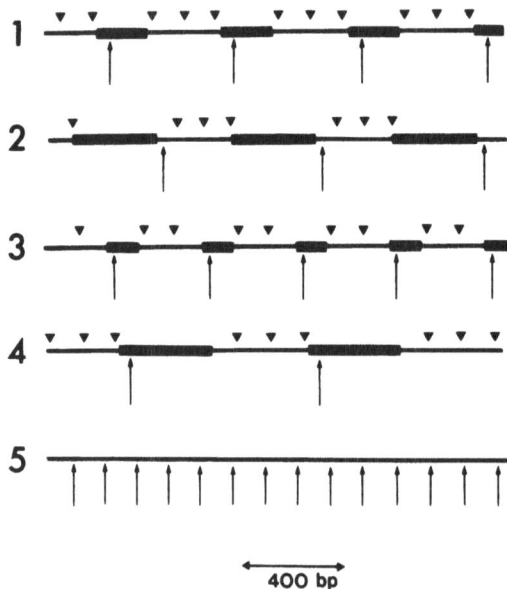

Fig. 7.   The repeating unit structure of five repeated sequence DNA
          families from rye. The repeating units are defined by the
          arrows (restriction enzyme sites) under each sequence array.
          The sub-repeats in each unit are defined by the sites for
          HaeIII restriction endonuclease shown above each array.
          The repeating unit of family 5 is closely related to the
          sub-repeats in families 3 and 4. These five families make
          up about 10% of the rye genome.

into arrays of repeated sequences, as suggested earlier, to create
non-repeated DNA interspersed with repeated DNA and (2) repeated
sequences become reamplified in new configurations as illustrated in
figures 3 and 6.

IMPLICATIONS OF REAMPLIFICATION FOR THE INTERPRETATION OF IN VITRO
RENATURATION EXPERIMENTS

     Families amplified from compound sequences as depicted in Figs.
3 and 6 have part homology with other families of repeated sequences.
Thus when DNA fragments are renatured in vitro, repeated sequences
which belong to arrays arising from different amplification events
will form duplexes. When such duplexes are treated with S$_1$ nuclease,
only short duplexes will remain because the arrays have only partial
homology. This result has often been used as evidence that repeated
sequences in the same family are often not in tandem arrays. However,
the alternative possibility of partial homology between arrays should
now be borne in mind.

A second consequence of the amplification of compound sequences containing an already repeated portion is that subcomponents of the new repeats will renature in vitro at different rates. Renaturation of one portion of the repeat will be driven by the related sequences in other families. This greatly complicates the interpretation of in vitro renaturation patterns. It would also result in slow renaturing sequences being seen on the same DNA fragments as fast renaturing sequences. This has been found in many renaturation analyses of plant DNAs but clearly does not necessarily imply the large scale dissipation of repeats from families of low copy number into families of high copy number or vice versa.

A GENERAL SCHEME FOR THE EVOLUTION OF THE CEREAL GENOMES

The models and ideas discussed in this paper for the evolution of much of DNA of cereal chromosomes can be summarised in a cyclical scheme (**fig.** 8). At any point in time there exists regions of the chromosomes which consist of short non-repeated sequences interspersed with repeated sequence DNA and other regions which consist of long stretches of repeated sequence DNA. With time, repeated sequences evolve into regions with interspersed non-repeats due to the accumulation of mutations and the insertion of non-repeated DNA. In the scheme they also become more complex by the insertion of other pieces of repeated DNA. New arrays of repeated sequences are generated in amplification events and non-repeated, non-repeated interspersed with repeated or repeated DNAs can be the substrate in these amplification events. One of the features of this scheme is that it does not separate into different categories those parts of the genome which have non-repeats interspersed with repeats from those parts which consist of tandem arrays of repeats. DNA sequences in both kinds of organisation belong to common cycles of DNA evolution. The scheme predicts that in diverging species, some different sets of sequences are in the amplification cycles and as divergence time increases, interspecies sequence homologies decrease for repeated and non-repeated DNAs. We would not expect the scheme to apply to those parts of the genome involved in the coding of polypeptides and carrying out sequence-specific roles. These sequences would be expected to evolve under substantial constraints.

Throughout this paper, we have discussed many events which lead to increases in total genome size. Undoubtedly, there are restraints on genome growth (Flavell et al., 1974) and loss of DNA must occur too. Where the accumulation of tandem arrays of repeats is selected against because of the effect on the total DNA content, then the proportion of the genome consisting of interspersed repeats and non-repeats should be higher. Plants with small DNA contents have a smaller proportion of their genome in long tandem arrays of repeats (Flavell et al., 1974). The non-repeated sequence regions in small

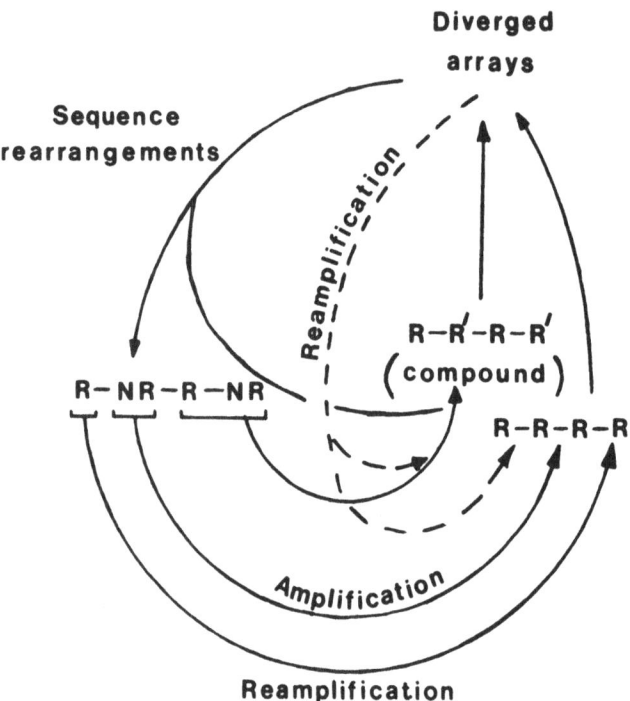

Fig. 8.   A scheme for the evolution of chromosomal DNA which can
          evolve without strong sequence-specific constraints.   The
          scheme has been devised, as described in the text, from
          knowledge of genome organisation, of how genomes have
          diverged in related species and of the structure of some
          families of repeated sequences.   NR = Non-repeat.
          R = repeat.

genomes are also often longer than 1000 bp (Walbot and Dure, 1976;
Thompson, W.J., personal communication).   This would result if the
generation of non-repeated DNA from repeated DNA continued with a
reduced rate of amplification.

ACKNOWLEDGEMENTS

    JRB was a recipient of a long term EMBO postdoctoral fellow-
ship.   JR was supported by Deutsche Forschungsgemeinschaft while
on leave from the Institut für Pflanzenzüchtung der Universität
Göttingen, 34 Göttingen, West Germany.

REFERENCES

Flavell, R. B., Bennett, M. D., Smith, J. B., and Smith, D.B., 1974, Genome size and the proportion of repeated nucleotide sequence DNA in plants. Biochem. Genet. 12:257-269.

Flavell, R. B., O'Dell, M., and Smith, D. B., 1979, Repeated Sequence DNA comparisons between Triticum and Aegilops species. Heredity - in press.

Flavell, R. B., Rimpau, J., and Smith, D. B., 1977, Repeated sequence DNA relationships in four cereal genomes. Chromosoma (Berl.) 63:205-222.

Flavell, R. B., and Smith, D. B., 1976, Nucleotide sequence organisation in the wheat genome. Heredity 37:231-252.

Murray, M. G., Cuellar, R. E., and Thompson, W. F., 1978, DNA sequence organisation in the pea genome. Biochemistry 17:5781-5790.

Rimpau, J., Smith, D. B., and Flavell, R. B., 1978, Sequence organisation analysis of the wheat and rye genomes by interspecies DNA/DNA hybridisation. J. Molec. Biol. 123:327-359.

Rimpau, J., Smith, D. B., and Flavell, R. B., 1979, Sequence organisation in barley and oats chromosomes revealed by interspecies DNA/DNA hybridisation. Heredity - in press.

Smith, D. B., and Flavell, R. B., 1977, Nucleotide sequence organisation in the rye genome. Biochim Biophys Acta 474:82-97.

Walbot, V., and Dure, L. S., 1976, Developmental biochemistry of cotton seed embryogenesis and germination. J. Mol. Biol. 101:503-536.

CHROMOSOME AND GENE STRUCTURE IN PLANTS: A PICTURE DEDUCED FROM ANALYSIS OF MOLECULAR CLONES OF PLANT DNA

J. Bedbrook, W. Gerlach, S. Smith[†], J. Jones and
R. Flavell
Department of Cytogenetics
Plant Breeding Institute, Cambridge CB2 2LQ, England
[†]Biological Sciences Dept. University of Warwick, UK

Molecular cloning technology has introduced exciting ways of studying eukaryotic chromosomes. The purpose of this paper is to illustrate ways in which molecular cloning has enabled us to study components of plant chromosomes.

By way of introduction we discuss properties of plant nuclear DNA which have proven important in approaching cloning experiments.

Some technical considerations

While molecular cloning experiments have proven straight-forward for chloroplast DNA (Bedbrook et al., 1978; Bedbrook et al., 1979) many workers have encountered difficulties in their attempts to clone nuclear DNA - (various pers. comm.).

We have discussed previously general factors, such as, repeated DNA content of plant genomes and 5 methyl cytosine content which may affect the efficiency of cloning plant nuclear DNA (Bedbrook et al., 1980). DNA purity is a major factor in determining the efficiency of cereal DNA cloning. We have described elsewhere the details of our plant DNA purification scheme (Bedbrook et al., 1979; Gerlach and Bedbrook, 1979) Table 1 shows results of experiments in which we compared the relative cloning efficiency of wheat and E. coli DNA. Our preliminary experiments showed that the use of intercalating dyes in CsCl density gradients was the best way of removing contaminants (probably polysaccharides) which inhibited the activity of restriction endonucleases and reduced cloning efficiency. In the experiments of Table 1 the efficiency of cloning of DNA purified either on

Table 1. The relative cloning efficiency of E. coli and wheat DNA
using plasmid pAC184 and E. coli strain HB101.

DNA sample	Number trans- formants per μg vector DNA	Frequency transformants per cell	Insertion frequency %
No DNA (control)	0	0	-
closed circular pAC184 DNA	1.0×10^6	1.0×10^{-3}	-
Eco R1 digested pAC184 DNA	1.2×10^3	1.2×10^{-6}	-
Ligated pAC184 DNA	1.0×10^5	1.0×10^{-4}	-
pAC184 DNA ligated to wheat DNA 1	2.5×10^4	2.5×10^{-5}	11
pAC184 DNA ligated to wheat DNA 2	4.0×10^4	4.0×10^{-5}	12
pAC184 DNA ligated to E. coli DNA 1	8.5×10^4	8.5×10^{-5}	13
pAC184 DNA ligated to E. coli DNA 2	1.2×10^5	1.2×10^{-4}	12

Wheat and E. coli sample 1 DNAs were purified by two cycles of
CsCl - ethidium bromide centrifugation and sample 2 DNAs were
purified by CsCl - ethidium bromide centrifugation and one cycle
of CsCl - actinomycin D centrifugation. 10 μg of each of these
DNAs were digested to completion with Eco R1 and ligated with
0.2 μg of Eco R1 digested pAC184 plasmid DNA. The ligated DNA
was mixed with 0.1 μg of untreated RSF1030 plasmid DNA. RSF1030
is compatible with pAC184 plasmid (33) and provided an internal
control which was used to standardise the transformation frequencies of the various DNA samples. The DNA samples were used to

Legend to Table 1 contd.

transform 10^9 CaCl$_2$ treated HB101 cells as described in methods.
As controls closed circular pAC184 DNA (0.2μg), EcoRI digested
pAC184 DNA (0.2μg) and pAC184 DNA digested with EcoRI and self
ligated were also used to transform HB101. The cells were plated
on media containing 30 μg/ml ampicillin, to select RSF1030 transformants, on media containing 10 μg/ml tetracycline to select
pAC184 transformants, and on non selective media to assay the total
number of viable cells per transformation. The number of pAC184
transformants was standardised for each transformation according
to the number of RSF1030 transformants. The frequency of transformation was standardised according to the viable cell titre.
The fraction of pAC184 transformants containing DNA fragments
inserted at the EcoRI site of the Cm^R locus in pAC184 are shown.

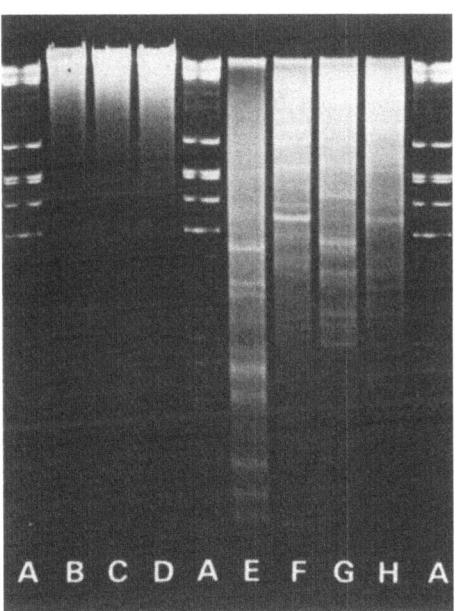

Fig. 1. Agarose gel electrophoresis of rye DNA digested with
various restriction endonucleases

Secale cereale DNA (5 μg) prepared from embryo nuclei by two cycles
of density gradient centrifugation in CsCl-ethidium bromide was
digested to completion with restriction endonucleases. Digests
were fractionated by electrophoresis through 1% agarose gels.
(A) Size markers (B) S. cereale DNA digested with SmaI, (C) HpaII
(D) PstI, (E) Hind III (F) BglII (G) Bam HI (H) EcoRI
Electrophoresis was from top to bottom.

CsCl-ethidium bromide or CsCl actinomycin D gradients was
investigated.

In these experiments EcoRI fragments of wheat DNA were cloned
in the plasmid pAC184. pAC184 closed circular DNA gave 1 x 10^6
transformants per µg of vector DNA. This number was reduced to
approximately 1 x 10^3 by EcoRI linearisation of the DNA. Self-
ligation of EcoRI digested pAC184 DNA restored the frequency to
1 x 10^5 transformants per µg. Ligation of pAC184 DNA to E. coli
DNA, prepared by two different methods (see legend to Table 1) did
not significantly reduce the transformation frequency. Ligation
of wheat DNA prepared by CsCl-EtBr density centrifugation to
pAC184 reduced its transformation frequency by a factor of 4 and
wheat DNA prepared by CsCl-EtBr plus CsCl-Act-D density centrifugation reduced transformation by a factor of 2. The frequency
of inserts was about the same for E. coli and wheat DNA. The
reduction in expected transformation frequency by ligation to wheat
DNA is a consistent feature. We presume, though we have no direct
evidence, that many "chimeric" plasmids formed with wheat DNA
either do not become established in E. coli or are unstable for
some reason. We have investigated the size of inserted wheat DNA
in plasmids. Plasmid DNA was prepared from 60 clones and the size
of the inserted fragments measured. The number average size of
wheat DNA EcoRI fragments inserted in pAC184 DNA was close to the
number average size of fragments produced by EcoRI digestion of
total DNA (Bedbrook et al., 1980).

The clones of wheat, rye and barley DNA which we have studied
in detail have shown no instability on storage and correspond in
size to fragments in the genome. In the following sections we
briefly describe results of the physical analysis of clones which
have enabled us to study some components of chromosomes.

Molecular clones of highly repeated DNA have enabled a description
of DNA sequences in telomeric heterochromatin of S. cereale

Much of plant DNA is in the form of highly repeated sequences
Flavell et al., 1974). These sequences have been studied in some
detail by solution hybridization studies (e.g. Rimpau et al., 1978)
S. cereale contains much repeated DNA in blocks of constitutive
heterochromatin at the telomeres. The amount of this heterochromatic material varies greatly in different Secale species.
(Bennett, Gustafson and Smith, 1977).

Figure 1 shows that restriction endonuclease digestion reveals
many repeated DNA sequences in rye DNA. Such sequences are seen
as bands in gel fractionated digests of nuclear DNA. We have found
(Bedbrook, Jones, Thompson and Flavell submitted) that DNA at the
limiting mobility of gels of such digests is enriched for

sequences located within telomeric heterochromatin of rye. We have used Hae III fragment clones of this highly repeated DNA to characterise the structure of these repeat DNA families in telomeric heterochromatin of rye. Figure 2 shows gel electrophoresis fractionation of plasmids containing a Hae III fragment which is a component of the most highly repeat sequence in rye DNA.

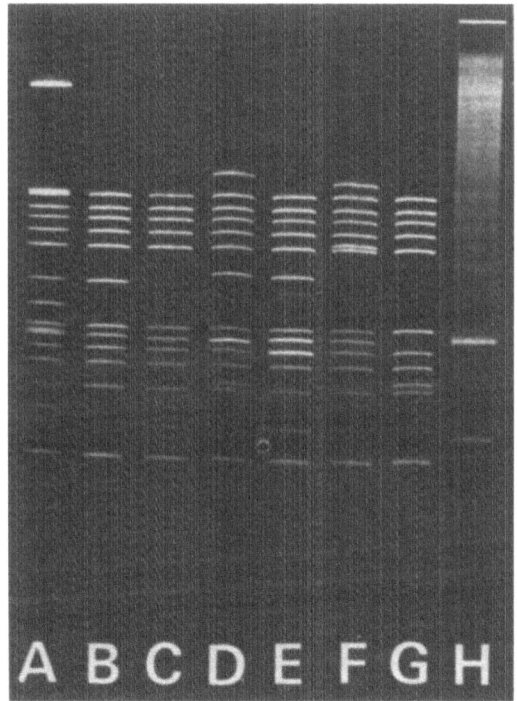

Fig. 2. Gel electrophoresis of molecular clones of Hae III fragments of S. cereale DNA

Plasmid DNA containing Hae III fragment inserts and total rye DNA were digested with Hae III and fractionated by gel electrophoresis on a 6% polyacrylamide gel. (A-F) Hae III digests of chimeric plasmid DNAs containing sequences which hybridized to the major repeat sequence of rye (G) pBR322 DNA digested with Hae III (H) S. cereale DNA digested with Hae III.

Importantly, all the clones contain a fragment of the same size as
the major Hae III fragment band in the genome. In situ hybridization (Fig. 3) of such clones show that the sequences in these
plasmids are predominantly telomeric. The cloned sequences have
enabled us to describe 4 major repeated sequence families in rye
DNA. Physical maps of these sequences have been deduced by
hybridizing the clones back to genomic DNA using the method of
Southern (1975). The chromosomal location of the cloned DNAs have
been investigated by in situ hybridization of probes prepared from
the clones to rye chromosomes.

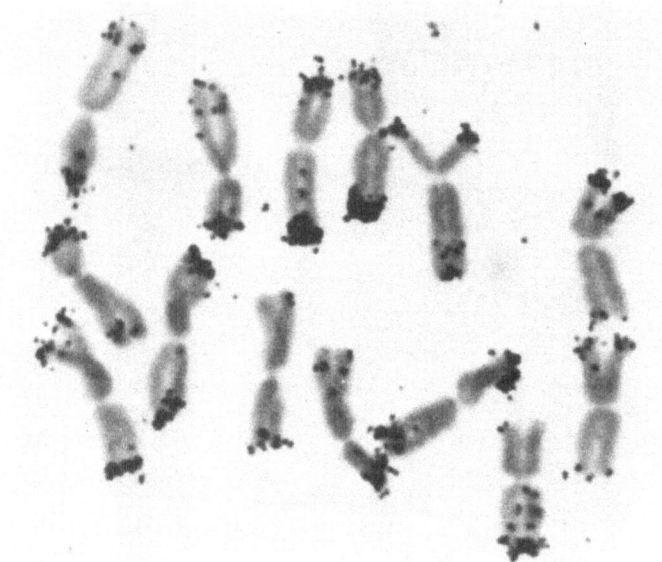

Fig. 3. In situ hybridization of a probe prepared from cloned
         DNA containing part of the major repeated sequence family
         in S. cereale

In situ hybridization of a cloned major repeat to metaphase
chromosomes from root tips of 2-3 day old seedlings was as
described in Bedbrook, Jones, Thompson and Flavell, submitted
and Appels et al., 1978.

Family 1 sequences are simple tandem repeats (illustrated in
Figure 4A) approximately 120 bp long as defined by Hae III
digestion. This family comprises approximately 2.25% of the
S. cereale chromosome and is present in high copy number in all the
Secale species we have studied. This sequence is also present in
wheat. Family 2 is a tandem repeat sequence of 480 bp defined by
Mbo II and Taq I endonucleases. Analysis with the enzyme Hae III
shows this family to contain a subrepeating unit with the arrangement diagrammed schematically in Figure 4B. This family is
present only in Secale species with large heterochromatic blocks
and is not detected in wheat chromosomes. This sequence comprises
approximately 6% of the total rye DNA. Family 3 which represents
approximately 2.7% of rye DNA has a basic repeating unit of 610 bp
defined by the enzyme Taq I. This repeat family also has a complex
structure with a subrepeating unit which is revealed by Hae III
digestion. This family is also found only in Secale species with
large amounts of telomeric heterochromatin; it is absent from wheat.
A fourth family of telomeric repeated sequences has three different
basic repeating length units of 120, 356 and 630 bp. Together
these three repeats represent about 0.5% of rye DNA. The 120 bp
repeat is present in all the Secale species tested but the 356 and
630 bp repeats have been detected only in Secale cereale.

The structure and distribution of these repeats has led us to
propose that complex repeats of the type shown in Fig. 4B may be
derived from simple repeats (Fig. 4A) by the insertion of DNA
elements in simple repeats followed by amplification (Bedbrook,
Jones, Thompson and Flavell, submitted; Rimpau, Smith and Flavell
1978).

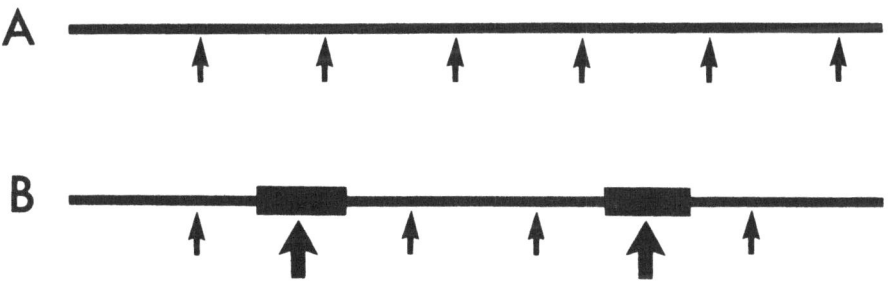

Fig. 4.   Schematic representation of the major repeated sequence
          families in S. cereale telomeric heterochromatin

Horizontal lines represent chromosomal DNA. Vertical arrows indicate the length periodicity of a repeating unit. The solid
reactangles in B represent sequences which interrupt the repeat
units defined by vertical arrows. The large vertical arrows
define the repeat unit of subrepeat plus insert.

Molecular cloning of repeated genes: the ribosomal RNA genes of
wheat and barley

    Molecular cloning of repeated gene sequences permits an
analysis of the types of structural variation found in different
individual examples of a particular repeat gene family.

    Our preliminary experiments with "southern hybridization" of
25 + 18S ribosomal RNA to wheat and barley DNA showed that the
genes for these RNAs were tandem arranged repeats of approximately
9 Kb as defined by EcoRI and that there was slight length heterogeneity in the repeat unit. In order to determine the basis of
this length heterogeneity and to obtain a detailed picture of rDNA
in wheat and barley we cloned rDNA from these species (Gerlach and
Bedbrook, 1979). We first enriched for rDNA using CsCl-Actinomycin
D density gradients (Hemleben and Grierson, 1978). EcoRI fragments
of this DNA were cloned in the plasmid pAC184, Table 2 shows the
results. The number of plasmids obtained containing full length
repeats was approximately that expected on the basis of the proportion of the starting DNA that was rDNA. Gel electrophoresis of
all the rDNA plasmids from the two species showed that 17 of the
18 clones obtained contained full length repeats. One rDNA
plasmid contained a deletion involving both the rDNA and the vector.

Table 2. Cloning of rDNA from wheat and barley

rDNA	Fraction rDNA[1]	Transformation[2] efficiency	Insert frequency	Fraction of clones with full length rDNA repeats
Wheat	1.12%	6×10^4	4.9%	1.8%
Barley	2.01%	2.4×10^4	12.1%	3.6%

Wheat DNA (7.5 µg) and barley DNA (11.6 µg) enriched for rDNA was
digested with EcoRI and ligated to 0.5 µg of pAC184 DNA. Ligated
DNA was used to transform E. coli HB101. Transformants were
selected for resistance to tetracycline and clones containing
inserts were screened by sensitivity to chloramphenicol. Clones
containing sequences complementary to rDNA were screened using the
procedure described by Grunstein and Hogness (1975).

(1) Number average fraction of EcoRI digests in form of length
     rDNA repeats

(2) Number of transformants per µg of vector DNA

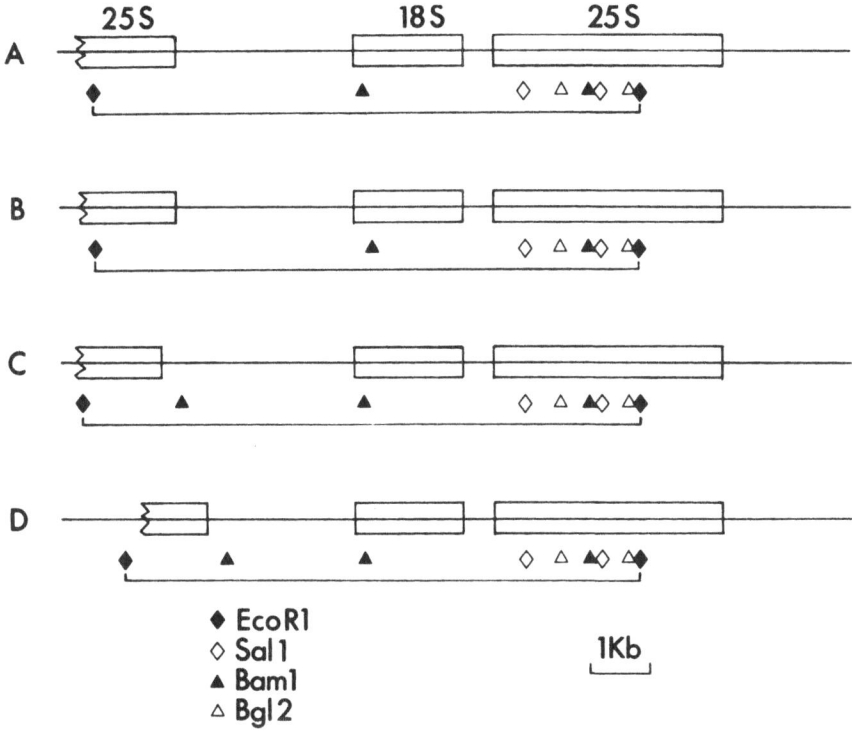

Fig. 5. Physical maps of the rDNA repeats of barley and wheat
         using restriction endonuclease targets as markers

A) map of the major wheat rDNA repeat   B) map of a minor wheat
rDNA repeat   C) map of the longer barley rDNA repeat
D) map of the shorter barley rDNA repeat

The various clones enabled us to define three length variants of
the rDNA repeat in wheat and two length variants of the rDNA
repeat in barley. Figure 5 shows diagrams of the repeats for
which we have cloned examples. It was found that all the length
heterogeneity is confined to the "non-transcribed" spacer. In
wheat the major class of repeat has a non-transcribed spacer of
3Kb and two minor length variants with spacers of 3.15 and 3.3 Kb.
In barley two major length variants exist whose non transcribed
spacers differ in length by approximately 1 Kb.

Molecular cloning of unique sequences - Isolation of cDNA clones
for the small subunit of ribulose bisphosphate carboxylase

        The probability of obtaining a clone of a specific unique
sequence from many higher plants via clone libraries of complete

hexanucleotide restriction digests is in the order of 1 x 10^{-6}.
Various schemes for reducing the number of clones required using
partial restriction enzyme digests or sheared fragments have been
presented (Maniatis et al., 1978). A requirement for obtaining such
clones from the genome is to have a pure source of sequence to
screen clone libraries for the presence of the required sequence.
Purified mRNA could be used for this purpose though it is technic-
ally difficult to purify to homogeneity even the most abundant mRNA.
Maniatis et al., (1976) have demonstrated that cDNA prepared from
RNA can be incorporated and amplified in bacterial plasmids. cDNA
clones can be used as pure probes for the selection of genomic
clones.   In this section we discuss experiments in which we have
cloned cDNA of polyA containing RNA from pea leaves in bacterial
plasmids and used the hybridization assay of Grunstein and Hogness
(1975) to select clones likely to contain sequences of the small
subunit of ribulose bisphosphate carboxylase (SU). We have used
the hybridization "release translation" assay of Smith et al.,
(1979) to positively identify SU cDNA clones. Details of the methods
of our cDNA cloning procedure are found in the legend to Figure 7.
Our estimates indicate that the total polyA containing RNA of pea
leaves is somewhat less than 2% by weight of the total RNA preparation. We took advantage of two facts in the identification of SU
cDNA clones. Firstly, following the observation of Highfield and
Ellis (1978) that green pea leaves contain considerably more
translatable mRNA for SU than do etiolated leaves, (see Figure 6)
and secondly that preparative gel electrophoresis could be used to
prepare RNA fractions differentially enriched for SU mRNA (Fig. 6).
RNA preparations from dark grown leaves and from illuminated
leaves and size fractions of polyA containing RNA were labelled
in vitro using polynucleotide kinase and 32p phosphate   labelled
ATP and hybridized to cDNA clones of total polyA + RNA from peas.
Figure 7 shows the results of such analysis. Specific clones
hybridized RNA from light grown leaves more strongly than RNA from
dark grown leaves. These clones also hybridized size fractions of
RNA enriched for SU mRNA more strongly than fractions containing
less SU mRNA. The clones hybridizing in this manner were used to
prepare plasmid DNA the plasmid DNA was bound to filter discs of
diatotized cellulose and total polyA RNA from pea leaves hybridized to DNA by the method of Smith et al. (1979). Unhybridized
RNA was removed by washing and the specifically hybridized RNA
eluted. This RNA was then used to promote protein synthesis in an
in vitro translation system derived from wheat germ. All the
clones identified as SU possibles by the "Grunstein-Hogness"
(1975) assay of Figure 7 hybridized a mRNA which gave on in vitro
translation a product which co-migrated with the protein precursor
of SU. That this protein was in fact SU precursor was confirmed
by the in vitro processing assay described by Highfield and Ellis
(1978). Figure 8 gives a few examples of the translation and
processing results.

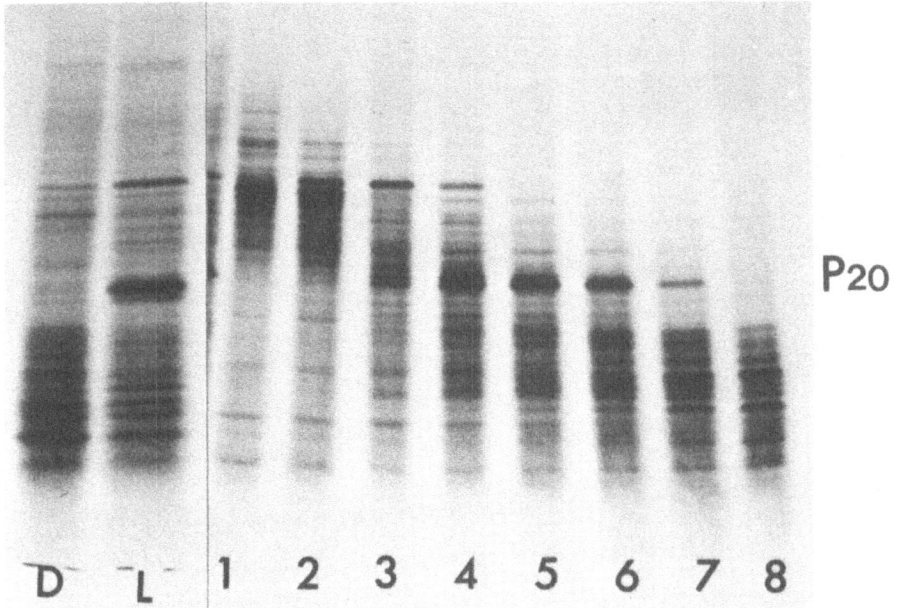

Fig. 6. Gel fractionation of translation products of polyA + RNA
         from peas containing varying amounts of p20 mRNA

Polysomal polyadenylated RNA prepared from pea apices was translated in a wheat-germ protein synthesising system and the products
analysed as described previously (Highfield and Ellis 1978;
Smith and Ellis 1979). The figure shows an autoradiograph of SDS
polyacrylamide gel electrophoresis of translation products of
(D) RNA prepared from peas grown in darkness, (L) RNA prepared
from peas grown 9 days in darkness followed by 48 hours continuous
illumination

1-8 RNA from peas grown as for (L) and fractionated by gel
    electrophoresis

1 is higher molecular weight RNA and 8 is lower molecular weight RNA

P20 = precursor to the small subunit of Ribulose bisphosphate
      carboxylase

    We are currently using SU cDNA clones to obtain the primary
sequence of the SU precursor mRNA and to select from clone banks
genomic clones containing the SU coding sequence.

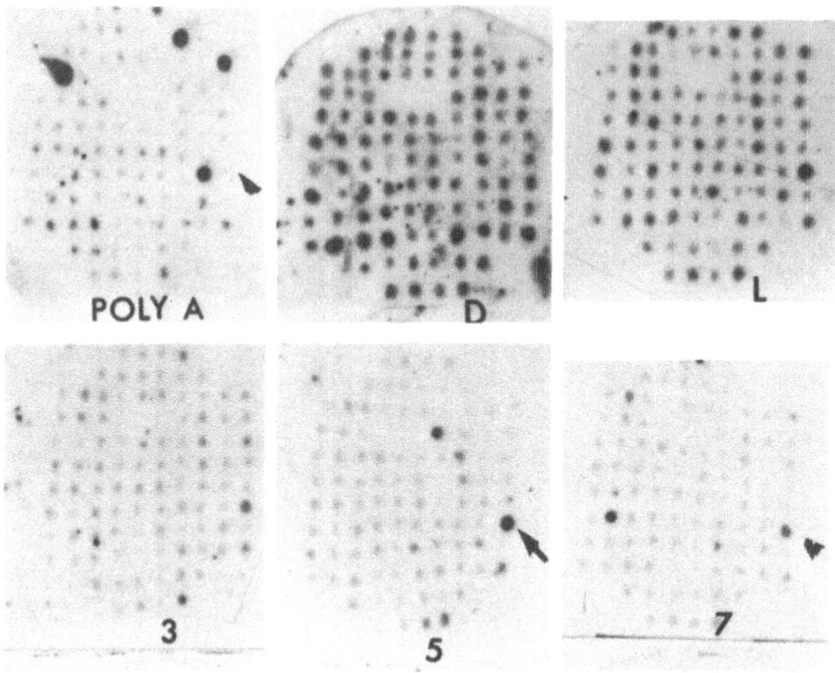

Fig. 7. Hybridization screen of cDNA clones with RNA probes
         containing a differential amount of P20 mRNA

Polyadenylated RNA from peas was used to prepare double stranded
cDNA as described by Maniatis et al. (1977). Double stranded cDNA
was ligated to synthetic linkers containing the recognition target
for the restriction endonuclease Hind III. After digestion with
Hind III the cDNA was ligated to the bacterial plasmid pBR322
linearised with Hind III. The ligated DNA was used to transform
the E. coli strain HB101. 300 transformants were screened by the
hybridization assay of Grunstein and Hogness (1975), with RNA
probes described in the legend to Fig. 6. (D)=Hybridization with
RNA from dark grown leaves (L)=hybridization with RNA from
illuminated leaves 3, 5, 7 hybridization with RNA of various size
classes (see legend to Fig. 6).

polyA = Hybridization with polyA.
Arrows point to a colony with the hybridization pattern expected
for a SU clone.

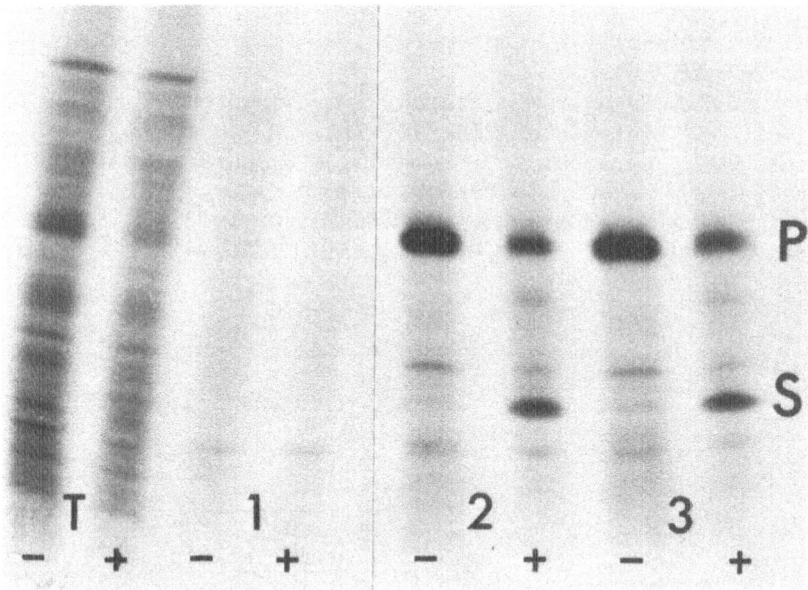

Fig. 8. Identification clones containing DNA encoding the precursor to the small subunit of RBP Case

Plasmid DNA, from clones hybridizing to polyA RNA rich in p20 mRNA, was prepared by the method of Clewell 1972. The DNA was linearised with EcoRI denatured and bound to 11 mm diameter discs of diazobenzylomethyl-paper. (Smith et al., 1979) PolyA RNA from peas was hybridized to DNA discs using the method of Smith et al., (1979). Hybridized RNA was eluted, precipitated and translated in a wheat germ derived in vitro translation system. Half of the translation products were subjected to the in vitro processing procedure described by Smith and Ellis (1979).

T. translation of unfractionated RNA. (−) non processed.
(+) post-translational processed with chloroplasts
P = precursor of small subunit of bulose bisphosphate carboxylase
S='mature' small subunit of ribulose bisphosphate carboxylase

1 Control filter minus DNA
2 & 3 Filters containing plasmid DNA from putative SU clones
(see Fig. 7).

Acknowledgments

J.R.B. was a recipient of a long term EMBO postdoctoral fellowship. All recombinant DNA experiments described here were carried out under Class II containment as defined by GMAG UK.

REFERENCES

Appels, R., Driscoll, C. and Peacock, W.J. 1978. Chromosoma
    70: 67-89.
Bedbrook, J.R., Coen, D.M., Beaton, A.A., Bogorad, L. and Rich, A.
    1979. J. Biol. Chem. 254: 905-910.
Bedbrook, J.R., Gerlach, W., Thompson, R., Jones, J., and
    Flavell, R. 1980. In Press.
Bedbrook, J.R., Kolodner, R. and Bogorad, L. 1977. Cell 11: 739-749.
Bennett, M.D., Gustafson, J.P. and Smith, J.B. 1977. Chromosoma
    61: 149-174.
Clewell, D.B. 1972. J. Bact. 110: 667-676.
Coen, D.M., Bedbrook, J.R., Bogorad, L. and Rich, A. 1978.
    Proc. Natl. Acad. Sci. USA 74: 5487-5491.
Flavell, R.B., Bennett, M.D., Smith, J.B. and Smith, D.B. 1974.
    Biochemmical Genetics 12: 257-269.
Gerlach, W. & Bedbrook, J. 1979. Submitted Nucleic Acids Research.
Grunstein, M. and Hogness, D.S. 1975. Proc. Natl. Acad. Sci. USA
    72: 3961-3965.
Hemleben, V. and Grierson, D.1978. Chromosoma 65: 363-368.
Highfield, P.E., Ellis, R.J. 1978. Nature 271: 420-424.
Maniatis, T., Hardison, R.C., Lacy, E., Lauer, J., O'Connell, C.,
    Quon, D., Sim, G.K. and Efstratiadis, A. 1978. Cell 15:
    687-701.
Maniatis, T., Kee, G.G., Estratiadis, A., Kafatos, F.C. 1976.
    Cell 8: 163-182.
Rimpau, J., Smith, D. and Flavell, R. 1978. J. Mol. Biol. 123:
    327-359.
Smith, S.M. and Ellis, R.J. 1979. Nature 278: 662-664.
Smith D.F., Searle, P.F. and Williams, J.G. 1979. Nuc. Acids Res.
    6: 487-506.
Southern, E.M. 1975. J. Mol. Biol. 98: 503-517.

A MODEL FOR A MOLECULAR CLONING SYSTEM IN HIGHER PLANTS: Isolation

of Plant Viral Promotors

Richard B. Meagher[+†*] and Thomas D. McKnight[+*]

University of Georgia

Departments of Microbiology[+] and Botany[†], Program in
Genetics[*],Athens, Georgia 30602

Several different technical approaches could be taken in the
development of a genetic engineering system in higher plants. Our
laboratory is planning to construct molecular cloning vehicles in
E. coli which can be used in higher plant cells and in regenerated
higher plants. By combining isolated transcriptional "promotors"
from plant DNA and known bacterial genes, we hope to obtain selectable genetic markers for plant cells. Fragments of rDNA and other
repeat sequence DNA from host plants are being isolated as a homologous sequence for recombination with the host genome. The
rationale of our plan is outlined herein and our results on the
isolation of promotor sequences from a plant DNA virus are summarized.

The most versatile bacterial plasmid vehicles contain numerous
selectable drug resistance markers for the identification of transformants (Bolivar, 1978). When foreign DNA is inserted into a gene
encoding drug resistance, cells containing these recombinant molecules can be screened for the loss of this drug resistance charac-
ter. With lambda phage vehicles an even stronger selection can be
made for recombinant molecules. Vehicles will not form plaques on
host strains unless they contain inserted foreign DNA (Williams and
Blattner, 1979). These genetic selections have been of absolute
importance in the majority of genetic engineering experiments in E.
coli.

Critical experiments have demonstrated the molecular cloning
of DNA in mammalian culture cells. For example, rabbit globin cDNA
in a recombinant SV40 vehicle (Mulligan et al., 1979) and any
foreign DNA such as procaryotic DNA (Wigler et al., 1979) can be
manipulated into animal cells. In the SV40 system cells which have

taken up DNA and are transformed can be selected on proper media.
In the system of Wigler et al. (1979) transformation of the foreign
DNA fragments is dependent upon the coselection of a homologous
thymidine kinase genetic marker.

The most highly developed of the eucaryotic molecular cloning
systems is found in yeast (Hinnen et al., 1978). It has not been
possible to detect transformation of yeast auxotrophic mutants with
total DNA from prototrophic yeast cells. Two technical advances
have made molecular cloning in yeast possible. First of all, recombinant plasmids constructed in E. coli containing a complementing
yeast biosynthetic gene can be amplified in large quantity. A 20 µg
sample of recombinant plasmid contains approximately 100 times as
much of the complementing yeast gene as is contained in a 100 µg
sample of total prototrophic yeast DNA. If this amount of recombinant plasmid DNA is used in a transformation, approximately 20 to
200 transformants can be detected. Genetic and biochemical analysis
of the transformants show that the entire recombinant plasmid is
integrated at the site of the auxotrophic mutant gene in the yeast
chromosome (Hicks and Fink, 1977). Establishing the genetic selection for transformed cells was the critical step in developing a
yeast transformation system.

Secondly,if the complementary recombinant plasmid also harbors
a multicopy yeast gene, such as the yeast ribosomal DNA, then a 10-
to 100-fold increase is obtained in the total number of transformants
per experiment (Szostack and Wu, 1979). The entire recombinant
plasmid can integrate into any of the numerous sites on the yeast
chromosomes containing the rDNA gene. Apparently, increasing the
target for DNA integration increases the total number of stable
transformants.

Additional improvements in the molecular cloning vehicles used
in the yeast system have made it possible to obtain from 10,000 to
20,000 transformants per experiment. However, these new vehicles,
which replicate independently of host chromosomes, do not yield a
high percentage of stable transformants.

The data from yeast suggest two focused areas of research for
the development of a molecular cloning system which should yield
stable genetically altered higher plants. First, a selectable
genetic marker for plant cells must be isolated, molecularly cloned
and amplified in E. coli. Secondly, a mechanism for integration
of this genetic marker into the plant host genome must be provided.

It has been shown by numerous methods that plant protoplasts
can take up DNA (Lerquin and Kado, 1977). However, the stable
transformation of a foreign genetic marker in plant cells has not
yet been detected. In no case have even the minimum requirements

of the yeast transformation system been met. There has not been a strong selectable genetic marker amplified in a large quantity on a vehicle and there has not been a known sequence homology between the host genome and the vehicle which would allow integration.

Insertion of DNA Into the Host Plant Genome

Yeast tranformation was greatly enhanced by inserting recombinant plasmids into a homologous multicopy gene in the yeast genome. It may be possible to use the large amount of repeat sequence DNA contained in higher plant genomes as a target for gene integration in plant cell transformation experiments. Most higher plants contain at least as high a percentage of ribosomal DNA as does yeast. Therefore, integration of recombinant plasmids into the rDNA of plants also may be productive. Other less well defined but more highly repeated plant DNA sequences could easily be isolated and used as targets for homologous recombination.

Other approaches exist which may provide useful gene integration in plants. (1) The Ti plasmid from Agrobacterium tumafaciens must carry specific sequences for stable integration into a higher plant's genetic complement (Chilton et al., 1977). It is not known if these sequences are functionally linked to the information for transformation of plant cells. Once isolated the DNA responsible for integration of Ti could be used to construct vehicles with a wide host range. The Ti plasmid will certainly be used as a vehicle for molecular cloning of foreign DNA in plant cells. This vehicle should provide the first powerful genetic selection applied to plants. Cells can be transformed from normal hormone-requiring callus to hormone independent tumor tissue which grows on hormone-free media. (2) Inverted repeat sequence DNA in plants may be responsible for gene movement comparable to that demonstrated in procaryotes (Kleckner, 1977). Once the mechanisms for insertion of inverted repeat sequences are known it may be possible to design a plant molecular cloning vehicle with inverted repeat sequences which aid in insertion of this DNA into the genome. (3) Another approach which should aid in the replication of foreign DNA would be the construction of autonomous replicating elements of DNA in plants analogous to bacterial plasmids, and to the 2 μm circle (Gerbaud et al., 1979) and mini-chromosomes in yeast (Szostack and Wu, 1979). However, the unstable nature of these autonomous genetic elements may make these vehicles less desirable, particularly for transformation of plants with agriculturally important genetic characters.

Selectable Genetic Markers for Plant Cell Genetic Manipulation

Complementation of plant auxotrophic mutants with normal plant DNA should provide a straight forward selection for transformation.

This method has been very effective in other eucaryotic systems.
From our knowledge of plant biochemistry, there is every reason to
believe that plant auxotrophic mutants requiring amino acids, nucleoside bases and vitamins, can be isolated. However, there are only
a few such plant auxotrophs identified for which the biochemical
requirement is known to be linked to loss of a structural gene
function (Bourgin, 1978). This has not been for want of effort on
the part of numerous plant biologists and biochemical geneticists.
Many theories have been put forth to explain this problem although
none are completely supported by positive scientific data. It may
be that many structural genes for these biosynthetic markers are
not single copy genes, even in the haploid plant genome. A number
of eucaryotic structural genes have been found to be in multicopies
in the haploid complement of DNA. Among these are the dihydrofolate
reductase genes in hamster (Shauske et al., 1978), the actin gene in
Dictostelium (Kindle and Firtel, 1978), and zein gene in corn (Burr
and Burr, manuscript in preparation) and the leghemoglobin gene in
soybeans (Baulcombe and Verma, 1978). If the plant structural genes
for many plant biosynthetic enzymes are contained in multiple copies
the probability of isolating auxotrophic mutations in these genes
would be infinitely small.

     It is likely that positively selected genetic markers which
can be added to a normal prototrophic genetic background will be
most useful in plant systems. In particular, in the application
to agricultural problems it will be necessary to use prototrophic
healthy plant cells as recipients in transformation experiments.
Ideal genetic markers should involve the expression of a single
gene which will allow a plant to grow normally in the presence of
a toxic compound and without which the cell could not survive. At
best, such a gene function should inactivate the toxic compound by
a chemical cleavage or modification reaction.

     Herbicides could be used in genetic selections in plants in
much the same way as drugs have been used in procaryotes. Pichloram
resistant mutants of tobacco callus cells have been selected and
resistant plants regenerated (Chaleff and Parsons, 1978). This
resistance marker shows Mendelian inheritance indicating that it is
a chromosomal gene. If the gene or genes responsible for this
resistance were available for genetic manipulation into sensitive
plants, expression of this DNA might provide an adequate genetic
selection for a plant transformation system. In addition to
pichloram many other potent herbicides are available and could be
examined.

     Drug resistance genes in E. coli have allowed numerous E. coli
strains of different genetic backgrounds to be used without the need
for specific auxotrophic mutations which the yeast system requires.
Some of the same drugs can be used in genetic selections in plants.

Most of the drugs that effect both eucaryotes and procaryotes are known to effect eucaryotic organelles. The isolation of herbicide, drug, analogue and other resistance mutants of higher plant cells is well reviewed by Malaga (unpublished manuscript).

Assuming that plant cells resistant to the compound in question are isolated, the native plant genes encoding these resistance functions could be used as genetic markers. However, isolation of the genes encoding these traits from an enormous plant genome could be very difficult. In general, the method used would be as follows. The enzyme responsible for inactivating the toxic compound would have to be identified and isolated. Using a sensitive radioimmune assay for the protein and in vitro translation of RNA the mRNA encoding this protein would have to be at least partially purified. A cDNA copy of this mRNA once molecularly cloned in E. coli and isolated can be used as a probe in order to molecularly clone the actual plant structural gene, which should then be a potential genetic marker for genetic engineering in plants. Such an approach is extremely laborous. It may be possible to shorten this procedure by cloning and selecting plant genes directly in yeast. However, there is still the problem that such resistance functions are generally uncharacterized in plants.

If bacterial resistance markers can be expressed in plant cells, this will greatly facilitate the development of strong selectable markers for plant systems. For example,   the major protein responsible for divalent mercury resistance in E. coli is mercuric reductase (Summers and Silver, 1978). The polypeptide which composes this enzyme is the product of a single gene. Mercuric reductase uses NADPH as an electron donor to reduce toxic divalent mercury to less toxic volatile metalic mercury. The enzyme, if present in plant cells, should render cells resistant to mercuric ion. Thus, cells containing this gene could be selected for on media containing mercuric ion. As another example, if functional genes for chloramphenicol acetyltransferase were introduced into plant cells they should render plant organelles chloramphenicol resistant (Shaw, 1971). The complete nucleotide sequence of the chloramphenicol acetyltransferase gene has been determined (Vapnek and Alton, personal communication). Sequence information will be invaluable to the in vitro reconstruction of this gene. However, the expression of procaryotic DNA in eucaryotic cells has not yet been examined. Problems relating to the expression of eucaryotic DNA in procaryotes have been discussed elsewhere (Meagher et al., 1977b). Some of the translational differences between eucaryotes and procaryotes which might create barriers to the expression of heterologous genetic information have recently been reviewed (Kozak, 1978).

Control of Foreign DNA via Homologous Promotors

By inserting foreign DNA adjacent to a bacterial promotor on a plasmid, high levels of expression of foreign eucaryotic genes have been obtained in E. coli. Synthetic eucaryotic genes for the two chains of insulin (Goeddel et al., 1970) and for somatostatin (Itakura et al., 1977), cDNA copies of the mRNAs encoding ovalbumin (Fraser and Bruce, 1978), and restriction fragments of hepatitis B virus (Burrell et al., 1979) have directed the synthesis of correct or nearly correct polypeptides in E. coli under the direction of bacterial promotors. Using a similar strategy, it may be possible to design vehicles containing strong plant "promotors" which can ensure the expression of foreign DNA and in particular bacterial DNA in plant cells. Procaryotic genes may have distinct advantages over any eucaryotic source of genetic information. Selectable bacterial genes are easily identified and isolated for genetic manipulation. They do not contain introns which must be processed ● correctly from the transcribed RNAs. In vitro manipulation of the number of promotors controls the levels of gene expression in E. coli (Backman and Ptashne, 1978) and multiple promotors are not uncommon in highly transcribed native E. coli genes (Musso et al., 1977; Young and Steiz, 1979). By placing a plant and bacterial promotor in tandem controlling the expression of a bacterial resistance gene it should be possible to obtain expression for the same marker in both bacteria and plants.

We plan to isolate selectable bacterial genes, such as the previously mentioned gene for mercuric reductase, and place this DNA under the control of a strong plant viral promotor isolated from cauliflower mosaic virus. By this method we hope to ensure that the bacterial gene could be transcribed and translated in plants. We have attempted to locate RNA polymerase binding sites on cauliflower mosaic virus, and we have evidence that these sequences act as promotors in E. coli. To facilitate insertion of foreign DNA into the plant genome, we plan to include in the cloning vehicle some repetitive DNA, such as the rDNA genes, to provide a region of homology with the plant genome. This method has been successfully employed with yeast transformation. Initial data on our isolation of soybean rDNA and other soybean DNA fragments will be presented elsewhere at this workshop.

A schematic of the type of vehicle we hope to construct for molecular cloning in plant cells is shown in Figure 1. We believe that this type of vehicle can be constructed in E. coli by making each desired addition on an existing E. coli plasmid vehicle such as pBR325 (Bolivar, 1978) which contains numerous sites for the insertion of foreign DNA.

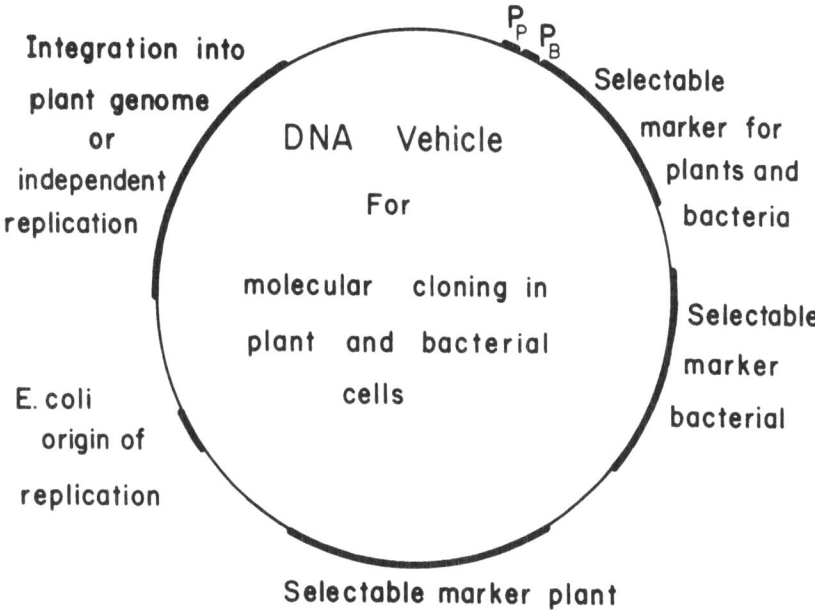

Figure 1.   Proposed DNA Vehicle for Molecular Cloning in Plants and
Bacteria.   Numerous genetic markers are required for selection in
plant and bacterial hosts.   Inactivation of these markers after the
in vitro insertion of foreign DNA can be used to screen for newly
constructed recombinant molecules. P_p and P_b are the plant and
bacterial promotor sequences, respectively, controlling the single
marker for both plant and bacterial hosts.   A ColE1 origin of
replication is included for replication and amplification in E.
coli.   DNA sequences allowing the stable integration into the plant
host genome or autonomous replication in plant cells is also provided.

Locating "Promotor" Sequences on Cauliflower Mosaic Virus DNA

A great deal is known about the binding of E. coli RNA polymerase to E. coli structural gene promotors (Pribnow, 1975).   E. coli
holo-RNA polymerase binds correct promotors several orders of magnitude more strongly than transcribed regions.   However, much less
is known about the binding of eucaryotic RNA polymerases to their
substrates.   The majority of data published on eucaryotic RNA polymerase II "promotors" describes the location of the 5' termini of
eucaryotic mRNAs on their DNA substrates (Dhar et al., 1977).   Binding studies are difficult in eucaryotic systems due to the very
small differences between the binding affinity of eucaryotic RNA
polymerase II at correct promotors relative to nonspecific DNA
binding.   DNA sequences resembling bacterial promotors have been

observed for mammalian genes but the functional role of these
sequences is yet to be demonstrated (Gannon et al., 1979).

When fragments of cauliflower mosaic virus are cloned via recombinant plasmids in E. coli three major polypeptide products are
synthesized from the CaMV DNA. These polypeptides of 42K, 40K and
37K daltons represent nearly 50% of the coding capacity of the virus
when reading the 4.8×10^6 dalton genome in a single frame. It has
been shown that the information for initiation of transcription and
translation of these polypeptides comes entirely from within the
cloned CaMV DNA (Meagher et al., 1977b). The very high levels of
synthesis observed for these polypeptides relative to other plasmid
polypeptides argues that a high level of transcription must occur
from within the cloned fragments. These observations have not been
made for any other higher eucaryotic DNA (Meagher, 1977). Because
of these indications that CaMV contains strong promotors and because
most bacterial viruses contain strong promotors, cauliflower mosaic
virus should be a good source of plant "promotor" sequences for
molecular cloning work in plants.

Protection from Endonuclease Cleavage by E. coli and Wheat Germ RNA Polymerase

If E. coli holo-RNA polymerase is bound specifically to an E.
coli promotor it will protect this site from nuclease cleavage
(Jones et al., 1977). Non-specifically bound polymerase will not
inhibit cleavage until the polymerase to DNA ratio is very high. We
have defined a restriction endonuclease cleavage site as protected
if 80% of the molecules containing this site are left uncleaved when
cleavage at other control sites for the same enzyme contained in the
same reaction mixture is virutally complete (95%). When about five
E. coli holo-RNA polymerase molecules are present per sequence of
cauliflower mosaic virus at least four restriction endonuclease
cleavage sites out of the 40 we have examined so far are protected
(McKnight and Meagher, 1979). These are sites for EndoR·HpaI-HincII,
HindIII, BglII and XbaI. Although our mapping data is not complete
the last three of these cleavage sites may lie within 200 base pairs
of each other. Thus, only two regions of the DNA may be involved in
binding polymerase. In our initial experiments with wheat germ RNA
polymerase II (Jendrisak and Burgess, 1975) these same sites and one
additional HindIII site appear to be protected. ·Protection with
wheat germ polymerase is not as complete as it is with E. coli polymerase.

Do the regions of DNA which bind RNA polymerase function as
promotors; that is, do they promote transcription in one direction?
This question can be tested in E. coli by making use of a plasmid
molecular cloning vehicle derived from pBR322 (amp^{r}tet^r) (Bolivar
et al., 1977). This new plasmid, pHB1 (amp^{r}tet^s) has the majority

of its promotor for the tetracycline resistance gene deleted and
replaced by a synthetic EcoRI endonuclease site (Rodriquez et al.,
1980). If sequences containing promotor activity are cloned in the
correct orientation in this site the tetracycline resistance gene
can be transcribed and translated and cells containing the recombinant plasmid are tetracycline resistant.

CaMV DNA is cleaved by EcoRI* producing at least 45 fragments
(McKnight, Heyneker and Meagher, 1980). These fragments have an
average molecular weight of 120,000 daltons and range in size from
50,000 to 500,000 daltons. The EcoRI* activity of EcoRI endonuclease
has a four base recognition sequence homologous with the internal
four bases of the normal six base EcoRI sequence. Both activities
yield a common 5'AATT single stranded tail and, therefore, EcoRI*
DNA fragments can be molecularly cloned in an EcoRI site (Polisky
et al., 1975).

In order to select for viral sequences functioning as promotors
in E. coli, the EcoRI* DNA fragments of the virus are first ligated
into the EcoRI site of pHB1. The recombinant plasmids are transformed into E. coli and transformants are selected for various levels
of tetracycline resistance. Resistant colonies are screened for
their plasmid DNA content and selected fragments are mapped by
Southern blot hybridization back to the restriction map of the virus.
Thus far, only transformants as tetracycline resistant as the parent,
pBR322 plasmid, have been analyzed. Out of the 16 independent transformants chosen as the most highly tetracycline resistant, we found
that nine contain a common 220,000 dalton fragment and that all
transformants could be accounted for by a total of four small fragments with promotor activity. The most highly tetracycline resistant clone, pMM38, has a fragment of 0.34×10^6 daltons which contains
the HpaI site. This clone also contains an adjacent EcoRI* star
fragment from the viral genome evidently the result of a partial
EcoRI* cleavage (McKnight, Heynecker and Meagher, 1980). The other
cloned CaMV fragments which are functional as promotors in pHB1 are
being analyzed further.

Is it possible that sequences contained at the ends of these
fragments have fortuitously recreated a promotor-like sequence when
combined with the remaining promotor sequence contained in pHB1?
Or is the promotor activity coming entirely from within the cloned
fragments? If the sequence containing the protected HpaI site within
pMM38 is functional as a promotor then it should function as a promotor when cloned within other restriction fragments of the virus.
This site is contained on a HindIII fragment of 0.3×10^6 daltons in
CaMV (H3i-H3j, Meagher et al., 1977). The HindIII site of pBR322
is within a few base pairs of the EcoRI site of pHB1 relative to
the tetracycline gene. Insertion into this HindIII site in pBR322
separates the tetracycline resistance promotor from the structural

gene. Unlike seven other HindIII fragments from CaMV, when this
HindIII fragment (H3i-H3j) is inserted in pBR322 the recombinant
plasmid is resistant to 20 µg per ml tetracycline. Initial data
also indicate that the EcoRI* and HindIII fragments containing this
HpaI site are functional in only one orientation as would be
expected for a true promotor sequence.

Verification of the in vivo function of these presumptive CaMV
"promotor" sequences will depend upon a detailed in vivo transcription map of the whole viral genome.

The key to all flexible molecular cloning systems in bacteria,
yeast or animal cells has been strong selectable genetic markers.
It is likely that the lack of such markers for plant cells will
limit the development of a genetic engineering system in plants.
The isolation of active plant "promotor" sequences may allow selectable gene markers to be constructed from existing bacterial genes,
which should function in both plant and bacterial cells.

Acknowledgements: The authors would like to thank Barbara Rutledge
for her thoughtful suggestions on the final stages of the manuscript
and R. A. Lansman for numerous stimulating discussions relating to
this paper. This research was supported by National Science Foundation Grants PCM-7715011 and PCM-7900833. T.D.M. is supported as an
N.I.H. predoctoral trainee in genetics at the University of Georgia.

References

Backman, K., and Ptashne, M., 1978, Maximizing gene expression on
    a plasmid using recombination in vitro, Cell,13:65.
Baulcombe, D., and Verma, D.P.S., 1978, Preparation of a complemen-
    tary DNA for leghaemoglobin and direct demonstration that
    leghaemoglobin is encoded by the soybean genome, Nuc. Acid Res.,
    5:4141.
Bolivar, F., 1978, Construction and characterization of new cloning
    vehicles, III Derivatives of plasmid pBR322 carrying unique
    EcoRI sites for selection of EcoRI generated recombinants,
    Gene, 4:121-136.
Bolivar, F., Rodriquez, R. L., Greene, P. J., Betlach, M. C.,
    Heyneker, H. C., Boyer, H. W., Crosa, J. M., and Falkow, S.,
    1977b, Construction and characterization of new cloning
    vehicle, II A multipurpose cloning system, Gene 2:95:113.
Bourgin, J. P., 1978, Isolement de mutants a partir de cellules
    vegetales en culture in vitro, Physio. Veg. 16:339.
Burrell, C. J., MacKay, P., Greenaway, P. J., Hofschneider, P. H.,
    and Murry, K., 1979, Expression in Escherichia coli of hepatitis
    B virus DNA sequences cloned in plasmid pBR322, Nature 2479:43.
Chaleff, R. S., and Parsons, M. F., 1978a, Direct selection in vitro
    for herbicide-resistant mutants of Nicotiana tabacum, Proc.
    Natl. Acad. Sci. USA, 75:5104.

Chaleff, R. S., and Parsons, M. F., 1978b, Isolation of a glycerol-utilizing mutant of Nicotiana tabacum, Genetics 89:723.

Chilton, M. D., Drummond, M. H., Merlo, D. J., Sciaky, D., Montoya, A. L., Gordon, M. P., and Nester, E. W., 1977, Stable incorporation of plasmid DNA into higher plant cells: the molecular basis of crown gall tumorigensis. Cell 11:263.

Dhar, R., Subramanian, K. N., Par, J., and Weissman, S. M., 1977, Nucleotide sequence of a fragment of SV40 DNA that contains the origin of DNA replication and specifies the 5' ends of "early" and "late" viral RNA IV, J. Biol. Chem. 252:368.

Fraser, T. H., and Bruce, B. J., 1978, Chicken ovalbumin is synthesized and secreted by Escherichia coli, Proc. Natl. Acad. Sci. USA 75:5936.

Gannon, F., O'Hare, K., Perrin, F., LePennec, J. P., Benoist, C., Cochet, M., Breathnech, R., Royal, A., Garapin, A., Cami, B., and Chambon, P., 1979, Organization and sequences at the 5' end of a cloned complete ovalbumin gene, Nature 278:428.

Gerbaud, C., Fournier, P., Blanc, H., Aigle, M., Heslot, H., and Guerineau, M., 1979. High frequency of yeast transformation by plasmids carrying part or entire 2-µm yeast plasmid, Gene 5:233.

Goeddel, D. V., Kleid, D. G., Bolivar, F., Heyneker, H. L., Yansura, D. G., Crea, R., Hirose, T., Kraszewski, A., Itakura, K., and Riggs, A. D., 1979, Expression in Escherichia coli of chemically synthesized genes for human insulin, Proc. Natl. Acad. Sci. USA, 76:106.

Hicks, V., and Fink, G. R., 1977, Identification of chromosomal location of yeast DNA from hybrid plasmid pYeleu 10, Nature, 269:265.

Hinnen, A., Hicks, J. B., and Fink, G. R., 1978, Transformation of yeast, Proc. Natl. Acad. Sci. USA 75:1929.

Itakura, K., Hirose, T., Crea, R., Riggs, A. D., Heyneker, H. L., Bolivar, F., and Boyer, H. W., 1977, Expression of Escherichia coli of a chemically synthesized gene for the hormone somatostatin, Science 198:1056.

Jendrisak, J. J., and Burgess, R. R., 1975, A new method for the large scale purification of wheat germ RNA polymerase II, Biochemistry 14:4639.

Jones, B. B., Chan, H., Rothstein, S., Wells, R. D., and Reznikoff, W. S., 1977, RNA polymerase binding sites in λ plac 5 DNA, Proc. Natl. Acad. Sci. USA 74:4914.

Kindle, K. L., and Firtel, R. A., 1978, Identification and analysis of Dictostelium actin genes, a family of moderately repeated genes, Cell 15:763.

Kleckner, N., 1977, Translocatable elements in procaryotes, Cell, 11:11.

Kozak, M., 1978, How do eucaryotic ribosomes select initiation regions in messenger RNA?, Cell 15:1109.

Lurquin, D. F., and Kado, C. I., 1977, E. coli plasmid pBR313 insertion into plant protoplasts and into their nuclei, Molec. gen. Genet, 154:113.

Malaga, P., 1979, Resistance mutants and their use in genetic manipulation (unpublished manuscript, Inst. Plant Phys., Hungarian Acad. Sci., Szeged, Hungary).

McKnight, T. D., Heyneker, H. L., and Meagher, R. B., 1980, Isolation and characterization of DNA sequences from cauliflower mosaic virus which function as promotors in E. coli,(manuscript in preparation).

McKnight, T. D., and Meagher, R. B., 1979, Isolation and location of potential promotors of cauliflower mosaic virus, Plant Phys. 63:33, Abst. No. 183.

Meagher, R. B., 1977, The development of a molecular cloning system in higher plants, In, Genetic Engineering for Nitrogen Fixation, Ed. by A. Hollaender, et al., Plenum Corp., N.Y., N.Y.

Meagher, R. B., Shepherd, R. J., and Boyer, H. W., 1977a, The structure of cauliflower mosaic virus. 1. A restriction endonuclease map of cauliflower mosaic virus DNA, Virology 80:362.

Meagher, R. B., Tait, R. C., Betlach, M., and Boyer, H. W., 1977b, Protein expression in E. coli mini-cells by recombinant plasmids, Cell 10:521.

Mulligan, R. C., Howard, B. H., and Berg, P., 1979, Synthesis of rabbit β-globin in cultured monkey kidney cells following infection with SV40 β-globin recombinant genome, Nature 277: 108.

Musso, R. E., DiLauro, R., Adhya, S., and deCrombrugghe, B., 1977, Duel control for transcription of the galactose operon by cyclic AMP and its receptor protein at two interspersed promotors, Cell 12:847.

Pribnow, D., 1975, Bacteriophage T7 early promotors: Nucleotide sequences of two RNA polymerase binding sites, J. Mol. Biol. 99:419.

Poliski, B., Greene, P., Garfin, D. E., McCarthy, B. J., Goodman, H. M., and Boyer, H. W., 1975, The specificity of substrate recognition by the EcoRI restriction endonuclease, Proc. Natl. Acad. Sci. USA, 72:3310.

Rodriquez, R. L., West, R. W., Heyneker, H. L., Bolivar, F., and Boyer, H. W., 1980, Characterizating wild-type and mutant promotors of a tetracycline resistance gene in pBR313, (manuscript in preparation).

Shanske, S., Melera, P. W., and Biedler, J.L., 1978, Overproduction of dihydrofolate reductase by antifolate resistant chinese hamster cells, Cell Biol. 79:345.

Shaw, W. V., 1971, Comparative enzymology of chloramphenicol resistance, Ann. N.Y. Acad. Sci. 182:234-242.

Szostack, J. W., and Wu, R., 1979, Insertion of a genetic marker into the ribosomal DNA of yeast, Plasmid, in press.

Summers, A. O., and Silver, S., 1978, Microbial transformations of metals, Ann. Rev. Microbiol. 32:637.

Wigler, M., Street, R., Sim, G. K., Wold, B., Pellicer, A., Lacy, E., Maniatis, T., Silverstein, S., and Axel, R., 1979, Transformation of mammalian cells with genes from procaryotes and eucaryotes, Cell 16:777.

Williams, B. G., and Blattner, F. R., 1979, Construction and characterization of the hybrid bacteriophage lambda charon vectors for DNA cloning, J. Virol. 29:555.

Young, R. A., and Steiz, J. A., 1979, Tandem promotors direct E. coli ribosomal RNA synthesis, Cell 17:225.

Zain, S., Sambrook, J., Roberts, R. J., Keller, W., Fried, M., and Dunn, A. R., 1979, Nucleotide sequence in a copy of adenovirus 2 fiber mRNA, Cell 16:851.

PURIFICATION, STRUCTURES AND FUNCTIONS OF THE NUCLEAR RNA POLY-

MERASES FROM HIGHER PLANTS

Jerry Jendrisak

Department of Botany
University of Minnesota
St. Paul, Minnesota U.S.A.   55108

INTRODUCTION

Eukaryotic organisms contain three classes of nuclear DNA-dependent RNA polymerases (EC 2.7.7.6) which can be distinguished 1) chromatographically, 2) by their differential sensitivities to inhibition by the fungal toxin α-amanitin, 3) by their subunit structures, 4) by their intranuclear localizations, and 5) by the types of genes which they transcribe[1].

The following properties are characteristic of RNA polymerases from a wide variety of higher eukaryotes: 1) The enzyme classes are named I, II and III according to their order of elution from DEAE-Sephadex columns developed with a linearly increasing salt gradient--usually ammonium sulfate. RNA polymerases I, II and III elute at approximately 0.1 M, 0.2 M and 0.3 M ammonium sulfate from this anion exchange resin. 2) RNA polymerase I is unaffected by α-amanitin, RNA polymerase II is inhibited by very low concentrations of α-amanitin, and RNA polymerase III is inhibited by relatively high concentrations of α-amanitin. 3) The subunit structures of the three enzymes are distinctly different and very complex. The enzymes superficially resemble each other in that they are all composed of two or three high molecular weight subunits and many low molecular weight subunits. 4) RNA polymerase I is localized in the nucleolus and RNA polymerases II and III are nucleoplasmic. 5) RNA polymerase I transcribes the ribosomal RNA precursor, RNA polymerase II transcribes messenger RNA precursors, and RNA polymerase III transcribes low molecular weight RNAs such as transfer RNA precursors and 5S ribosomal RNA.

Control mechanisms of eukaryotic gene transcription are largely unknown. Three enzyme classes with different transcriptive functions permit the independent regulation of the transcription of different RNAs by modulation of the levels or activities of the three RNA polymerases. In order to regulate the transcription of various messenger RNAs by RNA polymerase II or of transfer RNAs and 5S RNA by RNA polymerase III may require mechanisms operative at the level of the template. Current theory suggests that chromatin structural modifications govern the accessibility of sites for transcription by RNA polymerases. However, other control mechanisms can be envisioned which operate through specific RNA polymerases and affect their ability to recognize certain genes and not others. Whatever the mechanisms of regulation, their elucidation will require the establishment of in vitro reconstituted transcription systems which are capable of mimicking in vivo RNA synthesis. Highly purified and well defined components will be required, including the three nuclear RNA polymerases. Well defined RNA polymerases implies knowledge of their physical and chemical properties. Detailed studies on the enzymes will require that ample quantities be readily available for analysis. Readily available enzymes are also needed for in vitro transcription studies which will be useful for assessing the selectivity of transcription by RNA polymerases I, II and III. Those interested in other aspects of RNA polymerases such as structure-function relationships of the various RNA polymerase subunits will also require large quantities of highly purified material. Consequently, many laboratories have been involved in the purification of the multiple RNA polymerases from a wide variety of eukaryotic organisms. This feat is not trivial for several reasons: 1) The concentration of RNA polymerases in most tissues is extremely low such that purifications of several thousand to one hundred thousand fold are sometimes required to achieve homogeneity, 2) the instability of enzymes and the poor recoveries at various purification steps have plagued investigators, and 3) rather low amounts of pure enzyme are often obtained due to limitations inherent in purification procedures. Selection of tissues which are rich in RNA polymerases and where large amounts of starting material are available, and the development of efficient purification techniques are important considerations. Finally, systems where in vivo RNA synthesis can be studied or characterized, and from which highly active in vitro transcription systems can be constructed are also important considerations. The wheat embryo system possesses many of these desirable traits. The utility of this system for in vitro protein synthesis has been amply demonstrated in the literature. It is anticipated that the wheat embryo system will serve a similarly useful role for in vitro transcription studies.

In this report is presented chromatographic evidence for three RNA polymerases in ungerminated wheat embryos. The enzymes were

characterized with respect to their α-amanitin sensitivities. A
large-scale procedure which allows the simultaneous purification
of all three enzymes is briefly described. The resulting enzymes
were characterized with respect to their subunit structures.
Finally, changes in the subunit structure of RNA polymerase II were
detected during wheat embryo germination. These will be described
and related to changing rates of nuclear RNA synthesis during
germination.

RESOLUTION OF RNA POLYMERASES I, II AND III FROM UNGERMINATED
WHEAT EMBRYOS

    RNA polymerases are usually separated by chromatography of
cellular, nuclear or chromatin extracts on DEAE-Sephadex. Figure
1a illustrates the separation of wheat embryo RNA polymerases on
DEAE-Sephadex. A small peak of RNA polymerase I elutes at about
0.1 M ammonium sulfate, a large peak of RNA polymerase II elutes
at 0.2 M ammonium sulfate, and an almost undetectable peak of RNA
polymerase III elutes at about 0.3 M ammonium sulfate. If RNA

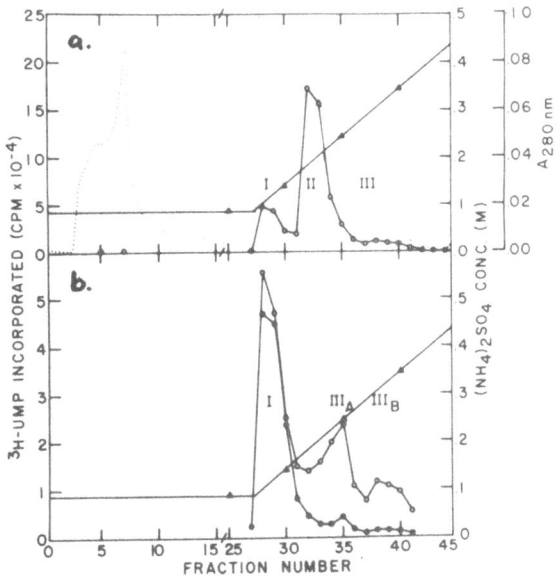

FIGURE 1. Resolution of wheat embryo RNA polymerases I, II and III
by DEAE-Sephadex chromatography. a) RNA polymerase assays of column
fractions were done in the absence of α-amanitin. RNA polymerase
activity (o); ammonium sulfate concentration (Δ); protein (...).
b) RNA polymerase activities were determined in the presence of
1 ug/ml α-amanitin (o) or 100 ug/ml α-amanitin (●).

polymerase assays are done in the presence of low concentrations of
α-amanitin (1 ug/ml), the profile seen in Figure 1b is obtained
(open circles). The ordinate scale has been magnified 10 fold to
aid in the presentation of RNA polymerase I and III activities.
Note that the RNA polymerase II peak has been completely inhibited.
RNA polymerase III which has been partially obscured by the large
amount of RNA polymerase II appears as two peaks called IIIA and
IIIB. When assays are conducted in the presence of a higher concentration of α-amanitin (100 ug/ml), the profile represented by
closed circles in Figure 1b is obtained. RNA polymerases IIIA
and IIIB are similarly inhibited by the higher dose of α-amanitin
but the RNA polymerase I activity is largely unaffected by even
100 ug/ml α-amanitin. Heterogeneity in RNA polymerase III has
been demonstrated in animal systems as well[1]. The physical basis
for the heterogeneity in wheat embryo RNA polymerase III is unknown.

α-AMANITIN INHIBITION PROPERTIES OF WHEAT EMBRYO RNA POLYMERASES
I, II AND III

From Figure 1 can be gained an appreciation of the relative
sensitivities of RNA polymerases I, II and III to α-amanitin. This
is more quantitatively presented in Figure 2 in which the activities
of the enzymes are determined in the presence of various levels of
α-amanitin. RNA polymerase I is refractory to inhibition even by
100 ug/ml α-amanitin, RNA polymerase II is 50% inhibited by 0.05
ug/ml α-amanitin, and RNA polymerase III is 50% inhibited by 5 ug/ml
α-amanitin. Animal and plant RNA polymerases exhibit remarkably
similar α-amanitin inhibition properties [1,2].

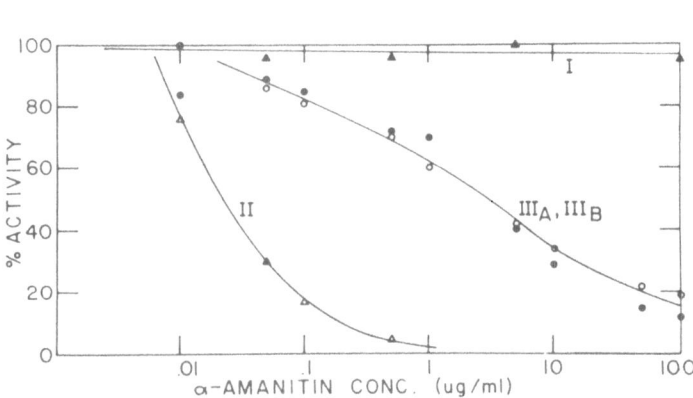

FIGURE 2. α-amanitin inhibition of RNA polymerases I, II, IIIA
and IIIB from wheat embryos. Enzymes were individually assayed
in the presence of various concentrations of α-amanitin.

TRANSCRIPTIONAL ROLES OF RNA POLYMERASES I, II AND III

Information concerning the transcriptional roles of the multiple plant RNA polymerases is still rather incomplete. To elucidate the roles in animal systems, isolated nuclei have been employed for in vitro RNA synthesis in the presence of various concentrations of α-amanitin[1]. It has been established that in nuclei, RNA chains that were initiated in vivo are elongated in vitro. RNA polymerases I and II appear to be unable to reinitiate RNA chains in vitro in nuclei, but RNA polymerase III appears to reinitiate chains repeatedly if assay conditions are controlled[3]. With no α-amanitin present in an assay, the activity of all three RNA polymerases is expressed. With low concentrations of α-amanitin (1 ug/ml), RNA polymerase II activity is completely inhibited and only RNA polymerases I and III are expressed. With high concentrations of α-amanitin present, both RNA polymerases II and III are inhibited and only RNA polymerase I activity is expressed. In vitro labelled RNA products synthesized in the presence of various concentrations of α-amanitin are analyzed by competition hybridization experiments against 18 and 28S ribosomal RNA, cDNA copies of purified messenger RNAs, viral RNAs, transfer RNAs and 5S ribosomal RNA. In the case of RNA polymerase III, complete transcripts are made in vitro and the RNA is small enough such that size analysis by polyacrylamide gel electrophoresis is also useful in assigning a role for RNA polymerase III.

Our information in plant systems is incomplete since few studies have been done with isolated nuclei. Relevant studies have been done with soybean nuclei[4] which indicate that RNA products synthesized in the presence of 1 ug/ml α-amanitin are competed by 28S and 18S ribosomal RNA in hybridization experiments. RNA synthesized by RNA polymerase II on the other hand was not competed by these ribosomal RNAs. Transcription by RNA polymerase III was not considered in these studies. We therefore have no direct proof that plant RNA polymerase II synthesizes precursors to any messenger RNAs nor that RNA polymerase III synthesizes precursors to transfer RNAs or 5S ribosomal RNA. cDNA probes to plant messenger RNAs and nuclear systems displaying vigorous RNA polymerase III activity will be useful in establishing the roles of these enzymes in plants. Due to the remarkable similarity in properties between plant and animal enzymes, it is anticipated that RNA polymerase II will eventually be shown to synthesize precursors to messenger RNAs and RNA polymerase III to synthesize small cellular RNAs.

PURIFICATION OF RNA POLYMERASES I, II AND III

Two approaches have been used for the purification of eukaryotic RNA polymerases. One approach takes advantage of the fact that RNA polymerases can be isolated with nuclei[5] or chromatin, thus affording significant purification from the total protein in a crude extract. RNA polymerases are extracted from nuclei or chromatin by high salt extraction, sometimes coupled with sonication[1]. Nucleic acids are removed by high speed centrifugation and the soluble proteins are concentrated by ammonium sulfate precipitation. Further purification involves standard column chromatography on a variety of ion exchange resins and sizing by gel filtration or sedimatation in sucrose or glycerol gradients.

Another approach takes advantage of the fact that a portion of the RNA polymerase activity (in some instances the bulk of the enzyme activity) is not tightly associated with nuclei or chromatin and can be readily extracted from tissues by simple homogenization in buffers of low or moderate ionic strength[6,7]. A disadvantage of this method is that the specific activity of the RNA polymerases in such extracts is very low compared to activities obtained from nuclear or chromatin extracts (perhaps causing some investigators to ignore such soluble enzymes). Treatment of crude extracts with polyethyleneimine (Polymin P®) however results in a very efficient initial purification and concentration of RNA polymerases[7]. Polyethyleneimine, a polycation, precipitates acidic macromolecules including chromatin, RNA, ribosomes and RNA polymerases from crude extracts while leaving a bulk of the cellular protein soluble and eliminated by centrifugation. RNA polymerases can be extracted from the polyethyleneimine precipitate by suspending it in a buffer of higher ionic strength. Nucleic acids remain precipitated along with a bulk of the protein, also eliminated by centrifugation. RNA polymerases are concentrated by ammonium sulfate precipitation which also affords further purification and removes residual polyethyleneimine. Enzymes are further purified by standard column chromatography. The polyethyleneimine method has been used for the purification of bacterial RNA polymerase[8,9], RNA polymerases I, II and III from wheat embryos and Acanthamoeba castellanii[10,11,12], RNA polymerase II from soybean hypocotyls and embryos[13], cauliflower influorescences[2], maize mesocotyls[2], calf thymus[14], Physarum polycephalum[15], Agaricus bisporus[16], and RNA polymerases II and III from yeast[17].

One other recent advance in the purification of RNA polymerases is the use of heparin agarose chromatography[18]. Heparin, a sulfated polysaccharide, has long been known to be a very potent inhibitor of polymerases[19]. The enzymes bind to heparin very tightly. Presumably heparin mimicks the phosphodiester backbone of DNA. RNA polymerases can be applied to heparin agarose columns at very high ionic strength thus minimizing protein aggregation which sometimes hampers purification of RNA polymerases. Most other proteins do not bind to heparin agarose under these conditions and are readily purified from RNA polymerases. Little resolution of RNA polymerases I, II and III is achieved on heparin agarose and separation of the enzymes is delayed until a later step. Usually RNA polymerases are resolved on DEAE-Sephadex, but a number of other ion exchangers can be used for further purification and separation of RNA polymerases I, II and III. Table 1 lists the chromatographic elution positions of wheat embryo RNA polymerases I, II and III from several ion exchangers commonly used in the purification of these enzymes.

A purification scheme which we use for the purification of RNA polymerases I, II and III from wheat embryos and which takes advantage of the polyethyleneimine and heparin agarose steps is summarized in Figure 3.

Approximately 3, 30 and 2 mg of RNA polymerases I, II and III can be purified from 1 kg of starting material in 5 days each at a yield of 30-50%. RNA polymerases I, II and III are purified approximately 40,000; 4,000 and 80,000 fold. Milligram quantities of all three enzymes afford the opportunity to perform subunit structural analyses. The high degree of purity and large quantities of all enzymes make them useful for in vitro transcription studies as well as sources of antigens for antibody production.

SUBUNIT STRUCTURES OF WHEAT EMBRYO RNA POLYMERASES I, II AND III

Figure 4 shows dodecyl sulfate polyacrylamide gel electro-phoresis patterns of wheat embryo RNA polymerases I, II and III. Figure 4a presents a 5% gel which separates the high molecular weight subunits optimally, while the smaller subunits migrate with the marker dye. It can be seen that RNA polymerases I, II and III possess two or three high molecular weight subunits which are

Table 1.   Chromatographic Properties of Wheat Embryo RNA
          Polymerases I, II and III

Ion Exchange Resin	RNA Polymerase Class		
	I	II	III
DEAE-Sephadex	0.1 M	0.2 M	0.3 M
DEAE-Cellulose	0.1 M	0.2 M	0.1 M
Phosphocellulose	0.2 M	0.11 M	0.12 M

Values represent concentrations of ammonium sulfate required to elute RNA polymerases from the indicated ion exchangers.

different in the three RNA polymerases. Figure 4b presents a 15%
polyacrylamide gel which separates the low molecular weight subunits efficiently. It can be seen that the enzymes are also distinctly different concerning the smaller subunits. Subunit molecular weights for these enzymes are listed in Table 2.

COMMON SUBUNITS IN RNA POLYMERASES I, II AND III

From Table 2 and Figure 4b it can be seen that RNA polymerases
I, II and III from wheat embryos apparently share small subunits
in common with molecular weights of 20,000; 17,800 and 17,000.
Two dimensional mapping (isolectric focussing-SDS polyacrylamide
gel electrophoresis) and peptide mapping studies should confirm
these observations. Peptide mapping of several low molecular
weight subunits shared by yeast RNA polymerases I and II has been
done[20]. One dimensional polyacrylamide gel electrophoresis of
animal RNA polymerases I, II and III suggests that several low
molecular weight subunits may be shared in common between all
three enzyme classes[21].

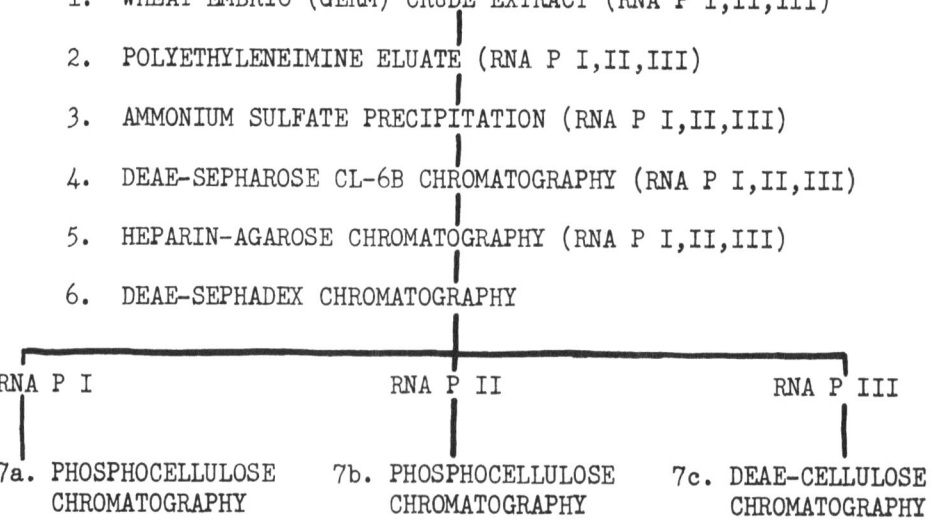

1. WHEAT EMBRYO (GERM) CRUDE EXTRACT (RNA P I,II,III)

2. POLYETHYLENEIMINE ELUATE (RNA P I,II,III)

3. AMMONIUM SULFATE PRECIPITATION (RNA P I,II,III)

4. DEAE-SEPHAROSE CL-6B CHROMATOGRAPHY (RNA P I,II,III)

5. HEPARIN-AGAROSE CHROMATOGRAPHY (RNA P I,II,III)

6. DEAE-SEPHADEX CHROMATOGRAPHY

RNA P I   RNA P II   RNA P III

7a. PHOSPHOCELLULOSE   7b. PHOSPHOCELLULOSE   7c. DEAE-CELLULOSE
    CHROMATOGRAPHY   CHROMATOGRAPHY   CHROMATOGRAPHY

FIGURE 3. Schematic for the purification of wheat embryo (germ)
RNA polymerases I, II, and III.

COMPARATIVE SUBUNIT STRUCTURES OF RNA POLYMERASE II FROM VARIOUS
PLANT SPECIES

Subunit structures for RNA polymerase II have been more firmly
established than those for RNA polymerases I and III. This enzyme
is far more abundant in most tissues than the other enzymes and
has consequently been purified from many plant species. Comparative
subunit structural studies have been done on one dimensional and
two dimensional polyacrylamide gel electrophoresis systems[2]. It
has been shown that analogous subunits can be identified on the
basis of charge and molecular weight in several poant species, a
fact which reinforces the contention that these are indeed RNA
polymerase subunits. Comparative subunit structural studies of

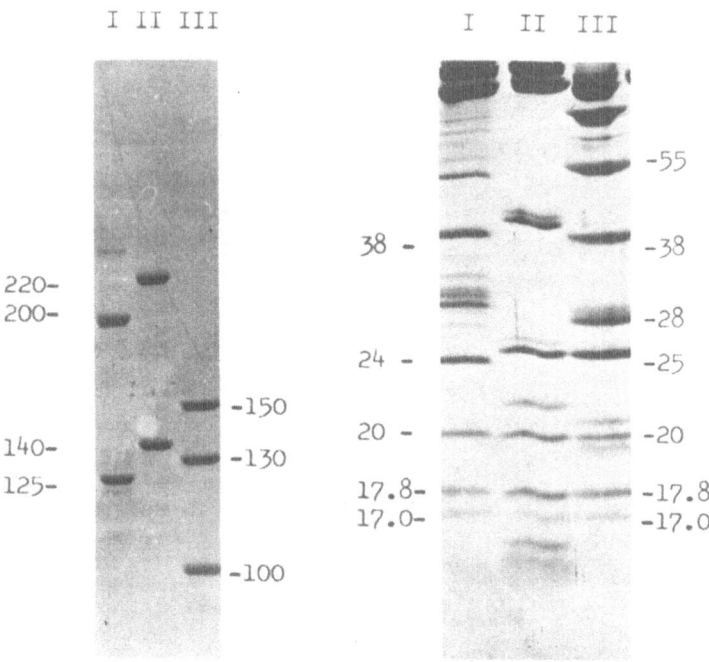

FIGURE 4. Dodecyl sulfate polyacrylamide gel electrophoresis of
wheat embryo RNA polymerases I, II and III. a) 1 ug quantities of
the indicated enzymes were applied to a slab gel containing 5%
polyacrylamide. b) 5 ug quantities of the indicated enzymes were
applied to a slab gel containing 15% polyacrylamide. Migration
was toward the anode from top to bottom.

RNA polymerases I and III should likewise more firmly establish
the subunit composition of these enzymes as well. The subunit
structures of RNA polymerases II from a variety of plant species
are also listed in Table 2.

Table 2. Subunit Structures of RNA Polymerases I, II and III
          From Higher Plants

| I | Subunit Molecular Weights x 10^{-3} | | | | III |
	II				
wheat embryo	wheat embryo	maize mesocotyl	soybean embryo	cauliflower inflorescence	wheat embryo
200	220	185	215	180	150
125	140	145	138	140	130
					100
38	42+40†	40	42	40	55
					38
24	27+25‡	27+26	27	25	28
	21	22	22	22	25
20	20	20	19	19	20
17.8	17.8	17.8	17.6	17.5	17.8
17.0	17.0	17.5	17.0	17.0	17.0
	16.3	16.3	16.2	16.2	
	16.0	16.1	16.1	16.0	
	14	14	14	14	

†Two subunits of approximately 40,000 mol. wt. are seen in wheat
RNA polymerase II which have an additive molar ratio of 1 relative to the rest of the subunits.

‡Two subunits of approximately 25,000 mol. wt. are seen in wheat
and maize RNA polymerase II which have an additive molar ratio
of 2 relative to the rest of the subunits.

All of the other subunits in all of the rest of the enzymes are
present in a molar ratio of 1 except for the subunits with molecular weights of approximately 25,000. These are present at a
molar ratio of 2. Numbers in squares represent subunits believed
to be shared in common among wheat RNA polymerases I, II and III.

DIVERSITY IN THE SUBUNIT STRUCTURE OF RNA POLYMERASE II

As seen in Table 2, there is a major difference in the subunit structure of RNA polymerase II purified from several plant sources. RNA polymerase II purified from wheat and soybean embryos have a large subunit with a molecular weight of approximately 220,000. These are called IIA enzymes. RNA polymerase II purified from maize mesocotyls and cauliflower inflorescences have a large subunit of 180,000 molecular weight and none at 220,000 molecular weight. These are called IIB enzymes. The enzymes purified from quiescent tissues were all IIA and those from metabolically active tissues were all IIB enzymes. Presumably the IIA enzymes give rise to IIB enzymes by a specific proteolysis. The question arises whether the IIB enzyme is merely an artifact of enzyme purification due to the presence of relatively high amounts of proteases in metabolically active tissues. In yeast[20] and _Drosophila_ larvae[22], IIB enzymes are largely obtained. In most animal systems a mixture of IIA and IIB is generally obtained[1].

CONVERSION OF WHEAT EMBRYO RNA POLYMERASE IIA TO IIB DURING GERMINATION

In order to investigate the significance of RNA polymerases IIA and IIB further, RNA polymerase II was purified from germinating wheat embryos to see if conversion occurred during early stages of germination and to see how rapidly it occurred. Figure 5 illustrates the results from this study. It can be seen that the 180,000 molecular weight subunit increases during germination at the expense of the 220,000 molecular weight subunit. Half of the RNA polymerase II appears to have been converted 24 hours after inbibition. Peptide maps of the 220,000 and 180,000 molecular weight polypeptides are similar, indicating that the 220,000 molecular weight subunit is the precursor for the 180,000 molecular weight subunit (data not shown). Similar studies have been previously done with soybean RNA polymerases IIA and IIB[13].

A likely explanation for the appearance of the IIB enzyme is that it is merely an artifact produced during enzyme purification due to an increase in protease activity in the tissue during germination. To test this hypothesis, equal amounts of ungerminated and 36 hour germinated embryos were combined and RNA polymerase II was purified from the combined tissues. If no in vitro proteolysis occurred, RNA polymerase IIA should be recovered in equal amounts with RNA polymerase IIB. If extensive proteolysis occurred, very little or no IIA should be recovered, having been converted to the IIB enzyme during purification. The results in Figure 6 indicate that equal amounts of RNA polymerases IIA and IIB were recovered suggesting that extensive proteolysis did not

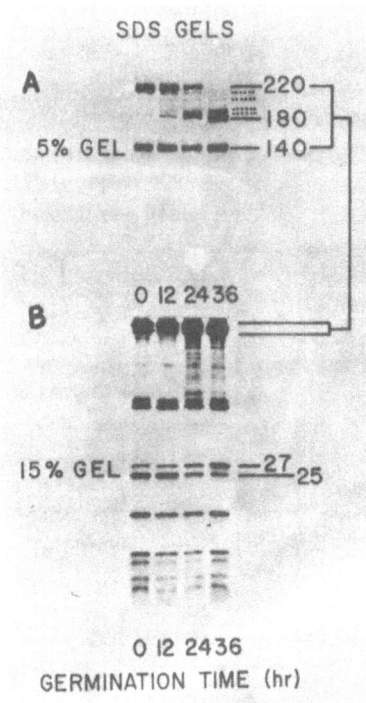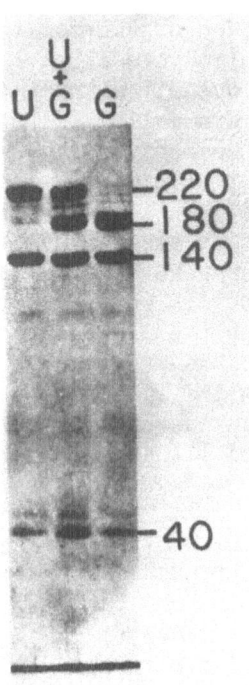

FIGURE 5. Conversion of wheat embryo RNA polymerase IIA to IIB
during germination. RNA polymerase II was purified from wheat
embryos germinated for 0, 12, 24 and 36 hours. Enzymes were
applied to A) a 5% polyacrylamide slab gel which contained dodecyl
sulafate, or B) a 15% polyacrylamide slab gel containing dodecyl
sulfate. Numbers to the right of the gels indicate polypeptide
molecular weights x 10^{-3}. Migration was toward the anode from
top to bottom.

FIGURE 6. Extent of proteolysis of wheat embryo RNA polymerase IIA
during purification. A mixing experiment was performed by combining
one part 36 hour germinated embryos with one part ungerminated
embryos. RNA polymerase II was purified from the combined tissues.
Enzymes were analyzed by dodecyl sulfate 7.5% polyacrylamide slab
gel electrophoresis. One ug amounts of enzyme from ungerminated
(U) embryos; from the mixing experiment containing ungerminated
and 36 hour germinated embryos (U+G); and from 36 hour germinated
embryos (G) were analyzed. Numbers to the right of the gel refer
to polypeptide molecular weights x 10^{-3}. Migration was toward the
anode from top to bottom.

occur and that the IIB enzyme may not be merely an artifact. Similar mixing experiments have been reported previously with soybeans[13] and the results agree with those obtained here.

The appearance of the IIB enzyme correlates with the activation of RNA synthesis during embryo germination. The nuclear RNA polymerase II activity increases 13 fold after 36 hours of germination compared to the activity in nuclei isolated from ungerminated embryos, although the total amount of RNA polymerase II in the embryo appears to be constant over this same time period. A hypothesis has been put forth that the IIA enzyme may be a storage or precursor enzyme which is activated for transcription by conversion to a IIB enzyme[13]. In vitro reconstitution-transcription experiments may shed further light on this hypothesis.

ADDITIONAL MODIFICATION IN WHEAT EMBRYO RNA POLYMERASE II DURING GERMINATION

Examination of the subunit structure of RNA polymerase II gels of lower porosity (Figure 4b) indicates that another subunit structural modification has occurred. As mentioned in Table 2, two subunits are observed in wheat with molecular weights of 27,000 and 25,000 with molar rations of approximately 0.4 and 1.6 which add to 2.0. As germination proceeds, the quantity of the 27,000 molecular weight subunit increases at the expense of the 25,000 molecular weight subunit while still maintaining the additive stiochiometry of 2.0. The results are more clearly shown in Figure 7 which shows the low molecular weight subunits of ungerminated and 36 hour germinated embryos. The molecular basis of this transition is unknown. The modification does not appear to be due to phosphorylation or ribosyladenylation since treatment of RNA polymerases IIA or IIB with bacterial alkaline phosphatase or snake venom phosphodesterase did not alter the proportions of the 27,000 or 25,000 molecular weight polypeptides. When embryos were germinated in the presence of ^{32}P-phosphate, no label was incorporated into either of these two polypeptides although label was incorporated into the 220,000 molecular weight subunit (data not shown). Both the 25,000 and 27,000 molecular weight polypeptides behave as very basic proteins as indicated by isotachophoresis (non-equilibrium pH gel electrophoresis) and they appear to be related in gross overall composition (data not shown). Peptide mapping studies should further clarify the relationship of the two polypeptides.

CONCLUSIONS

Methodology has been presented for the purification of RNA polymerases I, II and III from wheat embryos which have been

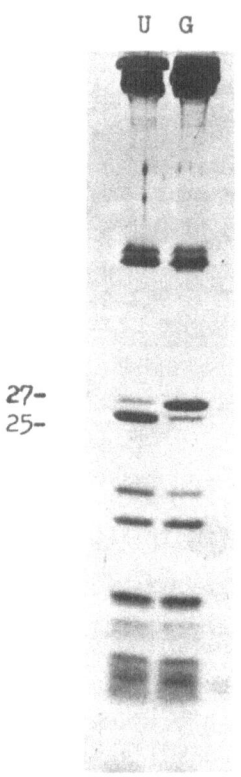

FIGURE 7.   Alteration in low molecular weight subunits of RNA
polymerase II during germination.   Enzymes purified at various
stages of germination were subjected to dodecyl sulfate 15% polyacrylamide slab gel electrophoresis (Figure 5).   Shown here are
RNA polymerase II enzymes purified from ungerminated (U) and 36
hour germinated (G) embryos.   Numbers to the left of the gel
refer to polypeptide molecular weights x 10^{-3} which change during
germination.   Migration was toward the anode from top to bottom.

characterized with respect to their chromatographic properties,
α-amanitin sensitivities and subunit structures.   Alterations
in the subunit structure of RNA polymerase II were detected during
the course of embryo germination when the nuclear RNA synthetic
rates were increasing.

REFERENCES

1. R. G. Roeder, Nuclear RNA Polymerases, in "RNA Polymerase",
   R. Losick and M. Chamberlin eds., Cold Spring Harbor
   Laboratory, New York (1976).
2. J. Jendrisak and T. J. Guilfoyle, Eukaryotic RNA Polymerases:
   Comparative Subunit Structures, Immunological Properties,
   and α-amanitin Sensitivities of the Class II Enzymes from
   Higher Plants, Biochemistry 17:1322 (1978).
3. R. Weinmann, T. G. Brendler, H. J. Raskas, and R. G. Roeder,
   Low Molecular Weight Viral RNAs Transcribed by RNA Polymerase III during Adenovirus 2 Infection, Cell 7:557 (1976).
4. W. B. Gurley, C. Y. Lin, Y. M. Chen, R. T. Nagao, and J. L.
   Key, Synthesis of Ribosomal RNA by Soybean Chromatin-bound
   RNA Polymerase, Biochim. Biophys. Acta 418:344 (1976).
5. S. B. Weiss and L. Gladstone, A Mammalian System for the Incorporation of Cytidine Triphosphate into Ribonucleic Acid,
   J. Amer. Chem. Soc. 81:4118 (1959).
6. R. J. Mans and G. D. Novelli, Ribonucleotide Incorporation by
   a Soluble Enzyme from Maize, Biochim. Biophys. Acta 91:186
   (1964).
7. J. J. Jendrisak and R. R. Burgess, A New Method for the Large-
   Scale Purification of Wheat Germ DNA-Dependent RNA Polymerase II, Biochemistry 14:4639 (1975).
8. W. Zillig, K. Zechel, and H. Halbwachs, A New Method of Large
   Scale Preparation of Highly Purified DNA-Dependent RNA
   Polymerase from E. coli, Hoppe-Seyler's Z. Physiol. Chem.
   351:221 (1970).
9. R. R. Burgess and J. J. Jendrisak, A Procedure for the Rapid,
   Large-Scale Purification of Escherichia coli DNA-Dependent
   RNA Polymerase Involving Polymin P Precipitation and DNA-
   Cellulose Chromatography, Biochemistry 14:4634 (1975).
10. S. R. Spindler, G. L. Duester, J. M. D'Alessio, and M. R.
    Paule, A Rapid and Facile Procedure for the Preparation of
    RNA Polymerase I from Acanthamoeba castellanii, J. Biol.
    Chem. 253:4669 (1978).
11. J. M. D'Alessio, S. R. Spindler, and M. R. Paule, DNA-Dependent
    RNA Polymerase II from Acanthamoeba castellanii, J. Biol.
    Chem. 254:4085 (1979).
12. S. R. Spindler, J. M. D'Alessio, G. L. Duester and M. R. Paule,
    DNA-Dependent RNA Polymerase III from Acanthamoeba castel-
    lanii, J. Biol. Chem. 253:6242 (1978).
13. T. J. Guilfoyle and J. Jendrisak, Plant DNA-Dependent RNA Poly-
    merases: Subunit Structures and Enzymatic Properties of
    the Class II Enzymes from Quiescent and Proliferating
    Tissues, Biochemistry 17:1860 (1978).

14. H. G. Hodo III and S. P. Blatti, Purification Using Poly-
    ethyleneimine Precipitation and Low Molecular Weight
    Subunit Analyses of Calf Thymus and Wheat Germ DNA-De-
    pendent RNA Polymerase II, Biochemistry 16:2334 (1977).

15. S. S. Smith and R. Braun, A New Method for the Purification of
    RNA Polymerase II (or B) from the Lower Eukaryote Physarum
    polycephalum, Eur. J. Biochem. 82:309 (1978).

16. A. C. Vaisius and P. A. Horgen, Purification and Characteri-
    zation of RNA Polymerase II Resistant to α-amanitin from
    the Mushroom Agaricus bisporus, Biochemistry 18:795 (1979).

17. G. I. Bell, P. Valenzuela, and W. J. Rutter, Phosphorylation
    of Yeast DNA-Dependent RNA Polymerases in Vivo and in Vitro,
    J. Biol. Chem. 252:3082 (1977).

18. H. Sternbach, R. Engelhardt, and A. G. Lezius, Rapid Isolation
    of Highly Active RNA Polymerase from Escherichia coli and
    Its Subunits by Matrix-Bound Heparin, Eur. J. Biochem.
    60:51 (1975).

19. G. Waster, W. Zillig, P. Palm and E. Fuchs, Initiation of DNA-
    Dependent RNA Synthesis and the Effect of Heparin on RNA
    Polymerase, Eur. J. Biochem. 3:194 (1967).

20. A. Sentenac, S. Dezelee, F. Iborra, J-M. Buhler, J. Huet,
    F. Wyers, A. Ruet, and P. Fromageot, Yeast RNA Polymerases,
    in "RNA Polymerase", R. Losick and M. Chamberlin eds., Cold
    Spring Harbor Laboratory, New York (1976).

21. V. E. F. Sklar, L. B. Schwartz, and R. G. Roeder, Distinct
    Molecular Structures of Nuclear Class I, II, and III DNA-
    Dependent RNA Polymerases, Proc. Nat. Acad. Sci. 72:348
    (1975).

22. A. L. Greenleaf and E. K. F. Bautz, RNA Polymerase B from
    Drosophila melanogaster larvae: Purification and Partial
    Characterization, Eur. J. Biochem. 60:169 (1975).

RNA POLYMERASES AND TRANSCRIPTION DURING DEVELOPMENTAL

TRANSITIONS IN SOYBEAN

Tom Guilfoyle, Neil Olszewski, and Linda Zurfluh

Department of Botany
University of Minnesota
St. Paul, Minnesota 55108

INTRODUCTION

We have chosen two developmental transitions in soybean
(Glycine max), (1) germination of embryonic axes and (2) auxin-
induced cellular proliferation in mature hypocotyl, to investigate
the effects of these transitions on DNA-dependent RNA polymerases
and nuclear transcription. In both of these developmental transitions, relatively quiescent tissues are induced to proliferate.
Our experimental approach has been to examine the levels and
possible structural modification of RNA polymerases and attempt to
correlate these with the rate of transcription in isolated nuclei
and chromatin during various stages of growth and development in
soybean.

GERMINATION OF SOYBEAN EMBRYONIC AXES

Germination of soybean embryonic axes presents a developmental
system where highly quiescent, dehydrated tissue undergoes a transition to rapid growth following the imbibition of water under
appropriate environmental conditions. There is evidence from
studies with several plant species that with the onset of germination and growth of the embryonic axis, nucleic acid synthesis
rises in a relatively synchronous manner (1, 2, 3). We have
investigated whether the levels and/or modification of RNA polymerases play a major role in regulating transcription during germination of soybean embryonic axes.

During the first 48 hours of germination, RNA polymerase
activity assayed in isolated nuclei increases many-fold (Figure 1).

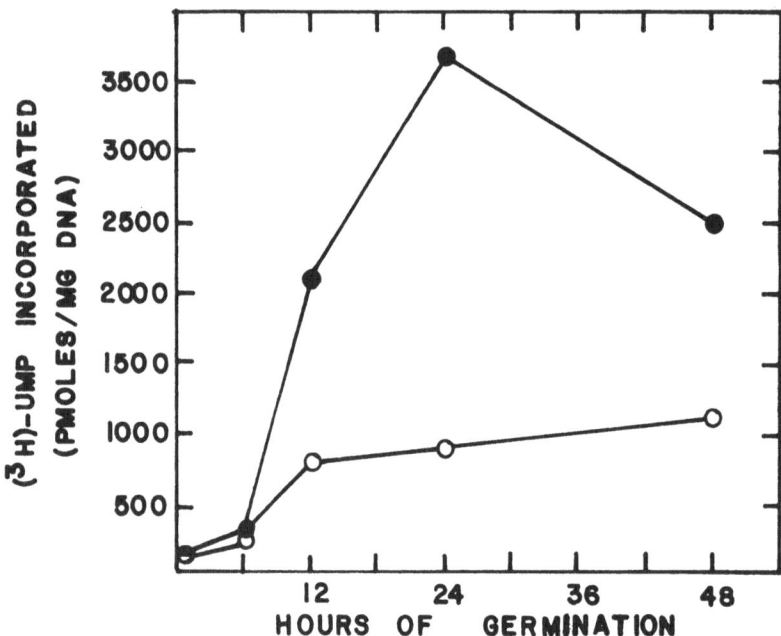

Fig. 1.   RNA polymerase activities associated with isolated nuclei
at various stages during the first 48 hours of germination
of soybean embryonic axes. Soybean seeds were germinated
in the dark at 30C in moist vermiculite and embryonic axes
were excised at the time indicated. Nuclei were isolated
by the method of Chen et al. (4) and assayed for α-amanitin-resistant (●——●) and α-amanitin-sensitive
(○——○) activities as described by Guilfoyle et al. (5).

Both α-amanitin-resistant (RNA polymerases I and III) and α-amanitin-sensitive (RNA polymerase II) RNA polymerase activities
increase in isolated nuclei during the first 24 hours of germination and subsequently level off or decline within the next 24 hour
period. During this same germination period, the major type of
RNA polymerase (RNA polymerase II) present in the embryonic axes
remains at a fairly constant level (Table 1). A high level of
RNA polymerase is largely restricted to the embryonic axes. The
amount of RNA polymerase II found in cotyledons is about 10-fold
less than that found in embryonic axes (Table 1).

We have further investigated the possible modification of RNA
polymerase II during germination since we had suggested earlier
that activation of nuclear-associated RNA polymerase II activity
might be related to the conversion of RNA polymerase IIA to IIB
(6, 7). Figure 2 shows the subunit structures of RNA polymerase II
purified from soybean embryonic axes at various stages during the
first 48 hours of germination. There is a rapid conversion of RNA

Table 1. Summary of purification of RNA polymerase II from 1 kg amounts of soybean cotyledons and embryonic axes germinated for 0, 24, and 48 hours

Fraction[a]	Protein (mg) Cot. 0[b]	Protein (mg) Embryonic Axes 0	Protein (mg) Embryonic Axes 24	Protein (mg) Embryonic Axes 48	Specific Activity (units/mg)[c] Cot. 0	Specific Activity (units/mg)[c] Embryonic Axes 0	Specific Activity (units/mg)[c] Embryonic Axes 24	Specific Activity (units/mg)[c] Embryonic Axes 48
1. Crude extract	132,000	238,000	21,000	7,500	0.012	0.048	0.052	0.050
2. (NH4)2SO4 precip.	3,720	11,550	915	320	0.34	0.83	0.87	0.88
3. DEAE cellulose	90	126	9.3	3.5	13	69	72	75
4. Phospho-cellulose	4.3	17.7	1.5	0.6	230	380	395	415
5. DNA Agarose	1.2	10.2	1.0	0.3	525	550	525	535
	(42)[d]	(49)	(48)	(42)		(0.057)[e]	(0.055)	(0.050)

[a] Fraction 2 represents the (NH4)2SO4 precipitate of the Polymin P eluate (8).
[b] Time of germination in hours.
[c] A unit is 1 nmole of UMP incorporated into RNA per 30 min at 30C using the method of Guilfoyle et al. (5).
[d] % recovery.
[e] Units of activity per embryonic axis. Average weights per embryonic axis for 0, 24, and 48 hours of germination were 5, 50, and 135 mg, respectively. Values are corrected for % recovery of the enzymes.

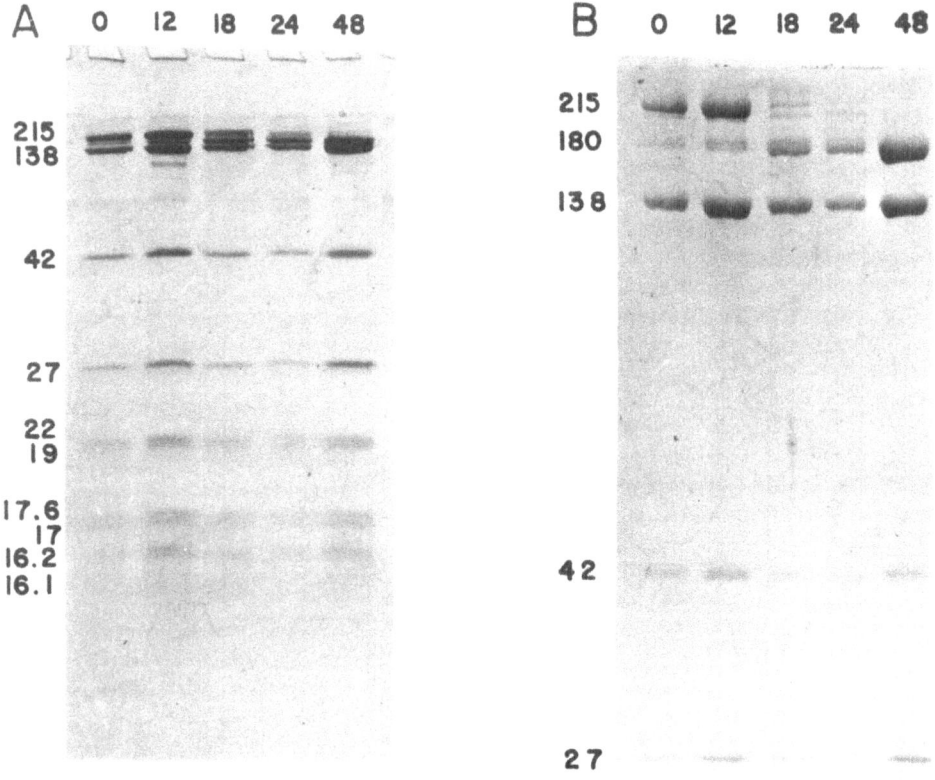

Fig. 2. Subunit structures of RNA polymerase II purified from soybean embryonic axes during various stages of germination.
RNA polymerases were purified according to the scheme
summarized in Table 1 and subjected to electrophoresis in
the presence of dodecyl sulfate (7). Numbers above each
well indicate the time of germination in hours. Numbers
to the left or right of each figure indicate molecular
weights of RNA polymerase II subunits in kilodaltons (7).
A: 15% polyacrylamide gel; B: 7.5% polyacrylamide gel.

polymerase IIA to IIB within the first 24 hours of germination.
This corresponds to a period of germination when nuclear-associated
activity rises dramatically (Figure 1). Thus there is some correlation between the rise in RNA polymerase activity associated with
isolated nuclei and the conversion of RNA polymerase IIA to IIB.
This conversion of RNA polymerase IIA to IIB should be interpreted
cautiously, however, since this conversion may be artifactual in
nature. Immunoprecipitates of partially purified enzymes during
24 and 48 hours of germination contain significant quantities of
RNA polymerase IIA while enzymes purified to homogeneity are almost
exclusively IIB enzymes (data not shown). This suggests that at

least a part of the conversion of RNA polymerase IIA to IIB occurs
during purification of the enzymes eventhough mixing experiments
indicate that conversion is not an artifact of purification (7).

Several conclusions can be drawn from the results we have
obtained with germinating soybean embryonic axes. (1) There is a
large amount of RNA polymerase associated with ungerminated soybean
embryos. This large quantity of enzyme is restricted to the embryonic
axes and is not found in large quantities in cotyledons. (2) During
the first 24 hours of germination, there is a rapid rise in nuclear
DNA transcription with little or no increase in total cellular RNA
polymerase. This suggests that the RNA polymerases may be dissociated
from the nuclear DNA template prior to germination and as germination
progresses, there is an increase in template engaged enzymes. (3)
There is a rapid conversion of RNA polymerase IIA to IIB during the
first 24 hours of germination as determined by purification of the
enzymes to homogeneity; however, immunoprecipitation of cruder
fractions of RNA polymerase II indicate that at least a portion of
RNA polymerase IIB may arise by an artifact of purification.

AUXIN-INDUCED CELLULAR PROLIFERATION IN MATURE SOYBEAN HYPOCOTYL

Application of auxin (2,4-dichlorophenoxyacetic acid) to etiolated 3-day old soybean seedlings results in large increases in cell
division, protein and nucleic acid synthesis, and chromatin-bound
RNA polymerase activity (reviewed by 9, 10). We have continued to
investigate the mechanisms underlying the activation of RNA synthesis
in mature soybean hypocotyl tissue induced by auxin. In contrast
to the germinating soybean embryonic axis where levels of RNA polymerase remain relatively constant, during auxin-induced cellular
proliferation there is a large increase in the levels of RNA polymerase I (Table 2) and RNA polymerase II (Table 3). Investigation
of possible subunit modification during auxin-induced cellular
proliferation has been completed with RNA polymerase II, but
no modification of has been detected in the subunit structure of
the Class II enzymes following auxin treatment (Figure 3). Preliminary results also indicate that the subunit structure of RNA
polymerase I is identical in untreated and auxin-treated hypocotyls
(data not shown).

We have labeled rootless soybean seedlings (11) with ^{35}S-
methionine to determine whether auxin results in increased de novo
synthesis of RNA polymerase subunits. Auxin application results
in a several-fold increase of ^{35}S incorporation into RNA polymerase
I (Figure 4) and RNA polymerase II (Figure 5) subunits. However,
auxin application does not result in changes in the level of ^{35}S
incorporation into any one subunit relative to other RNA polymerase
polypeptides.

Table 2. Summary of purification of RNA polymerase I from 1 kg amounts of soybean hypocotyls treated with auxin for 0, 12, 24, and 48 hours

Fraction	Protein (mg) Time after auxin-treatment (hours)				Specific Activity (units/mg)[a] Time after auxin-treatment (hours)			
	0	12	24	48	0	12	24	48
1. Solubilized[b] chromatin	15.3	31	49	125	0.7	0.95	1.3	1.7
2. Heparin[c] Sepharose	1.0	3.0	5.2	12.2	8.0	8.2	8.7	12.0
3. DEAE cellulose	0.3	0.7	1.5	4.6	23.5	24.1	25.0	29.0
4. Phospho-cellulose	0.03	0.07	0.12	0.4	195	200	215	220
	(54)[d]	(48)	(40)	(41)				

[a] Specific activity is defined in Table 1.
[b] RNA polymerase I was solubilized from chromatin as described by Guilfoyle et al. (12) in chromatography buffer (7) containing 0.5 M $(NH_4)_2SO_4$. The solubilized RNA polymerase was precipitated with 0.35g/ml of solid $(NH_4)_2SO_4$, and the precipitate was suspended in chromatography buffer (7) and assayed for RNA polymerase activity.
[c] Heparin Sepharose chromatography was conducted similar to the method described by Spindler et al. (13).
[d] % recovery.

Table 3. Summary of purification of RNA polymerase II from 1 kg amounts of soybean hypocotyls treated with auxin for 0, 12, 24, and 48 hours

Fraction[b]	Protein (mg)				Specific Activity (units/mg)[a]			
	Time after auxin-treatment (hours)				Time after auxin-treatment (hours)			
	0	12	24	48	0	12	24	48
1. Crude extract	5,236	7,110	12,950	30,500	0.060	0.067	0.080	0.078
2. $(NH_4)_2SO_4$ precip.	119	158	259	628	3.9	4.5	5.3	5.0
3. DEAE cellulose	1.4	2.0	3.2	7.8	210	225	256	250
4. Phospho-cellulose	0.31	0.45	0.78	1.8	510	517	515	520
	(34)[c]	(33)	(29)	(30)				

[a] Specific activity is defined in Table 1.
[b] Purification procedures are identical to those described by Guilfoyle and Jendrisak (7).
[c] % recovery. Fraction 2 represents 100% since this fraction has more total units than Fraction 1.

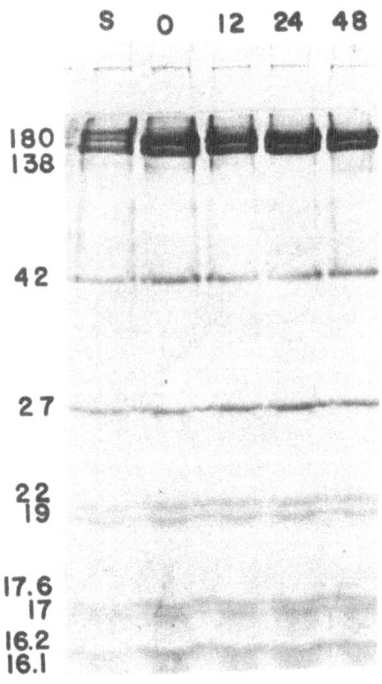

Fig. 3.   Subunit structures of RNA polymerase II purified from
          mature soybean hypocotyls during various times after
          auxin-treatment.   RNA polymerases were purified according
          to the scheme summarized in Table 3, and 3 μg amounts
          of each enzyme were subjected to electrophoresis in the
          presence of dodecyl sulfate (7).   Gels are 15% acrylamide.
          Numbers above wells indicate time after auxin application.
          Numbers to the left of the figure indicate subunit molecular
          weights in kilodaltons.   S is RNA polymerase II purified
          from embryonic axes germinated for 15 hours; this enzyme
          contains equal quantities of the 215,000 and 180,000 dalton
          subunits.

      We have also investigated the phosphorylation of RNA polymerase
subunits in untreated and auxin-treated rootless seedlings incubated
in the presence of ^{32}P-orthophosphate.   Although there is increased
incorporation of ^{32}P into the 215,000 and 42,000 dalton subunits
of RNA polymerase II following auxin application, the labeling
pattern is identical in untreated and auxin-treated tissues (data
not shown).

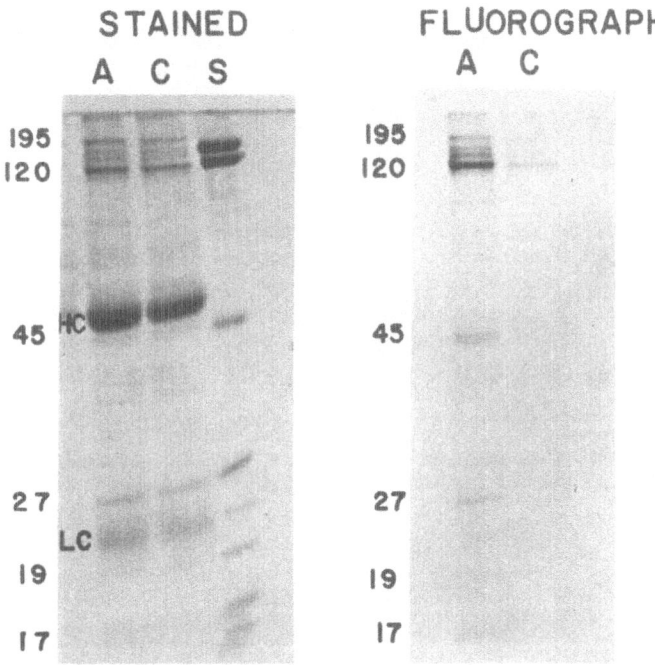

Fig. 4.   Incorporation of ^{35}S into RNA polymerase I subunits.
Soybean rootless seedlings were labeled for 24 hours in
the presence of ^{35}S-methionine. The labeled mature
hypocotyls (10 g) were combined with 200 g of unlabeled
48 hour auxin-treated hypocotyls and RNA polymerase I was
purified through the DEAE cellulose fraction as described
in Table 2. Approximately 10 units of C and A were
immunoprecipitated (14) with antibodies to soybean RNA
polymerase I. The immunoprecipitates were solubilized in
dodecyl sulfate and subjected to electrophoresis on 10-16%
polyacrylamide gels in the presence of dodecyl sulfate.
C is untreated and A is auxin-treated. ^{35}S cpm immunoprecipitated from 10 units of C and A were 50 and 520 cpm,
respectively.

From the results presented above and previous studies (10),
the following conclusions can be made about the effects of auxin-
induced cellular proliferation on RNA polymerases. (1) Application
of auxin (2,4-D) results in large increases in RNA polymerase I and
II in mature soybean hypocotyl. Within 48 hours after application

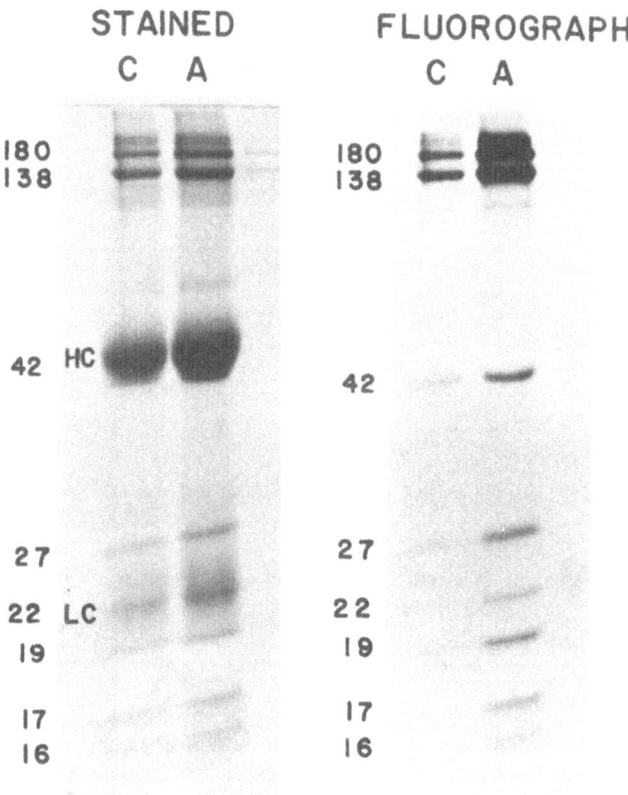

Fig. 5.   Incorporation of ^{35}S into RNA polymerase II subunits.
Details on labeling are found in Fig. 4. RNA polymerase
II was purified through the DEAE cellulose fraction as
described in Table 3. Approximately 50 units of C and A
were immunoprecipitated (14) with antibodies to soybean
RNA polymerase II and analyzed as described in Fig. 4.
C is untreated and A is auxin-treated. ^{35}S cpm immunoprecipitated from 50 units of C and A were 275 and 1770
cpm, respectively. Numbers to the left of the figures
indicate subunit molecular weights in kilodaltons. HC
and LC are the heavy and light chains of IgG fraction.

of auxin, RNA polymerases I and II increase about 10-fold and
6-fold, respectively, when equivalent amounts of tissue are analyzed.
(2) Labeling with ^{35}S-methionine indicates that auxin application
results in enhanced de novo synthesis of RNA polymerases I and II;
labeling patterns are similar in untreated and auxin-treated
hypocotyls. (3) Subunit structures and phosphorylation patterns
of RNA polymerases are identical in untreated and auxin-treated
hypocotyls. (4) Although enhanced RNA synthesis may be partially

attributed to increased levels of RNA polymerases I and II in auxin-treated hypocotyls, increased levels of in vitro transcription observed with isolated nuclei or chromatin are largely due to enhanced RNA polymerase I activity in auxin-treated hypocotyls (5, 10). Much of the enhanced RNA synthesis observed with chromatin isolated from auxin-treated hypocotyls appears to result from increased rates of RNA chain propagation catalyzed by RNA polymerase I (15, Olszewski and Guilfoyle, unpublished).

ACKNOWLEDGEMENTS

     The authors wish to acknowledge the technical assistance of Sandra Malcolm. This research was supported by Public Health Service Grant number GM 24096.

REFERENCES

1. S. Spiegel, R. L. Obendorf, and A. Marcus, Transcription of ribosomal and messenger RNAs in early wheat embryo germination, Plant Physiol. 56: 502 (1975).
2. Z. Grzelczak and J. Buchowicz, A comparison of the activation of ribosomal RNA synthesis during germination of isolated and non-isolated embryos of Triticum aestivum L., Planta 134: 263 (1977).
3. V. Walbot, RNA metabolism during embryo development and germination of Phaseolus vulgaris, Develop. Biol. 26: 369 (1971).
4. Y. M. Chen, C. Y. Lin, H. Chang, T. J. Guilfoyle, and J. L. Key, Isolation and properties of nuclei from control and auxin-treated soybean hypocotyl, Plant Physiol. 58: 78 (1975).
5. T. J. Guilfoyle, C. Y. Lin, Y. M. Chen, R. T. Nagao, and J. L. Key, Enhancement of soybean RNA polymerase I by auxin, Proc. Nat. Acad. Sci. U. S. A. 72: 69 (1975).
6. T. J. Guilfoyle and J. L. Key, The subunit structures of soluble and chromatin-bound RNA polymerase II from soybean, Biochem. Biophys. Res. Commun. 74: 308 (1977).
7. T. J. Guilfoyle and J. J. Jendrisak, Plant DNA-dependent RNA polymerases: subunit structures and enzymatic properties of the class II enzymes from quiescent and proliferating tissues, Biochemistry 17: 1860 (1978).
8. J. J. Jendrisak and R. R. Burgess, A new method for the large-scale purification of wheat germ RNA polymerase II, Biochemistry 14: 4639 (1975).
9. J. L Key, Hormones and nucleic acid metabolism, Ann. Rev. Plant Physiol. 20: 449 (1969).
10. T. J. Guilfoyle and J. L. Key, Purification and characterization of soybean DNA-dependent RNA polymerases and the modulation of their activities during development, in: "Nucleic Acids and Protein Synthesis in Plants," L. Bogorad and J. H Weil, eds., Plenum Press, New York (1977).

11. R. E. Holm and J. L. Key, Inhibition of auxin-induced deoxy-
     ribonucleic acid synthesis and chromatin activity by 5-
     fluorodeoxyuridine in soybean hypocotyl, Plant Physiol. 47:
     606 (1971).
12. T. J. Guilfoyle, C. Y. Lin, Y. M. Chen, and J. L. Key, Purific-
     ation and characterization of RNA polymerase I from a higher
     plant, Biochim. Biophys. Acta 418: 344 (1976).
13. S. R. Spindler, G. Duester, J. M. D'Alessio, and M. R. Paule,
     A rapid and facile procedure for the preparation of RNA
     polymerase I from Acanthamoeba castellanii, J. Biol. Chem.
     253: 4669 (1978).
14. J. M. Buhler, F. Iborra, A. Sentenac, and P. Fromageot, The
     presence of phosphorylated subunits in yeast RNA polymerases
     A and B, FEBS Lett. 71: 37 (1976).
15. T. J. Guilfoyle and J. B. Hanson, Greater length of ribo-
     nucleic acid synthesized by chromatin-bound polymerase from
     auxin-treated soybean hypocotyls, Plant Physiol. 53: 110
     (1974).

ANALYSIS AND RESOLUTION OF mRNA POPULATIONS

J.O. Bishop, P.M. Clissold, J.A. Davis and P. Mason

Department of Genetics
University of Edinburgh
Edinburgh, EH9 3JN

Populations of polyadenylated messenger RNA (mRNA) sequences can conveniently be characterised by studying the kinetics of hybridisation between the mRNA population and a cDNA copy population which is usually synthesised by using reverse transcriptase.[1,2] The most common method is to anneal unlabelled mRNA with radio-actively labelled cDNA, and to measure how much of the latter has become duplexed at different times by treating successive samples of the annealing mixture with nuclease S1. A description of the reaction and its analysis is given in references 1-4.

In some cases the objective of the experiment is to define the total number of different mRNA sequences that are present in a particular cell-type, in a tissue or even in a whole organism. The poly(A)mRNA of yeast contains about 3,500 sequences[5], that of the L3 line of Drosophila cells about 6,000,[6,7] and mouse fibro-blasts contain about 16,000. During its lifetime an individual Drosophila may express 10,000 sequences. Based on a comparison of three mouse tissues[3] we can estimate that the mouse expresses at least 20,000, and the true number may be very much larger. The measurements are not absolute, but they are important in suggesting that during eukaryote evolution there have occurred not only changes in preexisting functions, but also increases in the number of different functions which have come to be, so to speak, at the disposal of more highly evolved organisms.

Before this conclusion can be firmly drawn, however, it is important to ask whether the polyadenylated RNA sequences that we measure are actually mRNAs, in the sense of being translated within the cell. High resolution two dimensional gel electro-phoresis of proteins so far resolves perhaps 2,000 different

polypeptides[8]. In the case of a mammalian cell, nucleic acid
reassociation methods suggest that this may be about 10% of the
total complement of polypeptides if all or most polyadenylated RNA
sequences are translated. Some of the evidence bearing on this
question will be discussed below.

Reassociation experiments between mRNA and cDNA are also
carried out in order to study the most abundant mRNA sequences
in the population. These are often of interest because they are
tissue-specific, and some at least code for known proteins. The
reassociation experiments can tell us how many abundant sequences
are present and how much of the total population they comprise.
They can also be useful in following the purification of a
particular sequence. The abundant sequences are of course the
most accessible to experiment. What we know about mRNA populations
as a whole is mainly based on studies of the abundant sequences.

Chick myoblast mRNA

Chick embryo myoblasts in primary culture fuse together to
form myofibrils over a period of days. When translated in a
cell-free system, poly(A)mRNA from myofibrils causes the synthesis
of seven prominent proteins, among them actin and myosin. These
are seen against a background of many minor proteins, and are
present if at all in greatly reduced amounts in the cell-free
translation products of myoblast poly(A)mRNA.

Reassociation experiments between myofibril mRNA and cDNA show
an abundant component comprising about 20% of the total mass and
representing about 6 distinct sequences. These sequences do not
form a distinct abundance component in myoblast mRNA, but they
are nevertheless present in myoblasts in a lower quantity.

Paterson[9] annealed myofibril mRNA to a low $R_o t$ value with
homologous cDNA, and translated the hybrid mixture in a cell-
free system. Translation of the abundant muscle-specific
proteins was strongly inhibited, but it was restored if the
hybrid mixture was first heated to dissociate duplexes. This
procedure (now called hybrid-arrest translation or HART)
clearly demonstrated a correspondence between abundant
sequences found in reassociation experiments and the proteins
synthesised in abundance when the same mRNA was translated.

Mouse liver mRNA

Hastie and Bishop[3] estimated that mouse liver poly(A)mRNA
contains 9 abundant mRNA sequences, together making up about 22%
of the total population. On average, these sequences are present
in 12,000 copies per liver cell, while the rarest sequences are
present in only 15 copies. Hastie and Held[10] observed a number

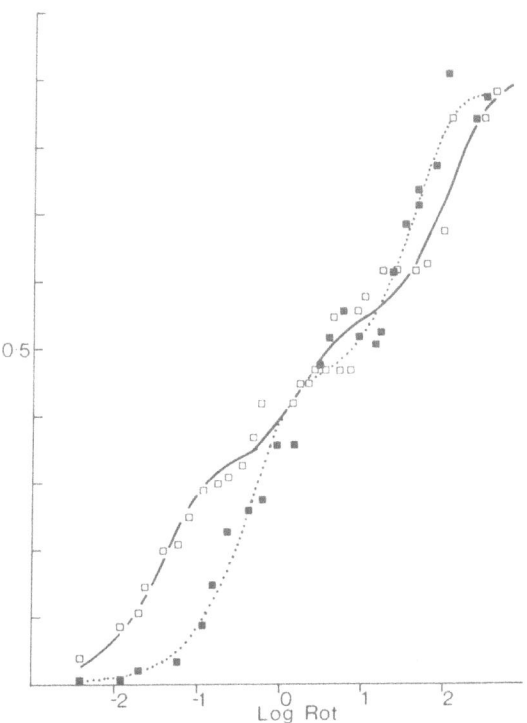

Fig.1.   Reassociation analysis of liver poly(A)mRNA fractions.
Continuous line - mRNA from ER polyribosomes x homologous cDNA.
Broken line - mRNA from free polyribosomes x homologous cDNA.
The ordinate shows the proportion of cDNA resistant to nuclease S1.

of abundant liver proteins by SDS-PAGE.   By annealing liver mRNA
with an excess of cDNA (HART), Hastie and Held were able to show
that abundant mRNA sequences are the templates for synthesis of
the abundant proteins[10].

     More recently, we have found that most of the abundant liver
mRNA sequences are preferentially associated with the endoplasmic
reticulum (Fig.1).   The overall complexities of the mRNA populations of free reticulum-bound polyribosomes are very similar, and
in fact there are no qualitative differences between the two
populations.   The reticulum population, however, contains a high
abundance component not found in the free polyribosomes.   Upon
translation of the two populations in vitro very different patterns
of synthesis of abundant proteins are observed (Fig.2).   Precipitation of the in vitro translation products with antibody against
whole mouse serum shows that most of the abundant proteins are
exported from the liver (data not shown).   In this case, then, we

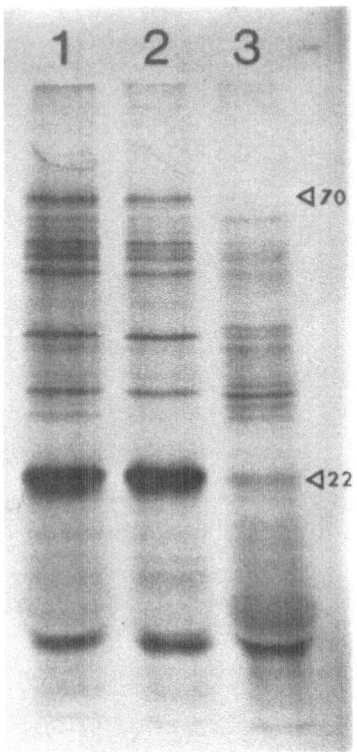

Fig.2. SDS-PAGE of translation products of liver polyribosomal
poly(A)mRNA fractions. 1, from ER; 2, from nucleus-associated ER;
3, from free polyribosomes. The fractions were translated in the
reticulocyte primer-dependent lysate.[11]

observe a good correlation in the abundant class between the
reassociation measurements and the number of different species of
mRNA, as defined by the criterion of translation. The reassociation
experiments point to the existence of 12 abundant sequences in ER
mRNA while 10 abundant proteins are observed by one dimensional
SDS-PAGE.

Drosophila melanogaster mRNA populations

     In Drosophila melanogaster, we come across a striking example
of a sequence which scores as a messenger sequence in reassociation
experiments, but which turns out not to be a messenger after all.
This is the larger (14S) mitochondrial rRNA, which binds to oligo
(dT) cellulose and is an oligo(dT)-dependent template for reverse
transcriptase[7].   Poly(A)RNA prepared from total cytoplasmic
RNA contains large amounts of 14S RNA, but it is also found in
significant amounts in poly(A)RNA prepared from sucrose gradient

polyribosomes.

    Total cytoplasmic poly(A)RNA preparations from the L3 cell
line and from embryos (14h) contain about 5300 and about 3500
sequences, respectively.   Each population shows three distinguishable abundance classes with similar characteristics. The most
abundant class comprises the 14S rRNA and, at least in the case of
the L3 cell line, the copia sequence (refs. 12-14, and see below).
A poly(A)RNA preparation made from embryo cytoplasmic RNA was
depleted of 14S RNA by pooling sucrose gradient fractions sedimenting before and after the 14S peak[7].   Double-stranded cDNA was
prepared and cloned in plasmid pCM2[15] by the poly(dA)-(dT) method[16].
A number of clones were selected on the basis of size (>500 bp) and
screened by filter hybridisation with kinase-labelled embryo mRNA.
Table 1 shows some results. The hybrid plasmid pEE 2328 showed the
strongest reaction, and proved to contain the 14S mitochondrial rRNA
sequence. The other plasmids contain cDNA sequences corresponding
to mRNA sequences that differ in abundance by at least a factor of
50. In itself, this is a direct confirmation of the existence of
differences in the abundance of mRNA sequences. Analysis of these
clones and of the corresponding nuclear DNA sequences is in
progress.

Table 1. Screening Drosophila Embryo cDNA-Plasmid Clones by
          Filter Hybridisation with End-labelled RNA.   The $R_0 t$
          was about 18.

DNA Bound to filter	RNA Bound (c.p.m.)	Δ	Saturation (%)
Blank	318		
pCM2	516	(0)	
pEE2313	879	363	12.1
2319	1348	832	27.7
2321	784	268	8.9
2328	1758	1242	41
2331	635	119	4.0
2334	531	15	0.5
2340	594	78	2.6

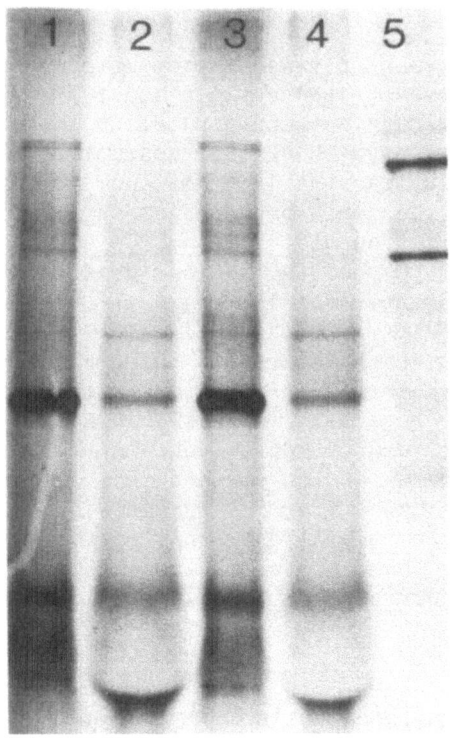

Fig.3. SDS-PAGE of translation products of fat-body poly(A)RNA.
1,3, with ^{35}S-methionine; 2,4, with ^{3}H-leucine; 5, heat-shock
markers, 84, 70, 26 and 22k, respectively.

    The fat body of the late third instar larva synthesises a
small number of particularly abundant proteins, as well as a large
number of others (Fig.3). The abundant proteins include LSP-1,
LSP-2 and ADH[17,18]. Fat body total cytoplasmic poly(A)RNA shows
two transitions in reassociation experiments with homologous cDNA
(Fig.4). The first has the characteristics of a single sequence
and is due to the 14S RNA. The second can be calculated to
represent about 100 sequences.

    With a view to the screening of cDNA clones, hybrid plasmids
were constructed between <u>D.melanogaster</u> mitochondrial DNA fragments and PstI-cleaved pBR322. The plasmid selected for further
use, pMD417, is shown in Fig.5. It contains much of the mitochondrial DNA sequence coding for 14S RNA, and in particular, the
3'-end of the 14S RNA is contained within a 0.77 kb Eco.RI fragment that can readily be isolated by agarose electrophoresis. This
fragment reacts with the appropriate fragments of pEE2328 in
Southern blot[19] experiments (see above).

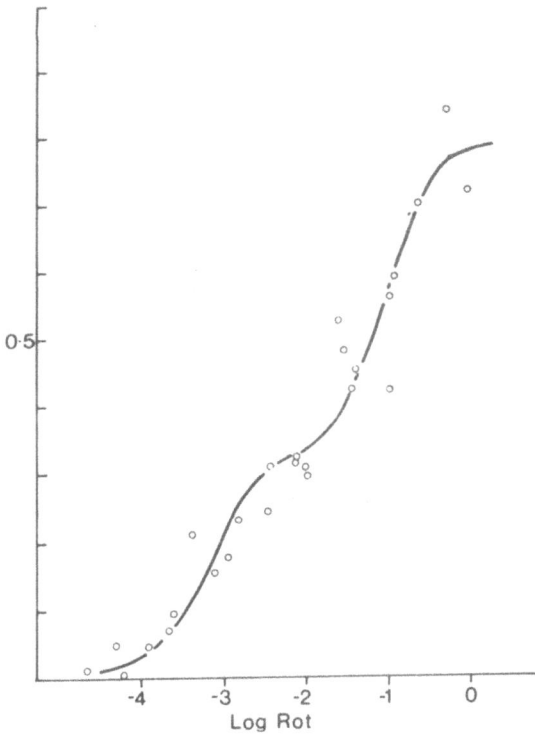

Fig.4. Reassociation of fat-body mRNA with homologous cDNA. The
ordinate shows the proportion of cDNA resistant to nuclease S1.

    Double-stranded cDNA was prepared against poly(A)RNA made
from the total cytoplasmic RNA of 3rd instar larvae. Two parallel
preparations were made, one labelled with ^{32}P-dCTP and the other
with ^{3}H-dCTP. The preparations were run in parallel on an agarose
gel and transferred to a nitrocellulose filter. The ^{3}H-containing
lane was annealed with the 0.77 kb fragment of pMD417, nick-translated with ^{32}PdCTP, and autoradiographically exposed alongside the
^{32}P-containing lane. The results are shown in Fig.6 in the form
of densitometric scans of the X-ray film. It is clear that some
portions of the preparation are enriched in 14S cDNA, while others
are not.

    Hybrid plasmids were constructed by inserting double-stranded
3rd larval instar fat body poly(A)RNA into plasmid pCM2. After
size-screening, a number of plasmids were screened by filter
hybridisation against the 0.77 kb fragment of pMD417 and
separately against fat-body cDNA. The results show (Table 2)
that about half of the hybrid plasmids contain sequences complementary to 14S RNA. Of the remainder, some are complementary to
more abundant, and some to rarer mRNA sequences. Among the more

Table 2. Screening Fat-body-specific cDNA Plasmids

Plasmid	Description	Reaction with pMD417 fragment	Reaction with fat-body cDNA[a]	Eco.RI site in insert[b]	Conclusion[c]
pCM2	Vector	76	48		
pMD417	Mitochondrial	6480	2350	+	
pFB132	Fat body cDNA	2080	3210	+	Mit
147		177	824	-	mRNA
151		1429	832	+	Mit
153		385	1000	-	mRNA
163		438	1245	+	Mit
172		10160	4260	-	mRNA
177		34	560	-	mRNA

a. Filter hybridisation, expressed as cts/min hybridized
b. Agarose electrophoresis of Eco.RI digests
c. Mit = mitochondrial 14S RNA sequence

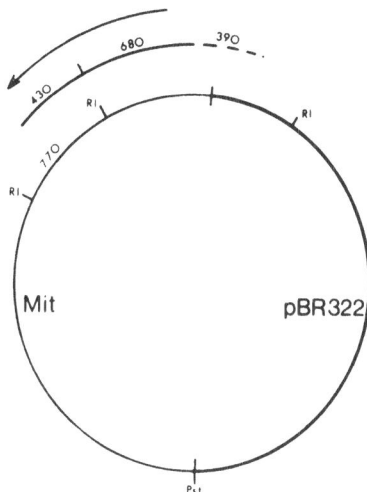

Fig.5. Map of hybrid plasmid pMD417, which consists of a piece
of D.melanogaster mitochondrial DNA cloned in pBR322. The region
coding for 14S RNA is shown, the broken line representing a
region which may or may not be present in pMD417. The direction
of transcription is also shown.[20]

abundant, we have identified pFB177, by HART[21], as being complementary to LSP in mRNA.

Heat-shocked Drosophila and Drosophila cell lines synthesise
a small number of proteins (probably 8) in response to the preferential synthesis of a small number of mRNA sequences (possibly 6)
and the displacement of the normal mRNA sequences from the polyribosomes (review, ref.22). When the reassociation of heat-shock
cDNA (from L3 cells) with mRNA from normal cells[23] was compared
with the reassociation of cDNA and mRNA from normally-grown L3
cells[7], it seemed likely that the copia sequence is a prominent
component of poly(A)RNA from the polyribosomes of heat-shocked
cells. This was demonstrated by means of a reassociation reaction
between heat-shock-specific cDNA and a copia sequence cloned in
plasmid pPW220[13] (Table 3). The identity of the reactive sequence
was further confirmed by a Southern transfer[19] experiment using
pPW220 DNA cleaved with KpnI and Hind III and heat-shock-specific
cDNA as the probe. As shown in Table 3, the heat-shock cDNA
also contains sequences complementary to 14S RNA. Together,
14S RNA and copia RNA account for the heat-shock sequences found
to be abundant in normal L3 cell RNA.

To what extent the same is true of normal cellular mRNA
populations remains to be seen. Furthermore, what we make of this
situation will depend very much on whether copia and the αβ sequ-

Table 3.   Reaction of Heat-shock cDNA with Hybrid Plasmid
           DNA

Plasmid	Description	Reaction of total heat-shock cDNA[a] (%)
pMB9	Vector	3.7
pPW220	Copia	20.8
pCM2	Vector	3.8
pEE2328	14S	7

a. Filter hybridisation

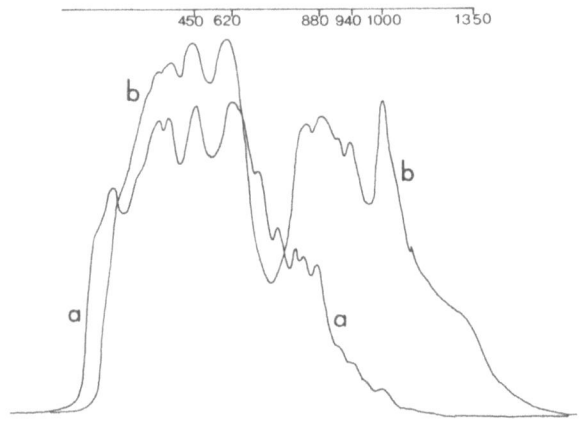

Fig.6. Analysis of double-stranded fat-body poly(A)RNA prior to
cloning. For a description, see the Text. a. ^{32}P-labelled cDNA.
b. ^{3}H-labelled cDNA after annealing with ^{32}P-labelled 14S-
specific probe.

ences have a function. If, as has been suggested, they are regulatory RNA sequences, then the DNA sequences that code for them
are as much 'genes' as are the sequences that code for proteins.
If, on the other hand, they have no function, and if such sequences
are common in normal RNA populations, then our gene-number
estimates may be quite wrong.

REFERENCES

1. J.O. Bishop, J.F. Morton, M. Rosbash and M. Richardson,
   Three abundance classes in HeLa cell mRNA. Nature 250:199
   (1974).
2. G.D. Birnie, E. MacPhail, B.D.Young, M.J.Getz and J.Paul, The
   diversity of the mRNA population in growing Friend cells.
   Cell Differentiation 3: 221 (1974).
3. N.D. Hastie and J.O. Bishop, The expression of three abundance
   classes of mRNA in mouse tissues. Cell 9: 761 (1976).
4. J.O. Bishop, J. Beckman, M.S. Campo, N.D. Hastie, M.Izquierdo
   and S. Perlman, DNA-RNA hybridisation. Phil.Trans.Roy.Soc.B
   272: 147 (1975).
5. L.M. Hereford and M. Rosbash, Number and distribution of
   polyadenylated RNA sequences in yeast. Cell 10L 453 (1977).
6. B. Levy W and B.J. McCarthy, Messenger RNA complexity in
   Drosophila melanogaster. Biochemistry, 14: 2440 (1975).
7. M. Izquierdo and J.O. Bishop, An analysis of cytoplasmic RNA
   populations in Drosophila melanogaster, Oregon R. Biochemical
   Genetics (1979)
8. P.Z. O'Farrell, H.M. Goodman and P.H. O'Farrell, High resolution
   two-dimensional electrophoresis of basic as well as acidic
   proteins. Cell 12: 1133 (1977).
9. B.M. Paterson and J.O. Bishop, Changes in the mRNA population of
   chick myoblasts during myogenesis in vitro. Cell 12: 751
   (1977).
10.N.D. Hastie and W.A. Held, Analysis of mRNA populations by cDNA-
   mRNA hybrid-mediated inhibition of cell-free protein synthesis.
   Proc.Nat.Acad.Sci., U.S. 75: 1217 (1978).
11.H.R.B. Pelham and J.Jackson, An efficient mRNA-dependent translation system from reticulocyte lysates. Eur.J.Biochem. 67:
   247 (1976).
12.D.J. Finnegan, G.M. Rubin, M.W.Young and D.S. Hogness, Repeated
   gene families in Drosophila melanogaster. Cold Spring Harbor
   Symposia Quant.Biol. 42: 1053 (1978).
13.S.S.Potter, W.J. Brorein, P.Dunsmuir and G.M. Rubin, Transposition of elements of the 412, copia and 297 dispersed
   repeated gene families in Drosophila. Cell 17:415 (1979).
14.E.Strobel, P. Dunsmuir and G.M. Rubin, Polymorphisms in the
   chromosomal locations of elements of the 412,copia and
   297 dispersed repeated gene families in Drosophila. Cell
   17: 429 (1979).

15. J.O. Bishop, A DNA sequence cleaved by restriction endo-
    nuclease R.Eco RI in only one strand. J.Mol.Biol. 128: 545
    (1979).
16. P.C. Wensink, D.J. Finnegan, J.E. Donelson and D.S. Hogness,
    A system for mapping DNA sequences in the chromosomes of
    Drosophila melanogaster. Cell 3: 315 (1974).
17. M.E. Akman, D.B. Roberts, G.P. Richards and M. Ashburner,
    Drosophila: the genetics of two major larval proteins.
    Cell 13: 215 (1978).
18. H. Ursprung and J. Leone, Alcohol dehydrogenase: a polymorphism
    in Drosophila melanogaster. J.Exper.Zool. 110:147 (1965).
19. E.M. Southern, Detection of specific DNA sequences among DNA
    fragments separated by electrophoresis. J.Mol.Biol. 98:503
    (1975).
20. C.K. Klukas and I.B. Dawid, Characterization and mapping of
    mitochondrial ribosomal RNA and mitochondrial DNA in
    Drosophila melanogaster. Cell 9: 615 (1976).
21. B.M. Paterson, B.E. Roberts and E.L. Kuff, Structural gene
    identification and mapping by DNA-mRNA hybrid-arrested cell-
    free translation. Proc.Nat.Acad.Sci., U.S. 74: 4370 (1977).
22. M. Ashburner and J.J. Bonner, The induction of gene activity
    in Drosophila by heat shock. Cell 17: 241 (1979).
23. D.J. Bower, Ph.D. Thesis, University of Edinburgh (1979).
24. J.T. Lis, L. Prestidge and D.S. Hogness, A novel arrangement
    of tandemly repeated genes at a major heat shock site in
    Drosophila. Cell 14: 901 (1978).

STRUCTURAL GENE EXPRESSION IN TOBACCO

Robert B. Goldberg

Department of Biology
University of California
Los Angeles, California   90024

INTRODUCTION

Cellular and developmental processes in higher plants differ
significantly from those of other eukaryotes. Although plant
biology plays a central role in food production and agriculture,
our knowledge of the molecular mechanisms which control gene
expression in plants remains scant. This paper briefly reviews
current information regarding DNA sequence organization and expression in the tobacco plant. In addition, new data are summarized which show that mRNA sequence sets are strikingly regulated
in major plant tissues and that post-transcriptional selection
mechanisms play a central role in the regulation of gene expression in plants.

DNA SEQUENCE ORGANIZATION

The tobacco genome, like that of most other plants and animals, has repetitive and single-copy sequences which are organized
in a short period interspersion pattern (Davidson et al., 1973).
A simple experiment which illustrates this point is shown in
Figure 1. At Cot 100 approximately 70% of short DNA fragments
have bound to hydroxyapatite as compared to 94% of those which are
long. Since few single-copy sequences have reacted at this Cot,
short repetitive elements must be contiguous to >90% of the
single-copy sequences in the tobacco genome. A quantitative
summary of a large number of experiments on tobacco genome
organization (Zimmerman and Goldberg, 1977) is diagrammed in the
inset to Figure 1. At present, the physiological significance of

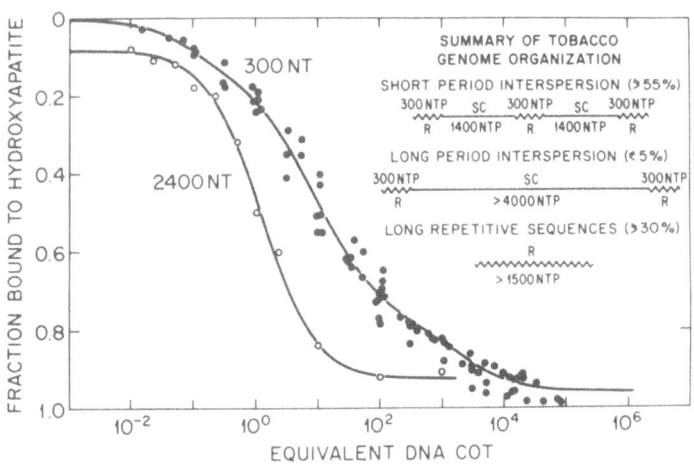

Fig. 1.   Reassociation cf Short and Long DNA Fragments (Zimmerman
          and Goldberg, 1977). Unlabeled 350 and 2400 nucleotide
          DNA fragments were separately annealed at a criterion
          equal or equivalent to 0.18 M Na$^+$, 60°C (Britten,
          Graham, and Neufeld, 1974) and the fraction of fragments
          containing a duplex region was assayed by hydroxyapatite
          chromatography. The open circles show the reassociation
          of 2400 nuceotide fragments while the closed circles that
          of 350 nucleotide fragments. The curves represent least
          squares solutions to the data points, using one second-
          order component for the 2400 nucleotide data and four
          second-order components for the 350 nucleotide data. The
          rate constant for the long DNA fragments was 0.82 M^{-1}
          sec^{-1}. The fraction of DNA fragments, rate constants,
          and average reiteration frequencies for the putative com-
          ponents describing the 350 nucleotide reassociation data
          were: repetitive-1, 15%, 9.5 M^{-1} sec^{-1}, 20,000;
          repetitive-2, 40%, 0.21 M^{-1} sec^{-1}, 450; repetitive-3,
          22%, 0.03 M^{-1} sec^{-1}, 65; single-copy, 18%, 0.00048
          M^{-1} sec^{-1}, 1. The inset depicts the average patterns
          of DNA sequence arrangement in the tobacco genome. S-1
          nuclease studies have shown that about 60% of tobacco DNA
          sequences are repeated as compared to 40% which are
          single-copy.

short period interspersion in plant genomes remains obscure.
Britten and Davidson (1969) have speculated that interspersed
repetitive sequence elements function in the regulation of structural gene expression. However, direct experimental verification
of this proposition is not yet at hand (see Davidson and Britten,
1979).

DNA SEQUENCE REPRESENTATION IN MESSENGER RNA

Figure 1 shows that the majority of DNA mass in tobacco chromosomes is repeated. Of considerable functional significance,
therefore, is whether structural gene sequences are represented in
the repeated or single-copy DNA class. Experiments which demonstrate that >95% of the diverse structural genes are single-copy
DNA sequences are shown in Figure 2. After correcting to 300
nucleotide fragment length, ^{3}H-cDNA which was transcribed from
leaf poly(A) mRNA, hybridized to excess leaf DNA with a rate close
to that of purified ^{3}H-single-copy DNA (7 x 10^{-4} M^{-1} sec^{-1}
vs. 4.8 x 10^{-4} M^{-1} sec^{-1}). This result demonstrates that
transcripts which comprise most of the mRNA mass are transcribed
from single-copy sequences. As shown below, however, most of the
mRNA mass includes only a small fraction of the mRNA diversity.
To show that, within the limits of measurement, all mRNA sequences
are transcribed from single-copy DNA, a different approach was
utilized. In this case, ^{3}H-DNA which was isolated from
mRNA/^{3}H-single-copy DNA hybrids was reassociated with excess
leaf DNA. Figure 2 shows that it too reacts with a rate equal to
that of single-copy DNA sequences. Hence, in the tobacco leaf at
least, most of the mRNA mass as well as mRNA diversity is transcribed from single-copy DNA sequences.

ABUNDANCE DISTRIBUTION OF LEAF MESSENGER RNA SEQUENCES

While the genes which code for leaf mRNAs are present as
single copies in tobacco chromosomes, the mRNA transcripts vary
considerably in cellular concentration. Figure 3 shows that the
hybridization kinetics of leaf ^{3}H-cDNA with excess poly(A) mRNA
are heterogenous, indicating that the concentration of individual
mRNA species may vary by as much as three orders of magnitude.
Approximately 60% of the poly(A) mRNA mass contains relatively few
diverse mRNA sequences (500-1500), which are represented by up to
several thousand molecules each per cell. These abundant mRNAs
code for proteins which are readily visualized in electrophoretic
patterns. On the other hand, 40% of the poly(A) mRNA mass contains species which are represented by only a few molecules each
per cell. Collectively, this rare mRNA class contains the over-
whelming majority of the mRNA diversity. Since proteins which are
coded for by rare class mRNAs cannot as yet be identified, their
cellular functions are not known.

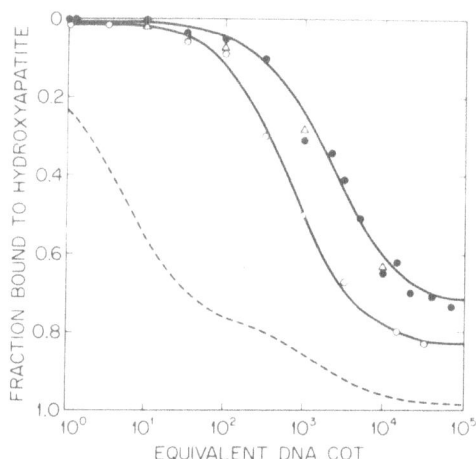

Fig. 2. Reassociation of Leaf ^{3}H-cDNA and ^{3}H-Single-Copy DNA
with Total DNA (Goldberg et al., 1978). Trace amounts of
640 nucleotide ^{3}H-cDNA (open circles) or 300 nucleotide
^{3}H-single-copy DNA (closed circles) were reassociated
with a >20,000 fold mass excess of unlabeled leaf DNA
and the extent of reassociation was measured by hydroxyapatite chromatography. The ^{3}H-cDNA was transcribed
from leaf poly(A) mRNA using reverse transcriptase while
the ^{3}H-single-copy DNA was labeled in vitro by "gap-
translation" using E. coli DNA polymerase I (Galau et
al., 1976). The solid curves represent least squares
solutions to the tracer data using a single second-order
component while the dashed curve depicts the driver
reaction (Figure 1). Attempts to fit the tracer data to
more than one component were unsuccessful. The rate constant of the ^{3}H-cDNA data was 0.0015 M^{-1} sec^{-1}
while that of the ^{3}H-single-copy DNA was 0.00048 M^{-1}
sec^{-1}. The triangles represent the reassociation of
^{3}H-single-copy DNA which was isolated from mRNA/^{3}H-
single-copy DNA hybrids.

SEQUENCE COMPLEXITY OF LEAF MESSENGER AND NUCLEAR RNAs

     To directly measure the sequence diversity of leaf mRNA
(i.e., of the rare mRNA class), ^{3}H-single-copy DNA (Figure 2)
was hybridized to excess leaf total polysomal RNA. The results of

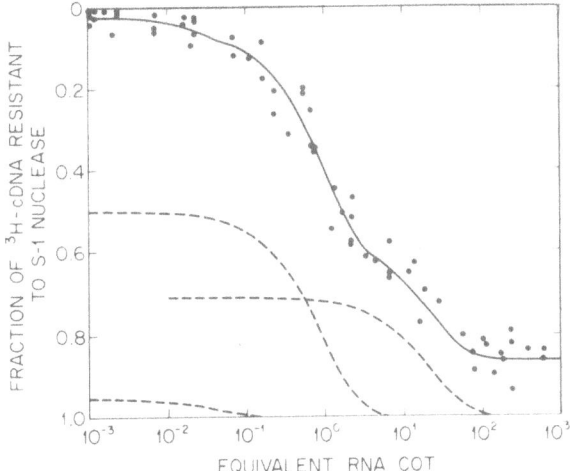

Fig. 3. Hybridization of leaf ^{3}H-cDNA to Poly(A) mRNA (Goldberg
         et al., 1978). Trace amounts of ^{3}H-cDNA were
         hybridized to a >1000 fold mass excess of template mRNA
         at a criterion equivalent to 0.18 M Na+, 60°C and the
         degree of hybridization assayed by S-1 nuclease (Leong et
         al., 1972). The curve through the data points represents
         one least squares solution using three pseudo first-order
         components. This solution should not be considered
         unique but only as a model approximation of the abundance
         distribution of leaf poly(A) mRNA sequences. The per-
         cent poly(A) mRNA mass, rate constant corrected to 100%
         purity, number of diverse 1240 nucleotide sequences, and
         average number of molecules per cell per sequence for
         these components were: class-1, 9%, 156 M^{-1} sec^{-1},
         10, 4500; class-2, 52%, 1.9 M^{-1} sec^{-1}, 770, 340;
         class-3, 39%, 0.13 M^{-1} sec^{-1}, 11,200, 17.

this experiment, shown in Figure 4, indicate that about 2% of the
labeled single-copy DNA hybridizes to the mRNA. After correcting
for asymmetric transcription and single-copy tracer reactivity
with total DNA, this amount of hybridization is equivalent to 5.2%
of the single-copy DNA or 27,000 average-sized mRNA sequences
(assuming that rare class messages average 1240 nucleotides in
length).

    That transcripts present in the mRNA constitute only a frac-
tion of the RNA sequence diversity in the leaf cell is also shown
in Figure 4. When ^{3}H-single-copy DNA was reacted with nuclear
RNA about 8% hybridized, four times that which was obtained with
mRNA. This experiment demonstrates that 75% of the diverse RNA

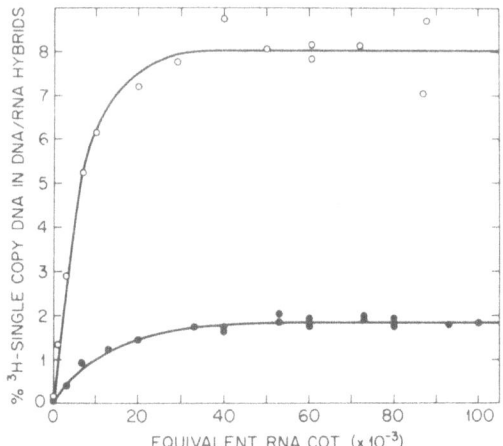

Fig. 4.   Hybridization of ^{3}H-Single-Copy DNA to Leaf Nuclear and
Messenger RNAs (Goldberg et al., 1978).   Trace amounts of
[3]H-single-copy DNA were hybridized to a 2,000 fold
mass excess of nuclear (open circles) or total polysomal
RNA (closed circles) and the extent of hybridization was
measured by hydroxyapatite chromatography. Each reaction
was assayed for both DNA/DNA reassociation and DNA/RNA
hybridization (Galau, Britten, and Davidson, 1974). The
curves represent least squares solutions for a single
pseudo first-order component. After correcting for asymmetric transcription and tracer reactivity approximately
18.7% and 5.2% of the single-copy DNA was found to be
represented in the nuclear and polysomal RNA populations,
respectively. These values correspond to 1.10×10^8
and 3.33×10^7 nucleotides of diverse nuclear and mRNA
transcripts. Each nuclear RNA transcript is present an
average of 2 times per cell while those of mRNA an
average of 12 times per cell. The latter transcripts
form the rare mRNA class (class-3, Figure 3) which constitute only 20% of the mRNA mass but >95% of the mRNA
diversity.

sequences in the leaf cell remain in the nucleus while only 25%
are transported to the cytoplasm to be translated on polysomes.

REGULATION OF MESSENGER RNA SEQUENCE SETS

     Genes which code for abundant mRNAs have been demonstrated to
be developmentally regulated in plants. i.e., their proteins are

PREPARATION OF LEAF mDNA AND LEAF NULL mDNA

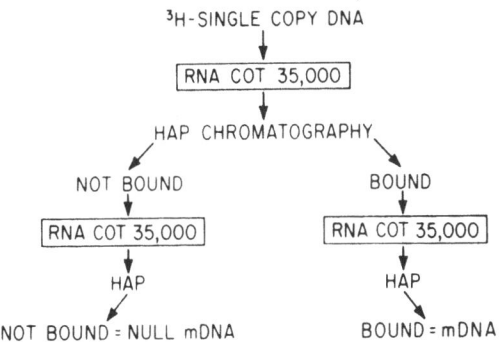

Fig. 5. Isolation of leaf mDNA and Null mDNA (Kamalay and
Goldberg, 1979). Two labeled single-copy DNA fractions
were prepared by reacting ^{3}H-single-copy DNA with
>2,000 fold mass excess of leaf total polysomal RNA and
then separating the DNA fragments which did and did not
hybridize by hydroxyapatite chromatography (Galau et al.,
1976). The fraction that hybridized, termed leaf
[3]H-mDNA, contained 2% of the labeled single-copy DNA
(see Figure 4), and was enriched 24 fold for structural
gene sequences represented in leaf mRNA. The non-
hybridized fraction, designated ^{3}H-null mDNA, contained
98% of the labeled single-copy DNA, and was depleted of
structural gene sequences represented in leaf mRNA.

tissue specific (e.g., leghemoglobin, α-amylase, carboxylase,
seed proteins). However, as shown above, the vast majority of
structural genes which are expressed in plant cells code for rare
class mRNAs. Do these genes show tissue specific patterns of
expression or are they simply those which are commonly expressed
in all plant cells (i.e., housekeeping genes)?

Experiments which relate to this important issue were carried
out by Kamalay and Goldberg (1979) and will be presented in detail
elsewhere. However, the overall strategy and a summary of results
are presented in Figures 5 and 6. Several important conclusions
can be made. First, each tissue expresses the same number of
structural genes. That is, about 25,000-30,000 diverse mRNAs are
present on the polysomes of each tissue. Second, rare mRNA
sequence sets are strikingly regulated in plant tissues. Assuming
that petal and leaf are homologous organs, each tissue possesses a
unique set of mRNAs, which correspond to thousands of diverse
structural genes. Each mRNA sequence set correlates with the
specific form and function of a complex organ system. Hence,
structural genes which code for rare class mRNAs do not simply

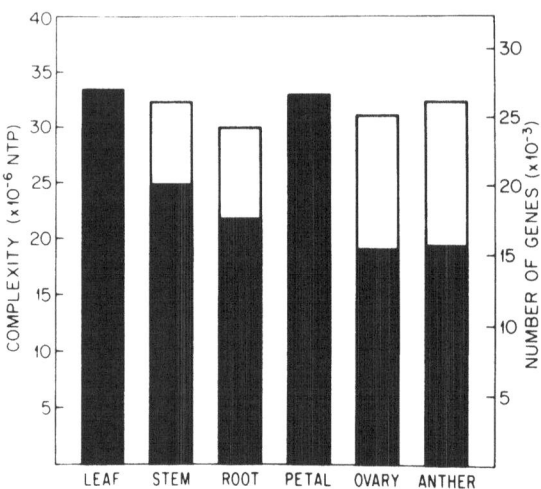

Fig. 6.   Messenger RNA Sequence Sets Present on Polysomes of Major
          Organ Systems (Kamalay and Goldberg, 1979). Leaf
          ^{3}H–mDNA and ^{3}H–null mDNA were separately hybridized
          to an excess of leaf, stem, root, ovary, anther, and
          petal polysomal RNAs and the extent of hybridization was
          assayed by hydroxyapatite chromatography. Hybridization
          with leaf ^{3}H–mDNA measured the degree of overlap
          between the leaf mRNA sequence set and that of other
          tissues, while hybridization to ^{3}H–null mDNA measured
          mRNA sequence sets unique to other tissues but absent
          from leaf polysomes. The solid portion of the bars
          summarizes the results of the ^{3}H–mDNA reactions, while
          the open portion those of the ^{3}H–null mDNA reactions.
          The height of each bar represents the total mRNA sequence
          complexity present on polysomes of each tissue expressed
          as nucleotide pairs of single-copy sequences or as the
          number of 1240 nucleotide pair structural genes. The
          leaf mRNA complexity was set at 3.33 x 10^7 nucleotide
          pairs (Figure 4).

fulfill a housekeeping role, but appear to be of central importance to the establishment and maintenance of specific differentiated states in plant cells. Third, overlap does occur in the
mRNA sequence sets. The minimum number of mRNAs shared by all
tissues is approximately 8,000, or the equivalent of about 1.5% of
the single-copy DNA. Finally, taking these results as a whole at
least 60,000 diverse structural genes are expressed in the entire
tobacco plant during the dominant phase of its life cycle. These
genes comprise 11% of the single-copy DNA.

Table 1.   Hybridization of Leaf ^{3}H—mDNA to Nuclear and
           Polysomal RNAs[a]

Tissue	Reaction with Polysomal RNA[b]	Reaction with Nuclear RNA[b]
Leaf	100	100
Stem	75	100

[a]Kamalay and Goldberg (1979)
[b]Leaf ^{3}H—mDNA (Figure 5) was hybridized to >50,000 fold
   mass excess of nuclear and polysomal RNAs from the leaf and
   stem. The degree of hybridization was measured by hydroxyapatite chromatography. Values represent the percentages
   of ^{3}H—mDNA hybridization which were normalized to the
   reaction with leaf RNAs.

MESSENGER RNA SEQUENCE SETS ARE REGULATED POSTTRANSCRIPTIONALLY

The large differences in structural gene expression found in
plant tissues could be a reflection of transcriptional and/or
post-transcriptional control mechanisms. If regulation occurred
primarily at the transcriptional level, genes which were not
utilized in a given tissue would be expected to be inactive or
repressed. On the other hand, if post-transcriptional control
mechanisms dominated, all structural genes, irrespective of their
expression in a given tissue, might be expected to be represented
by transcripts in the nuclear RNA (i.e., constitutively
transcribed).

An experiment which distinguishes between these two alternatives is summarized in Table 1, the details of which will be
described elsewhere (Kamalay and Goldberg, 1979). While only 75%
of leaf mRNA sequences are represented on polysomes of the stem,
all of the leaf mRNAs are present in the stem nuclear RNA
population. Thus, leaf structural genes, which are not utilized
to produce proteins in the stem, are nevertheless transcribed into
stem nuclear RNA. Clearly, the major conclusion which can be
drawn from this result is that post-transcriptional processing
and/or selection mechanisms play a central role in the regulation
of plant gene expression.

ACKNOWLEDGEMENTS

I express deep gratitude to my associates Lynn Zimmerman, Joe Kamalay, and Gisela Hoschek for their tireless efforts in contributing to the experiments reported here. These investigations were supported by NSF grants PCM 76-24593 and PCM 78-22321.

REFERENCES

Britten, R.J. and Davidson, E.H., 1969, Science, 165:349-368.
Britten, R.J., Graham, D.E. and Neufeld, B.R., 1974, in "Methods in Enzymology," 29E, L. Grossman and K. Moldave, eds., pp. 363-418.
Davidson, E.H. and Britten, R.J., 1979, Science, 204:1052-1059.
Davidson, E.H., Hough, B.R., Amenson, C.S. and Britten, R.J., 1973, J. Molec. Biol., 77:1-23.
Galau, G.A., Britten, R.J. and Davidson, E.H., 1974, Cell, 2:9-21.
Galau, G.A., Klein, W.H., Davis, M.M., Wold, B.R., Britten, R.J. and Davidson, E.H., 1976, Cell, 7:487-505.
Goldberg, R.B., Hoschek, G., Kamalay, J.C., and Timberlake, W.E., 1978, Cell, 14:123-131.
Kamalay, J.C. and Goldberg, R.B., 1979, in preparation.
Leong, J., Garapin, A., Jackson, N., Fonshier, L., Levison, W. and Bishop, J.M., 1972, J. Virol., 9:891-902.
Zimmerman, J.L. and Goldberg, R.B., 1977, Chromosoma, 59:227-252.

MESSENGER RNA DOMAINS IN THE EMBRYOGENESIS AND GERMINATION OF

COTTON COTYLEDONS

L. S. Dure, III, A. M. Capdevila and S. C. Greenway

Department of Biochemistry
University of Georgia
Athens, GA   30602   U.S.A.

INTRODUCTION

We have begun the task of describing the development of
cotton cotyledons through embryogenesis and germination in terms
of the mRNA subsets that, through their appearance and disappearance, determine this developmental sequence. This, of course,
represents an attempt to describe a developmental sequence in
terms of sequential gene activity.

Initially, we wish to establish a fundamental molecular basis
for this developmental sequence so that its regulation takes on a
more definitive biochemical perspective. Ultimately, we hope to
isolate mRNAs representative of discrete developmental stages and,
through them to isolate their genes, so as to examine their
unique features that govern their expression.

Our approach to this description has been, in part, to
delineate cotyledon development in terms of the changing protein
populations that determine the ontogenic progression. Due to the
limitations of the techniques available for studying protein populations only the rather abundant proteins of the total population can be catalogued. The total genetic complexity of the mRNA
population existing at a given developmental stage, and the extent
to which it is included in the population from other stages, is
being determined by other means. These measurements will not be
presented today. What we wish to present here are some preliminary findings concerning the mRNA subsets active during cotyledon
development as described by catalogs of their protein products.

CATALOGS OF PROTEINS DURING DEVELOPMENT

We have begun the description of the abundant members of mRNA subsets by compiling the following protein catalogs at selective points during the developmental sequence.

1. Catalog of extant proteins . . Coomassie staining of 2D gels
2. Catalog of proteins synthesized in vivo . . . fluorography of 2D gels
3. Catalog of proteins synthesized in vitro from purified RNA fractions . . . fluorography of 2D gels.
4. Determination of RNA complexity and sequence overlap during development. . . RNA:cDNA hybridization.

The first of the catalogs involves extracting protein from cotyledons in various solvent systems at time points during development and separating them by electrophoresis in one dimension by the discontinuous sodium dodecyl sulfate (SDS) system of Laemmli[1] or by the two dimensional electrophoretic system of O'Farrell[2] in which the proteins are first separated on cylindrical gels by migration to their respective isoelectric points and then electrophoresed in the presence of SDS into a discontinuous slab gel where they separate by virtue of their differences in molecular weight. The separated proteins are visualized by staining with Coomassie Brilliant Blue R-250. This technique displays only the more abundant proteins of the tissue, and the thousands of enzymes whose concentration in the tissue is likely to be low are not observed. Furthermore, such a display of proteins is not a direct measure of mRNA population since many proteins are likely to be stable and to accumulate and persist after their mRNAs have disappeared. Likewise rapidly turning over proteins would under-represent their mRNA concentration. Nevertheless, such displays do allow for a developmental description of the tissue in terms of extant proteins.

The second catalog presents a measure of the mRNA population present at a given developmental point in terms of the proteins being synthesized in vivo during the time interval that the tissue is exposed to radioactive amino acids. Here, the tissue is exposed to the labelled amino acids for brief periods at points during ontogeny after which the proteins are extracted and separated by the two dimensional electrophoretic system. The location on the slab gel of the proteins synthesized during the incubation period (radioactive proteins) are displayed by the fluorographic techniques of Lasky and Mills[3]. In order to follow the processing or degradation of certain proteins, the cotyledons are further incubated in the absence of radioactive isotopes in some instances (pulse-chase labelling).

The third catalog involves purifying RNA fractions from cotyledons at discrete developmental points and allowing these fractions to direct in vitro translation using both the rabbit reticulocyte and wheat germ systems[4,5]. The radioactive protein products are separated and visualized as are the in vivo synthesized
products.

DEVELOPMENTAL STAGES CATALOGUED

The basis for choosing particular time points in the ontogeny
of cotton cotyledon tissue for study stems from earlier work on
the developmental biochemistry of this tissue in our laboratory[6,7,8]. This developmental framework is depicted diagrammatically in Figure 1. The upper half of the figure indicates the
developmental points at which the extant proteins are extracted
for the stain catalog and at which cotyledons are pulse-labelled
for the catalog of the proteins being synthesized in vivo. The
lower half shows that at the same developmental points RNA is
extracted for translation in the cell-free systems to establish
the mRNA populations directly by in vitro protein synthesis.

Cotyledons from embryos 50 mg in weight are in the midst of
their logarithmic phase of growth, and the outer seed tissues are

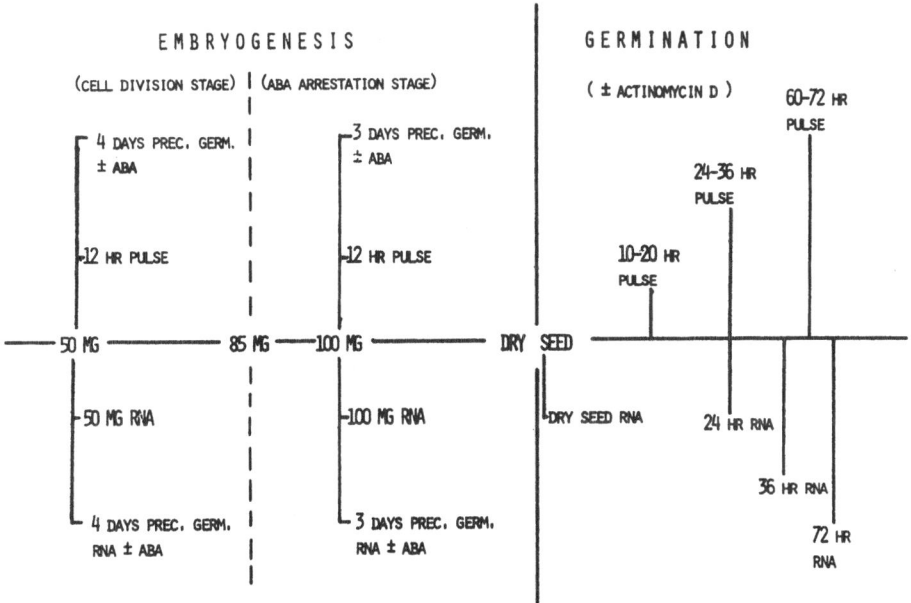

Fig. 1.   Diagrammatic presentation of the developmental points
          studied. Top half indicates preparations made for stain
          and in vivo synthesis catalogs.   Bottom half indicates
          preparations of mRNA made for in vitro synthesis catalog.

in vascular connection with the mother plant. Cell number is increasing at this point and the storage nutrients are accumulating. ABA levels are very low at this point. If embryos are removed from the boll at this point and placed on moist filter paper they will precociously germinate[7]. However, removal brings about an immediate cessation of cell division in the cotyledons which causes the resultant seedling to have half-sized cotyledons. An ABA solution of 10^{-6}M applied to the filter paper totally prevents this precocious germination. For this reason the catalogs have been extended to include the possible changes in mRNA subsets related to the cessation of cell division, to precocious germination or to ABA arrestation. All of these phenomena have occurred by the 4th day after removal of embryos from the seed tissue.

When the embryos have reached 100 mg in wet weight they have entered the maturation phase of embryogenesis. (Final embryo wet weight is about 125 mg). The vascular connection between the mother plant and the seed has atrophied and cell division has stopped. ABA levels have risen markedly, presumably to prevent "vivipary"[8]. These embryos precociously germinate readily when excised from the seed tissues, and, again this is prevented if the immature embryos are placed on filter paper moistened with an ABA solution. Again the catalogs have been extended to include measurements of changes occurring during precocious germination. Cotyledons from excised embryos at this stage of development incubated in the ABA solution are of interest since these embryos have been removed from a high ABA environment in the seed.

The dessicated dry seed represents a special developmental stage. The mature cotton seed has no form of dormancy to overcome. Its cotyledons contain mRNApoly(A)[9], and it is of interest to know if this RNA is still functional after several hours of germination or whether it is simply residual embryonic RNA that is rapidly degraded as has been suggested by earlier experiments[9].

With germination many new enzyme activites are required in cotyledons, and many of these activities are known to result from the de novo synthesis of both mRNA and protein during this period[10,11,12]. Other new enzyme activities appear as the result of de novo protein synthesis but without requiring concomittant RNA synthesis[8,13]. For this reason the catalogs include data from cotyledons germinated in the presence of actinomycin D so as to determine those proteins that may arise from pre-existing but unexpressed mRNA (stored mRNA).

Finally, during the 3rd day of germination, proteins coded for by the chloroplast genome become detectable, and, since their mRNAs are not likely to have poly(A) sequences, they may be made apparent by the fact that all of the in vitro cell-free translation is carried out with poly(A)-RNA as well as polyA+ RNA at all the developmental stages.

With this developmental scheme in mind, it is possible to hypothesize the existance of certain mRNA subsets at the outset; each subset under different regulatory influences as to time of synthesis and disappearance but whose concerted, integrated and overlapping functioning insures cotyledon embryogenesis and germination. It is anticipated that the catalogs will indicate the members of such subsets. A rather small number of such putative subsets are listed below.

CONCEIVABLE mRNA SUBSETS

Embryogenesis   Representative protein
1. functional only during cell. . . . . . actins
   division stage

2. functional only during ABA
   arrestation stage (ABA induced?) . . .

3. functional throughout embryogenesis. .

4. non-functional during ABA arrestation. stored mRNA derivatives

5. constitutive (independent of develop-
   mental events). . . . . enzymes of intermediary
                                            metabolism

Germination
1. Residual . . . .

2. Stored   . . . . carboxypeptidase C

3. Newly synthesized, unique
   to germination. . . . . asparagine synthetase,
                                            glyoxylate cycle enzymes

4. Newly synthesized, constitutive . . . enzymes of
                                            intermediary metabolism

LIMITATIONS AND POTENTIAL ARTEFACTS OF TECHNIQUES

Before presenting some of our data it is well to emphasize the constraints and limitations of the techniques used and some of the sources of artefact in these procedures so that the data may be viewed with the proper perspective.

First, as has been mentioned, only the 200-400 most abundant proteins are made readily visible in these procedures. Furthermore, only those proteins whose pIs are between pH 7.5 to 4.5 are focused in the first dimension. Many proteins have pIs beyond this range as indicated by the pile up of stain and radioactivity often seen on the edges of the slab gels. Technique for visualizing these proteins have been developed ("NEPHGE gels"[14]), but have not yet been incorporated into these studies. Next, quite a different display of proteins is observed when radioactive methionine is used as the protein precursor in comparison with radioactive leucine or amino acid mixtures. Proteins with inordinate amounts of methionine attain very high specific radioactivity that is not a true indication of their rate of synthesis in vivo or in vitro. The presence of the N-terminal methionine on in vitro labelled proteins, but its likely absence on these labelled in vivo contributes to this distortion of synthesis.

A further difficulty is encountered when in vivo and in vitro synthesis patterns are compared. Not a great deal of match-up is observed. This makes identifying identical proteins detected by the two methods hazardous. The failure of the two displays to indicate common radioactive proteins is likely to be due to the absences of the processing events in the in vitro translating systems. Large differences in the intensity of identical proteins between the two types of measurements may also indicate a rapid turnover of certain proteins in vivo (failure to accumulate although their mRNA is in high concentration) or simply result from the fact that translation of certain mRNAs may be regulated in some fashion in vivo, whereas such regulation is missing in in vitro translation.

RESULTS

Our catalogs are far from complete at present. Much of the in vivo synthesis catalog during germination remains to be completed. Much of the data is yet to be analyzed in detail. However, from the data collected to date, several interesting observations have accrued, and it is these observations we wish to present here.

PROCESSING OF THE PRINCIPAL STORAGE PROTEINS

Figure 2A presents a one dimensional view of the proteins present in the cotyledons of a mature cotton seed. Early on we found it convenient to divide the total protein of this tissue to that which is readily soluble (extracted in 0.1 M NaCl, pH 8.3) from that which requires high salt (0.5 M NaCl) or 6 M urea or 2% SDS or 2% deoxycholate to make soluble. The left hand gel well

contains total cotyledon protein (extracted 2% SDS), the center
well contains the readily soluble fraction (0.1 M NaCl extract)
and the right hand well contains the protein that can be pelleted
at low speed from the 0.1 M NaCl homogenate. The two principal
storage proteins are seen to be in the pellet fraction and to comprise together about 25% of the total cotyledon protein.

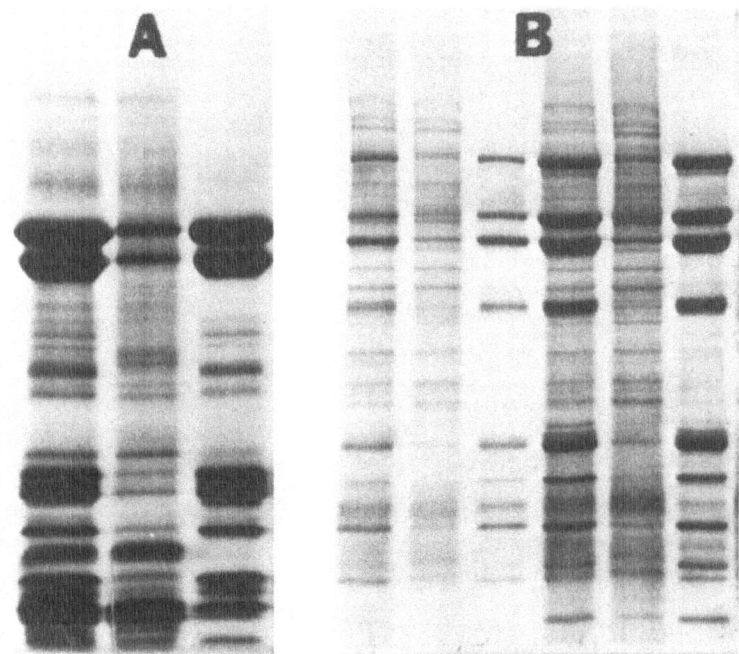

Fig. 2. Display of cotyledon proteins separated by discontinuous
        gel electrophoresis in SDS and stained for protein. In A,
        proteins from dry seed cotyledons; left well = total pro-
        tein, center well = readily soluble protein, right well =
        pellet protein. In B, proteins from cotyledons of immature
        embryos. Left three wells are total, readily soluble and
        pellet proteins from cotyledons of 50 mg embryos. Right
        three wells are the same preparations from cotyledons of
        100 mg embryos.

   Figure 3 shows the one dimensional electrophoretic separation
of total protein extracted from cotyledons during points in embryogenesis and germination. Several interesting phenomena are
observed in this gel. Two principal storage proteins have apparent molecular weight of 53,000 and 48,000 daltons. It is
apparent that the smaller of the two storage proteins appears
first and accumulates faster initially than does the larger storage protein. It is also degraded much more rapidly in germination.

Further, several proteins that are abundant in embryogenesis vanish during the last days of embryogenesis and are undetectable in dry seed cotyledons. Two of these have molecular weights of about 70,000 and 40,000 daltons and are seen in Figure 2B to be confined to the pellet fraction.

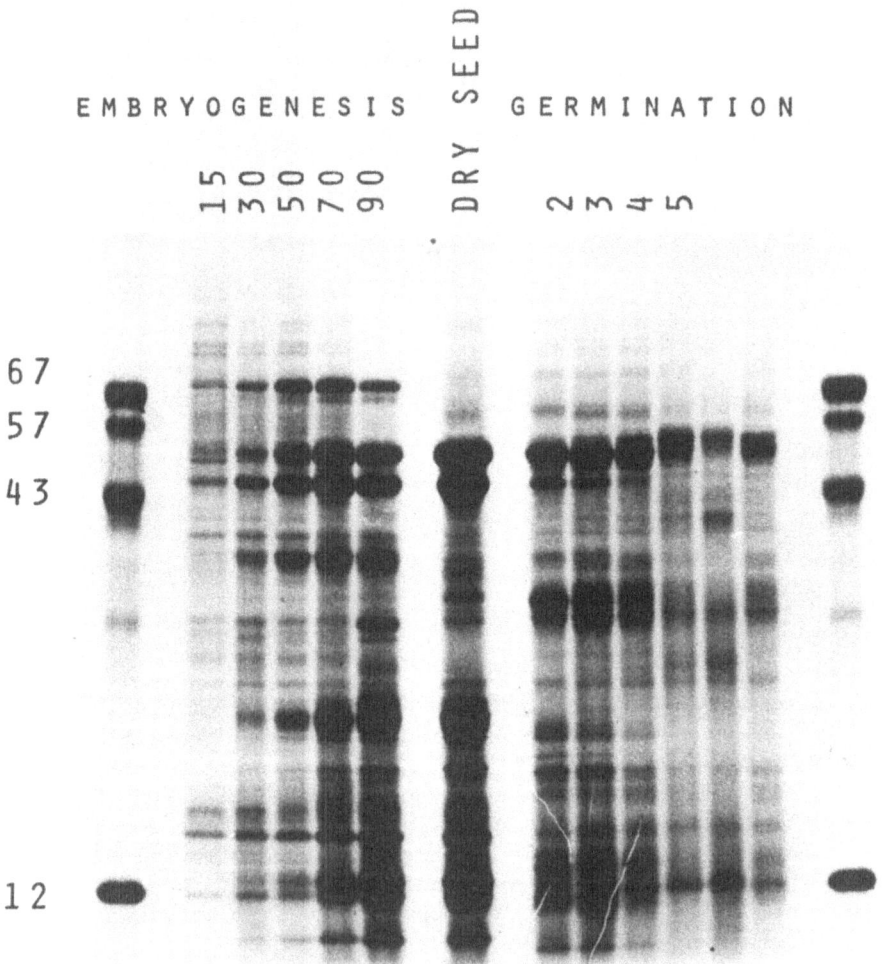

Fig. 3. Display of total cotyledon protein extracted at points during development. Electrophoresis as in Fig. 1. Numbers under EMBRYOGENESIS refer to the size in mgs of the embryos used. Numbers under GERMINATION refer to the days that the embryos were germinated. Numbers on the left margin are the molecular weights in kilodaltons of reference proteins.

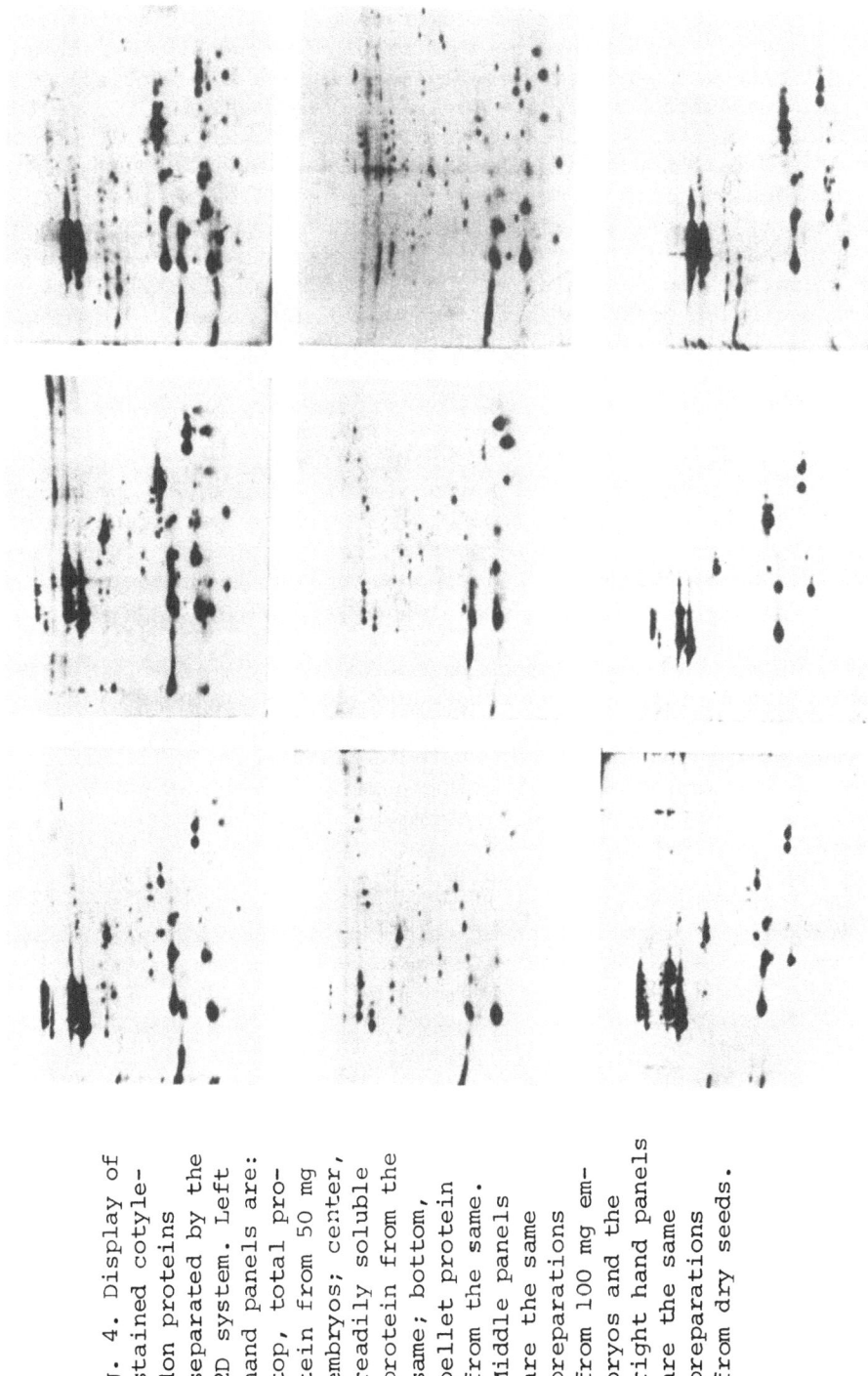

Fig. 4. Display of stained cotyledon proteins separated by the 2D system. Left hand panels are: top, total protein from 50 mg embryos; center, readily soluble protein from the same; bottom, pellet protein from the same. Middle panels are the same preparations from 100 mg embryos and the right hand panels are the same preparations from dry seeds.

Figure 4 is the two dimensional display of proteins from cotyledons
from 50 mg, 100 mg and dry seed embryos. The principle storage
proteins are found to have a range of pI heterogeneity, which subsequent data will show is probably not introduced during solubilization or electrophoresis artefactually. Again the 70,000 dalton
protein is seen to disappear as the tissue matures as well as does
the 40,000 dalton protein. The same pI heterogeneity found in
the storage proteins is observed in the 70,000 dalton protein.
The readily soluble fraction (supernatant fraction) is seen to have
fewer abundant proteins than does the pellet fractions, and these
are of rather low molecular weight. A number of changes are seen
in this display of extant proteins during embryogenesis, but only
a few of these have been studied in detail to date.

A number of changes occur in these populations during cotyledon embryogenesis; however, we have concentrated principally on the
changes in the dominant storage proteins and the two disappearing
proteins. Our initial in vivo translation studies for the second
catalog yielded rather intriguing results which are presented in
Figure 5. Here, pellet proteins from 100 mg cotyledons that have
been incubated 6 hours in ^{14}C amino acids have been electrophoresed
in two wells in the Laemmli one dimensional gel system. An autoradiogram of these two wells is on their right. (The gel is
highly overloaded to increase its radioactivity). The two principal storage proteins are grotesquely obvious in the stained
wells. The autoradiograph shows extensive synthesis of the smaller
storage protein, but no apparent radioactivity in the larger.
This was curious at the time, since, at this point in embryogenesis, the larger protein is accumulating faster than the
smaller one (Figure 3). Furthermore, the figure shows extensive labeling of the 70,000 dalton protein, which, although an
abundant protein at this developmental stage, is in the process
of vanishing from the tissue at this time. Obviously its synthesis
continues into the period of its degradation or processing.

A reasonable conjecture at this point would be that the
larger storage protein is synthesized initially as the 70,000
dalton species which is cleaved in a rather slow process. To test
this idea we carried out a series of pulse-chase incubations of
50 mg cotyledons in which they were pulsed for 6 hours, and incubated thereafter for varying time periods without isotopes before
being harvested. These and other data are presented in Figure 6.
At the top left of this composite is shown a stained two dimensional gel of pellet protein from 50 mg cotyledons. At the top
right is an identical gel stained for carbohydrate by the PAS method[15]. The larger storage protein and its 70,000 dalton putative
precursor are found to be glycoproteins, and to be the only glycoproteins demonstratable in the pellet fraction by this method.
In the middle of the left panel a fluorograph of pellet protein

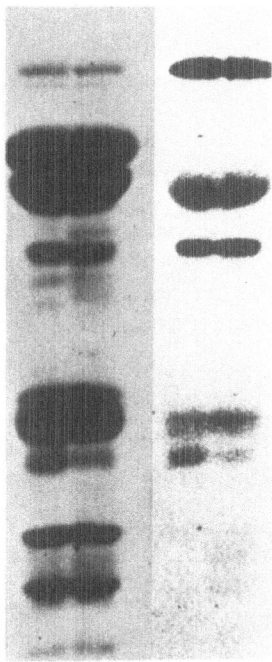

Fig. 5. In vivo protein synthesis in cotyledons of 100 mg embryos.
        Left two wells are the electrophoretic separation of pellet
        proteins. The right two wells are autoradiographs of the
        left wells.

from 50 mg cotyledons that have been pulsed 6 hrs and chased 12
additional hours is presented. Radioactivity is seen here in the
larger storage proteins, but the various pI isomers do not have
the same specific radioactivity. To the right of this gel is
given an expanded view of fluorographs of the storage protein
region of the two dimensional gel from gels containing protein
from cotyledons pulsed 6 hrs and chased 6, 9, 15 and 24 hrs.
From this it is apparent that the larger storage proteins gain
radioactivity with time, and the 70,000 dalton proteins appear to
lose radioactivity. These data reinforce the precursor-product
idea; however, to firmly establish this relationship we utilized
immunochemical techniques.

    First, the large storage proteins were purified and antibodies
against them prepared in rabbits by conventional means. Figure
7, left side, shows a typical titration of the antibody preparation by the "rocket" technique[16]. Shown in the right hand side of
this figure is the result obtained when total cotyledon protein
from 50 mg cotyledons, separated by one dimensional electrophoresis, is electrophoresed into the antibody containing region of the

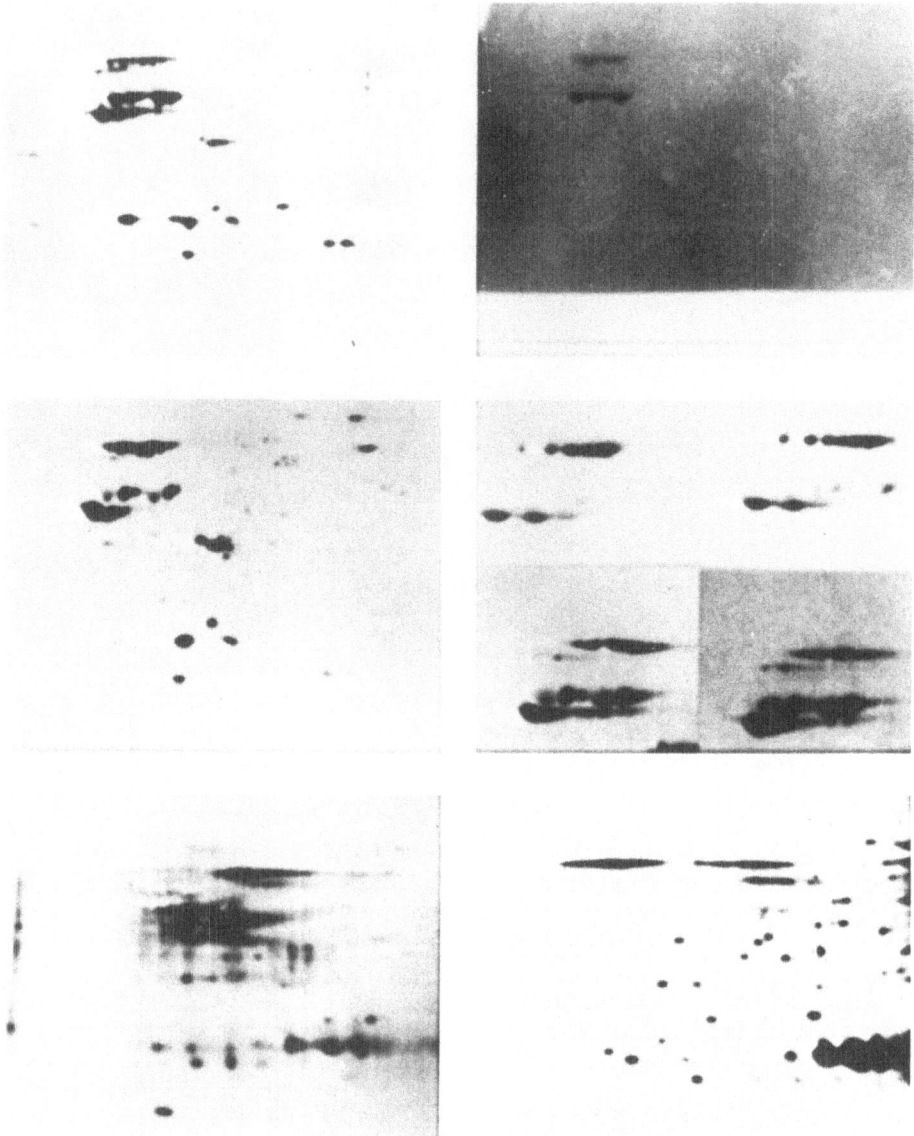

Fig. 6. Top, left: stained pellet protein from 50 mg embryos. Top, right: same preparation stained for carbohydrate. Middle, left: Fluorograph of the same preparation incubated in radioactive amino acids for 6 hrs and chased for 12 hrs. Middle, right: expanded views of fluorographs of storage protein area of gels. Bottom panels: In vitro synthesized proteins from total RNA from 50 mg cotyledons. Left = ^{3}H leu; right = ^{35}S met.

gel by the method of Chua and Blomberg[17]. Two regions of the gel
containing the separated cotyledon proteins are seen to contain
proteins that immunoprecipitate the antibodies, and these regions
contain the 70,000 dalton protein and the larger storage protein
which served as the antigen. From this it is apparent that the
larger storage protein species are derived from the 70,000 dalton
species.

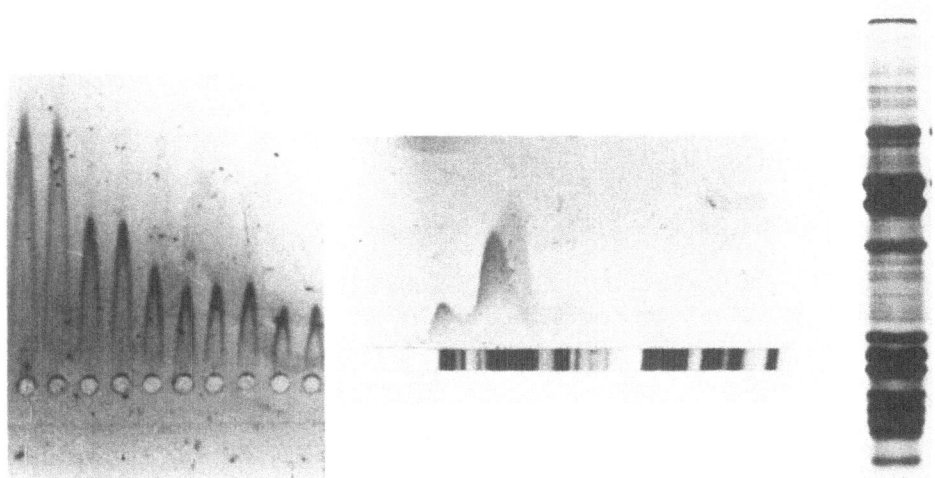

Fig. 7. Left: Rocket immunoelectrophoresis of antigen (large storage
        protein) vs. rabbit antiserum. Stained gel. Middle and
        right: Total protein from 50 mg cotyledons was electro-
        phoresed as shown on far right. A longitudinal section
        of the ID gel was electrophoresed into antiserum to the
        large storage protein[17]. Both are stained gels.

     Data gathered for the third catalog (in vitro translation
products) allowed us to attempt to identify the initial translation products that give use to the 70,000 dalton species.
Since the processing of initial translation products into glycoproteins probably does not occur in the wheat germ systems, the
initial products of the mRNA for the 70,000 dalton pI isomers
would not be expected to co-migrate with these proteins. The
bottom left of Figure 6 is the fluorograph obtained when total
RNA from 50 mg cotyledons is translated in the wheat germ system
and its products electrophoresed. In this instance ^{3}H leucine
was used as the radioactive precursor. A band of proteins of
about 60,000 molecular weight and showing pI heterogeneity is
found to the acidic side of the storage proteins and could be
considered the unglycosylated precursor of the 70,000 dalton

proteins. (In this fluorograph, non-radioactive protein from dry
seed cotyledons has been included in the preparation separated by
the two dimensional system. The mature unlabelled storage proteins
can be seen on the fluorograph as white areas since very abundant
proteins tend to exclude other proteins from their region of the
gel). However, when the experiment is carried out with ^{35}S
methionine (bottom right-hand panel) other potential precursor
bands show up. These proteins obviously have a different ratio
of methionine to leucine than does the band seen on the left hand
panel. We have not yet resolved the question of the identity of
the precursor of the 70,000 dalton proteins. There does not appear
to be an easily demonstrable immunochemical relationship between
the larger and smaller storage proteins. However, the glycosyl
residues of the larger storage proteins may have played a deter-
minative role in antibody production, which might leave the unglycosylated smaller storage proteins unreactive.

PROTEIN CHANGES DURING PRECOCIOUS GERMINATION

As previously mentioned our catalogs include changes in
cotyledon proteins occurring during the precocious germination of
50 and 100 mg embryos. This portion of the catalogs also includes
protein changes that may take place in cotyledons from these
dissected embryos whose precocious germination is artifically
arrested by ABA. Figure 8 shows the stained catalog of these
changes. The left hand column shows from top to bottom, total
protein from cotyledons of ungerminated 50 mg embryos, from these
cotyledons after 4 days of precocious germination and from cotyledons incubated 4 days in ABA. In both sets of dissected embryos
the 70,000 dalton precursor has vanished. The 40,000 dalton protein is also disappearing in both sets. Some loss of the two
principal storage proteins is evident in the precociously germinating cotyledons as well as of the many other abundant proteins.
This loss has not occurred in the ABA treated cotyledons. The
middle column shows the same sequence for the 100 mg embryo cotyledons. In this case precocious germination ±ABA was carried
out for 3 days. Again the 70,000 dalton and 40,000 dalton proteins
disappear in the germinating cotyledons and are decreased in those
treated with ABA. When these two sets of embryos are incubated
for 2 additional days (bottom two panels of the right hand column
of the figure) the changes are more apparent. The germinating
cotyledons show a great loss of storage protein and other abundant
proteins, whereas those treated with ABA have lost the 70,000 and
40,000 dalton proteins, but none of the others.

Of particular interest in this figure is the accumulation of
several new proteins in the ABA-treated cotyledons (enclosed in
pen). Some of these proteins are not apparent in 50 or 100 mg
embryos nor in those precociously germinating. However, they are

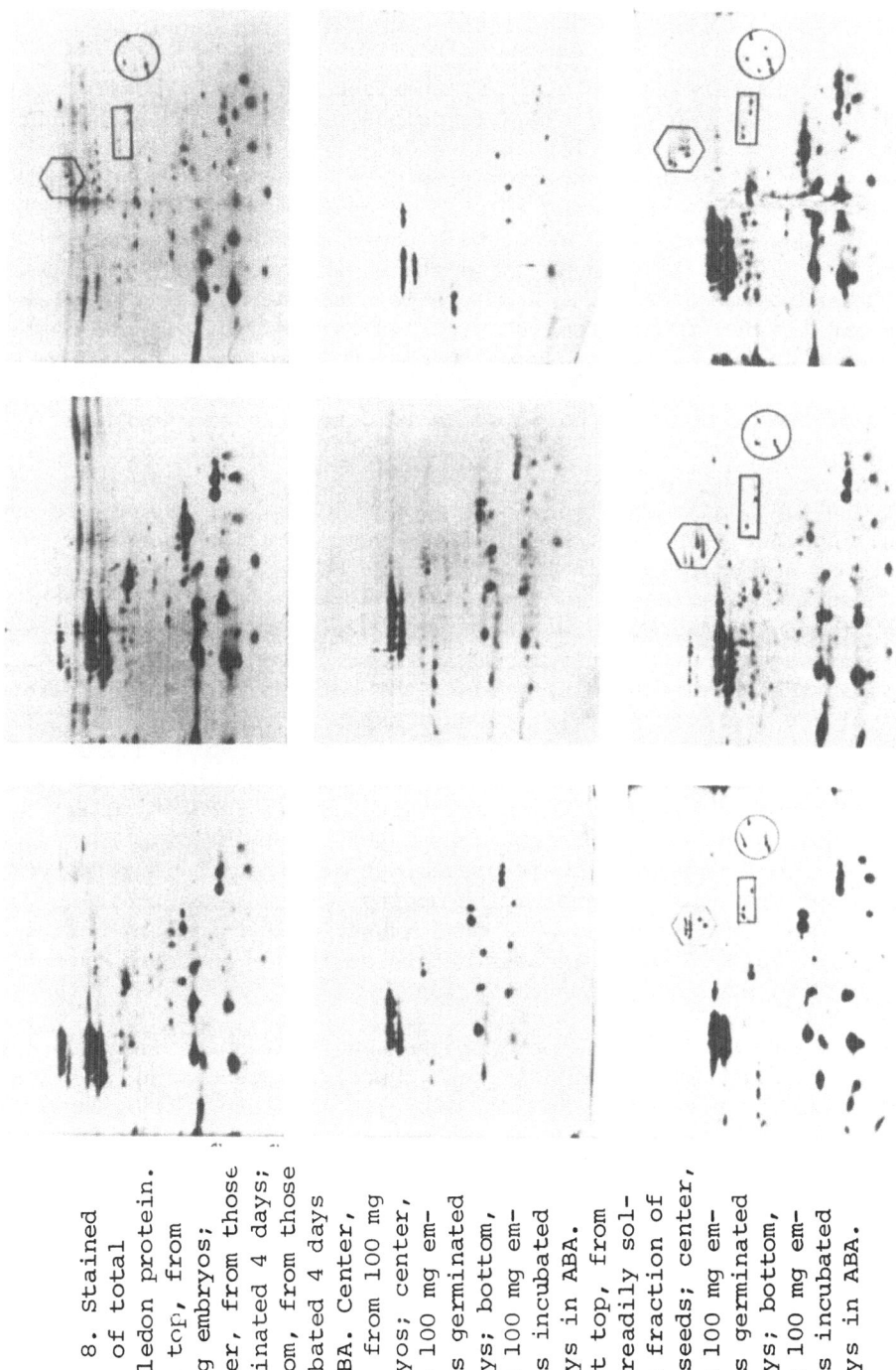

Fig. 8. Stained gels of total cotyledon protein. Left top, from 50 mg embryos; center, from those germinated 4 days; bottom, from those incubated 4 days in ABA. Center, top, from 100 mg embryos; center, from 100 mg embryos germinated 3 days; bottom, from 100 mg embryos incubated 3 days in ABA. Right top, from the readily soluble fraction of dry seeds; center, from 100 mg embryos germinated 5 days; bottom, from 100 mg embryos incubated 5 days in ABA.

found in the mature seed as shown by the upper right hand panel
which is the readily soluble protein from dry seed cotyledons.

Figure 9 shows the protein synthesized in vivo by cotyledons
from 50 and 100 embryos during precocious germination ±ABA. The
50 mg sequence is on the left hand side. From top to bottom are
shown total radioactive proteins from cotyledons of embryos transferred to isotope containing solutions for 6 hours immediately
after dissection and then harvested (these show the proteins
being synthesized in situ at these stages in development) followed
by the radioactive proteins from precociously germinating cotyledons and finally those from ABA arrested cotyledons. In the latter
two cases the 50 mg embryos were incubated 4 days and the 100 mg
embryos 3 days. They were exposed to the isotope for the last 12
hours of the incubation period so as to determine the proteins
being synthesized during the last phase of incubation.

The top panels show that both 50 and 100 mg cotyledons are synthesizing the 70,000 and 40,000 dalton proteins, although this synthesis is somewhat less in the older cotyledons. By the end of
the incubation period, all of this synthesis has stopped in both
sets of cotyledons, even in those treated with ABA. Many new
proteins are being synthesized in the unarrested cotyledons; however, in the ABA arrested cotyledons the pattern is quite different.
Much of the radioactivity is found in those proteins that were
found to accumulate in late embryogenesis in the stain catalog.
The salient question here is are any/all of these proteins induced
by ABA? Some of these proteins are being synthesized in 50 mg
cotyledons in situ when the endogenous ABA level is low. Their
synthesis continues when their precocious germination is arrested
by ABA but ceases when germination begins. Thus these proteins
may be considered products of a mRNA subset that functions during
the latter half of embryogenesis before and after ABA arrestation
and that disappeared upon germination.

Some of the proteins found in the ABA treated cotyledons,
however, are not synthesized in the younger cotyledons and may indeed result from exposure to ABA, appearing in late embryogenesis
in situ and upon exposure of young embryos to ABA after dissection.
These proteins we feel are the products of the mRNA subset induced
by ABA and whose function is to prohibit immature embryos from
germinating before seed maturation is complete. It has been known
for some time that ABA inhibition of the synthesis of several
developmentally important enzymes is, itself, a phenomenon that
requires continued RNA synthesis[8,18].

We have said very little about the 40,000 dalton protein that
disappears upon the dissection of immature embryos and disappears
normally in late embryogenesis. We believe this protein to be one

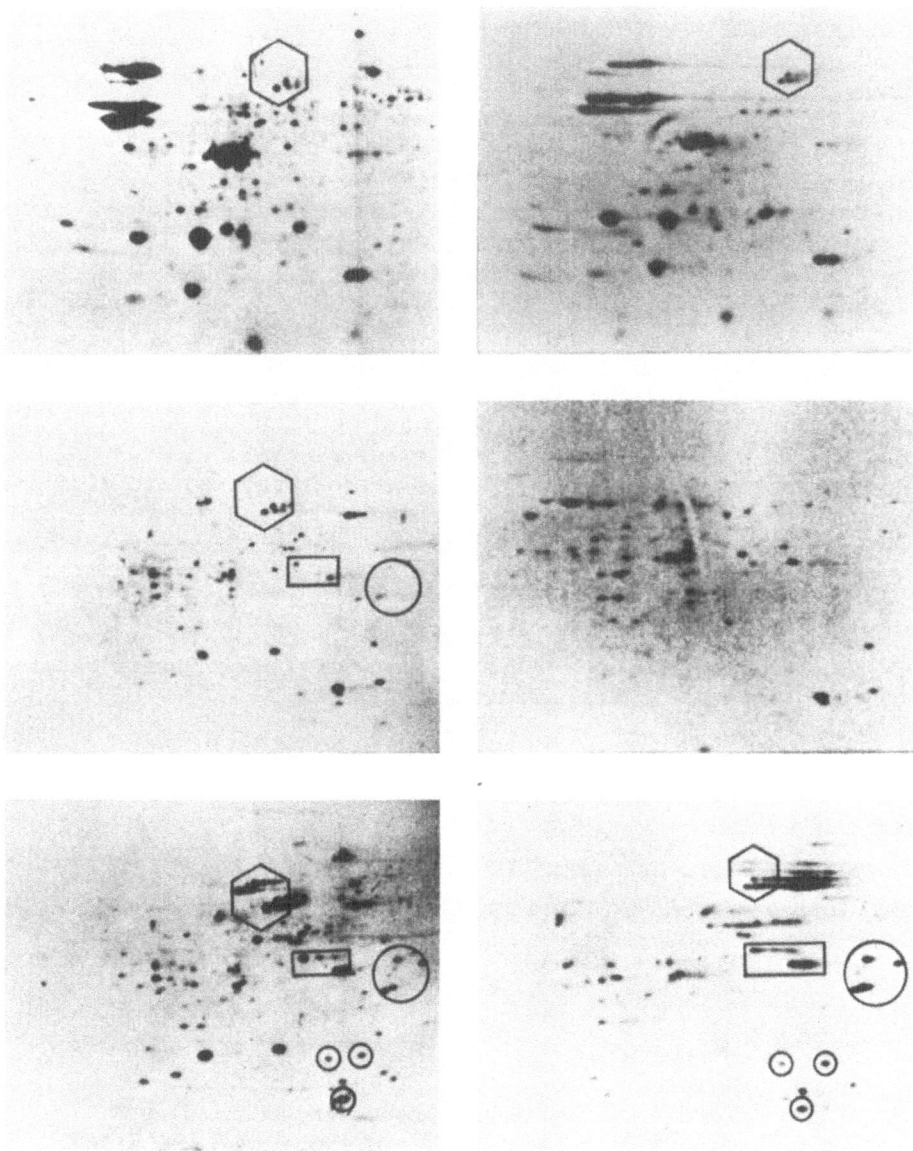

Fig. 9. Fluorographs of proteins synthesized in vivo. Left panels
        are from cotyledons of 50 mg embryos. Top, from embryos
        labelled briefly after dissection; middle from those
        labelled after 4 days germination; bottom, from those
        labelled after 4 days in ABA. Right panels are from coty-
        ledons from 100 mg embryos. Treatments are the same, ex-
        cept germination and incubation was for 3 days.

of the plant actins whose presence is linked to cell division.
Our rationale for this belief is summarized as follows:

PROPERTIES OF PURATIVE COTTON ACTIN

Chemical
1. Molecular weight = about 40,000 daltons
2. pI = 6.2 in 8 molar urea
3. Insoluble in dilute salts or Triton X-100
4. Soluble in SDS, DOC, 6 M urea, half molar salts
5. Solubilized by millimolar ATP, Ca^{++}
6. Precipitated by the addition of KCl at 0.1 M
7. Binds weakly to DNAse I-Sepharose

Biological
1. Abundant protein during cell division phase of embryogenesis.
2. Disappears during maturation phase.
3. In vivo synthesis demonstratable.
4. Synthesis ceases (as does cell division) and protein disappears
   when very young embryos are removed from seed.
5. Synthesis not maintained by ABA

SUMMARY

     These data from both the stain and in vivo synthesis catalogs
can be diagrammatically summarized as follows:

PROTEIN CHANGES IN COTYLEDON EMBRYOGENESIS

at 50 mg stage
1. low endogenous ABA
2. cell division occurring
3. storage protein synthesis and processing from 70K dalton
   precursor occurring
4. actin synthesis occurring

at 100 mg stage
1. high endogenous ABA
2. no cell division
3. storage protein synthesis and processing continuing
4. actin synthesis decreasing, actin disappearing

upon precocious germination
1. cell division stops prematurely in 50 mg embryos.
2. storage protein synthesis stops, degradation begins
3. processing of 70K dalton precursor continues until complete
4. actin synthesis stops, actin disappears
5. synthesis of embryogenic proteins stcps, they disappear

6. germination proteins appear

upon incubation in exogenous ABA
1. cell division stops prematurely in 50 mg embryos
2. storage protein synthesis stops but no degradation occurs
3. processing of 70K dalton precursor continues until complete
*4. actin synthesis stops, actin disappears
**5. many embryogenic proteins continue to be synthesized
***6. ABA proteins appear

in the mature seed
1. all storage protein processed, 70K dalton precursor gone
2. actin has disappeared
3. embryogenic proteins evident
4. ABA proteins evident

  *member of cell division mRNA subset which may depend on vascular
   connection
 **members of embryogenic mRNA subset which functions during and
   after cell division phase, and is not dependent upon vascular
   connection or high endogenous ABA, disappears when germination
   program begins
***members of ABA induced mRNA subset, functions to prevent initia-
   tion of the germination program.

REFERENCES

1. U. K. Laemmli. Cleavage of structural proteins during the
      assembly of the head of bacteriophage T4. Nature 227,
      680 (1970).
2. P. H. O'Farrell. High resolution two-dimensional electro-
      phoresis of proteins. J. Biol. Chem. 250, 4007 (1975).
3. R. A. Laskey and A. D. Mills. Quantitative film detection
      of ^{3}H and ^{14}C in polyacrylamide gels by fluorography.
      Eur. J. Biochem. 56, 335 (1975).
4. H. R. B. Pelham and J. Jackson. An efficient mRNA-dependent
      translation system from reticulocyte lysates. Eur. J.
      Biochem. 67, 247 (1976).
5. B. E. Roberts and B. M. Paterson. Efficient translation of
      tobacco mosaic virus RNA and rabbit globin 9S RNA in a
      cell-free system from commercial wheat germ. Proc. Nat.
      Acad. Sci. USA 70, 2330 (1973).
6. J. N. Ihle and L. Dure III. Synthesis of a protease in germ-
      inating cotton cotyledons catalyzed by mRNA synthesized
      during embryogenesis. Biochem. Biophys. Res. Comm. 36,
      705 (1969).
7. J. N. Ihle and L. Dure III. Hormonal regulation of translation
      inhibition requiring RNA synthesis. Biochem. Biophys. Res.

Comm. 38, 995 (1970).

8.   J. N. Ihle and L. Dure III. The developmental biochemistry of
     cottonseed embryogenesis and germination. III. Regulation
     of the Biosynthesis of enzymes utilized in germination.
     J. Biol. Chem. 247, 5048 (1972).

9.   B. Harris and L. Dure III. Developmental regulation of cotton-
     seed embryogenesis and germination. IX Polyadenylation of
     stored mRNA. Biochemistry 17, 3250 (1978).

10. B. Hock and H. Beevers. Development and decline of glyoxylate
     cycle enzymes in watermelon seedlings. Effects of dactino-
     mycin and cycloheximide Z. Pflanzenphysiol. 55, 405 (1966).

11. W. M. Becker, C. J. Leaver, E. M. Weir and H. Rieman. Regu-
     lation of glyoxysomal enzymes during germination of Cucum-
     ber I. Plant Physiol. 62 , 542 (1978).

12. J. E. Lamb, H. Riezman, C. J. Leaver and W. M. Becker, Regu-
     lation of glyoxysomal enzymes during germination of Cu-
     cumber II. Plant Physiol. 62, 754 (1978).

13. V. Walbot, A. Capdevila and L. Dure III. Action of 3'd adeno-
     sine (cordycepin) and 3'd cytidine on the translation of
     stored mRNA of cotton cotyledons. Biochem. Biophys. Res.
     Comm. 60, 103 (1974).

14. P. Z. O'Farrell, H. M. Goodman and P. H. O'Farrell. High re-
     solution two-dimensional electrophoresis of basic as well
     as acidic proteins. Cell 12, 1133 (1977).

15. J. P. Segrest and R. J. Jackson. Molecular weight determina-
     tion of glycoproteins by polyacrylamide gel electrophoresis
     in SDS. in "Methods in Enzymology," XXVIII, Academic Press,
     New York (1972).

16. C. B. Laurell. Quantitative estimation of proteins by elec-
     trophoresis in agarose gels containing antibodies. Anal.
     Chem. 15, 45 (1966).

17. N. H. Chua and F. Blomberg. Immunochemical studies of Thyla-
     koid membrane polypeptides from spinach and Chlamydomonas
     reinhardtii. J. Biol. Chem. 254, 215 (1979).

18. D. T. Ho and J. E. Varner. Response of barley aleurone layers
     to Abscisic Acid. Plant Physiol. 57, 175 (1976).

HORMONAL AND GENETIC REGULATION OF α-AMYLASE SYNTHESIS IN BARLEY

ALEURONE CELLS

Tuan-hua David Ho

Department of Botany
University of Illinois
Urbana, Illinois 61801 U.S.A.

INTRODUCTION

The production and secretion of α-amylase and several other hydrolases by the aleurone cells of germinating barley seeds are regulated by two hormones, gibberellic acid (GA_3) and abscisic acid (ABA) (16,18). While GA_3 enhances the production of α-amylase, ABA is able to reverse the GA_3 effect. Yet, these two hormones do not directly compete with each other and they appear to have different mode of action (1,7). It has been demonstrated un-equivocally that the GA_3 enhancement on α-amylase production is essentially due to the de novo synthesis of α-amylase molecules (3,8). The GA_3 enhancement on α-amylase synthesis has a lag period of 3 to 4 hours and a fast and constant accumulation of this enzyme takes place after 12 hours of hormone treatment (Fig. 1). Potent transcription inhibitors, such as cordycepin (3'-deoxy-adenosine), added at the same time as GA_3 administration effective-ly abolish the hormonal effect on α-amylase synthesis (6). However, this inhibitory effect of cordycepin on α-amylase synthesis is progressively decreased when the time of its addition is delayed and cordycepin does not inhibit α-amylase at all if it is added 12 hours or later after hormone administration (5). Since the effectiveness of cordycepin on RNA synthesis (cordycepin apparently inhibits both transcription and poly(A) addition in barley aleurone cells) does not change appreciably when its addition is delayed, this observation indicates the following. First, α-amylase mRNA is formed during, or even before, the first 12 hours of GA_3 treatment. Second, the fast synthesis of α-amylase after 12 hours of GA_3 treatment is not under transcription-al control.

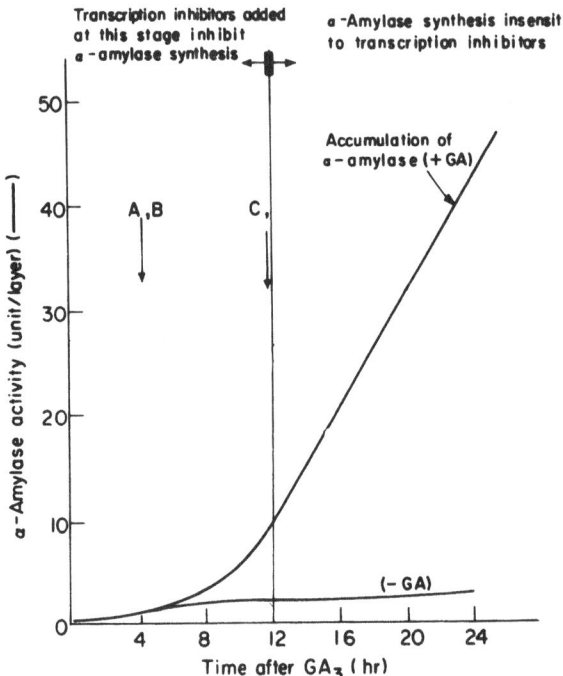

Fig. 1. Timecourse of the GA_3 enhancement on α-amylase production
         in barley aleurone cells. A, B and C are the starting
         points for membrane formation, total poly(A)+ RNA synthe-
         sis, and secretion of α-amylase, respectively.

EFFECT OF GA_3 ON THE FORMATION AND ACTIVITY OF α-AMYLASE mRNA

     α-Amylase is the most abundant protein synthesized in the
presence of GA_3. Thus, its synthesis can be easily quantitated
by simple SDS gel electrophoresis as shown in Fig. 2. Similarly,
the level of α-amylase specific mRNA can be estimated by employing
SDS gel to analyze proteins synthesized in vitro in wheat germ
extract primed with total RNA isolated from aleurone cells (Fig.
2). Aleurone cells without GA_3 treatment do not contain
significant amounts of active mRNA for α-amylase. Confirming the
report by Higgins et al (4) we have found that the level of α-
amylase mRNA started to increase after 3 to 4 hrs and leveled off
after 12 hrs of GA_3 treatment (Fig. 3). This timecourse correlates
well with that of cordycepin inhibition described above. Furthermore, it has been demonstrated that during the first 12 hours of
GA_3 treatment there is an increase in the synthesis of total mRNA
as measured by the incorporation of radioactive nucleosides into
poly(A)+ RNA (5,9). Although more vigorous demonstration such as
RNA:cDNA hybridization is needed, these observations tend to
suggest that α-amylase synthesis during this period is under
transcriptional control.

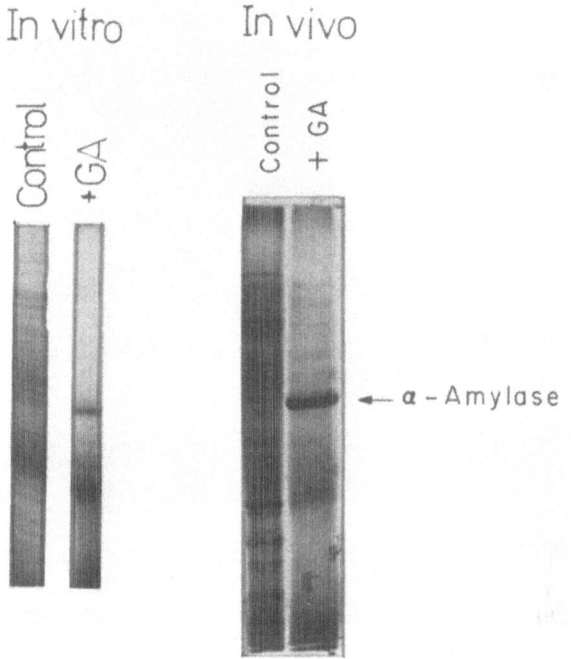

Fig. 2.   A. Profile of newly synthesized proteins from barley
          aleurone cells. Aleurone layers were treated with or
          without GA_3 (1 μM) before they were pulse-labeled with
          ^{35}S-methionine for 1 hr. The extracted proteins were
          analyzed by SDS gel electrophoresis and autoradiography.
          B. Profile of proteins synthesized in wheat germ cell
          free translation system primed with RNA isolated from
          barley aleurone cells. Total RNA was isolated from
          aleurone cells treated with or without GA_3 according
          to Higgins et al. (4). Ten μg of total ^{3}RNA was added
          to each 50 μl wheat germ reaction mixture with ^{35}S-
          methionine as the label precursor. The translation
          product were also analyzed by SDS gel electrophoresis
          and autoradiography.

     The action of GA_3 after the first 12 hours appears to be
different because the level of α-amylase mRNA remains constant
while the relative rate of α-amylase synthesis in vivo keeps on
increasing (Fig. 3), which is no longer subject to the inhibition
of cordycepin. How could a constant level of α-amylase mRNA lead
to a ever increasing rate of α-amylase synthesis? The α-amylase
mRNA is apparently translated more efficiently relative to other

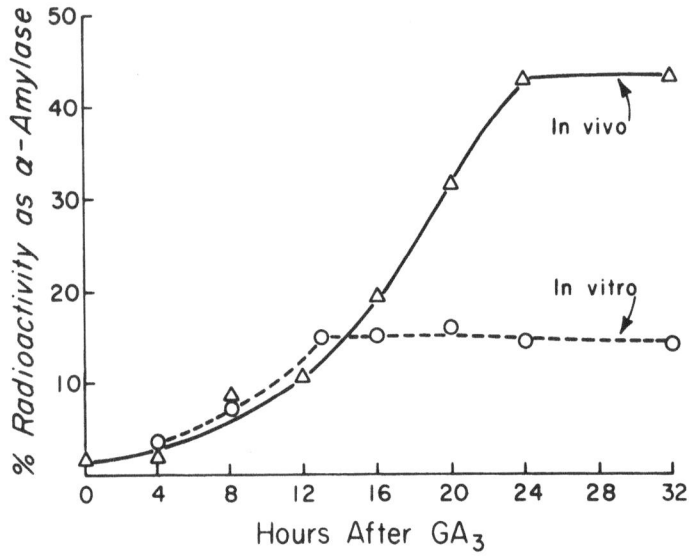

Fig. 3.   Timecourse of the GA₃ enhancement on level of translatable
          α-amylase mRNA in barley aleurone cells. O---O Relative
          abundance of α-amylase mRNA as determined by in vitro
          translation.   Δ---Δ Relative rate of α-amylase synthesis
          in vivo.% Radioactivity as α-amylase was determined by
          scanning the autoradiogram of SDS gel and integrating
          the area under the α-amylase peak.

messages.   What could be the mechanism?   In order to answer this
question we slowed down the overall elongation rate of protein
synthesis in barley aleurone cells by treating these cells with
suboptimal concentrations of an elongation inhibitor, cycloheximide.
As demonstrated by Lodish with α- and β-globin mRNA (12,13), if
the more efficient translation of α-amylase mRNA over other mRNAs
is due to the faster initiation rate of α-amylase mRNA, the advantage of α-amylase mRNA over other mRNAs will be minimized when
elongation, instead of initiation, becomes the rate limiting step
of protein synthesis.   As shown in Figure 4, the relative
abundance of α-amylase is 43% in the absence of cycloheximide.
However, this value decreases when elongation of protein synthesis
is slowed down by cycloheximide.   Cycloheximide at the concentration
of 2 μg/ml inhibits about 60% of total protein synthesis.   Under
this situation, elongation has to be the rate limiting step of
translation.   The relative abundance of α-amylase synthesized in
aleurone cells treated with cycloheximide at 2 μg/ml reaches 12%,
about the relative population of α-amylase mRNA (13%) in GA₃ treated aleurone cells.   This result suggests that α-amylase mRNA may be

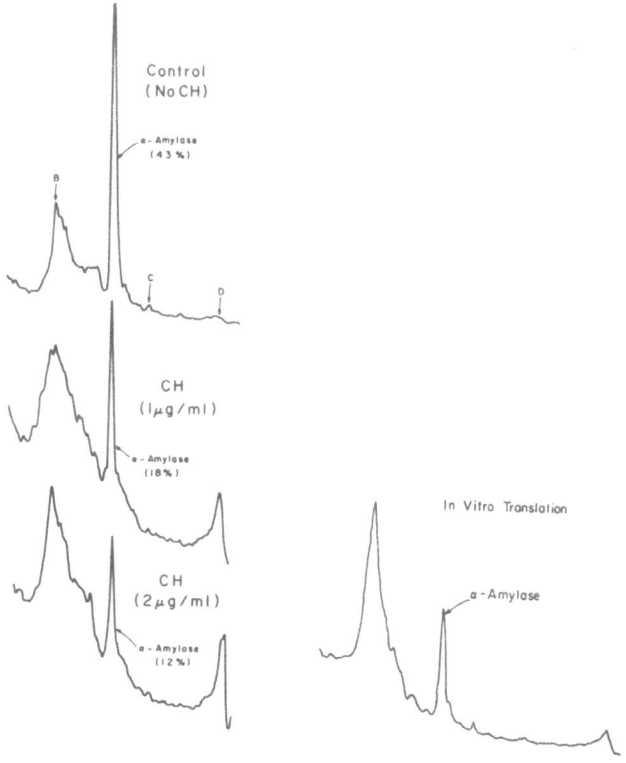

Fig. 4.   Effect of elongation inhibitor (cycloheximide) on the
          relative abundance of α-amylase synthesized in barley
          aleurone cells.   Aleurone cells were pretreated with GA₃
          (1 μM) for 18 hours before cycloheximide and ³⁵S-
          methionine were added.   After one hour labeling with ³⁵S-
          methionine proteins were extracted and analyzed by SDS
          gel electrophoresis.   Equal amounts of radioactivity
          were applied to each gel channel.   The tracings of the
          autoradiogram were presented and compared with the in
          vitro translation products of mRNA isolated from 18 hr
          GA₃ treated aleurone cells.

initiated faster than average mRNA in aleurone cells. When
elongation rate of protein synthesis is slowed down by cycloheximide, the advantage of α-amylase mRNA over other mRNAs, in
terms of initiation rate, decreases.

     In summary (Fig. 5), the regulation of GA₃ on the synthesis
of α-amylase appears to be complex.   During the first 12 hours of
GA₃ treatment membrane formation, as described by Jones and others
(10), and the increase of the level of α-amylase mRNA take place.
Subsequently, these two events lead to the synthesis of α-amylase

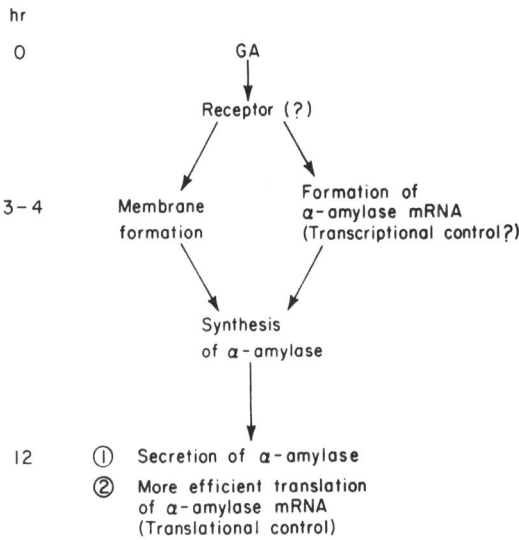

Fig. 5.   Summary of the effect of GA$_3$ in barley aleurone cells.

which is a secretory protein. After the first 12 hours of hormone treatment the α-amylase mRNA is apparently initiated faster than other messages in GA$_3$ treated tissue. However, whether this is due to the intrinsic properties of α-amylase mRNA or the presence of an α-amylase mRNA specific initiation cofactor is not yet clear.

SCREENING AND CHARACTERIZATION OF MUTANTS OF ALTERED HORMONE SENSITIVITY

Mutants have been intensively used to study genetic and bio-chemical regulation in microorganisms and to much less extent in higher organisms. Apparently, hormones have complex action in barley aleurone cells and there are several regulatory steps in-volved. It would be beneficial if mutants blocked in one of the regulatory steps could be obtained for comparative studies. Al-though barley, unlike bacteria, has a relatively long life cycle, genetic manipulations with various mutants are still possible. The fact that genetic studies in maize have been intensively carried out tends to support this view. High mutation frequency of barley can be induced by a potent base substitution mutagen, sodium azide (14,15). Using sodium azide, Nilan has observed very high frequency (>40%) of chlorophyll-deficient mutations in barley (14). Recently, nitrate reductase deficient mutants have been successfully isolated among azide mutagenized barley (9

nitrate reductase mutants among 6,000 plants screened) (17).   The
frequency of presumed single locus mutation, such as waxy endosperm and vine (gigas), has been shown to be 2.7 and 1.0 per
10,000 seeds, respectively (11).

In order to screen barley mutants with altered sensitivity to
GA_3, we cut the seeds produced by mutagenized barley into embryo
halves and endosperm halves. While the embryo half seeds are
stored the endosperm half seeds are placed on top of an agar plate
(in a petri dish) containing starch and GA_3. The petri dish is
incubated at 28° C for 30 hours. After incubation, half seeds are
taken off the plate and the plate is flooded with $KI-I_2$ solution.
The background of the plate is blue due to the reaction between
starch and I_2. However, transparent halos exist at the positions
where half   seeds are originally placed because α-amylase
secreted by the half seeds has digested the starch in the agar
plate (Fig. 6). If the half seed is a GA_3 insensitive mutant,
there will be no halo at the position where it is placed. In
order to recover the mutants, the embryos will be germinated when
the screening method indicates that their corresponding endosperm
halves do not respond to GA_3. For the ABA insensitive mutants
both GA_3 and ABA are included in the agar plate. While the wild
type does not produce and secrete α-amylase, the ABA insensitive
mutants cause the formation of transparent halos. Only ABA is included in the agar plates for the screening of constitutive
mutants in which the synthesis and secretion of α-amylase is not
subject to hormonal regulation. So far, we have been able to
isolate three homozygous GA_3 insensitive mutants and one GA_3
supersensitive mutant which are still stable after two generations
(Fig. 7). What is the potential of these mutants? To answer this
question I will describe the work we have done with a currently
existing wheat mutant (D 6899) which we obtained from C. O.
Qualset at UC-Davis.   D6899 is a dwarf wheat which produces only
5 to 10% α-amylase from its aleurone layers as that produced by
normal wheat.   The partial GA_3 insensitivity of D6899 is regulated
by a single gene ("Tom Thumb" gene) which is located on chromosome
4A (2).   Besides stem elongation and α-amylase production, we have
found that other GA_3 effects in aleurone cells, such as the production and secretion of protease, the release of phosphate ions,
and the secretion of sucrose also do not occur in D6899. This
indicates that the "Tom Thumb" gene in D6999 regulates an event which
is common to all these diversed secondary GA_3 effects. Furthermore,
since the receptor of GA_3 in aleurone cells has not been identified
and isolated, we have to use an indirect way to check the possibility
that D 6899 may have less GA_3 receptors than normal wheat. One
may expect that when the concentration of GA_3 is high ("saturat-
ing"), the rate of the binding between GA_3 and its receptor is
determined by the concentration of receptor. However, when the

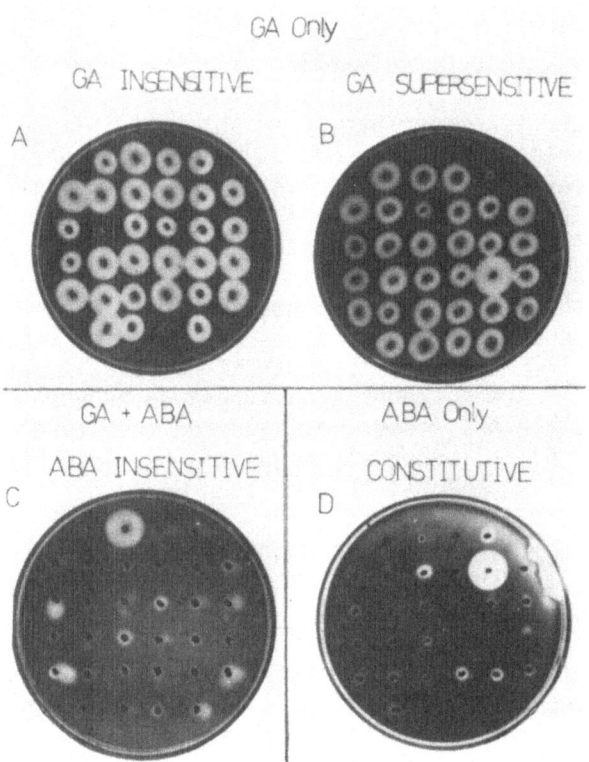

Fig. 6.   Screening of barley mutants with altered sensitivity to
          hormones. Barley half seeds were placed on 2% agar
          plates containing 0.15% starch and various combinations
          of hormones as described in the text. A. GA_3 in-
          sensitive mutants. B. GA_3 supersensitive mutant (3
          o'clock position on the plate). C. ABA insensitive
          mutant. D. Consititutive mutant. ABA (10^{-5} M) was
          included in the agar so that mutants with abnormal
          high endogenous level of GA_3 did not give positive
          response.

concentration of GA_3 is very low, the rate of hormone-receptor
formation is mainly determined by the concentration of hormone.
Under this condition, the difference in receptor concentrations
between two wheat varieties has very little impact on the synthesis of α-amylase. This is exactly what we observed when we
measured the difference of α-amylase produced between D6899 and
normal wheat at different concentrations of GA_3 (from 10^{-5} to
10^{-10} M). When the concentration of hormone decreases, the

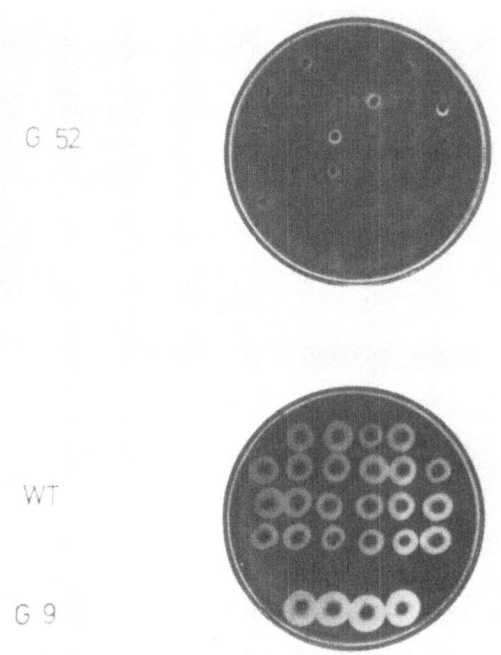

Fig. 7.   Stable barley mutants with altered GA_3 sensitivity.
          G52, GA_3 insensitive; G9, GA supersensitive.

difference between two wheat varieties is minimized, and in the
absence of exogenous GA_3 they produce exactly the same, although
very small, amount of α-amylase.

Mutants have been extremely helpful in the studies of
genetic and biochemical regulations in microorganisms. We believe that the barley and wheat mutants with altered hormone
sensitivity can be equally helpful when they are used to
supplement the molecular biology studies in order to better define
the mechanisms of hormonal regulation.

This work was supported by Jane Coffin Child Memorial Fund
and a grant from National Science Foundation (PCM78 16143).

LITERATURE CITED

1. Chrispeels, M. J. and J. E. Varner (1966). Nature 212, 1066.

2. Fick, G. N. and C. O. Qualset (1975). Proc. Natl. Acad. Sci. 72, 892.

3. Filner, P. and J. E. Varner (1967). Proc. Natl. Acad. Sci. 58, 1520.

4. Higgins, T. J. V., J. A. Zwar and J. V. Jacobsen (1976). Nature 260, 166.

5. Ho, T.-H. D. and J. E. Varner (1974). Proc. Natl. Acad. Sci. 71, 4783.

6. Ho, T.-H. D., R. A. B. Keates and J. E. Varner (1973). Plant Physiol. 51, s5.

7. Ho, T.-H. D. and J. E. Varner (1976). Plant Physiol. 57, 175.

8. Ho, T.-H. D. and J. E. Varner (1978). Arch. Biochem. Biophys. 187, 441.

9. Jacobsen, J. V. and J. A. Zwar (1974). Proc. Natl. Acad. Sci. 71, 3290.

10. Jones, R. L. (1969). Planta 88, 73.

11. Kleinhofs, A., R. L. Warner, F. J. Muelbauer and R. A. Nilan. Mutation Res., in press.

12. Lodish, H. F. (1971). J. Biol. Chem. 246, 7131.

13. Lodish, H. F. (1974). Nature 251, 385.

14. Nilan, R. A. (1974). Handbook of Genetics, Vol 2, p. 93.

15. Nilan, R. A., A. Kleinhofs and C. Sander (1975). in "Barley Genetics III", Proceedings of the third international barley genetics symposium, p. 113.

16. Varner, J. E. and Ho, T.-H. D. (1976). in "Molecular Biology of Hormone Action," ed. by J. Papaconstantinou, p. 173. Academic Press, NY.

17. Warner, R. L., C. J. Lin and A. Kleinhofs (1977). Nature 269, 406.

18. Yomo, H. and J. E. Varner (1972). "Hormonal control of
     secretory tissue." Chapter 4 in "Current Topics in Develop-
     mental Biology" vol. 6. Academic Press, New York.

AUXIN-REGULATED CELL ENLARGEMENT: IS THERE ACTION AT THE LEVEL OF

GENE EXPRESSION?

Larry M. Vanderhoef

Department of Botany
University of Illinois
Urbana, IL 61801 USA

INTRODUCTION

Auxin has been studied for fifty years as a regulating hormone in many developmental phenomena (1). Much of this work has been devoted to its effect on cell enlargement (2-6); however, our accumulated knowledge of auxin-regulated cell enlargement does not include an understanding of its primary mode of action. During the past two decades two hypotheses have, in turn, dominated and directed experimentation designed to solve this problem. In the 1960's the "gene expression" hypothesis of Key, and others (4) was popularly accepted. In the early 1970's, however, a new hypothesis, the "wall acidification" (7) hypothesis of Rayle and Cleland (5) and Hager et al. (8), was proposed, and it now directs the thinking of many researchers in this area. It is generally presumed that the tenets of these two hypotheses are incompatible.

Discovery and study of two separate elongation responses to auxin (9-11) has led me to believe that both hypotheses are tenable. While neither accounts for all of our knowledge of auxin-controlled cell enlargement, they can be combined to form a new and inclusive hypothesis.

THE GENE EXPRESSION HYPOTHESIS

The 1930's study of auxin at the California Institute of Technology brought together several investigators who would go on, by individual accomplishment, to dominate the field of plant hormone physiology. To this day, most workers in the field can

trace an influence in their training back to this group, which included Kenneth V. Thimann, H. E. Dolk, Folke Skoog, James Bonner,
Johannes van Overbeek, and Fritz W. Went. Much of what we know
about auxin today was discovered during that first half dozen
years. Many studies of hormone-induced elongation were based on
the simple straight-growth bioassay which consisted of the addition
of exogenous auxin to an elongating segment which had been excised from it endogenous auxin supply. Accumulated growth after
several hours, or a reestablished higher elongation rate of the
segment, was measured. Once formulated, the gene expression
hypothesis was supported by experiments much like the growth bioassay, but with the addition of transcription and translation
inhibitors such as actinomycin D and cycloheximide. In the absence
of RNA and/or protein synthesis, auxin was not able to induce the
higher rate of elongation. Thus, the hypothesis was supported
(4).

These experiments and their many sophisticated variations
were commonly reported and accepted for both plant and animal
hormone studies. It is especially clear in hindsight, though,
that their interpretation was difficult since it was always
possible that the inhibition of cell growth in the presence of
actinomycin D and/or cycloheximide was caused by events which
were not related to auxin action.

With no new approach to supporting the gene expression hypothesis forthcoming, it was chellenged in 1969 by Evans and Ray
(12). These authors presented an unusually inventive technique
for measuring minute-by-minute growth, experimental results that
measured a short 10 min lag time for auxin action, and an eloquent
discussion of the logic which led them to rule out gene expression
in auxin-promoted cell enlargement based on growth kinetics. While
the fast response to auxin had been known for many years (13),
its presentation and discussion in this new light caused its
graduation from a curiousity, to an important contribution to our
understanding of auxin action.

THE WALL ACIDIFICATION HYPOTHESIS

The Evans and Ray experiments initiated a series of debates
regarding the validity of the gene expression hypothesis, and its
popularity waned. The resulting void was quickly filled in 1971
when Cleland (2) and Hager et al. (8) independently suggested
that acidification of the cell wall may mediate in auxin action.
This hypothesis, attractive for its simplicity, stated that auxin
maintains the pH in the cell wall near 5, causing the increased
rate of elongation (Fig. 1). Two kinds of data supported the hypothesis. First, in most systems where exogenous auxin could
induce a higher rate of cell elongation, lowering the pH from 6

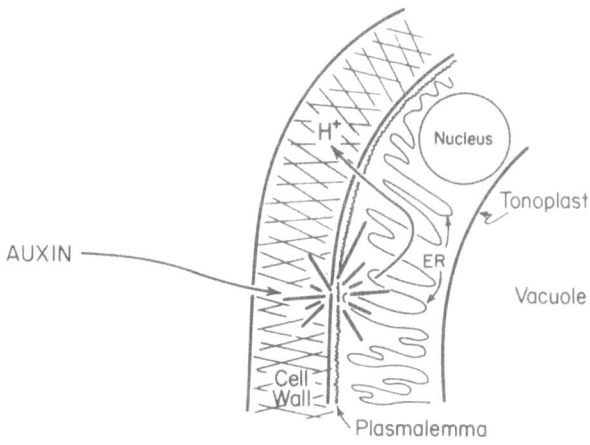

Fig. 1.   Wall acidification hypothesis.   Auxin causes wall
          loosening by inducing processes which lower the pH
          in the cell wall.

to about 4 could also induce a higher rate of cell elongation (9,
14-16).   Second, in certain circumstances auxin could be shown to
induce medium acidification by auxin-responsive elongating cells
(17).   Also, in one case (18) auxin was shown to induce responsive
cells to acidify the free space of pea stem segments and maize
coleoptile segments.

     The wall acidification hypothesis is currently being tested
by the research of many auxin physiologists.   Two problems have
emerged.   First, auxin-induced cell wall acidification was not
demonstrable in all auxin-responsive systems (19 and references
therein); this, however, may be a problem of methods and cannot
yet be assumed to constitute a major criticism of the hypothesis
(20).   Second, acid-induced elongation does not, in fact, mimic
auxin-induced elongation.   Cells, elongating in response to lowered pH, never attain a steady state rate.   The H^+-induced increase
in elongation rate is transient and over within 60 to 100 min, the
actual time depending on the species and experimental conditions
(5, 9, 15, 19, 21).   This is different from auxin-induced elongation, where a steady-state rate is attained and can be retained
for several hours.   Furthermore, acid damage is not responsible
for the transient nature of acid-induced elongation; segments
which have previously experienced the acid-induced burst of elongation are fully responsive to subsequent auxin addition (9, 19).

     This major problem with the acid-growth hypothesis suggested
that it should be expanded to include additional components.   In
fact, experiments which have distinguished two separate elongation
responses to auxin require such a new hypothesis.

THE SEPARATION OF TWO RESPONSES TO AUXIN

Early in our studies Catherine Stahl characterized the early
kinetics of auxin-induced elongation (9). The results were unusual.
Rather than simply rise to a higher steady-state rate, the rate
rose, then dropped, then finally rose again to a steady-state rate
(Fig. 2). Others had also seen unusual early kinetics (22, 23).

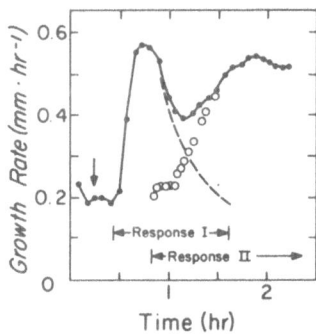

Time (hr)

Fig. 2. Rate kinetics show two separate elongation responses to
         auxin by soybean hypocotyl segments. Arrow marks auxin
         addition to auxin-depleted excised hypocotyl segments.

While they were not identical to our determinations, they were
similar. We now know that the common explanation for these early
fluctuations in the growth rate was that auxin induced elongation
in two separate ways. This concept, at first, was difficult to
accept. Initially, less complex explanations, such as variation
in turgor or osmoregulation (24), were more attractive. Also, if
one was selecting data so as to support a particular hypothesis,
other explanations would have been more palatable. Objective
appraisal of all the data, however, led to the conclusion that the
kinetics (Fig. 2) represented two overlapping elongation responses,
i.e., auxin induced a rapid (but transient) response, as well as a
more permanent response that began after a longer lag time.
Furthermore, it seemed quite likely (and has been confirmed) that
the chemical and physical processes that constitute the two responses were not identical.

This unusual explanation required, first, evidence that it
was correct. Second, if the explanation was correct, then the
two major hypotheses of auxin-regulated growth had to be re-
appraised. The supportive data for two separate elongation responses have appeared elsewhere, and are briefly summarized as
follows.

Evidence for separate elongation responses to auxin

Early in our studies of soybean stem elongation (25) we characterized the inhibition of auxin-induced cell elongation by cytokinin in long-term (6-12 hr) experiments (26). Cytokinin was a potent inhibitor, especially attractive because it is a plant hormone itself which appears to have a natural role in modulating auxin action in vivo. Study of this inhibitory effect of cyto-kinin in short-term experiments showed that while cytokinin was very effective in inhibiting the second response (and hence in inhibiting accumulated growth in previous long term experiments), it had no effect on the first response. This was true even when auxin addition to the stem segments was preceded by a 3 hr pre-incubation in cytokinin (9). The two responses thus could be cleanly separated by differential response to the elongation in-hibitor, cytokinin. A second kind of evidence came from our study of acid-induced elongation. As stated above a lowered pH will increase the rate of elongation of auxin-depleted stem segments. What is often overlooked, however, is the fact that acid-induced elongation is transient. In fact, the growth rate goes up and comes down without ever attaining a steady state rate, i.e., acid-induced elongation is equivalent only to the first response (9, 19). A third kind of evidence was discovered serendipitously. While studying the auxin analog 2-chloro-4-azidophenoxyacetic acid we noticed that the lag for the second response was increased but the lag for the first response was not. The two responses thereby were sometimes temporally separated (10). Separation of the two responses using another auxin analog, indoleacetamide, had been previously reported (23). A fourth kind of evidence came when we determined the half-life of the limiting protein(s) for each of the two responses (11). They were different, showing that the populations of proteins necessary for each of the two re-sponses were different, thus proving that the two responses were caused by different activities. The final kind of evidence for separate elongation responses to auxin, while not at all direct, was most convincing. This evidence came from summation con-structs. Recall that different people had found different kinds of early kinetics in response to auxin. Barkley and Leopold and others (22, 23) each found unusual, but different, early elon-gation kinetics in response to auxin. All of these different early kinetics can be simply explained: The summation of over lapping responses varies greatly with the amount of overlap (Fig. 3). So, for example, when auxin induced a high growth rate, which then dropped down to a medium steady-state rate of elongation in pea tissue (22, Barkley and Leopold), this was caused by a large amount of overlap of the two responses (Fig. 3). By com-parison, in soybeans and lupine beans the overlap is not as great and the two responses, though individually very similar to those in pea, gave a different kind of summation kinetics (Figs. 2, 3;

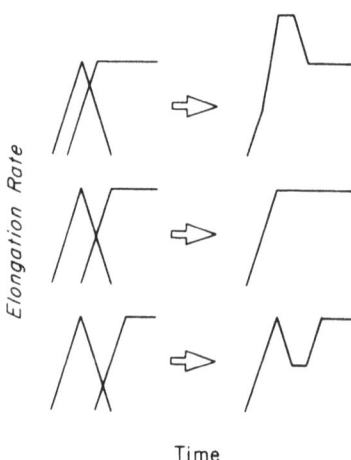

Fig. 3.   Summation response depends on amount of overlap of the
          separate elongation responses.

refs. 9, 10).   And so confusing results, i.e., unusual early
elongation kinetics in response to auxin that varied from species
to species, were explained by the concept of two elongation responses to auxin.

     Subsequent independent experiments in other laboratories have
confirmed the existance of two separate elongation responses to
auxin, and proton-independent growth has been identified in corn
(27).

A REAPPRAISAL OF THE HYPOTHESES OF AUXIN ACTION IN CELL ELONGATION

     The discovery and proof that auxin induces two separate
elongation responses has had an effect on our thinking regarding
auxin mode of action.   It was proposed (12) and widely accepted
(28) that the short lag time for auxin-induced elongation dis-
proved the gene expression hypothesis.   Now that was not at all
certain.   It presently is clear that the two hypotheses are
actually quite compatible.   In fact, all available data can be
accommodated in a "dual sites" hypothesis; AUXIN HAS TWO PRIMARY
ACTIVITIES IN ITS REGULATION OF CELL ELONGATION.   FIRST, IT MAIN-
TAINS THE WALL IN A CHEMICAL AND/OR PHYSICAL STATE WHICH ALLOWS
CONTINUED ELONGATION, AND SECOND, IT REGULATES, AT GENE EXPRESSION,
THE PRODUCTION OF COMPOUNDS NECESSARY FOR SUSTAINED CELL ELONGATION

Auxin's initial activity

The dual sites hypothesis states that auxin maintains the
wall in a chemical and/or physical state which allows continued
elongation. Supportive evidence for this action of auxin has been
described in discussion of the wall acidification hypothesis (5,
24). The research can be best evaluated if the normal events in
the intact seedling are described. Elongating stem cells, in a
soybean hypocotyl for example, are actively elongating in the
seedling for only a matter of hours. Cells just a few mm removed
are no longer elongating. Division and elongation of cells above
quickly move an elongating cell away from its auxin supply, and
wall extension in the long axis soon ceases; new events in the
cell eventually result in a cell wall that will not elongate in
response to auxin or lowered pH. It makes sense to assume that
when we excise actively elongating stem segments, we initiate
these same maturing events, for again we are moving the cells
away from their auxin supply (Fig. 4). The first events in the
wall which occur in response to the lack of auxin, however, are
clearly reversible; resupplying auxin puts the cells back into
the elongating mode. This reversible wall process has been best
studied by use of the instron apparatus (16).

Lowering the pH from 6 to 4 is another method for reversing
these first events. In this case, however, the cells do not go
back into a normal elongating mode. Rather, they show just a
burst of elongation which reaches a maximum rate very quickly,
then almost immediately begins to decline. The acid-induced
growth is over in 30 to 90 minutes in living tissue, even when
tension is applied to the elongating segment. While this brief
growth certainly does not mimic auxin-induced growth, it is
identical to the first response of auxin-induced elongation, as
described above.

These results and considerations support the conclusion that
one of auxin's first actions upon entering the cell is to cause
the wall to be maintained in a state which makes it receptive to
the processes necessary for continued elongation. Thus the components of the cell wall that must slide past each other and receive new materials during active elongation begin to bind to one
another when auxin disappears (Fig. 4). The reintroduction of
auxin in these cells causes the dissociation of these bonds (very
likely mediated by wall acidification) and a subsequent burst of
turgor-driven elongation. This elongation is transient; it lasts
30 to 90 min depending on conditions and species. Elongation
cannot continue by this first wall-loosening auxin action alone.
This very important point was first established with experiments
which employed cytokinin, an elongation inhibitor (9). Auxin must
have another activity in its regulation of cell elongation. Thus

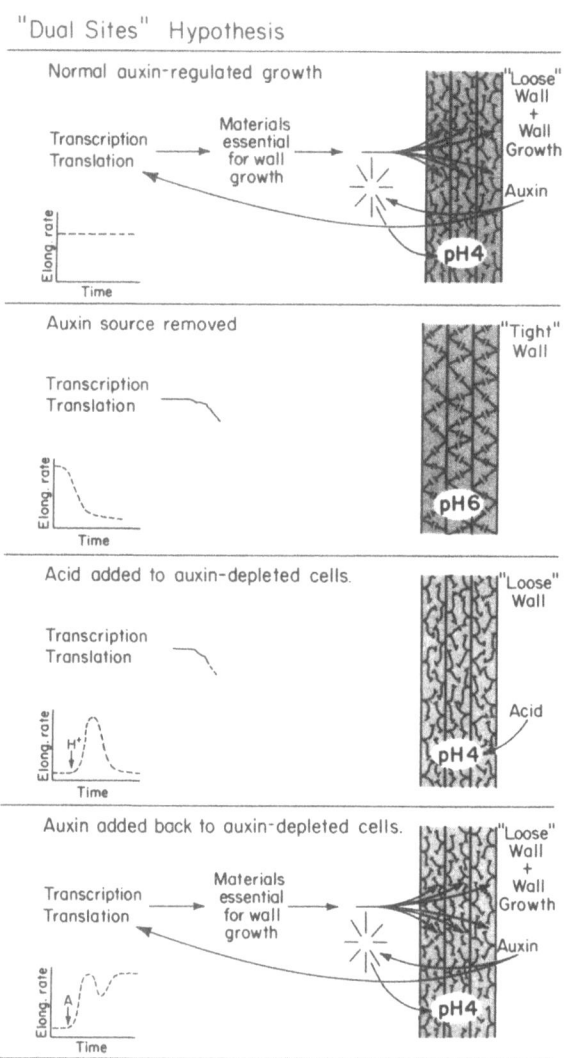

Fig. 4. The "dual sites" hypothesis. Auxin is postulated to regulate both wall loosening and wall synthesis.

the hypothesis further states that there is a second primary site
of auxin action, i.e., that auxin regulates, at gene expression,
the production of compounds necessary for sustained cell
elongation.

There is, in fact, no direct evidence for gene expression
regulation in any eucaryotic hormone system. However, this kind of auxin action must be considered most likely,
in light of several lines of indirect and circumstantial evidence.
First, it is now clear that the "fast response" experiments do not,
after all, refute the tenets of the gene activation hypothesis.
The arguments launched by Evans and Ray (12) are applicable to the
first elongation response only. This is best demonstrated by
replotting some data that were submitted in evidence of the contention that actinomycin D did not inhibit auxin-induced elongation.
While the first elongation response was not affected by actinomycin
D, the second elongation response was strongly affected by the
inhibitor (Fig. 5). The experiments of Key et al. and others (4)
which supported the gene expression hypothesis can not, after all,
be dismissed simply on the basis of the fast response.

A second line of evidence which supports a mediating role of
gene activation in auxin-induced elongation comes from studies of
auxin-affected protein synthesis in elongating cells. Recall that
experiments with estrogen are interesting since, like auxin,

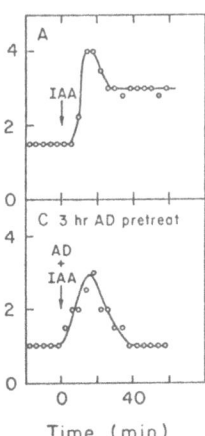

Time (min)

Fig. 5. Actinomycin D inhibits the second, but not the first
         elongation response in pea stem segments. Redrawn from
         figure 4 of ref. 22 (Barkley and Evans).

estrogen plays a role in cell enlargement. Estrogen, it is believed, induces the synthesis of a unique protein in uterine
development (29). Similarly, the studies of progesterone show the
binding of a receptor protein-hormone complex to a specific
chromosomal acidic protein (30). Hormone receptors also occur in

plant cells. While most auxin studies have concentrated on membrane-
bound receptors (31), soluble receptor proteins have been reported
(32), and preliminary results describe an auxin-protein complex
that binds to a chromosomal protein (32, Thompson and Slater).
Recently auxin has been shown to induce specific changes in the
pattern of protein synthesis in elongating soybean hypocotyl
segments (33). Certainly auxin activity at gene expression in
sustained cell elongation appears to be a real possibility.

It has been suggested that the first auxin-induced event
(causing the wall to be receptive to the active processes of
elongation) may be the only primary action of auxin; subsequent
events would be just secondary effects of this first event (5).
Most certainly this is possible. The idea is especially attractive
in light of the discovery that calcium can be a major determinant
in development. However, at the present time there are no data
which suggest that those activities that cause the second
elongation response are secondary or accessory processes.

TWO SITES FOR AUXIN ACTION: TEST AND PREDICTION

Most of the experiments cited above utilized a common technique. Rapidly elongating stem (or coleoptile) segments were
surgically removed from their endogenous auxin supply and incubated in an auxin-free pH 6 medium until the elongation rate was
low. Exogenous auxin was then added and the responses monitored.
This experimental technique, though not initially foreseen, was
absolutely required to distinguish the two major events of auxin-
regulated elongation. The first auxin-induced burst of elongation
is the result of the wall "breaking loose" when the reversible
bonds, formed in the absence of auxin, are broken. This first
response, therefore, can be observed only if the cells are allowed to incubate for a period of time in the absence of auxin.

The wall, made receptive by auxin to the insertion of new
wall materials, begins to show the first signs of irreversible
elongation about thirty-five minutes after auxin addition. A
steady state rate of elongation is established by 60 min after
auxin addition. This thirty to forty minute lag period can be
compared to recent results which show that soybean cells have a
distinct population of mRNA molecules which have half-lives near
0.5 hr (34).

The dual sites hypothesis allows an interesting prediction.
It requires that the pH of the cell walls of an elongating segment
rise when the segment is excised from its exogenous auxin supply.
Ostensibly it is this pH rise which causes the formation of elongation-resisting bonds. Exogenously-supplied auxin then, by

causing wall acidification, causes these bonds to break (the first
response). If this is all true, then a segment which is kept at
pH 4 and under tension from the moment of excision should show,
after depletion of endogenous auxin, only the second response when
exogenous auxin is applied. This experiment was performed and this
precise result was obtained, i.e., when the wall is not allowed to
form elongation-resisting bonds, auxin induced only the second
response (Fig. 6). These results are especially interesting because they also rule out a previously unmentioned alternative
explanation for the first elongation response to auxin. The first

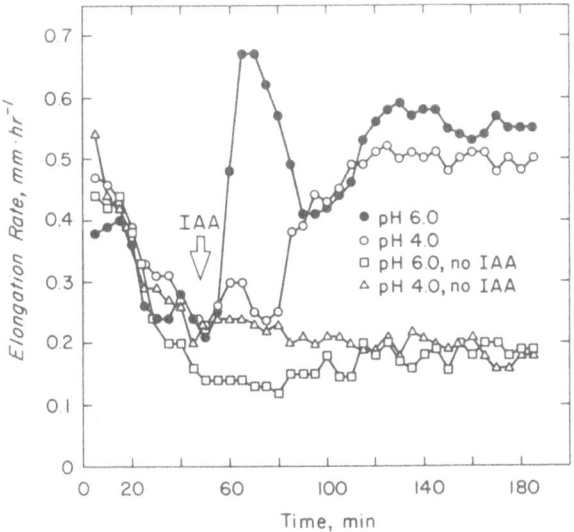

Fig. 6.   Auxin induces only the second response in acid-pretreated
          segments. The excised elongating segment was transferred
          immediately to pH 4 or pH 6 medium in the growth-measuring
          apparatus. Auxin, final concentration 45 µM, was added
          when the elongation rate decreased to less than
          0.2 mm per hr.

elongation response might represent the utilization of a backlog
of wall materials that could not be inserted while the wall was
temporarily, in the absence of auxin, at a high pH. However, if
this alternative explanation was correct, the 0 to 40 min growth
rate kinetics should have been measurably different at pH 4 and
6, i.e., those segments at pH 4 should have maintained a higher
growth rate while utilizing the proposed backlog of materials.
They did not, and the description of the first response suggested

by the dual sites hypothesis remains the more feasible explanation.

This dual-sites hypothesis is necessarily non-specific on several points, e.g., on the actual mechanisms of proton-mediated wall loosening, and gene expression regulation; the available data give us no substantial hints. On the other hand, the hypothesis is quite specific in omitting any consideration of osmoregulation. The experiments of Boyer and Wu (35) have made it clear, after years of debate, that auxin-induced changes in cell enlargement can not be attributed to changes in cell osmotic potentials or turgor.

Weakness of the Hypothesis

This model accommodates known data, and, as was shown above, it elicits testable predictions. Thus, our experiments are designed to disprove the basic tenets of the model, and to rigorously challenge the hypothesis upon which it rests. A major experimental weakness lies in the fact that we have failed to measure auxin-induced medium acidification in elongating soybean hypocotyl segments. This is not because we cannot, because of the cuticle for example, measure adjustment of the medium pH by the hypocotyl segments (19, Vanderhoef et al.). Indeed, these experiments have led us into a very interesting study of pH adjustment by hypocotyl cells (36). The hypocotyl cells (and presumably epicotyl and coleoptile cells, as well) very rapidly adjust the external pH to about 5.5. In hindsight this makes good sense, since these seedling tissue walls (a) are exposed to soil water, and (b) have wall enzymes and other components which have optimal pH requirements for maximal activity and/or proper conformation. These experiments, added to the fact that acid-induced growth has almost no lag in these segments, proved that the cuticle is not a limiting factor in our ability to measure tissue-mediated pH changes in the medium. In literally dozens of experiments we have never measured an effect of auxin on this regulation of wall pH. It may be relevant, though, that wounding by gentle abrasion causes a significant decrease in the equilibrium pH in the wall (36).

This leads me to an important concluding point. We in the field of plant hormones have seen many circumstances where hypotheses have been misused. Hypotheses are composed to guide further experimentation. Once proposed, a hypothesis should be rigorously tested. It should be mercilessly attacked and challenged, and its major proponent(s) should be leading the charge. This, in the long run, is how science will advance most rapidly. Unfortunately we sometimes cement our feet into our hypotheses; we see only data that supports our postulates. We develop convoluted reasoning and inconsistent arguments to explain away disagreeable data. We spend too much time defending, and not enough time attacking, hypotheses.

George Malacinski of Indiana University has some interesting thoughts on this. He points out that we who are attracted into science are likely to be inadequate in an important respect. We go into science because we want to bring order to chaos, and to explain the heretofore unexplained. Hence, in our heart of hearts we really want the hypothesis or model to be correct. (Even when it is not our own!). And so we may "see" only supportive data, and draw favoring conclusions before they are warranted. While this behavior may make us more comfortable, it surely hinders the advance of our knowledge.

Acknowledgments: Work supported by grants NSF BMS72-02496, NSF PCM77-14175, HEW PHS 07030, and by the University of Illinois Graduate Research Board.

References and Notes

1. K. Thimann, Hormone Action in the Whole Life of Plants.
    (U. Mass. Press, Amherst, 1977).
2. R. Cleland, Ann. Rev. Plant Physiol. 22, 197 (1971).
3. P. Ray, Adv. Phytochem. 7, 93 (1974); A. Trewavas, Progr.
    Phytochem. 1, 113 (1968).
4. J. Key, Ann. Rev. Plant Physiol. 20, 449 (1969).
5. D. Rayle and R. Cleland, Curr. Topics Dev. Biol. 34, 187
    (1977).
6. M. Evans, Ann. Rev. Plant Physiol. 25, 195 (1974).
7. The hypothesis has various names, e.g., the "acid growth"
    hypothesis, the "proton extrusion" hypothesis. I propose
    "wall acidification" hypothesis because it is more accurate-
    ly descriptive of the hypothesis, and it reflects the support-
    ing data.
8. A. Hager, H. Menzel and A. Krauss, Planta 100, 47 (1971).
9. L. Vanderhoef and C. Stahl, Proc. Nat. Acad. Sci. USA
    72, 1822 (1975).
10. _____, _____, C. Williams, K. Brinkmann and J. Greenfield,
    Plant Physiol. 57, 817 (1976).
11. _____, _____, and T-Y. Lu. ibid 58, 402 (1976).
12. M. Evans and P. Ray, J. Gen. Physiol. 53, 1 (1969).
13. T. Yamaki, Sci. Pap. Coll. Gen. Educ. Univ. Tokyo 4, 129
    (1954); K. Köhler, Planta 47, 159 (1956); P. Ray and A.
    Ruesink, Devel. Biol., 4, 377 (1962).
14. J. Bonner, Protoplasma 21, 406 (1934).
15. D. Rayle and R. Cleland, Plant Physiol. 46, 250 (1970); M.
    Evans, P. Ray and L. Reinhold, ibid. 47, 335 (1971).

16. D. Rayle, P. Haughton and R. Cleland, Proc. Nat. Acad. Sci. USA 67, 1814 (1970); D. Rayle and R. Cleland, Planta 104, 282 (1972).

17. R. Cleland, Proc. Nat. Acad. Sci. USA 70, 3092 (1973); D. Rayle, Planta 114, 63 (1973); J. Mentze, B. Raymond, J. Cohen and D. Rayle, Plant Physiol. 60, 509 (1977).

18. M. Jacobs and P. Ray, Plant Physiol. 58, 203 (1976).

19. L. Vanderhoef, J. Findley, J. Burke and W. Blizzard, Plant Physiol. 59, 1000 (1977); ____, T-Y. Lu, C. Williams, ibid. 59, 1004 (1977); D. Parrish and P. Davies, Plant Physiol. 60, 509 (1977).

20. The cuticle presents a special problem for workers intending to measure auxin-modified wall pH. If it is left intact it may slow the H^+ equilibration between the wall and external medium. If it is removed by abrasion or stripping extensive cell wounding occurs, which may cause artifactual pH changes (36). Additionally, pH microelectrodes to directly and accurately measure the wall pH are not available.

21. H. Kazama and M. Katsumi, Plant Cell Physiol. 17, 467 (1976).

22. G. Barkley and M. Evans, Plant Physiol. 45, 143 (1970); D. Penny, P. Penny, J. Munro and P. Bailey, Plant Growth Substances, ed. D. Carr (Springer-Verlag, N.Y., 1970), pp. 52-61; G. Barkley and A. Leopold, Plant Physiol. 52, 76 (1973).

23. P. Penny, D. Penny, D. Marshall and J. Heyes, J. Exp. Bot. 23, 23 (1972).

24. R. Cleland and D. Rayle, Bot. Mag. Tokyo (Special Issue) 1, 125 (1978).

25. L. Vanderhoef, C. Stahl, N. Seigel and R. Zeigler, Physiol. Plant. 29, 22 (1973).

26. These experiments usually measured only initial and final lengths of stem segments; hence, early growth rate kinetics (e.g., changes during the first 1 to 2 hr of growth) were not detectable.

27. H. Kazama and M. Katsumi, Plant Cell Physiol. 17, 467 (1976); N. Sakurai, D. Nevins, and Y. Masuda, ibid. 18, 371 (1977); M. Vesper, M. Evans and M. Cline, Plant Physiol. 63, S21 (1979).

28. For example Ray stated in his 1974 review, "The outcome of auxin research in the 1960's is that...regulation...of gene expression by IAA is probably involved in...auxin responses, such as cell division, differentiation, and morphogenesis [but] *some other more direct mode of action of IAA appears, in the view of most current workers, to be operating in cell enlargement.*" (italics added) P. Ray, Rec. Adv. Phytochem. 7, 93 (1974).

28. B. Katzenellenbogen and J. Gorski, J. Biol. Chem., 247, 1299 (1972).

30. B. O'Malley and W. Shrader, Sci. Amer. 234(2), 32 (1976).

31. P. Ray, U. Dohrmann and R. Hertel, Plant Physiol. 60,
    585 (1977); S. Batt, M. Wilkins and M. Venis, Planta 130,
    7 (1976); M. Venis, Biochem. Rev. 6, 325 (1978).
32. A. Wardrop and G. Polya, Plant Sci. Lett. 9, 155 (1977); M.
    Ihl, Planta 131, 223 (1976); J. Thompson and R. Slater,
    Intern. Bioch. Congress (Hamburg), Abstr. 08-3-290 (1976).
33. L. Zurfluh and T. Guilfoyle, Plant Physiol. 63, S143 (1979): T.
    Guilfoyle, this volume.
34. C. Silflow, Ph.D. Dissertation, U. Georgia, Athens, GA (1977);
    D. Baulcombe, this volume.
35. J. Boyer and G. Wu, Planta 139, 227 (1978).
36. A. Abu-Tabikh and L. Vanderhoef, Plant Physiol. 63, S20 (1979).

THE EFFECT OF AUXIN ON THE POLYADENYLATED RNA OF SOYBEAN

HYPOCOTYLS

David Baulcombe, Jarbas Giorgini,* Joe L. Key

Botany Dept.
University of Georgia
Athens, Georgia 30602   USA
*currently at: Faculdade De Filosofia, Ciencias
E Lettres De Riberirao Preto, Universidade De Sao
Paulo Brasil

INTRODUCTION

Studies with several plant systems have demonstrated that
environmental or developmental changes may cause an altered level
of specific mRNAs. For example, and considering only products of
the nuclear genome, it is known that leghaemoglobin synthesis and
mRNA levels increase during early soybean root nodule development
(Auger et al, 1979), that light stimulates production of phenylal-
nine ammonia lyase (PAL) and the level of PAL mRNA in parsley
leaves (Ragg et al., 1977), and that gibberellic acid induced
synthesis of α amlase in barley aleurone cells is a result of an
increased level of α amylase mRNA (Higgins et al., 1976). It is
not yet clear, however, to what extent the whole spectrum of gene
expression may shift during developmental or environmental changes.
In order to investigate this problem further we have examined the
effect of auxin (2,4-D) on the polyadenylated RNA populations of
soybean hypocotyls using nucleic acid hybridisation and in vitro
translation.

Treatment of young soybean hypocotyls with 2,4-D activates
cell division activity in the basal, non meristamatic region, after
a lag of 10-12 hrs., leading to massive tissue proliferation (Key et
al., 1964). Since auxin is probably an absolute requirement for
plant cell division growth (Thimann, 1969) an understanding of the
molecular events associated with the effect of auxin is essential
for an understanding of plant development. It has been known for
a number of years that synthesis of presumed mRNA is necessary for

175

many auxin effects on cell growth (reviewed by Jacobsen, 1977.) The
enhancement by auxin of cellulase mRNA level in pea epicotyls
(Verma et al., 1975) and competition hybridisation data with RNA
from auxin treated pea stem sections (Thompson and Cleland 1971)
further suggest that auxin-mediated cell growth may involve at
least some qualitative changes in gene expression. Our data have
extended these findings by showing that most polyadenylated RNA
sequences are common to treated and untreated hypocotyls, but that
a small number of abundant sequences have a drastically changed
abundance after hormone treatment.

    In this study of the extent of auxin-mediated changes in gene
expression in soybean hypocotyl only the total polyadenylated RNA
fraction has been used. The reason for this is mainly practical:
it is possible to isolate large quantities of high molecular weight
polyadenylated RNA almost free of rRNA (less than 5% contaminated
as shown by electrophoresis on 98% formamide gels). Non-messenger
polyadenylated RNA contaminants (mainly polyadenylated nuclear RNA)
are estimated, based on studies with soybean suspension culture
cells, to comprise a relatively small fraction of the total polyadenylated RNA mass (Silflow, 1977). Of course, this study would
not detect changes in non-polyadenylated mRNA sequences which may
represent a significant fraction of soybean RNA complexity (Silflow
et al., 1979).

RESULTS AND DISCUSSION

    a) Preparation of Complementary DNA and Single Copy DNA
Hybridisation Probes.

    Complementary DNA (cDNA) prepared using polyadenylated RNA
from either 2,4-D treated hypocotyls (TA+RNA) or untreated hypocotyls
(UA+RNA) was found by electrophoretic analysis in 98% formamide to
have a size distribution between 100 and 2000 nucleotides long with
a modal size of 900 nucleotides. The cDNA preparations were essentially free of foldback, having less than 1.5% S1 nuclease
resistance at a Cot of less than 10^{-5}.

    Single copy DNA was prepared by two cycles of hybridization
to a Cot of 250 followed by hydroxyapatite fractionation. The
single stranded, single copy DNA was annealed into hyperpolymers
(Galau et al., 1976) and labelled with ^{3}H-TTP by nick translation
to a specific activity of 10^7 cpm μg^{-1}.

    b) Comparison of TA+RNA and UA+RNA Sequence Complexities.

    In order to determine the complexity of the hypocotyl polyadeny-
lated RNA, ^{3}H-labelled single copy DNA was hybridised to saturation
with an excess of the RNA. At high Rot values both RNA populations
hybridised to a maximum of 4.6% of the single copy DNA (Figs. 1a, 1b).

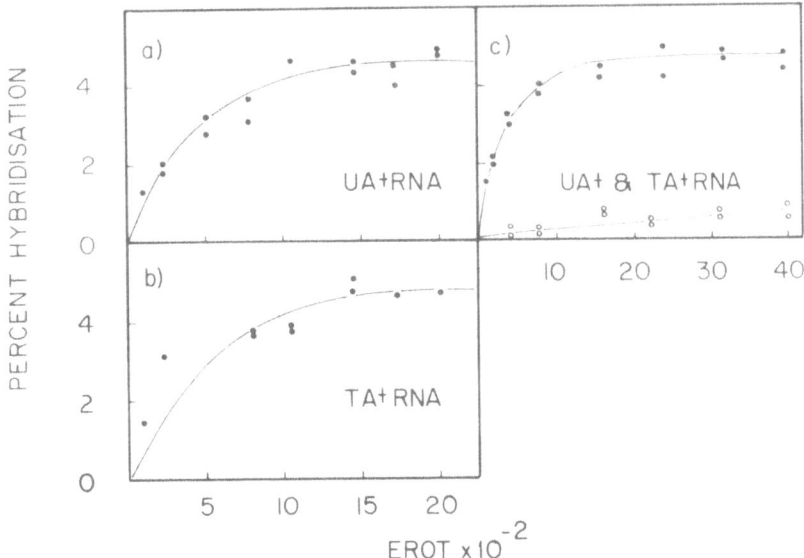

Figure 1 Determination of Hypocotyl RNA Sequence Complexity by
Saturation Hybridisation with Single Copy DNA.

[3]H-Single copy DNA was hybridised with a 1200 fold excess of
either UA+RNA (a) or TA+RNA (b) or a 2400 fold excess of an equimolar
mixture of UA+RNA and TA+RNA (c) in 0.3M NaCl, 10mM PIPES pH 6.4,
10mM EDTA at 64°c. Single copy DNA self reassociation is shown in
(c) (open symbols). The amount of hybridisation was assayed with
S1 nuclease and the rate corrected for salt concentration (Britten
et al., 1974).

After correction for 0.7% DNA self reassociation (Fig. 1c) and 70%
reactability of the single copy DNA and assuming asymmetric transcription, it is calculated that both TA+RNA and UA+RNA are complementary to 11.2% of the single copy DNA. The complexity of the
single copy fraction of the genome is equal to 5.55×10^8 nucleotide
pairs (Gurley et al., 1979). So, assuming that single copy DNA
is an accurate representation of the genomic single copy DNA, the
complexity of the hypocotyl polyadenylated RNA populations is
6.16×10^7 nucleotide pairs. This is equivilant to 44000 different
sequences of the length of hypocotyl polyadenylated RNA (1450
nucleotides, data not shown and Silflow et al., 1979).

    The complexity of soybean hypocotyl polyadenylated RNA measured
by saturation hybridisation of single copy DNA is in fair agreement
with other values determined from soybean suspension culture cells
(25,000 sequences Silflow et al., 1979) and from parsley root callus

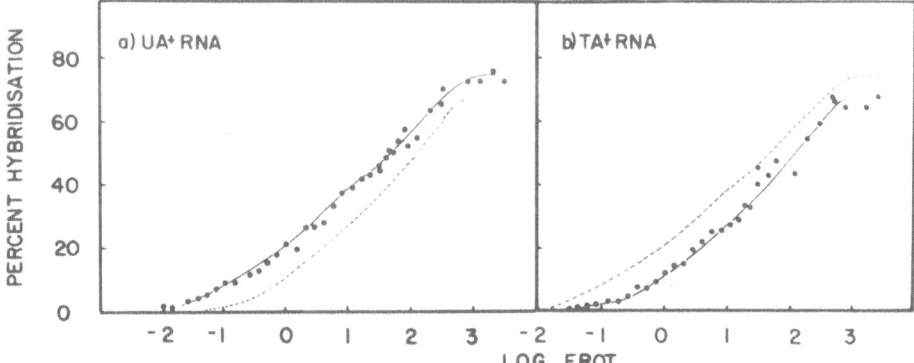

Figure 2.   The Abundance Distribution of Sequences in Hypocotyl
Polyadenylated RNA.

     UA+RNA (a) or TA+RNA (b) was hybridised in 3000 fold RNA excess
with the homologous cDNA type.   The dashed lines in a) and b) show
for comparison the opposite homologous reactions.   hybridisation
and assay conditions were outlined in the legend to Figure 1.

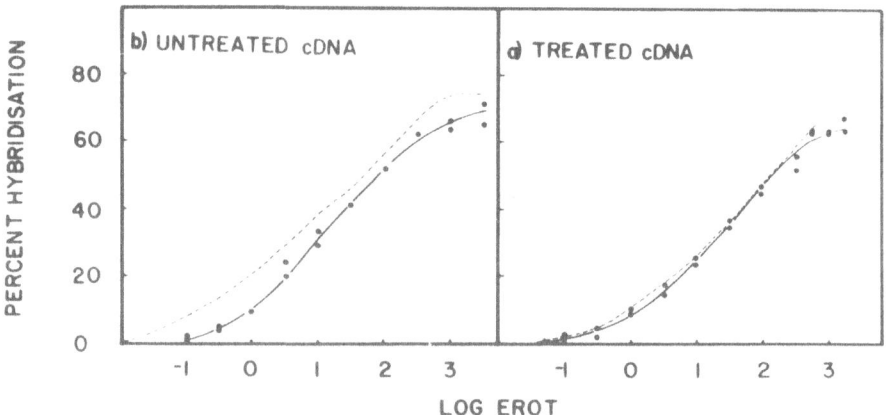

Figure 3.   The Effect of Auxin Treatment on the Abundance Distribution of Sequences in Hypocotyl Polyadenylated RNA.

     Complementary DNA of either TA+RNA (a) or UA+RNA (b) was
hybridised in 15000 fold RNA excess with the heterologous RNA
type.   The dashed lines show for comparison the reaction of the
indicated cDNA with the homologous RNA type (from Fig. 2).
Hybridisation and assay conditions were outlined in the legend
to Figure 1.

cells (35,000 sequences, Kiper et al., 1979). It has been suggested
by Kiper (1979) that these values may be artifactually high. He
points out that single copy DNA prepared from sheared DNA would be
depleted of any small single copy units which are interspersed
with repeated sequences. Consequently the single copy fraction
would be enriched for any long units which contain sequences coding
for polyadenylated RNA, and the proportion of single copy DNA coding
for polyadenylated RNA would be overestimated. While this is undoubtedly true, the number of small single copy units is probably
low in soybean (Gurley et al., 1979, Goldberg 1978) and the correction suggested by Kiper (1979) to compensate for enrichment of
coding sequences would be small in this particular case. Furthermore, since it is possible that small single copy units may code
for parts of larger "spliced" RNA molecules (Breathnach et al.,
1977) there cannot be certainty that depletion of single copy DNA
of small units is depletion of non-coding units. Perhaps it is better, in the light of these factors, to consider the data comparatively
rather than to generate absolute numbers. Unfortunately there is
no accurate independant test, since complexity estimates using
cDNA/RNA hybridisations may be varied severalfold without significantly affecting the fit of the data (e.g. Goldberg et al., 1978).

  c) Comparison of the Abundance Distribution of Sequences
Within TA+RNA and UA+RNA.

  It is known from a number of eukaryotic systems that different
polyadenylated RNAs exist at different abundance levels. We have
hybridised cDNA of either TA+RNA or UA+TNA with an excess of these
RNAs in either heterologous or homologous reactions in order to
detect any shifts in abundance of certain sequences.

  The homologous type cDNA/RNA reactions were clearly heterogeneous,
being spread over greater than 5 orders of magnitude of Rot (Figs
2a, 2b). This indicates that in soybean hypocotyl, as in soybean
suspension culture (Silflow et al., 1979) and other plant tissues
(Goldberg et al., 1978) the polyadenylated RNA contains sequences
at different levels of abundance. An obvious difference between
the TA+RNA and UA+RNA reactions is that the TA+RNA reaction was
slower than the UA+RNA reactions. This suggests that there is
a shift in the relative concentration of the rare and very
abundant sequences after auxin treatement.

  In the heterologous reactions the hybridisation of untreated cDNA
was slower than in homologous reactions, whereas there was no effect on the rate of hybridisation of the treated cDNA (Figs. 3a,
3b). This demonstrates that some UA+RNA sequences are rarer in
TA+RNA, but that the abundance of most TA+RNA sequences is unchanged
in UA+RNA. The plateau value of the heterologous reactions was
lowered only slightly relative to the homologous reactions, showing
that nearly all sequences are common to TA+RNA and UA+RNA. These
cDNA hybridisation data confirm the single copy DNA analysis and

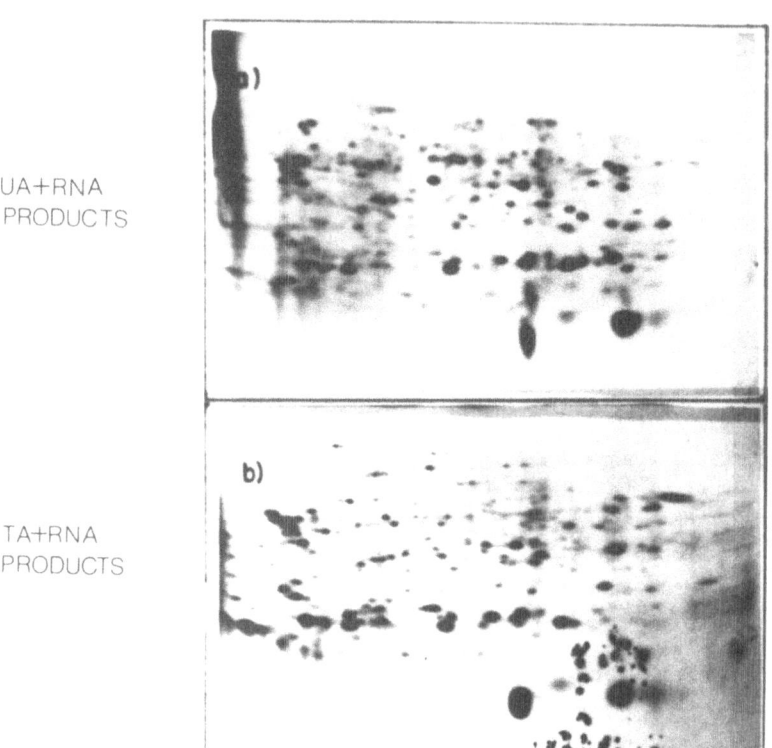

UA+RNA
PRODUCTS

TA+RNA
PRODUCTS

Figure 4.   Analysis of in vitro Translation Products of TA+RNA and
UA+RNA by 2 Dimensional Gel Electrophoresis.

further show that not only are most sequences common to TA+RNA and
UA+RNA, but also that most sequences are present at the same relative abundance. The major difference which is indicated by hybridisation data involves a decreased relative abundance in TA+RNA of the
5-10% most abundant UA+RNA sequences.

   In contrast to the averaged picture obtained with hybridisation,
the use of in vitro translation combined with product analysis on
2D gels allows a more detailed estimate of the change in concentration of certain mRNA species within a mixed population. Of course,
due to the limiting factor of resolution on gels, products of only
the most abundant mRNA sequences are detected using this technique.
There is also a bias towards the most efficiently translated
sequences. The translation products of TA+RNA and UA+RNA are shown
in Figure 4 and reveal that, within the gross shift in sequence
abundance revealed by the hybridisation data, there are sequences
being both increased and decreased in abundance. Approximately
220 distinct products are identifiable on the two gels. Of these
21 have a markedly higher concentration in TA+RNA products and 20

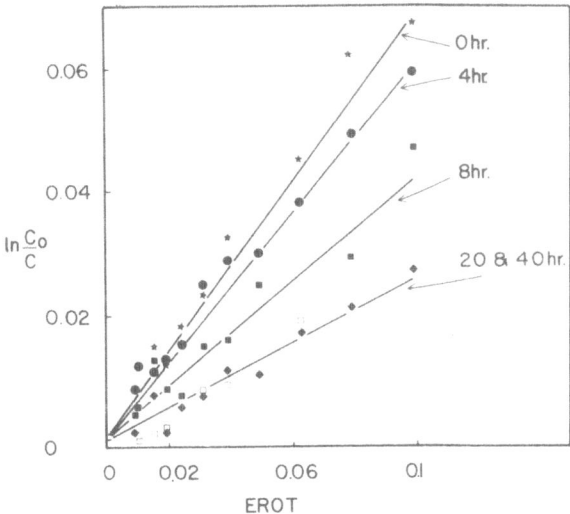

Figure 5.   The Time Course of the Effect of Auxin on Abundant
Polyadenylated RNA.

    Complementary DNA was prepared from total polyadenylated RNA
of soybean hypocotyls treated with 2,4-D for 0(◆), 4(●), 8 (■),
20(◆), or 40 hrs(□), and hybridised with the homologous RNA as
outlined in the legend of Fig 2.   The data have been linearized
for ease of presentation by plotting ln (C_0/C_{Rot}) v ERot where
C_0=proportion single stranded cDNA at Rot = 0 and C_{Rot}=proportion
single stranded cDNA at the given Rot value (Hereford and Rosbach,
1977).

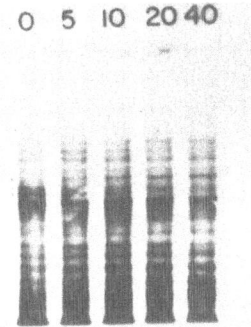

Figure 6. The Time Course of the Effect of Auxin on Translation
Products of Hypocotyl Polyadenylated RNA.

    The translation products of polyadenylated RNA which was
prepared from hypocotyls treated for varying times with 2,4-D
were analysed on single dimension SDS gels.

are higher in UA+RNA products. Components on the outer edges of
the gels are not considered in this analysis since these regions of
the gels are prone to variability between experiments.

d) The Timing of the Auxin Effect on Polyadenylated RNA.

An important aspect of the study of auxin stimulated developmental change is to determine how quickly the associated molecular
events are detectable. Events which are causal in initiating the
cell division would be detectable before the onset of cell division
which occurs in the soybean hypocotyl at 10→12 hrs. after application of 2,4-D (Key et al., 1964). We have examined the kinetics
of the abundant RNA/cDNA hybridisation of polyadenylated RNA prepared at 1, 4, 8, 20 and 40 hrs. after auxin treatment. The data
show that the slowed rate of hybridisation of the abundant component
(c.f. Fig. 3) was initiated by 4 hrs after treatment and was complete by 20 hrs (Fig. 6).

We also have examined the timing of the effect of auxin on
polyadenylated RNA by analysis of in vitro translation products.
It is clear, even from single dimension gels (Fig. 6) that the
concentration of at least 4 mRNAs or groups of mRNAs had begun to
change as quickly as 5-hrs after treatment. Thus both translation
and hybridisation data show that auxin-induced changes in the relative concentration of certain polyadenylated RNA sequences are
initiated within a few hours of auxin treatment, and well before
the commencement of cell division.

CONCLUSION

The hybridisation data presented here demonstrate that there
is a clear downward shift in the relative concentration of some
abundant UA+RNA sequences following auxin treatment. The in vitro
translation data also confirm that the abundance of some sequences
is changed, but reveal a more intricate situation in that some
sequences are increased and some decreased in relative concentration.
The timing of these changes shows that an auxin effect on the concentration of certain polyadenylated RNA sequences is an early
event and therefore probably required for division growth to
commence.

The difference between TA+RNA and UA+RNA is remarkably similar
in degree to the difference between polyadenylated mRNAs of dividing
and non dividing animal cells (Williams et al., 1977, Wilkes et al.,
1979). The small magnitude of this difference is hardly surprising
since, even with the development of highly differentiated animal or
plant cell types there is only a detectable change in the more
abundant polyadenylated RNA (e.g. Hastie and Bishop, 1976, Auger

et al., 1979). It should be noted however, that minute differences in the low abundant class of RNA, which could involve several hundred different sequences types are not detectable. The relatively few sequences involved in the difference between UA+RNA and TA+RNA may well determine the mode of hypocotyl growth and it will be important to determine the mechanisms regulating this difference and also the identity of any peptides (if any) encoded by these sequences. Several cDNA clones of auxin regulated polyadenylated RNA sequences have recently been isolated in our laboratory and will be used in a detailed analysis of these problems.

ACKNOWLEDGMENT

This work was supported by research grant CA11624 from the National Cancer Institute to J.L.K. We thank June Manderfield for skilled work on in-vitro translation and 2D electrophoresis.

ABBREVIATIONS

2,4-D: 2,4,dichlorophenoxyacetic acid, UA+RNA: polyadenylated RNA from untreated soybean hypocotyl, TA+RNA: polyadenylated RNA from 2,4-D treated soybean hypoctyls, Erot: equivalent Rot (mols, ℓ^{-1}, sec $^{-1}$).

REFERENCES

Auger, S., Baulcombe, D., Verma, D. P. S., 1979, Sequence complexity of the poly(A)-containing mRNA in uninfected soybean root and the nodule tissue developed due to the infection by Rhizobium, Biochim. Biophys. Acta, in press.

Breathnach, R., Mandel, J. L., Chambon, P., 1977, Ovalbumin gene is split in chicken DNA, Nature, 270:314-319.

Britten, R. J., Graham, D.E. Neufeld, B. R., 1974, Analysis of Repeating DNA Sequences by Reassociation, Methods in Enzymol., XXIX, 363-418.

Galau, G. A., Klein, W. H. Davis, M. M., Wold. B. J., Britten, R. J., Davidson, E. H., 1976, Structural Gene Sets Active in Embryos and Adult Tissues of the Sea Urchin, Cell 7:487-505

Goldberg, R. B., 1977, DNA Sequence Organization in the Soybean Plant, Biochem Genet., 16:45-68.

Goldberg, R. B., Hoschek, G., Kamaly, J. C., Timberlake, W. E., 1978, Sequence Complexity of Nuclear and Polysomal RNA in Leaves of The Tobacco Plant, Cell, 14:123-131.

Gurley, W. B., Hepburn, A. G., Key J. L., 1979, Sequence Organization in the Soybean Genome, Biochim. Biophys. Acta, 561:167-186.

Hastie, N. D., Bishop, J. O., 1976, The Expression of Three Abundance Classes of mRNA in Mouse Tissues, Cell, 9:7610774

Higgins, T. J. V., Zwar, J. A., Jacobsen, J. V., 1976, Gibberellic Acid Enhances the Level of Translatable mRNA for α-amylase in barley aleurone layers, Nature, 260:166-168.

Hereford, L. M., Rosbash, M., 1977, Number and Distribution of Polyadenylated RNA Sequences in Yeast, Cell, 10:453-462.

Jacobsen, J. V., 1977, Regulation of RNA Metabolism by Plant Hormones, Ann. Rev. Pl. Physiol. 28:537-564.

Key, J. L., Lin, C. Y., Gifford, E. M., Dengler, R., 1964, Relation of 2,4-D Induced Growth Aberrations to changes in Nucleic Acid Metabolism in Soybean Seedlings, Botan. Gaz. 127:87-94.

Kiper, M., 1979, Gene Numbers as Measured by Saturation Hybridisation are Routinely Overestimates, Nature, 278:279-280.

Kiper, M., Bartels, D., Herzfeld, F., Richter, G., 1979, The Expression of a Plant Genome in hnRNA and mRNA. Nucl. Acid. Res. 6:1961-1978.

Ragg, H., Schroder. J., Hahlbrock, K., 1977, Translation of Poly A-containing and Poly(A)-free mRNA for Phenylalanine Ammonia Lyase, a Plant Specific Protein, in a Reticulocyte Lysate, Biochim. Biophys, Acta. 474:226-233

Silflow, C., 1977, Ph.D. Thesis, University of Georgia.

Silflow, C., Hammet, J., Key, J.L., 1979, The Sequences Complexity of Polyadenylated RNA in Soybean Suspension Cultures, Biochemistry, in press.

Thimann, K. V., 1969, The Auxins, in The Physiology of Plant Growth and Development, M. B. Wilkins, ed., McGraw-Hill, New York.

Thompson, W. F., Cleland, R. E., 1971, Auxin and Ribonucleic Acid Synthesis in Pea Stem Tissue as Studied by DNA-RNA Hybridisation, Pl. Physiol, 48:663-670

Timberlake, W. E., Shumard, D. S., Goldberg R. B., 1977, Relationship between Nuclear and Polysomal RNA Populations of Achlya, Cell, 10:623-632.

Verma, D. P. S., Maclachlan, G. A., Byrne, H., Ewings, D., 1975, Regulation and Translation in vitro of mRNA for Cellulase from Auxin Treated Pea Epicotyls, J. Biol. Chem., 250:1019-1026.

Wilkes, P. R., Birnie G. D., Paul, J., 1979, Changes in nuclear and polysomal polyadenylated RNA sequences during rat-liver regeneration, Nucl. Acid. Res,, 6:2193-2208.

Williams, J. G., Hoffman, R., Penman, S., 1977, The Extensive Homology between mRNA Sequences of Normal and SV-40 Transformed Human Fibroblasts, Cell, 11:901-907.

The Role of Light in the Induction of mRNAs for
Phenylalanine Ammonia-Lyase and Related Enzymes in Plant
Cell Cultures

Klaus Hahlbrock, Susan E. Gardiner, Ulrich
Matern, Hermann Ragg and Joachim Schröder

Biologisches Institut II
Schänzlestraße 1
D-7800 Freiburg, W.-Germany

INTRODUCTION

Parsley cells (Petroselinum hortense Hoffm.) which
have been kept in suspension culture in complete darkness
for several years do not form chloroplasts in the light.
The only known response of these cell cultures to irradiation with visible and long-wave UV light is the synthesis and rapid accumulation of large amounts of flavone
and flavonol glycosides. These flavonoid glycosides are
also abundant in intact parsley plants under normal growth
conditions, but are not recognized by the human eye because of the intense colour of the chlorophyll of green
tissue. Dark-grown cell cultures are almost colourless,
and the accumulation of the greenish-yellow flavonoid
glycosides after irradiation is therefore easily visible.

Flavonoid glycoside formation in parsley cell cultures seems to have an absolute requirement for UV-light,
with the largest effect occurring around 290 nm (Wellmann,
1976).Sufficient long-wave UV light is transmitted through
Erlenmeyer flasks and glass vessels of fermenters to allow
the system to be fully induced with ordinary fluorescent
lamps. Thus, only simple and inexpensive equipment is
required for the induction of flavonoid biosynthesis in
parsley cell cultures under conditions, where large
amounts of material can be obtained (Knobloch and Hahlbrock, 1977). It was therefore possible to isolate and
purify enzymes and mRNAs in amounts sufficient for studies
of their properties.

SELECTIVE AND COORDINATED INDUCTION OF TWO GROUPS OF
ENZYMES

The flavonoids produced by irradiated parsley cell
cultures are glycosides of three flavones and three fla-
vonols, all of which are closely related with respect
to both their chemical structures and their biosynthesis.
A general scheme illustrating the sequence of biosynthetic steps which are involved in the formation of flavonoid glycosides in parsley is shown in Fig. 1. The first
sequence of three reactions is catalyzed by the enzymes
of general phenylpropanoid metabolism and leads to the
formation of 4-coumaroyl-CoA from phenylalanine. The
subsequent formation and modification of the flavonoid
skeleton involves approximately 10-12 enzymes of the
flavonoid glycoside pathway proper.

In addition to this formal distinction between two
groups of enzymes we have defined several operational
criteria for this classification. The two most important
criteria are (i) the induction of only group I, but not
of group II, in the absence of light upon subculturing
of the cells ("dilution effect"), and (ii) characteristic,
albeit small, differences in the shapes of the curves for
light-induced changes in the enzyme activities between
groups I and II. A schematic representation of these observations is given in Fig. 1.

Despite the significant difference between the two
groups with respect to the timing for maximal enzyme
activities, there appears to exist a high degree of coordination between both groups at the initial stage of
induction, particularly at the mRNA level (Schröder et
al., 1979). Furthermore, the induction by light of the
enzymes of the two groups is very selective. The induction does not include enzymes catalyzing the various reactions which supply substrates of flavonoid biosynthesis from intermediate metabolism, nor has a light effect
on any other enzyme activity in the cell cultures been
observed (Ebel and Hahlbrock, 1977).

METHODS FOR THE MEASUREMENT OF mRNA ACTIVITIES

One enzyme of group I, phenylalanine ammonia-lyase
(PAL), and two enzymes of group II, flavanone synthase
(FLS) and UDP-apiose synthase (UAS), have been purified
and used for the preparation of specific antisera. Each
of the three enzymes consists of subunits which are probably identical. UAS is associated with a second protein
(X) which is also composed of subunits and has no known

catalytic activity. The antiserum used for UAS contains
antibodies to both proteins.

     The procedure of measuring mRNA activity is out-
lined in Fig. 2. Immunoprecipitates were obtained with
appropriate mixtures of antisera after labelling of the
enzymes with ^{35}S-methionine either in vivo or in vitro.
Polyribosomal mRNA fractions were used for synthesis of
the enzymes in vitro in a rabbit reticulocyte lysate.
Details of the procedure have been described by Schröder
et al. (1979). Analysis of the immunoprecipitates by gel

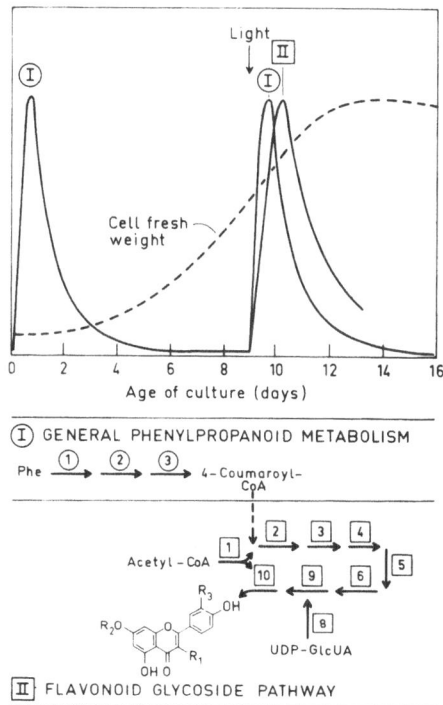

Fig. 1   Schematic representation of curves for changes
         in enzyme activities and of the sequence of re-
         actions involved in the formation of flavone
         glycosides from phenylalanine and acetyl-CoA,
         the two substrates providing the carbon atoms
         for the flavonoid skeleton. The numbering of
         the enzymes of the flavonoid glycoside pathway
         includes two reactions which are specific for
         flavonols (Nos. 7 and 11) and are not shown here.
         For details see Hahlbrock, 1976; Hahlbrock et
         al., 1976; Ebel and Hahlbrock, 1977.

electrophoresis was necessary even when the antisera
were used separately for each enzyme, because of the
relatively high contamination with non-specific radioactivity. The radioactivity associated with the enzyme
subunits was used to calculate the rates of enzyme synthesis. For the present experiments, mixtures of PAL
and FLS antisera and of PAL, UAS and X antisera were used.
It is important to note that essentially the same results
were obtained after labelling the enzymes either in vivo
or in vitro. The enzyme subunits synthesized in vitro
were indistinguishable under denaturing conditions on
polyacrylamide gels from those synthesized in vivo.

LIGHT-INDUCED CHANGES IN mRNA ACTIVITIES

        Two different irradiation programs for previously
dark-grown cell cultures, continuous irradiation and
short-term irradiation (2.5 h), were used for measuring
light-induced changes in mRNA and enzyme activities.
While continuous irradiation resulted in highest absolute values, the curves for changes in both mRNA and enzyme activities were more difficult to interprete than
those obtained after short-term irradiation.

Fig. 2
Procedure for the
determination of the
rates of synthesis
of (or mRNA activi-
ties for) various
proteins

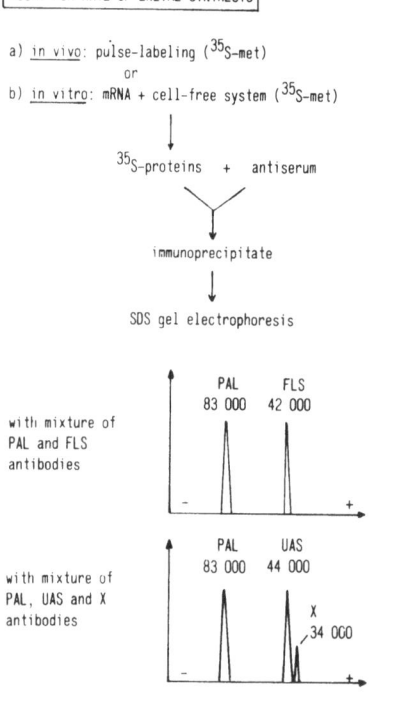

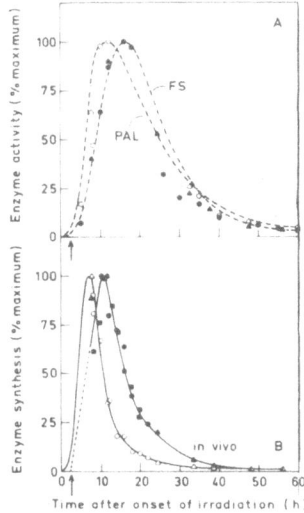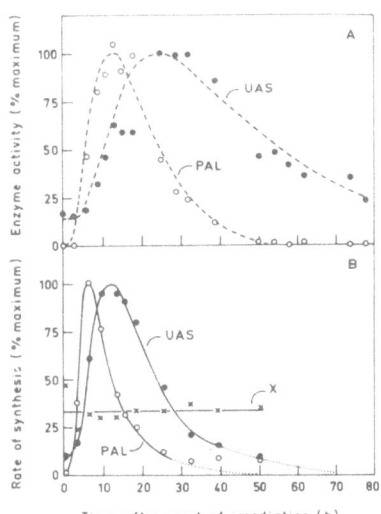

Fig. 3 Light-induced changes in (A) enzyme activities
and (B) rates of subunit synthesis for PAL and
FLS (left) and PAL, UAS and X (right) after
short-term irradiation for 2.5 h. The broken
lines were calculated using Equation (1). Data
partly from Schröder et al. (1979).

Figure 3 shows the results of two separate experiments in which the light-induced changes in the activities
PAL, FLS and UAS as well as in their rates of synthesis
in vivo were measured under the conditions of short-term
irradiation. In addition, the rate of synthesis of protein X was measured. The most striking observations were
the following:

    - Maximal activities for FLS and UAS occurred sig-
nificantly later than for PAL.
    - After a short period of low response, the rates
of synthesis in vivo (or mRNA activities in vitro) for
all three enzymes increased rapidly upon irradiation and
continued to increase for several hours beyond the time,
when the cells were returned to darkness.
    - The rates of synthesis (mRNA activities) ceased
to increase almost abruptly in all three cases after
6-12 hours and then declined exponentially. Maximal values for PAL synthesis were reached several hours earlier
than for the synthesis of FLS and UAS.
    - No significant light effect on the rate of synthe-
sis of the UAS-associated protein X was observed.

RELATIONSHIP BETWEEN mRNA AND ENZYME ACTIVITIES

     The expected changes in enzyme activity can be cal-
culated from the observed changes in the rate of synthesis, provided the rate of degradation is known and remains constant. In such a case, Equation (1) can be applied
(Betz et al., 1978) where enzyme activity (E) and the ze-
ro-order rate constant of enzyme synthesis (or mRNA activity, ^0k_s) change with time (t), while the first-order
rate constant of enzyme degradation (^1k_d) is independent
of time.

$$dE(t)/dt = {}^0k_s(t) - {}^1k_d \cdot E(t) \tag{1}$$

The value for ^1k_d can be calculated from the half-life
($\tau_{1/2}$) of the enzyme activity using Equation (2):

$$^1k_d = \ln2/\tau_{1/2} \tag{2}$$

     The half-lives of PAL, FLS and UAS activities were
about 7-10 h, 5-6 h and 17-20 h, respectively. The expected changes in these three enzyme activities were
then calculated from the data shown in Fig. 3B. The results (broken curves in Fig. 3A) are in good agreement
with the experimental data (symbols in Fig. 3A).

SIZE AND POLY(A)-CONTENT OF mRNAs

     Poly(A)-containing (poly(A)$^+$) mRNA was prepared by
passing polyribosomal RNA from irradiated cell cultures
twice through an oligo(dT)-cellulose column. The resulting, essentially rRNA-free mRNA preparation was fractionated by centrifugation on a sucrose gradient, and the positions of PAL and FLS mRNA activities on the gradient
were determined. Sedimentation constants of about 20-21 S
for PAL mRNA and 17 S for FLS mRNA were estimated from
these data, using 18 S and 25 S rRNAs as references. Similar results were obtained, when PAL and FLS mRNA activities were analyzed on polyacrylamide gels under denaturing conditions (Ragg, 1978; and Kreuzaler et al.,
1979).

     These data allowed us to calculate approximate mole-
cular weights of 1 x 10^6 for PAL mRNA and 0.6 x 10^6 for
FLS mRNA. These values are about 30-50% higher than those
calculated for polynucleotide sequences required to code
for the subunits of PAL (M_r = 83000) and FLS (M_r = 42000),
respectively. Thus, both mRNAs seem to contain large polynucleotide sequences which do not contain coding information for these enzymes.

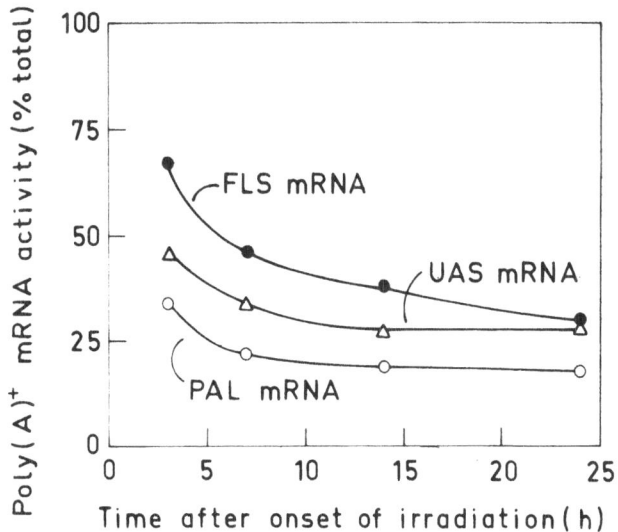

Fig. 4   Changes in poly(A)$^{+}$ mRNA activities for PAL, FLS
         and UAS mRNAs relative to the total amounts of
         mRNA activities, under continuous irradiation
         of a parsley cell culture. The onset of irradi-
         ation was at time zero. Data from Schröder et
         al. (1979) and unpublished results.

        According to the operational definition after chro-
matography on oligo(dT)-cellulose, mRNA activities for
all three enzymes, PAL, FLS and UAS, were found in both
poly(A)$^{+}$ and poly(A)$^{-}$ (poly(A)-free) mRNA fractions (Ragg
et al., 1977; Schröder et al., 1979; and unpublished results). Fig. 4 shows that the poly(A)$^{+}$ mRNA activity, relative to the total amount of specific mRNA activity, decreased in all three cases in a similar manner with time
after the onset of continuous irradiation.

CONCLUSIONS

        A high degree of coordination was observed for the
induction by light of mRNAs coding for enzymes of general
phenylpropanoid metabolism (PAL) and of the flavonoid glycoside pathway (FLS and UAS) in parsley cell cultures.
Coincidence of experimentally determined and calculated
curves for light-induced changes in the three enzyme activities indicated that these changes were mainly or
exclusively caused by large changes in the respective
mRNA activities. In contrast, the rates of degradation
of the enzymes appeared not to vary significantly during
or after irradiation.

Small, but significant differences between the shapes of the curves for changes in the individual enzyme activities can be attributed to two major causes: first, to slight differences in the lengths of time during which light is effective in the induction of mRNAs for the two groups of enzymes; and second, to different half-lives for the enzyme activities.

The finding that the molecular weights for PAL and FLS mRNAs were considerably higher than calculated for the minimal number of nucleotides required to code for the enzyme subunits is in agreement with similar observations with other eukaryotic mRNAs. The existence of polycistronic mRNAs which would explain the high degree of coordination in the induction of each of the two groups of enzymes is very unlikely for PAL and can be ruled out for FLS.

All three light-induced mRNAs investigated were only partially polyadenylated at various times after the onset of induction. A similar rate of decrease in the ratio of polyadenylated/non-polyadenylated mRNA activity is a further example of the great similarity in the regulation of these mRNAs in irradiated parsley cells.

Betz, B., Schäfer, E., and Hahlbrock, K., 1978, Arch. Biochem. Biophys., 190:126-135.

Ebel, J., and Hahlbrock, K., 1977, Eur. J. Biochem., 75:201-209.

Hahlbrock, K., 1976, Eur. J. Biochem., 63:137-145.

Hahlbrock, K., Knobloch, K.H., Kreuzaler, F., Potts, J.R.M., and Wellmann, E., 1976, Eur. J. Biochem., 61:199-206.

Knobloch, K.H., and Hahlbrock, K., 1977, Arch. Biochem. Biophys., 184:237-248.

Kreuzaler, F., Ragg, H., Heller, W., Tesch, R., Witt, I., Hammer, D., and Hahlbrock, K., 1979, Eur. J. Biochem., in press.

Ragg, H., 1978, Ph.D. Thesis, University of Freiburg.

Ragg, H., Schröder, J., and Hahlbrock, K., 1977 Biochim. Biophys. Acta, 474:226-233.

Schröder, J., Kreuzaler, F., Schäfer, E., and Hahlbrock, K., 1979, J. Biol. Chem., 254:57-65.

Wellmann, E., 1976, in:"Industrieller Pflanzenbau," Vol. 5, E. Bancher, ed., Wien.

FUNCTIONAL CHARACTERIZATION OF SOME RIBOSOMAL PROTEINS FROM

WHEAT GERM

A. B. Legocki, C. J. Mądrzak, D. Przybył,

M. Sikorski and U. Szybiak

Institute of Biochemistry, University of Agriculture
60-637 Poznań, Wołyńska 35, Poland

Detailed study of the topography and properties of ribosomal
proteins may greatly facilitate elucidation of the function of
ribosomes during the translational process. Although the basic
mode of action and the structural architecture of eukaryotic
ribosomes is similar to that of bacterial ribosomes, they are of
higher complexity and differ significantly in a number of their
structural components.

In comparison to mammalian ribosomes which have been intensely studied during the last few years (for review see Bielka and
Stahl[1]) much less is known about the ribosomes from higher plants.
In our recent studies we have estimated that purified 80S ribosomes
from wheat germ contain 79 proteins of which 35 are derived from
the smaller subunit and 44 from the larger subunit[2] (Fig. 1). This
report presents the identification of those ribosomal proteins
which interact with tRNA and poly(U) and describes some properties
of acidic proteins from wheat germ ribosomes.

I.   Identification of ribosomal proteins which interact with tRNA
     and poly(U).

Chromatography of ribosomal proteins on affinity adsorbents
containing RNA ligands appears to be a useful technique for studying the formation of specific complexes between either RNA or
polynucleotide fragments and bacterial or mammalian ribosomal
proteins.[5-7] We have applied this technique to detection of these
proteins of wheat germ ribosomes which are involved in the binding
of tRNA and synthetic template poly(U). As revealed by 2D electrophoresis (Fig. 2) there are five proteins from the smaller

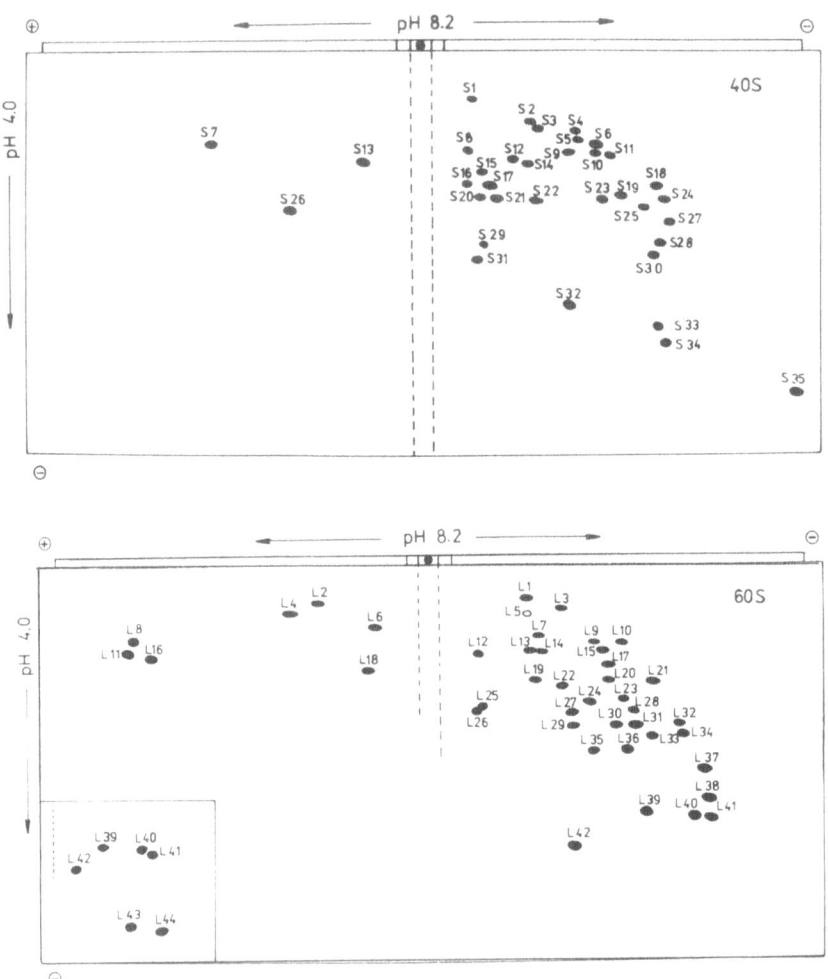

Fig. 1.   Schematic two-dimensional electrophoretic pattern of
          wheat germ 40S and 60S ribosomal proteins. Gel electro-
          phoresis was according to Kaltschmidt and Wittmann[3] and
          the nomenclature according to Sherton and Wool[4].

subunit (S4, S6, S9, S10, S28) and seven proteins from the larger
subunit (L1, L7, L13, L14, L19, L27, L35) which interact specifically with tRNA covalently coupled to Sepharose 4B matrix. Also
eight 40S proteins (S6, S15, S19, S23, S24, S25, S27, S30)˙ and
five 60S proteins (L17, L20, L22, L24, L30 in addition to four
minor proteins L33, L37, L40, L41) bind to immobilized poly(U).
A direct interaction of the above ribosomal proteins with tRNA or
poly(U) was also confirmed by the chromatography on RNA-cellulose
matrices. In this case RNA ligands were linked to cellulose by

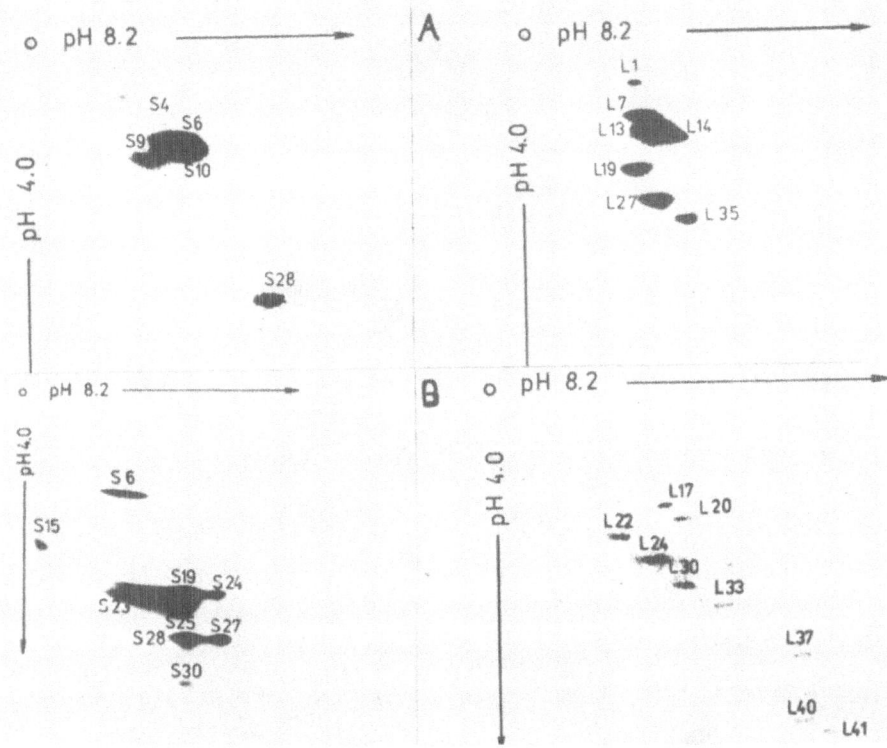

Fig. 2.   Two-dimensional gel electrophoresis of wheat germ 40S and
          60S ribosomal proteins bound to tRNA-Sepharose 4B (A) and
          poly(U)-Sepharose 4B (B). The ribosomal proteins were
          bound at 0.35M KCl and eluted with 1M KCl. In the control
          runs on Sepharose columns prepared without RNA no protein
          binding was observed.

irradiation with UV light according to a procedure developed earlier
for immobilization of native DNA[8]. It was found that the protein
patterns obtained with both types of affinity columns were quantitatively identical.

II.   Acidic proteins of wheat germ ribosomes.

     Wheat germ ribosomal acidic proteins represent a group of
proteins heterogeneous in size designated S7, S13, S26, L2, L4, L6,
L8, L11, L16 and L18 (Fig. 1). Five of these proteins (S13, **L2**,
L8, L11, L16) are phosphorylated in vitro in the presence of
$32P-\gamma$-ATP and yeast protein kinase 3 (Fig. 3). It is interesting
to note that acidic proteins are the only intensely labeled
proteins in the larger subunit. Two of the three proteins L8, L11
or 16 which migrate rapidly in the first dimension may correspond

to strongly acidic and phosphorylated bacterial proteins L7/L12.
Such a suggestion is consistent with the data from comparative
immunological analysis of ribosomal acidic proteins from various

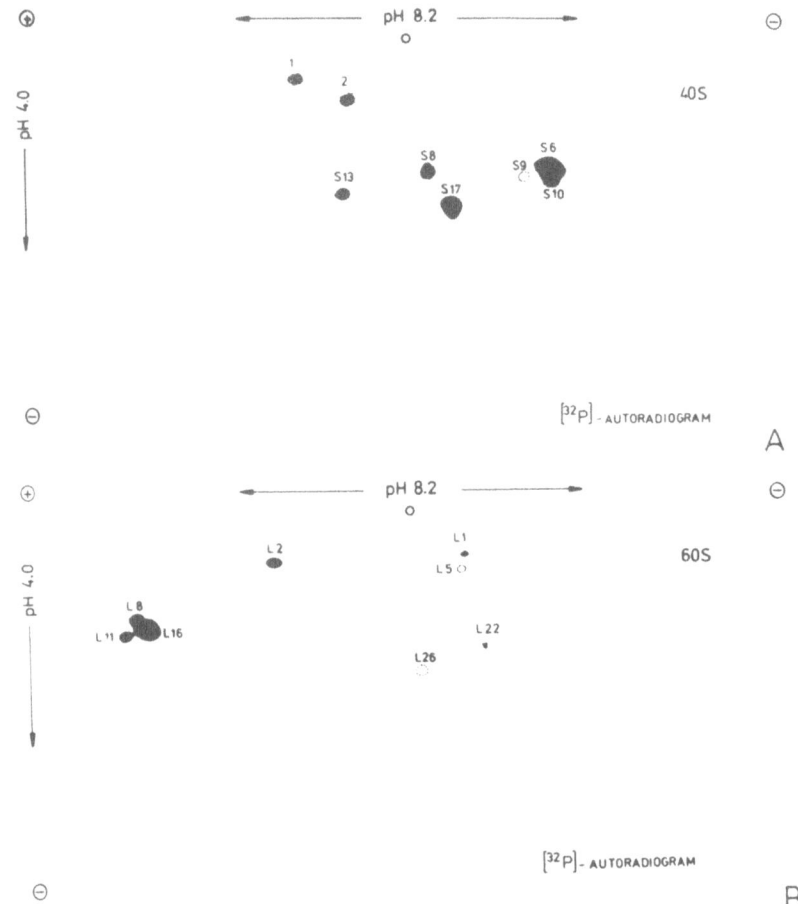

Fig. 3. Autoradiograms of ^{32}P-labeled wheat ribosomal proteins
         after separation on 2D gel electrophoresis. Each subunit
         was phosphorylated in vitro in the presence of yeast
         protein kinase 3 and ^{32}P-γ-ATP as described earlier[2].

species. These studies have shown that antibodies against the
L7/L12 proteins from E. coli cross-react with acidic proteins from
yeast[9], rat liver[10], Krebs ascites cells[11] and brine shrimp,
A. salina[12]. Since we have recently observed cross-reactivity
between E. coli L7/L12 and acidic proteins from wheat germ, this
also applies to plant ribosomal proteins. Structural homology
between acidic proteins from ribosomes of various species suggests

their high degree of evolutionary conservation.

As far as the biological function of ribosomal acidic proteins is concerned there are several lines of evidence that they are involved directly in various steps of polypeptide synthesis. In prokaryotes, protein S1 participates in the initiation[13] as well as in elongation[14] of the polypeptide chain, while proteins L7/L12 are involved in chain elongation through the hydrolysis of GTP (see review by Möller[15]).

Wheat germ ribosomal acidic proteins like bacterial protein S1 revealed high affinity to poly(U) as measured by retention to alkaline-treated Millipore filters (data not shown). Moreover, the addition of isolated acidic proteins to a cell-free system strongly affected polypeptide synthesis directed by either poly(U)

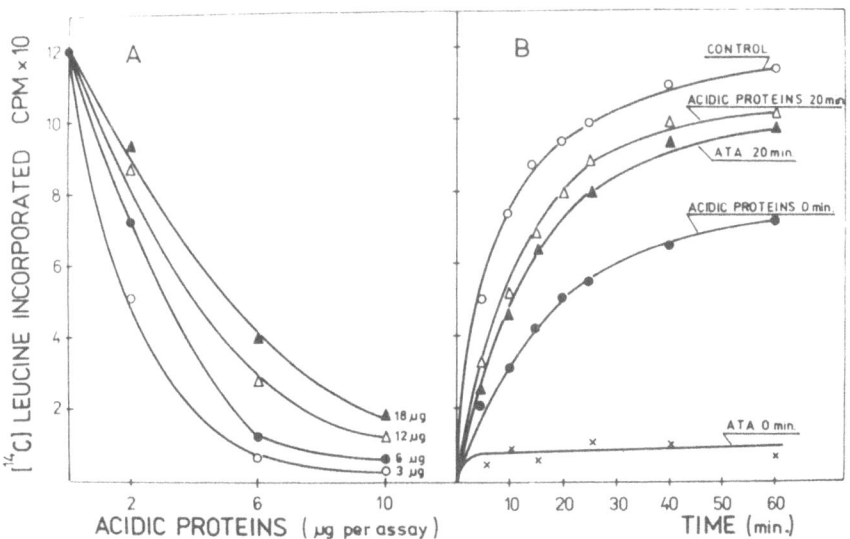

Fig. 4.   Effect of ribosomal acidic proteins on the translation of TMV-RNA in a wheat germ cell-free system. A. Inhibition as a function of the amount of acidic proteins added. B. Time kinetics. Standard incubation mixtures[16] contained 3-18 µg of TMV-RNA as indicated (A) or 6 µg (B) and 2-10 µg (A) or 2 µg (B) of acidic proteins. In B after the reaction was started 5 µl aliquots were withdrawn and assayed for ^{14}C leucine incorporation. Where indicated 2 µg of acidic proteins or 0.1 mM aurintricarboxylic acid (ATA) was added.

or TMV-RNA. Fig. 4 shows that the extent of the observed inhibition is a function of mRNA concentration. As the concentration of TMV-RNA in the system is increased, the inhibition decreased showing that the effect of acidic proteins is dependent on the ratio of mRNA/protein. Moreover, the inhibition was observed only when acidic proteins were added to a translational system prior to but not after 20 min of incubation. As is seen from Fig. 3 the kinetics of the inhibition are similar to that observed for aurintricarboxylic acid, a typical initiation inhibitor which prevents the binding of mRNA to 40S initiation complex.

The above results provide a preliminary indication that wheat germ ribosomal proteins added exogenously can modify the efficiency of mRNA translation; possibly on the level of initiation. This suggestion, however, requires further confirmation of experiments using the isolated individual proteins in a purified initiation system.

Acknowledgement

This study was carried out as Project 0.9.7 of the Polish Academy of Sciences.

References

1. H. Bielka and J. Stahl, Structure and Function of Eukaryotic Ribosomes, in: "Amino Acid and Protein Biosynthesis II", Vol. 18. Int. Rev. of Biochem. H. R. V. Arnstein, ed., University Park Press, Baltimore (1978).
2. M. M. Sikorski, D. Przybył, W. Kudlicki, J. Zając, T. Borkowski, E. Gąsior and A. B. Legocki, The Ribosomal Proteins of Wheat Germ. Distribution and Phosphorylation in vitro, Plant Sci. Letters, in press.
3. E. Kaltschmidt and G. H. Wittmann, Ribosomal Proteins. VII. Two-Dimensional Polyacrylamide Gel Electrophoresis for Fingerprinting of Ribosomal Proteins, Anal. Biochem. 36:401 (1970).
4. C. C. Sherton and I. G. Wool, Determination of the Number of Proteins in Liver Ribosomes and Ribosomal Subunits by Two-Dimensional Polyacrylamide Gel Electrophoresis, J. Biol. Chem., 247:4460 (1972).
5. H. R. Burrell and J. Horowitz, Binding of Ribosomal Proteins to RNA Covalently Coupled to Agarose, Eur. J. Biochem. 75:533 (1977).
6. M. Yukioka and K. Omori, Identification of Ribosomal Proteins with Affinity for tRNA Molecule by Affinity Chromatography on tRNA-Sepharose, FEBS Lett. 75:217 (1977).
7. M. Ustav, R. Villems, M. Saarma and A. Lind, The Interaction of Transfer Ribonucleic Acid with 50S Ribosomal Subunit Proteins, FEBS Lett. 83:353 (1977).

8. R. M. Litman, A Deoxyribonucleic Acid Polymerase from <u>Micro-coccus</u> <u>luteus</u> (<u>Micrococcus</u> <u>lysodeikticus</u>) Isolated on Deoxyri-bonucleic Acid-Cellulose, <u>J. Biol. Chem.</u>, 243:6222 (1968).

9. D. Richter and W. Möller, Properties and Functions of Ethanol-Potassium Chloride Extractable Proteins from 80S Ribosomes and their Interchangeability with the Bacterial Proteins L7/L12, <u>in</u>: "Lipmann Symposium: Energy, Regulation and Biosynthesis in Molecular Biology", D. Richter, ed., W. de Gruyter, Berlin (1974).

10. G. Stöffler, I. G. Wool, A. Lin and K. H. Rak, The Identification of the Eukaryotic Ribosomal Proteins Homologous with <u>Escherichia</u> <u>coli</u> Proteins L7 and L12, <u>Proc. Nat. Acad. Sci. USA</u> 71:4723 (1974).

11. D. P. Leader and A. A. Coia, The Acidic Ribosomal Phospho-protein of Eukaryotes and its Relationship to Ribosomal Proteins L7 and L12 of <u>Escherichia</u> <u>coli</u>, <u>Biochem. J.</u> 176:569 (1978).

12. W. Möller, L. I. Slobin, R. Amons and D. Richter, Isolation and Characterization of Two Acidic Proteins of 60S Ribosomes from <u>Artemia</u> <u>salina</u> Cysts, <u>Proc. Nat. Acad. Sci. USA</u> 72:4744 (1975).

13. G. Van Dieijen, C. J. Van der Laken, P. H. Van Knippenberg and J. Van Duin, Function of <u>Escherichia</u> <u>coli</u> Ribosomal Protein S1 in Translation of Natural and Synthetic Messenger RNA, <u>J. Mol. Biol.</u> 93:351 (1975).

14. R. Lipecky, J. Kohlschein and H. G. Gassen, Complex Formation Between Ribosomal Protein S1, Oligo- and Polynucleotides: Chain Length Dependence and Base Specificity, <u>Nucleic Acids Res.</u>, 4:3627 (1977).

15. W. Möller, The Ribosomal Components Involved in EF-G- and EF-Tu-Dependent GTP Hydrolysis, <u>in</u>: "Ribosomes", M. Nomura, A. Tissières and P. Lengyel eds., Cold Spring Harbor Labora-tory (1974).

16. A. Konieczny and A. B. Legocki, Isolation and <u>in</u> <u>vitro</u> Trans-lation of Leghaemoglobin mRNA from Yellow Lupin Root Nodules, <u>Acta Biochim. Polon.</u> 25:379 (1978).

MAIZE STORAGE PROTEINS: CHARACTERIZATION AND BIOSYNTHESIS

B. A. Larkins*, K. Pedersen*, W. J. Hurkman*,
A. K. Handa*, A. C. Mason*, C. Y. Tsai*, and M. A.
Hermodson[+], Department of Botany and Plant Pathology*,
and Department of Biochemistry[+]
Purdue University, West Lafayette, IN 47907

INTRODUCTION

Developing seeds provide plant molecular biologists with useful model systems for studying the physiological and genetic mechanisms regulating the synthesis of specific plant proteins, i.e. seed storage proteins. These studies have perhaps even greater significance considering the importance of seed proteins in human and livestock nutrition. Maize storage proteins are interesting from both these aspects, since maize is an economically important crop and mutations affecting both the quantitative and qualitative synthesis of maize storage proteins have been identified (Mertz et al., 1964; Nelson et al., 1965). Our research in the last several years has been devoted to the characterization of these storage proteins and the reactions regulating their biosynthesis.

The major storage protein fraction in maize consists of a group of alcohol-soluble proteins (prolamines) commonly called "zein". The term "zein" may be misleading, since this fraction actually consists of several proteins rather than a single one. Zein proteins are well known for their poor nutritional quality resulting from negligible contents of lysine and tryptophan.

Both environmental and genetic factors influence the synthesis of maize storage proteins. For example, application of nitrogen fertilizer increases the protein content of the seed. But this increase occurs primarily in the prolamine fraction, and as a consequence the proportion of essential amino acids is decreased. By screening maize genotypes for lysine content, Mertz et al. (1964) and Nelson et al. (1965), identified several endosperm

mutations, notably <u>opaque</u>-2 and <u>floury</u>-2, which improve nutritional
quality by decreasing the amount of prolamine. Unfortunately, the
reduction in seed dry matter associated with these mutations has
prevented their broad agricultural utilization. The problems of
improving the protein quality of the seed are complex, and an
understanding of the molecular mechanisms regulating storage protein
and starch synthesis may provide new insights for their solution.

Zein Isolation and Characterization

Zein proteins occur in the endosperm of the seed as deposits
called protein bodies. In developing endosperm, protein bodies
are spherical (1 to 2 microns in diameter) and are surrounded by
interconnecting rough endoplasmic reticulum (RER) membranes (see
arrows Fig. 1). In low prolamine mutants such as <u>opaque</u>-2 and
<u>floury</u>-2 the protein bodies are generally much smaller than in
normal endosperm.

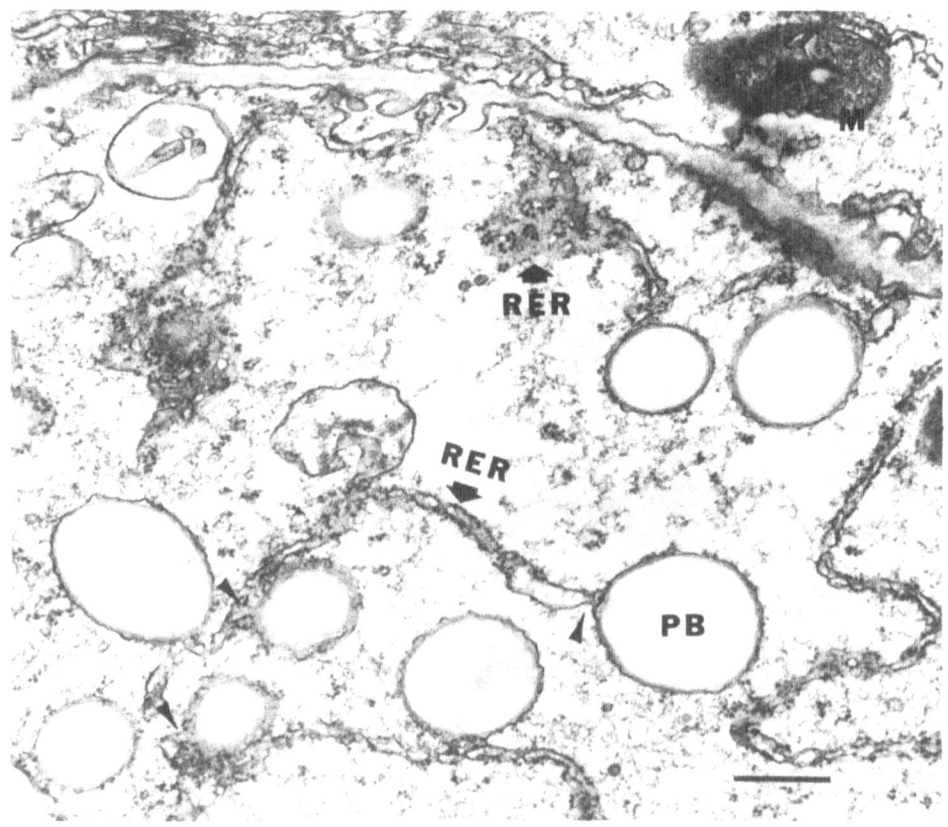

Fig. 1.   Electron micrograph of 19-day maize and endosperm cell.
          PB, protein body; RER, rough endoplasmic reticulum; M,
          mitochondrion. Scale bar = 0.5 um.

Zein polypeptides can be isolated by extracting endosperm meal or isolated protein bodies with 70% ethanol containing 1 mM 2-mercaptoethanol. SDS-polyacrylamide gel electrophoresis of these extracts reveals four different mol wt components (Fig. 2). The two major ones have mol wts of approximately 19,000 and 22,000; there are also small amounts of 15,000 and 10,000 mol wt polypeptides. Because these smaller mol wt proteins are present in lesser amounts, they often are not visible in photographs of stained gels.

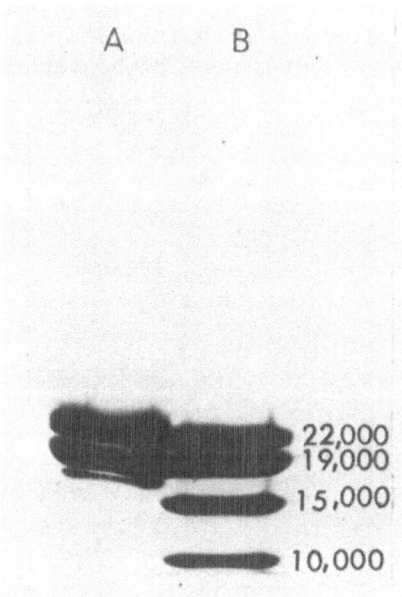

Fig. 2.   SDS-polyacrylamide gel electrophoresis of zein polypeptides. Zein proteins were dissolved in 0.01% phosphoric acid and separated by Sephacryl S-200 chromatography. Samples represent the leading (A) and trailing fractions (B) of the protein peak.

The different mol wt components have similar amino acid compositions. Each contains significant amounts of glutamine, leucine, alanine, and proline, and negligible amounts of lysine (Lee et al., 1976; Gianazza et al., 1977). The 15,000 mol wt component contains slightly larger amounts of methionine than the other proteins. There is apparently a significant amount of sequence homology between the major components since the peptides formed by trypsin/chymotrypsin digestion are very similar (Fraij and Melcher, 1978).

While the analysis of zein proteins on SDS gels suggests that this fraction is composed of a few polypeptides, analysis by iso-electric focusing (IEF) shows that substantial charge heterogeneity

exists within the different mol wt components (Righetti et al.,
1977). This heterogeneity does not appear to be entirely an
artifact of protein purification nor the IEF procedure. Righetti
et al. (1977) showed that there is significant variation in the
IEF patterns of zein proteins from different maize genotypes, and
certain of the IEF bands were shown to be inherited in a simple
Mendelian fashion (Soave et al., 1978). While these results suggest
that several different genes code for the same sized polypeptides,
whether each IEF band represents a different structural gene remains
to be demonstrated. IEF is a very sensitive technique and factors
such as aggregation, disulfide bonding, and deamidation might
contribute to the cha   heterogeneity of the proteins.

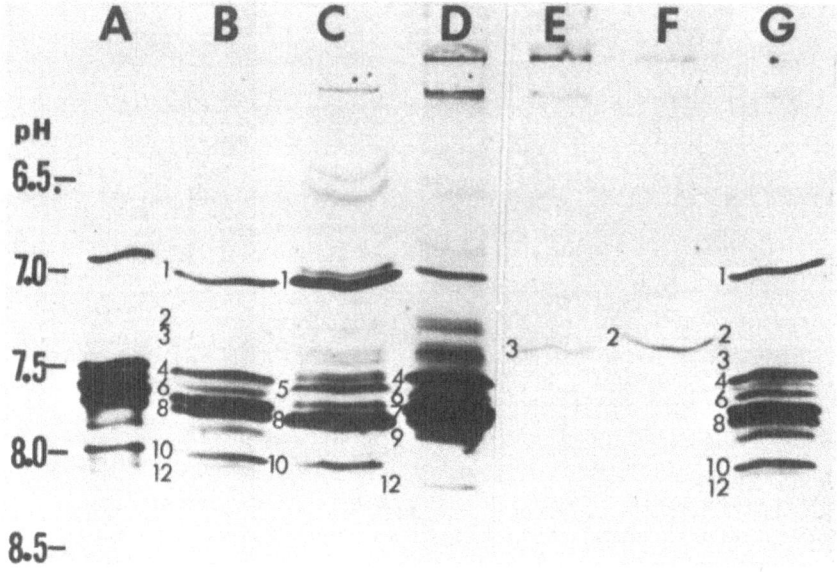

Fig. 3. Isoelectric focusing of zein polypeptides between pHs 6-9.
         Zein proteins from the maize inbred W64A were reduced and
         alkylated with 4-vinyl pyridine. Samples were dissolved
         in 1% SDS containing 20% 2-mercaptoethanol; after heating
         in boiling H_2O, Nonidet P-40 (final concentration 10%) was
         added and samples applied to the IEF gel. Individual zein
         components were isolated from preparative SDS-gels.
         Samples: (A) non-alkylated total zein; (B,G) alkylated
         total zein; (C) 22,000 Mr zein component; (D) 19,000 Mr
         zein component; (E) 15,000 Mr zein component; (F) 10,000
         Mr zein component.

    To determine if these factors contribute significantly to the
IEF pattern, we analyzed zein proteins from the inbred W64A. Zein
samples were reduced and alkylated prior to gel electrophoresis to
minimize potential protein interactions, and proteins were isolated

from SDS gels for analysis of individual mol wt components. Samples were dissolved in SDS to insure dissociation before IEF analysis. Samples B and G in Figure 3 show that 12 different protein bands were separated between pHs 6-9. A few additional minor bands were present, but their significance is questionable, since they were more prominant in samples that were not reduced and alkylated and applied in Nonidet P-40 (Fig. 3, sample A). The individual 22,000 and 19,000 mol wt components each accounted for 3-5 different IEF forms (Fig. 3, samples C and D, respectively). The 15,000 mol wt component (Fig. 3, sample E), and the 10,000 mol wt component (Fig. 3, sample F) each gave rise to one major IEF band.

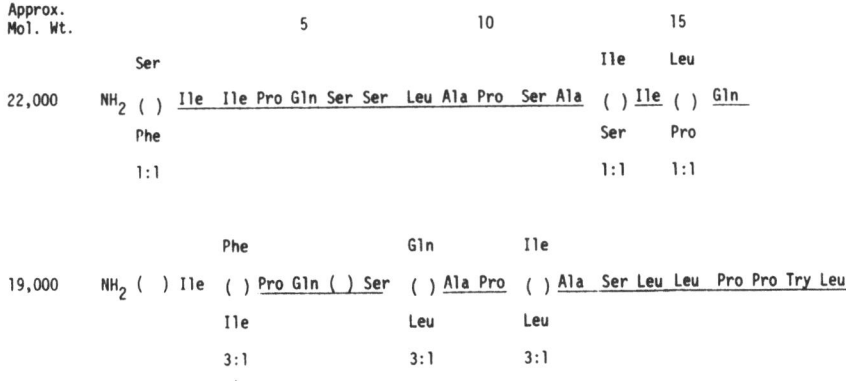

Comparison of the N-terminal Sequence of the Major Zein Components from W64A

Fig. 4. N-terminal sequence analysis of the 22,000 and 19,000 Mr zein components. Zein proteins were isolated from preparative SDS-polyacrylamide gels by electrophoretic elution. The N-terminal amino acid sequence was determined with a Beckman 890C sequencer. Identification of the amino acid phenylthiohydantions was by gas chromatography and high pressure liquid chromatography.

The charge differences detected within the major zein components could also be attributed to differences in their primary amino acid sequences. A minimum of two different but homologous N-terminal sequences were determined for each of the major components (Fig. 4). The 22,000 mol wt component had serine and phenylalanine as N-terminal amino acids in the ratio of 1:1. The sequences were identical thereafter except at positions 13 and 15 where isoleucine/serine and leucine/proline were in a 1:1 ratio.

The 19,000 mol wt component was found to contain several different sequences. It could be that there are two homologous sequences differing at positions 3, 8, and 11, although permutations that would result in a larger number of polypeptides are possible. These results provide a structural basis for at least part of the charge heterogeneity indicated by IEF.

In vivo and in vitro Synthesis of Zein

Electron micrographs of developing endosperm (Fig. 1) suggest that zein proteins are synthesized by membrane-bound polyribosomes, and there have been several reports demonstrating this (Larkins et al., 1976; Burr and Burr, 1976; Viotti et al., 1979; Melcher, 1979). Messenger RNAs that direct the synthesis of these proteins

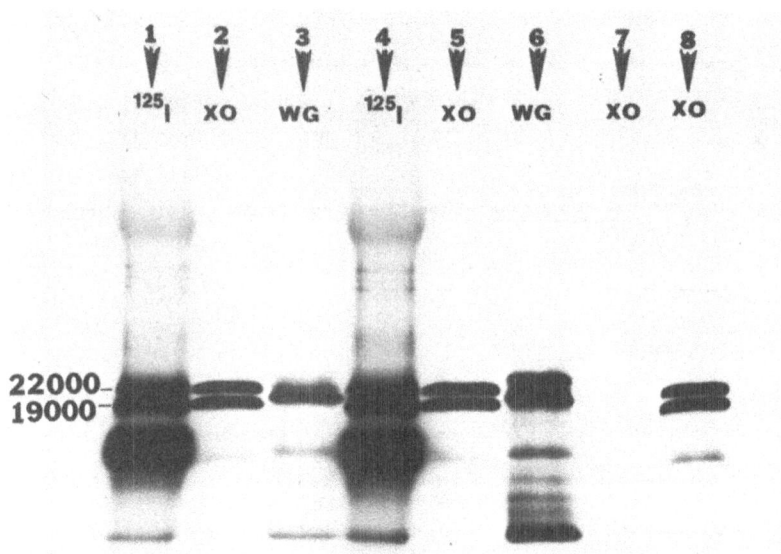

Fig. 5.   SDS-polyacrylamide gel analysis of 70% ethanol-soluble proteins synthesized in Xenopus oocytes. ^{125}I-labeled native zein proteins as well as normal and opaque-2 translation products from a wheat germ cell-free system were analyzed for comparison. Samples 1, 4: ^{125}I-labeled normal zein; 2, 5, 8: ^{3}H leucine-labeled ethanol-soluble proteins from oocytes injected with normal zein mRNAs; ^{3}H leucine-labeled products of normal and opaque-2 zein mRNA translation, respectively, from a wheat germ cell-free system; 7, ^{3}H leucine-labeled ethanol-soluble proteins from oocytes without injected zein mRNAs (number of oocytes equivalent to samples 2, 5, and 8).

in cell-free systems have been isolated from total RER (Larkins et al., 1976; Melcher, 1979) and the membranes surrounding protein bodies (Burr et al., 1978; Viotti et al., 1979). As has been found for other eukaryotic mRNAs, zein mRNAs are polyadenylated, monocistronic, and appear to be capped, since their translation is inhibited by the cap analogue [7]mGp (Burr et al., 1978; Larkins et al., 1978).

That these mRNAs direct the synthesis of zein polypeptides is indicated by the alcohol solubility, amino acid labeling pattern, and cyanogen bromide cleavage pattern of the products. Analysis of the cell-free translation products on SDS-polyacrylamide gels indicates that each of the different mol wt zein components are synthesized in vitro (Fig. 5, samples 3, 6). The cell-free products also exhibit a banding pattern similar to native zein proteins on IEF gels (Fig. 6).

Interestingly, each of the zein proteins synthesized in the cell-free systems are 2000 mol wt larger than native zein polypeptides (Fig. 5, cf. samples 3, 6, and 1, 4) suggesting that the mRNAs direct the synthesis of precursor polypeptides (Burr et

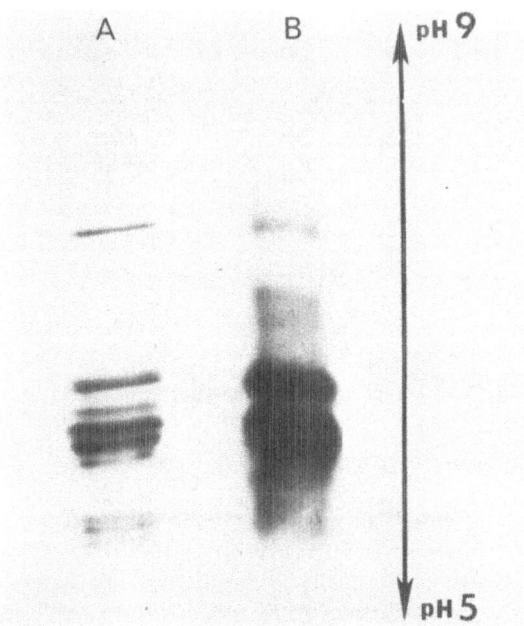

Fig. 6.   Isoelectric focusing of native zein proteins and zein proteins synthesized in a wheat germ-cell free system. Zein proteins synthesized in vitro were labeled with ^{3}H leucine, and the radioactive proteins were detected by fluorography.

al., 1978; Larkins and Hurkman, 1978). This result is not entirely
surprising since mRNAs for several animal secretory proteins direct
the synthesis of precursor proteins in cell-free systems. In the
case of the animal proteins a hydrophobic N-terminal sequence called
a "signal peptide" is responsible for directing the transport of the
polypeptide through the membrane. This signal sequence is normally
removed by proteolytic cleavage as the nascent polypeptide is passed
through the membrane (Blobel, 1976). Cell-free protein synthesizing
systems do not contain the membranes required for this post translational modification. However, cell-free systems supplemented with
RER membranes and Xenopus laevis oocytes will process signal
sequences and compartmentalize the proteins inside membrane vesicles
(Blobel and Dobberstein, 1975; Jilka et al., 1979).

Zein mRNAs injected into Xenopus oocytes are translated for
prolonged periods, and zein proteins are the only 70% ethanol-
soluble proteins synthesized in these oocytes (Larkins et al., 1979).
A comparison of the mol wts of ^{125}I-labeled native zein proteins with
those synthesized in the oocytes and the wheat germ cell-free system
is shown in Figure 5. Zein proteins synthesized in the oocytes have
the same mol wts as native proteins, and are 2000 mol wt smaller

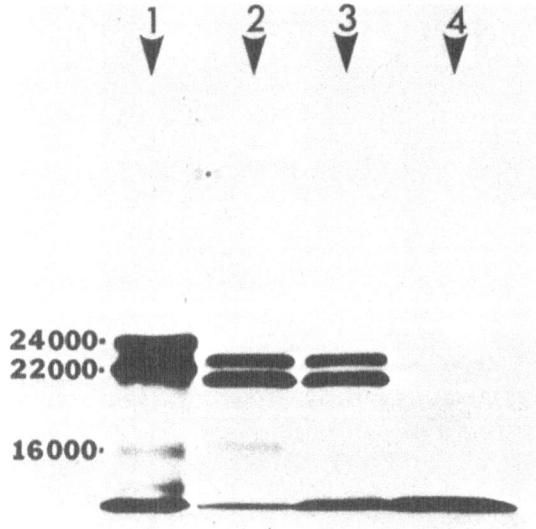

Fig. 7.   Evidence for compartmentalization of zein proteins in
          membranes of Xenopus oocytes. Sample 1, normal zein
          proteins synthesized in the wheat germ cell-free system;
          Sample 2, oocyte membranes extracted directly with 70%
          ethanol; Sample 3, oocyte membranes treated with 150 ug/ml
          protease K for 30 min and extracted with 70% ethanol;
          Sample 4, oocyte membranes treated with protease K and
          0.1% SDS for 30 min and extracted with 70% ethanol.

than the proteins synthesized in the wheat germ system. We also
found that zein proteins synthesized in the oocytes are in
protease-resistant membrane vesicles that have densities similar
to protein bodies from maize endosperm (Fig. 7).

Burr et al. (1978) previously reported that peptides associated
with signal sequences could be identified among the cyanogen
bromide cleavage products of zein polypeptides. We therefore
compared the mol wts of peptides generated by cyanogen bromide
cleavage of zein proteins synthesized in the oocyte and wheat
germ systems to determine if differences in mol wt could be
detected. To aid in the identification of fragments derived from
specific zein components, proteins whose synthesis was directed
by mRNAs from the normal and opaque-2 versions of W64A were
compared. As previously reported (Larkins et al., 1978), mRNAs
from the opaque-2 mutant do not synthesize detectable amounts of
the larger zein component. By comparing the cleavage products
of normal with those of opaque-2 it should be possible to determine
which fragments were derived from different mol wt zein components.

A comparison of the proteins synthesized in the wheat germ
and oocyte systems and their cyanogen bromide cleavage patterns is
shown in Figure 8. Although zein proteins treated with cyanogen
bromide showed a complex pattern, by comparing products of
opaque-2 mRNA translation with normal it was possible to determine
the origin of some of the fragments. Contrary to previous results
(Burr et al., 1978; Melcher, 1979) cyanogen bromide treatment
resulted in cleavage of some of the 19,000 mol wt component.
These peptides (Fig. 8, a, b) were present among the cleavage
products of opaque-2 as well as normal samples. Two distinct
polypeptides at positions e and f were among the cleavage products
of normal mRNA. Although both of these were presumably derived
from the larger mol wt component in normal samples, the similarity
of their mol wts in both oocyte and wheat germ samples precludes
concluding that either were associated with signal peptides.
While the cyanogen bromide cleavage patterns show relationships
between the zein proteins synthesized in oocytes and wheat germ
extracts, it is not possible to conclude which, if any, of the
peptides fragments were associated with signal sequences.

Zein mRNA Sequence Complexity and Gene Frequency

We have synthesized DNAs complementary to zein mRNAs to
analyze zein mRNA sequence complexity, and the frequency of zein
genes in the maize genome (Pedersen et al., 1979). As illustrated
in Fig. 9, zein cDNA/mRNA hybridization shows greater sequence
complexity than ovalbumin which was used as a standard. The
ovalbumin cDNA/mRNA hybridization followed a normal pseudo-first-
order reaction with 90% of the hybridization occurring within 1.5
decades of Rot. The zein hybridization encompasses 2.5 decades of

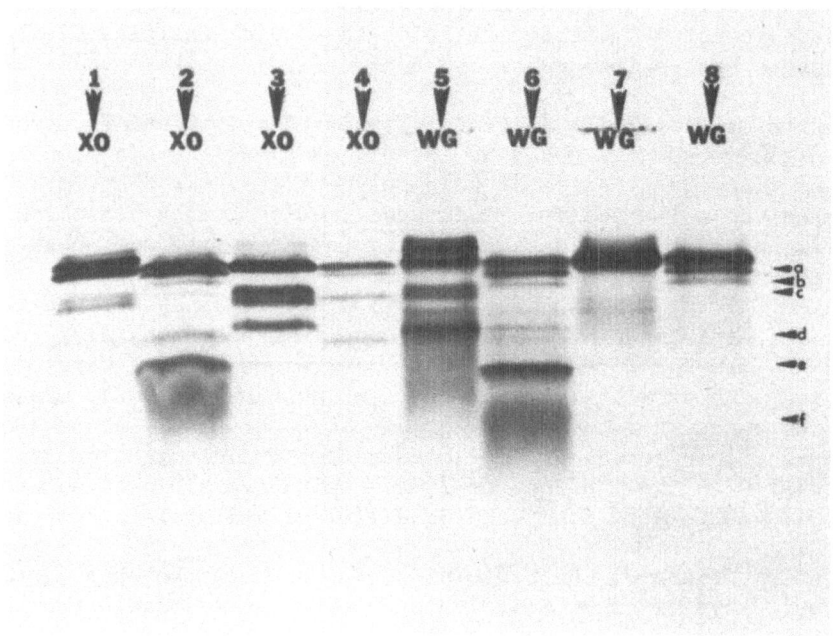

Fig. 8.   SDS-polyacrylamide gel analysis of cyanogen bromide
cleavage products.   Each sample contained approximately
75,000 cpm of ^{3}H leucine.   Samples 1, 2: normal zein
mRNA translation products from oocytes before and after
cyanogen bromide cleavage; 3, 4: opaque-2 mRNA
translation products from oocytes before and after
cyanogen bromide cleavage; 5, 6: normal zein mRNA
translation products before and after cleavage; 7, 8:
opaque-2 mRNA translation products from wheat germ before
and after cleavage.

Rot, and while it is possible to force the data to fit a single
component, a lower root mean square is obtained with a two component
fit. Based on the two component analysis, the first component has
a complexity of 190 and the second component has a complexity of
3753. Since the zein mRNAs have a complexity around 1100
nucleotides, this results suggests that the second component may
consist of about 4 different mRNAs having different sequence
complexities. The lower complexity component may be due to some
sequence redundancy within the mRNAs, however, we must point out
that this is only one possible explanation of the data.

By analyzing the hybridization of these cDNAs with maize DNA
we determined the frequency of zein genes within the genome.
Figure 10 shows that maize DNA reassociation can be divided into
three components which correspond to the repetitive, middle

repetitive, and single copy components of the DNA. The reassociation
of B. subtilis DNA was monitored as an internal standard to correct
for any factors that might alter the rate of DNA reassociation
(Fig. 10B).   Under the conditions used in this reaction the reassociation of B. subtilis DNA was retarded by a factor of 1.45.   Using
this factor to correct the reassociation rate of the maize single
copy component, and comparing the rate and complexity of B. subitilis
DNA, we calculate a complexity of 5.0×10^9 nucleotide pairs for the
haploid maize genome.   This value is equivalent to 5.3 pg of DNA
per haploid nucleus, which compares favorably with a reported value
of 5.0 pg (Phillips et al., 1974).

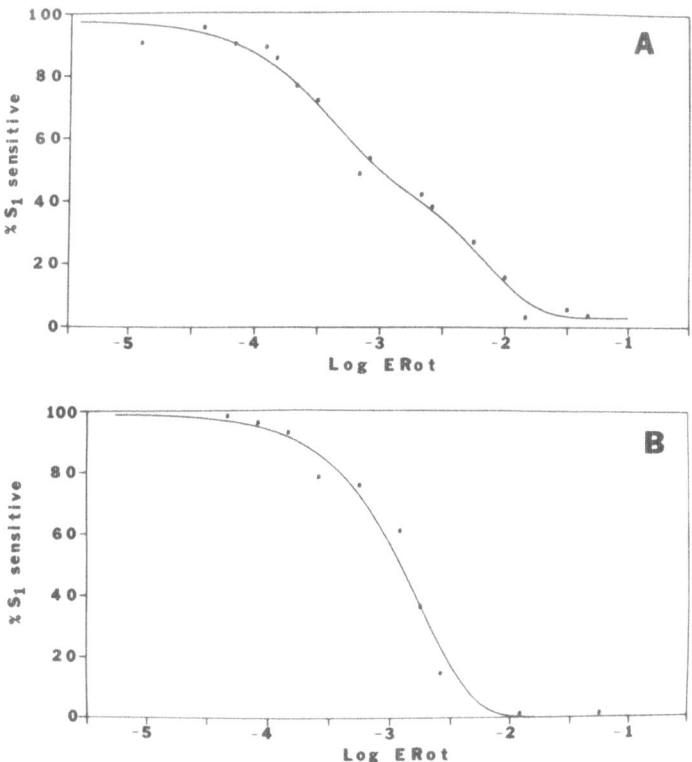

Fig. 9.   Comparison of zein and ovalbumin mRNA/cDNA hybridizations.
           Messenger RNAs were purified as described (Pederson et al.,
           1979), and cDNAs synthesized with AMV RNA-dependent reverse
           transcriptase.   Hybridization reactions contained 1000 cpm
           of [3]H cDNA and excess zein or ovalbumin mRNAs.   The curves
           represent the best least-squares fit for pseudo-first-
           order reactions.   (A) zein mRNA/cDNA fit to two components.
           (B) ovalbumin mRNA/cDNA fit to one component.

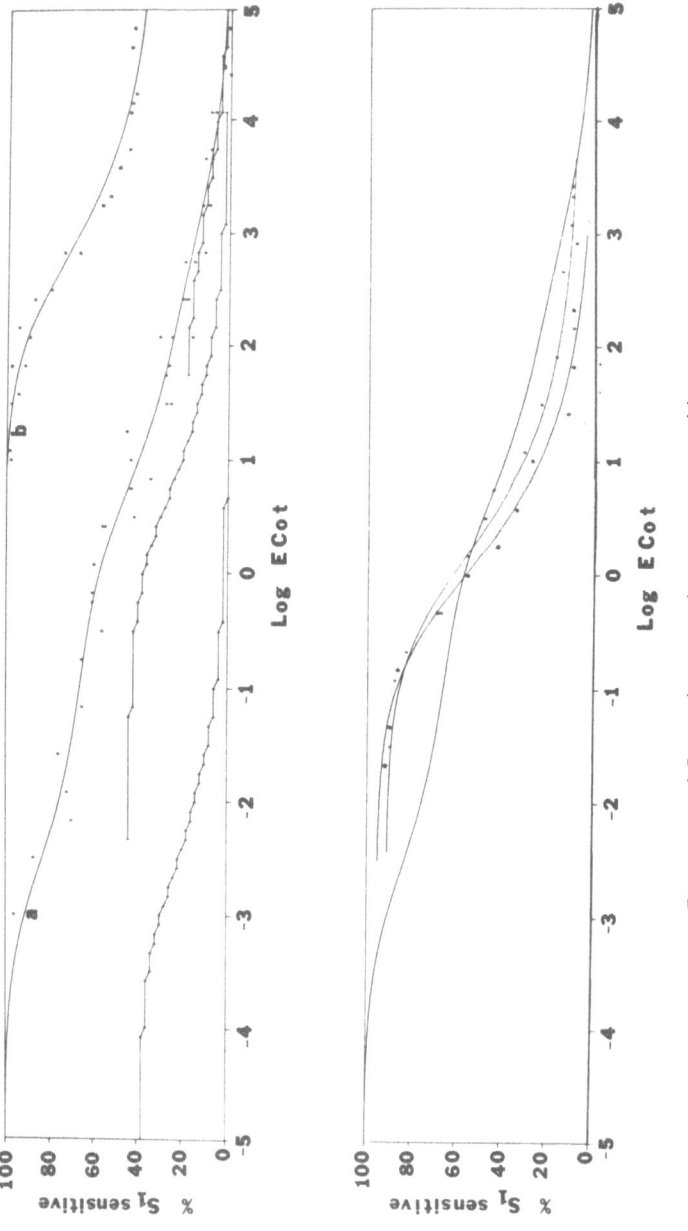

Figure 10.   Legend on preceding page

The cDNA prepared from zein mRNAs hybridize with the genomic DNA at a $Cot_{1/2}$ of 1036 which is half the value of 2317 determined from the reassociation of the single copy DNA component. This result indicates that there are possibly two, but probably no more than five copies of zein genes present in the genome; although large quantities of zein are synthesized during endosperm development, the zein genes are single copy rather than repetative genes.

Fig. 10.   Analysis of maize DNA reassociation and zein cDNA/DNA hybridization kinetics. DNA isolated as described (Pedersen et al., 1979), was banded in CsCl gradients and sheared to a single strand length of 300 nucleotides. The length of cDNAs was also 300 nucleotides. A: Reassociation of maize DNA in 0.6M NaCl. The curve through the data points (a) represents the best least-squares solution using three second order components (-.-.-.-.-). The second curve (b) shows the hybridization of zein cDNAs with the DNA. Opaque-2 cDNAs (500 cpm, 26 pg) were hybridized in 10-12 mg/ml DNA providing a driver sequence excess of 3:1. B: B. subtilis DNA was labeled with ^{32}P by nick translation, and used as a tracer in a B. subtilis reassociation and as an internal standard in the high ECot range of the maize DNA reassociation. The reassociation of B. subtilis tracer was retarded by a factor 1.45 in maize DNA compared to B. subtilis alone.

Summary

When maize storage proteins were analyzed by SDS-polyacrylamide gel electrophoresis two major components of 19,000 and 22,000 mol wt, and two minor components of 15,000 and 10,000 mol wt, were resolved. Each of the major components is composed of several homologous polypeptides of slightly different charge. Based on the N-terminal sequence analysis we concluded there was a minimum of two different polypeptides within each of the major components of the inbred W64A. However, additional polypeptides may also be present, since each component shows three of four bands on the IEF gels. The 19,000 mol wt component contained two polypeptides cleaved by cyanogen bromide as well as some protein that was not cleaved, indicating it consisted of at least three polypeptides.

The structural heterogeneity detected among zein proteins was reflected in the analysis of zein mRNA sequence complexity. The reaction of zein mRNAs and cDNAs was more complex than a single mRNA such as ovalbumin. Based on computer analysis of the reassociation kinetics of zein mRNAs, we calculated that there was sufficient sequence complexity to account for five different zein mRNAs. The similarity of the $Cot_{1/2}$ of the cDNA/DNA hybridization and the $Cot_{1/2}$ of the single copy DNA component indicated that zein genes are present in relatively few copies in the maize genome.

Translation of zein mRNAs in cell-free systems produced proteins that are 2000 mol wt larger than native zein polypeptides. However, translation of the mRNAs in vitro on RER membranes or in Xenopus oocytes produced proteins of the same mol wt as native zein polypeptides. These results suggested that zein mRNAs directed the synthesis of prezein proteins containing "signal sequences", similar to those which have been demonstrated for animal secretory proteins. The prezein peptides appear to be processed as the protein enters the RER, since processed polypeptides were recovered from protease-resistant membrane vesicles in Xenopus oocytes.

This research was supported in part by grant PCM 77-24210, National Science Foundation and GM 2504501, National Institutes of Health.

REFERENCES

Blobel, G., and Dobberstein, G., 1975, Transfer of proteins across membranes. II. Reconstitution of functional rough microsomes from heterogeneous components, J. Cell Biol. 67, 852.

Blobel, G., 1976. Synthesis and segregation of secretory proteins: The Signal hypothesis. in International Cell Biology. B. R. Brinkley and K. R. Porter, eds. The Rockefeller Univ. Press. p. 318.

Burr, B. and Burr, F. A., 1976, Zein synthesis in maize endosperm by polyribosomes attached to protein bodies, Proc. Natl. Acad. Sci., USA, 73, 515.

Burr, B., Burr, F. A., Rubenstein, I., and Simon, M. N., 1978, Purification and translation of zein messenger RNA from maize endosperm protein bodies. Proc. Natl. Acad. Sci., USA, 75, 696.

Fraij, B. and Melcher, U., 1978, Proteins of the zein extracts of corn, Plant Physiol. 61, 40.

Gianazza, E., Viglienghi, V., Righetti, P. G., Salamini, F., and Soave, C., 1977, Amino acid composition of zein molecular components, Phytochemistry 16, 315.

Jilka, R. L., Familletti, P., and Pestka, S., 1979, Synthesis and processing of mouse MOPC-321 k chain in Xenopus laevis oocytes, Arch. Biochem. Biophys. 192, 290.

Larkins, B. A., Jones, R. A., and Tsai, C. Y., 1976, Isolation and in vitro translation of zein messenger ribonucleic acid, Biochemistry 15, 5506.

Larkins, B. A. and Hurkman, W. J., 1978, Synthesis and deposition of zein in protein bodies of maize endosperm, Plant Physiol. 62, 256.

Larkins, B. A., Pearlmutter, N. L., and Hurkman, W. J., 1978, The mechanism of zein synthesis and deposition in protein bodies of maize endosperm. in The Plant Seed: Development Preservation, and Germination. Rubinstein, Phillips, Green, and Gengenbach, eds. Academic Press, N. Y. (in press).

Larkins, B. A., Pedersen, K., Hurkman, W. J. and Smith, L. D.,

1979, Synthesis and processing of maize storage proteins in Xenopus laevis oocytes. Natl. Acad. Sci., USA. (submitted)

Lee, K. H., Jones, R. A., Dalby, A., and Tsai, C. Y., 1976, Genetic regulation of storage protein content in maize endosperm, Biochem. Genet. 14, 641.

Melcher, U., 1979, In vitro synthesis of a precursor to the methionine-rich polypeptide of the zein fraction of corn, Plant Physiol. 63, 354.

Mertz, E. T., Bates, L. S., and Nelson, O. E., 1964, Mutant gene that changes protein composition and increases lysine content of maize endosperm, Science 145, 279.

Nelson, O. E., Mertz, E. T., and Bates, L. S., 1965, Second mutant gene affecting the amino acid composition of maize endosperm proteins, Science 150, 1469.

Pedersen, K., Bloom, K., Anderson, J. N., Glover, D. V., and Larkins, B. A., 1979, Analysis of the complexity and frequency of zein genes in the maize genome, Biochemistry (submitted).

Phillips, R. L., Weber, D. F., Kleese, R. A., and Wang, S. S., 1974, The nucleolus organizer of maize (Zea mays L.): Tests for ribosomal gene composition or magnification, Genetics 77, 285.

Righetti, P. G., Gianazza, E., Viotti, A., and Soave, C. 1977, Heterogeneity of storage proteins in maize, Planta, 136, 115.

Soave, C., Suman, N., Viotti, A., and Salamini, F., 1978, Linkage relationships between regulatory and structural gene loci involved in zein synthesis in maize, Theoret. Appl. Genetics. 52, 263.

Viotti, A., Sala, E., Alberi, P. and Soave, C., 1978, Heterogeneity of zein synthesized in vitro, Plant Sci. Lett., 13, 365.

RECENT EVIDENCE CONCERNING THE GENETIC REGULATION OF ZEIN SYNTHESIS.

Carlo Soave*, Angelo Viotti*, Natale di Fonzo**, Fran-

cesco Salamini**

*Laboratorio Biosintesi Vegetali, C.N.R. Milano, Italy
and
**Istituto Sperimentale Cerealicoltura,Bergamo, Italy

Recently it has been proposed that certain groups of functionally related RNAs or proteins may be encoded by families of genes
which are physically clustered and evolutionarily related[1]. The
18S, 28S and 5 S RNA genes, together with those for immunoglobulins,
histones, and, possibly, actins, keratins and silkmoth chorion proteins are convincing examples of this type of gene family [1-4].
Also the reserve proteins accumulated in the cereal endosperm could
be encoded by multigene families.

Zein, the major storage protein of maize endosperm, consists
of a group of polypeptides which can be partially resolved by gel
electrophoresis [5]. On SDS gels we obtain three to five major clas
ses with molecular weight ranging from about 23000 to 19000 and on
IEF gels from 8 to 15 bands in the pH range 6-9 (Fig.1). Using two-
dimensional analysis up to 25 components are detectable [6,7]. Moreo
ver, a large variability in the zein electrophoretic patterns exists among different maize genotypes [8]. The evidence suggests that
the zein polypeptides are the products of non identical,but homolo
gous, structural genes. For example, differences between the amino
acid composition of IEF and SDS zein components were observed [9];
the products synthesized in vitro by polysomes engaged in zein syn
thesis exhibit the same SDS or IEF heterogeneity as native zein[10];
the zein messenger RNAs is small sized and appear to be monocistro
nic [11]. Furthermore, in crosses between inbreds with different zein
patterns, the seeds of the two reciprocal F1s show additive patterns
and the amount of each component correlates to the gene dose present

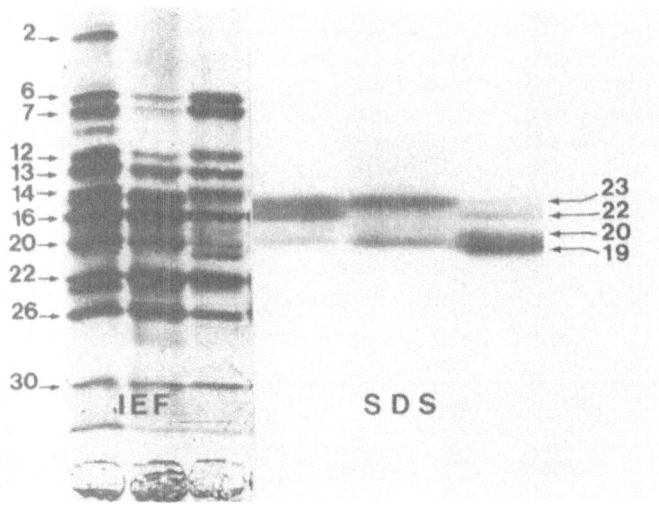

Fig. 1 - Isoelectric-focusing (IEF) and SDS polyacrylamide electro-
         phoresis of zein extracted from three different maize ge-
         notypes. The numbers on the left indicate the position of
         some IEF zein polypeptides, while those on the right repre-
         sent the molecular weight of the major zein subunite.

When seeds from the F2 generation are analyzed,simple Mendelian se
gregations of zein IEF bands are observed [12]. Moreover for a more
detailed characterization of the individual members of the zein fa
mily, the tryptic-chimotryptic peptide maps from different zein IEF
bands can be considered. Total zein produces more than 40 spots on
bidimensional analysis, though, judging from its amino acid composition, only about 20 peptide bonds should be enzymatically cleaved
to give rise to a maximum of 21 spots [13]. Furthermore, in our maps
not all spots have a similar intensity indicating that some peptides are present in less than molar amount per mole of polypeptide.
When we compare the chromatographic peptide maps obtained from 12
zein IEF bands, we observe that some spots are always present,whi-
le others are not and that some differ in intensity. The results
indicate that the zein polypeptides vary to a certain extent in their
primary sequence and thus must be the products of different genes.
    In favour of the zein multigene family hypothesis there are
also the experiments on hybridization studies among zein cDNA and
zein mRNA or total maize DNA. Figure 2 reports the hybridization
kinetics of zein cDNA to an excess of zein mRNA, in parallel to
those of α and β human globin cDNA/mRNA. The insert shows the products synthesized in vitro by a cell-free translation system direc

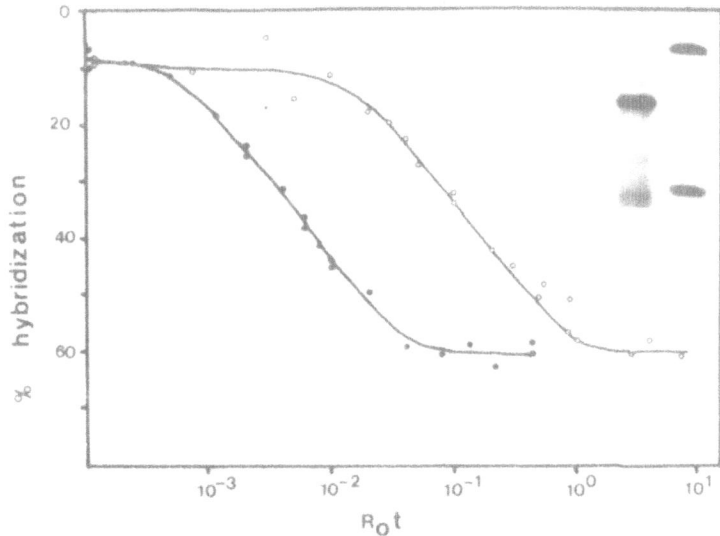

Fig. 2 - Hybridization of zein mRNAs (o) and rabbit α-β globin
mRNAs (●) to their respective cDNAs. The reaction was
carried out in 50% formamide, 0.5 NaCl, 25mM HEPES,0.5
mM EDTA, pH 6.8 at 43°C . The hybrid formed was tested
as S1 resistent material. The insert show the products
synthesized in vitro by a wheat germ cell-free system
supplemented with purified zein mRNA and analyzed by
SDS electrophoresis and fluorography. Concanavalin A
was used as molecular weight marker(right).

ted by our zein mRNA. Only zein proteins are synthesized indicating
the purity of the mRNA preparation. The observed Rot 1/2 of 8×10^{-2}
for the zein system, compared with that of 5×10^{-3} for the globin
system, indicates the presence in zein mRNA of about 13-16 non cross
reacting sequences. Furthemore, the sharp transition obtained is
characteristic of a mRNA population in which the different species
are represented roughly in equal quantities. The results of a saturation experiment among increasing amounts of zein cDNA and embryo-
DNA are shown in Fig. 3. At saturation, the quantity of cDNA hybridized to 3 μg od DNA is about 33 pg, a value which, taking into account the maize genome complexity (2.7×10^{12}) and the molecular
weight of the cDNA (1.8×10^5), indicates the presence of about 120
genes per haploid genome.
    By formal genetic analysis we demonstrated that the majority
of the genes which codify for zein IEF components are located on
the short arms of chromosome seven and four, and some of them are
clustered [12-14]. We report here new data concerning the genes loca-

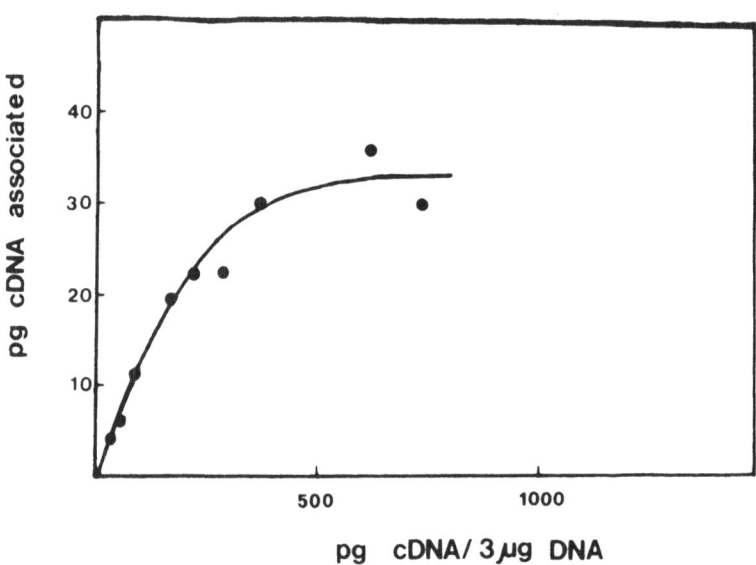

Fig. 3 - Saturation analysis of the association between zein cDNA
and maize embryo DNA. The association was carried out at
62°C in 0.12M NaPO4 and 1 mM EDTA buffer pH 6.8 with a
DNA concentration of 2.5 mg/ml for 1800 min. The hybrid
was tested by the S1 assay. A background of 1.2% of S1
resistance was subtracted from each point.

ted on chromosome 7. The inbred A69Y normal, containing the zein
IEF bands N° 1, 2, 3, 21, and 29 (genotype Zp1, Zp2, Zp3, Zp21 and
Zp29, Zp is for zein polypeptide) was crossed with A69Y o2 vp9
which lacks these IEF bands. The endosperm of 454 kernels from the
segreganting F2 ears were classified with respect to the normal ,
opaque-2 and viviparous-9 phenotype and, then, zein analyzed by IEF
for the presence absence of· the bands N° 1, 2, 3, 21, 29. At all
times the bands N°1, 2, 3 showed a coordinated presence or absence
as also did bands 21 and 29, indicating a close linkage between the
genes Zp1, Zp2, Zp3 and Zp21 and Zp29. Table 1 reports the linkage
relationships between the zein genes studied and the genetic markers o2 and Vp9 (chr.7 position 16 and 25 respectively). The genes
Zp21 and Zp29 are strictly linked to o2, while the genes Zp1, Zp2
and Zp3, which were previously located at about 5.5 map units from
o2 12 are more closely linked with the locus Vp9. Then, at least
two clusters of zein genes are located on chr.7, one near o2 and
the other near vp9. From other crosses, additional zein genes were
associated with these two clusters: the genes Zp6 and Zp16 are lin
ked to the clusters Zp21-29 and Zp 1-2-3 respectively.

Table 1.   Linkage relationships among the genes $\underline{Zp1}$, $\underline{Zp2}$, $\underline{Zp3}$,
           $\underline{Zp21}$, $\underline{Zp29}$ and the genes $\underline{O2}$ and $\underline{Vp9}$ from the cross
           A69Y normal x A69Yo2vp9[a].

Segregating genes	cM±s.e
O2-Zp1,Zp2,Zp3	12.8+1.6
O2-Vp9	9.0+1.4
Vp9-Zp1,Zp2,Zp3	3.7+0.9
0.2-Zp21,Zp29	0
Vp9-Zp21,Zp29	9.0+1.4
Zp1,Zp2,Zp3-Zp21,Zp29	12.8+1.6

[a]Data from 454 F2 seeds.

A particular feature of the zein system is the synchronous accumulation of all the zein polypeptides during the endosperm development. This coordinate expression is under the control of an integrated regulatory system. We have studied in detail the action of
the mutant alleles at four loci ($\underline{o2}$, $\underline{o7}$, $\underline{f12}$, De*30) which influence the rate of accumulation of the zein components in a specific
way: $\underline{o2}$ and $\underline{De*30}$ preferentially repress the heavier zein polypeptides, $\underline{o7}$ the lighter and f12 all the subunits to the same extent
6,15.

     The interaction of $\underline{f12}$ and $\underline{o7}$ with $\underline{o2}$ in complete dosage com-
bination were investigated. The $\underline{f12}$ repression of zein accumulation
is correlated to the amount of mutant allele present in the endosperm. However, in the presence of three doses of the recessive
allele of the $\underline{o2}$ locus, $\underline{f12}$ action is completely over-shadowed and
only the $\underline{o2}$ zein phenotype is maintained. Then $\underline{o2}$ appears to be
epistatic on $\underline{f12}$. $\underline{o2}$ and $\underline{o7}$ are both recessive: in the double mutant combination; both the alleles are active reducing zein additi-
vely, the first acting on the heavier and the second on the lighter
zein subunits. Then, it seems that at least two regulatory path -
ways, related to the control of the synthesis of the higher and
lower molecular weight zein subunits are active in maize endosperm.
$\underline{o2}$ and $\underline{o7}$ are involved respectively in the first and second pathways. $\underline{f12}$ seems instead to belong to the first pathway acting down-
stream or upstream to the $\underline{o2}$ locus.
     A major question, however, remains, $\underline{i.e.}$, how these regulato-
ry loci effect zein synthesis. Some data are now emerging on the
action of the $\underline{O2}$ locus. Fig. 4 reports the ratio between the SDS
zein subunits of 23-22000 and 20-19000 daltons during the develop-

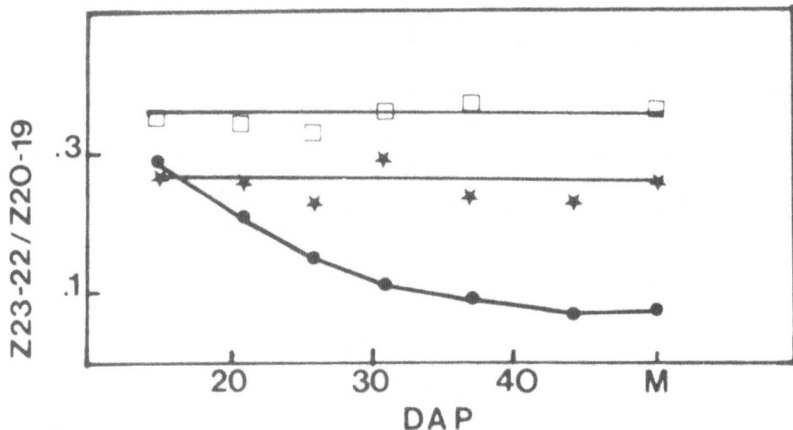

Fig. 4 - Zein extracted from normal (*), floury-2 (□) and opaque-2
        (●) endosperms at 15,21,26,31,37,45 days after pollination
        and at maturity (M) were electrophoresed in SDS-gels. After
        scanning of the gels at 600nm, the peak area of the zein
        SDS subunits 23+22000 and 20-19000 was calculated by trian
        gulation and their ratio reported in the graphic.

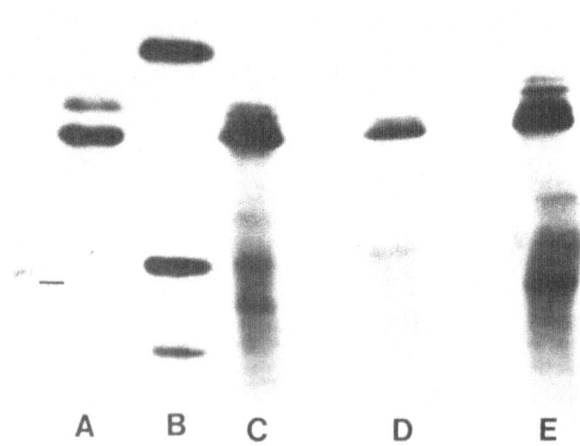

Fig. 5 -SDS-polyacrylamide gel analysis of zein synthesized in vi-
        tro by protein bodies-bound polyribosomes from normal (C),
        opaque-2 (D) and fluory-2 (E) 21 days old endosperms.
        Authentic zein (A) labelled with radioactive dansyl chlori
        de and ^{3}H-acetylated Concavavalin-A (B) were used as mar-
        kers.

Table 2.   Amount zein cDNA hybrid   :o
                                           a
             different total RNA pre   ons

	RNA	lybridization (%)
21 days	normal	77
	opaque-2	74
	floury-2	76
31 days	normal	72
	opaque-2	47
	floury-2	69

[a]Saturating amount of total RNA ha   added to a constant
quantity of cDNA transcribed from   normal zein mRNA . Incubation was carried out until cc   .on of the reaction.
The hybrid was measured as S1 nuc   resistant material.

ment of the endosperm in normal, f12   2 genotypes. In the normal and f12 samples the ratio does n   nge during development
indicating a synchronous accumulati   the two zein classes.In
o2 in the early phases of zein synth   from 15 to 20 days after
pollination) the ratio is similar to   found in the normal but
after that time progressively lowers   ating a decreasing synthesis of the heavier zein subunits.   her words, it means that
the action of the o2 allele progress   becomes more evident during endosperm development. Moreover   we analyse the products
synthesized in vitro by polysomes en;   in zein synthesis from
normal o2 and f12 endosperms collecte   n the repression of the
heavier components becomes evident ii   we note that all the zein
subunits are present in normal and f   ples, but only the lighter
ones in o2 (Fig.5). Thus from the 21:   following pollination
the o2 polysomes are no longer able   tain the zein 23-22000
synthesis. This inability is not due   absolute deficiency in
zein mRNA since by hybridizing zein   ranscribed from normal
zein mRNA) to a total RNA preparatior   normal, o2, or f12 21
day old endosperms, the same value of   idization is observed for
the three genotypes (Table 2). It is   ided that at this stage
all the zein mRNA sequences are prese   the o2 cells. Only at

later stages in development(31 days) the level of hybridizable cDNA
is lowered in o2, indicating the absence of some mRNA sequences.
These lacking sequences could be those responsible for the synthe
sis of the heavier zein polypeptides. Different models can fit the
above reported data, in particular the presence in o2 at 21 days
of all the zein mRNAs sequences when the synthesis of the heavier
zein polypeptides is already inhibited. One possibility could be
that of a translational block of the 23-22000 subunits followed by
a degradation of the untranslated mRNAs.

     A final point merits some attention. Apparently,in the segment
of chr.7 between the O2 and Vp9 loci, two clusters of zein genes
lie near two regulatory elements: O2 and D*30 which was recently
mapped at about 5 map units from O2 and is strictly linked with
Zpl, Zp2, Zp3. It is difficult to believe that these associations
are casual, especially if we consider that the zein genes located
on chr.4 and 10 lie respectively near the two other regulatory ele
ments f12 and o7.

REFERENCES

(1) Hood,L., Campbell,J.H. and Elfin,S.C.R., (1975)
     Ann. Rev. Genet. 9, 305-353.
(2) Mckeown,M., Taylor,W.C., Kindle,K.L., Firtel,R.A., Bender,W.
     and Davidson,N. (1978) Cell 15, 789-800.
(3) Fuds,E., Green,H., (1978) Cell 15, 887-897.
(4) Goldsmeith,M.R., Basehoar,G., (1978) Genetics 90, 291-310.
(5) Righetti,P.G., Gianazza,E., Viotti,A., Soave,C., (1977)
     Planta 136, 115-123.
(6) Soave,C., Viotti,A., Di Fonzo,N., Salamini,F., (1979)
     Seed Protein Improvement in Cereals and Grain Legumes. FAO IAEA
     eds., Vienna, 165-174.
(7) Miflin,B.J., Shewry,P.R., (1979) Seed Protein Improvement in
     Cereals and Grain Legumes. FAO-IAEA eds., Vienna, 137-158.
(8) Nucca,R., Soave,C., Motto,M., Salamini,F., (1978)
     Maydica 23, 239-249.
(9) Gianazza,E., Viglienghi,V., Righetti,P.G., Salamini,F., Soave,
     C., (1977) Phytochemistry 16, 315-317.
(10) Viotti,A., Sala,E., Alberi,P., Soave,C., (1978)
     Plant. Sci. Lett. 13, 365-375.
(11) Wienand,V., Feix,G., (1978) Eur.J.Biochem. 92, 605-611.
(12) Soave,C., Suman,N., Viotti, A., Salamini, F., (1978) Theoret.
     Appl. Genet. 52, 263-267.
(13) Fraj,B., Melchers,V., (1979). Biochim. Biophys. Acta, in press.
(14) Valentini,G., Soave,C., Ottaviano,E., (1979) Heredity,in press.
(15) Salamini,F., Di Fonzo,N., Gentinetta,E., Soave,C., (1979) Seed
     Protein Improvement in Cereals and Grain Legumes. FAO-IAEA eds.,
     Vienna, 97-108.

THE CLONING OF ZEIN SEQUENCES AND AN APPROACH TO ZEIN GENETICS

Frances A. Burr and Benjamin Burr

Biology Department
Brookhaven National Laboratory
Upton, New York   11973

In addition to the cell biology of zein synthesis[1,2], we are interested in the genetics of zein structural genes. Initially we had hoped that there might be as few as two gene products. This was based on the fact that when reduced zein polypeptides are separated on the basis of molecular weight on sodium dodecyl sulfate-polyacrylamide gels two bands are resolved at Mr 19,000 and 22,500[2]. We have recently separated these two components preparatively and determined the amino terminal sequence for each[3]. While the two chains have about 30% conserved homology there was no indication of sequence heterogeneity within the first 35 residues of the light chain or the first 25 residues of the heavy chain. In contrast to these results, zein has long been reported to be heterogeneous with respect to charge[4]. Our own results with isoelectric focusing in the presence of urea and reducing agent indicate that there are usually 3 major components and a varying number of minor ones which differ with respect to charge. However, when the major charge components were separated preparatively on thin layer isoelectric focusing and re-examined on SDS-polyacrylamide gels both heavy and light chains were found in each of the three major bands. Twenty-one percent of the amino acid side chains of zein are amides[5] so it seemed reasonable to postulate that at least some of the charge heterogeneity could be caused by random deamidation. This would have a pronounced effect on zein chains separated on the basis of charge since there are relatively few charged side chains. Furthermore, zein, being extremely hydrophobic presents some problems for analytical separation since aggregates can form when the protein is not completely denatured. Another complication is that zein undergoes at least two post-translational modifications[1] which might further introduce heterogeneity. Despite these difficulties Soave and his colleagues[6] have demonstrated that there are at least three unlinked groups of electromorphs.

We have conducted a rather extensive search for zein structural gene mutations. We reasoned that whereas charge changes can be introduced post-synthetically, frame shift or nonsense mutations would reduce the molecular weight or condition the absence of a polypeptide when analyzed on SDS-polyacrylamide gels and undeniably represent alterations in the structural gene. Some opaque and floury mutants, none of which are thought to be zein structural gene mutations, can cause their endosperm phenotype by reducing the amount of zein accumulated. Structural gene mutations might also reduce the amount of zein in the kernel and thereby also produce an opaque phenotype. Kernel and plant phenotype mutants of maize are easily induced by treating pollen with ethyl-methane sulfate or N,N-nitrosoguanidine and examining segregating M2 seed for recessive mutations[7]. Dr. Jerry Neuffer allowed us to examine his collection of a large number of M2 ears resulting from EMS mutagenesis. We chose some 300 independently derived opaques for further study. Zein was prepared from these and examined on SDS-polyacrylamide gels. Two of these samples had a slightly altered pattern on SDS electrophoresis and were chosen for further study. One was found to be associated with seedling lethality in the homozygous state and the second is being used to attempt to map a zein gene to chromosome arm with BA translocations.

We have also embarked on another approach to identify the zein structural genes: molecular cloning permits the preparation of homogeneous sequences in quantity for use as molecular probes. Quantitative hybridization will give an accurate estimate of the number of homologous sequences and the use of monosomic or trisomic stocks will allow the identification to chromosome.

We have made two preparations of double-stranded DNA complementary to zein mRNA which was subsequently inserted into a bacterial plasmid[8]. In the first instance zein mRNA was purified to apparent homogeneity from protein bodies of developing endosperm[9] and double-stranded DNA was synthesized and inserted by A/T tailing into the bacterial plasmid pMB9 by Tom St. John in the laboratory of Ron W. Davis at Stanford University. More recently in our laboratory, a second indosperm RNA fractionated on a denaturing sucrose gradient. Synthetic Eco RI linkers were attached and the DNA was inserted into the Eco RI site of the plasmid pBR325. Rapid lysates were prepared to identify plasmids with inserts. Plasmids containing zein mRNA sequences were identified by hybridizing endosperm RNA and translating the specifically bound mRNA. Linearized DNA was denatured and bound to diazotized aminobenzoyl cellulose[10]. RNA was hybridized with the immobilized DNA in formamide and the eluted RNA was translated in the wheat germ cell-free system. Zein cDNA clones were identified by those which bound light or heavy chain zein mRNAs. Similar results have also been reported by Udo Weinand and colleagues[11].

To show that the cloned sequences were homologous to zein mRNA for most or all of their length, experiments similar to those described by Berk and Sharp[12] were conducted. mRNA was hybridized with plasmid DNA under conditions which favor RNA-DNA hybrids. The single-stranded nucleic acids were removed with S1 nuclease and either a) the RNA-DNA duplexes were displayed on a neutral agarose gel (Fig. 1) or b) the protected DNAs were separated on an alkaline agarose gel, blotted on nitrocellulose paper, and detected by hybridization with nick translated cloned sequences. Both methods indicated that most or all of the cloned sequences are complementary to zein mRNA.

Examination of internal restriction endonuclease sites of the cloned sequences indicates that there are possibly a minimum of three heavy chain sequences and four light chain sequences. These results are being confirmed by hybridization to restriction cut DNA immobilized on filters, and more extensive restriction mapping of the cloned sequences themselves.

If these preliminary results are borne out, the consequences for mutation breeding to improve endosperm protein quality by intervention at the zein structural gene loci become problematical. If there are multiple genes, detection of amino acid substitutions at the various loci and recombination to create multiple substitutions at a given locus would be nearly impossible by conventional techniques. Furthermore, should a method be found to accomplish this goal, and the multiple genes turn out to be unliked, as seems likely, their uniform transfer to inbred backgrounds for the production of hybrid seed adapted to various environments would be very difficult to achieve.

Despite this there are some interesting problems to be investigated. It will be important to see how nucleic acid sequences have diverged rather rapidly relative to the protein seuqences and to examine the divergence at the gene level by using the cloned cDNAs as probes to isolate genomic clones.

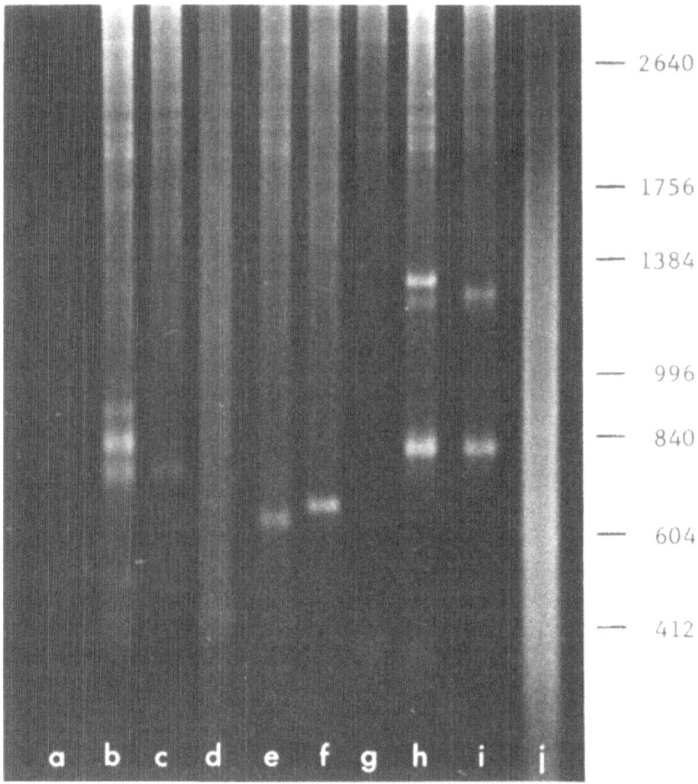

Fig. 1.   RNA-DNA duplexes formed by hybridizing zein mRNA to
recombinant plasmid DNA in 80% formamide buffer and
treated with S1 nculease.   The remaining oligonucleotides were separated on a neutral 1.6% agarose slab gel
and visualized by ethidium bromide staining.   Channels
are DNA from clones: a) A4, b) B59, c) B54, d) B59, e)
B41, f) B40, g) B36, h) B19, i) A30, j) pMB9.   Nine out
of 16 of the plasmids gave duplex bands apparently 70 to
300 base pairs longer than the length of the inserts
including A/T tails.   The remaining 7 gave duplex 145 to
270 base pairs shorter than the total insert length.
However, it is not known whether duplex RNA-DNA hybrids
prepared from a possible heterogeneous RNA population
can be readily sized relative to duplex DNA standards.

Supported by U. S. Department of Energy, by Grant GM 24057 from the National Institutes of Health, and by the Science and Education Administration of the U. S. Department of Agriculture under Grant No. 5901-0410-8-0023-0 from the Competitive Research Grants Office.

1. F. A. Burr and B. Burr, Molecular basis of zein protein synthesis in maize endosperm, in: "The Plant Seed: Development, Preservation, and GErmination," I. Rubenstein, ed., Academic Press, New York (1979).

2. B. Burr and F. A. Burr, Zein synthesis in maize endosperm by polyribosomes attached to protein bodies, Proc. Nat. Acad. Sci. U.S.A. 73:515 (1976).

3. L. Silver, M. Elzinga, F. A. Burr and B. Burr, Zein storage protein mRNA translation and processing in Xenopus oocytes (in preparation).

4. J. Mossé, Alcohol-soluble proteins of cereal grains, Fed. Proc. 25:1663 (1966).

5. J. S. Wall, Cereal proteins, in "Proteins and Their Reactions, Symposium on Foods," H. W. Schultz and A. F. Anglemeir, eds., Avi Publishing Co., Westport, Conn. (1964).

6. C. Soave, N. Suman, A. Viotti and F. Salamini, Linkage relationships between regulatory and structural gene loci involved in zein synthesis in maize, Theor. Appl. Genet. 52:263 (1978).

7. M. G. Neuffer, Chemical mutagens in mineral oil very effective on corn pollen, Maize Genet. Coop. Newsletter 42:124 (1968).

8. B. Burr, F. A. Burr, T. P. St. John, M. Thomas, and R. W. Davis, Corn storage protein DNA coding sequences cloned in bacteria (in preparation).

9. B. Burr, F. A. Burr, I. Rubenstein, and M. N. Simon, Purification and translation of zein messenger RNA from maize endosperm protein bodies, Proc. Nat. Acad. Sci. U.S.A. 75:696 (1978).

10. B. F. Noyes and G. R. Stark, Nucleic acid hybridization using DNA covalently coupled to cellulose, Cell 5:301 (1975).

11. U. Weinand, C. Brüschke, and G. Feix, Cloning of double stranded DNAs derived from polysomal mRNA of maize endosperm: isolation and characterization of zein clones, Nuc. Acids. Res. 6:2707 (1979).

12. A. J. Berk and P. A. Sharp. Sizing and mapping of early adenovirus mRNAs by gel electrophoresis of S1 endonuclease-digested hybrids, Cell 4:721 (1977).

THE SYNTHESIS OF BARLEY STORAGE PROTEINS

Benjamin J. Miflin, Jayne A. Matthews, Shirley R.
Burgess, Audrey J. Faulks, and Peter R. Shewry

Rothamsted Experimental Station
Harpenden,
Herts, AL5 2JQ

THE PROTEINS

Characterisation

The storage protein fraction of barley, called hordein, is classified as a prolamin. It constitutes some 35-55% of the proteins of the endosperm depending on the variety and N nutrition of the plant[1]. The prolamins are characterized by their extractability and solubility in alcoholic solvents which increases in the presence of reducing agents, their amino acid composition and their presence in protein bodies (see[2] for a review). In our laboratory hordein is routinely extracted in 55% propan-2-ol and 2% 2-mercaptoethanol and the cysteine residues alkylated with 4-vinyl pyridine [3,4] before further characterisation. The component polypeptides are then separated by either sodium dodecylsulphate polyacrylamide gel electrophoresis (SDS-PAGE), urea-PAGE or isoelectric focusing (IEF) in one dimension or by IEF/SDS PAGE or urea-PAGE/SDS-PAGE in 2-dimensional systems [3,4,5,6]. Based on electrophoretic separations hordein was originally separated into $\alpha, \beta, \gamma, \delta$ and ε fractions[7,8] although now this is usually simplified into A, B and C fractions[9,5]. The A fraction is relatively high in lysine and is probably not a true storage protein (see discussion in[10]). The B and C fractions have been separated and purified and various of their characteristics determined[11] (Table 1). Perhaps the most noticeable points are the large content of proline and the virtual absence of S-amino acids in the C fraction and the inconsistency in the molecular weight determinations between SDS-PAGE and equilibrium ultracentrifugation; this has been noted for other cereal proteins[12] and since the latter method is the one of choice indicates the unreliability of SDS-PAGE for mol. wt. determination of these types of protein.

Table 1

Comparison of 'B' and 'C' Hordein Fractions

	'B' hordein	'C' hordein	Ref
Mol. Wt. (Kd)			
1. SDS-PAGE	45–60	68–80	29
	30–51	67–86	5
2. G200 Sephadex	52,66	101	8
also	190–350[c]		
3. Equilibrium Ultracentrifugation	31.9	52.6	11

Amino Acid Composition

	a	b	a	b	11
Asx	1.4	3.8	1.0	4.4	
Glx	35.4	95.7	41.2	182.4	
Pro	20.6	55.8	30.6	135.6	
Lys	0.5	1.5	0.2	0.7	
Cys	2.5	6.8	t	–	
Met	0.6	1.5	0.2	0.7	
Phe	4.8	13.1	8.8	38.9	

[a]Asn + Gln	33.0	38.9
[d]Asn + Gln	89.8	92.3
Residues/mole	271	439

N terminal Amino Acids

	blocked	Arg	11

[a]expressed as mole %; [b]expressed as residues/mole; [c]probably an aggregated form; [d]expressed as % of total recovered Asp + Glu

Two dimensional IEF/SDS-PAGE of hordein resolves the protein into about 20 polypeptides depending on variety[2,6]. There have been various criticisms of the validity of the heterogeneity of polypeptides in IEF[13] but in our view these are largely unsubstantiated. We have investigated the structural homology of the 'B' hordein polypeptides by two-dimensional mapping and gradient SDS-PAGE of the CNBr cleavage products . Fig. 1a,b shows 2-D maps of B hordein polypeptides from 2 varieties, Jupiter and Proctor. In both varieties a number of polypeptides differing in pI and mol. wt. are present. Analysis of a mixture of 'B' hordein from these 2 varieties (Figures 1c,d) shows that of 17 major polypeptides present, eight (Nos. 2,3,6,10,11,13,14,15) occur only in Proctor, five (Nos. 4,5,7,8,9) only in Jupiter and four (Nos.1,12, 16,17) in both varieties. The areas containing the major polypeptides were excised from the 2-D gels, homogenised with CNBr in 70% formic acid and the cleavage products separated by gradient SDS-PAGE. Figure 1e,f shows the resulting peptide maps of six major polypeptides from Jupiter. Although polypeptides 8, 9 and 5 had very similar patterns, minor differences in band intensity and migration distance occurred. These are shown clearly for 8 and 5 in Figure 1f. Polypeptide 12 had a similar pattern to polypeptides 8,9 and 5, but two peptides of intermediate mol. wt. were absent (arrowed in Figure 1e). Polypeptides 1 and 16 had peptide maps which were distinctly different to each other, and to those of the other four polypeptides (Figure 1e). We have also mapped polypeptides 1,2,3,6,10,12 and 13 from Proctor. Polypeptides 1-3 are of slightly higher mol. wt., and all had maps very similar to polypeptide 1 of Jupiter. Polypeptides 6,10,12 and 13 are of slightly lower mol. wt. and had maps similar to those of polypeptide 12 of Jupiter. As in Jupiter, the peptide maps of Proctor showed minor differences in relative band intensity and in migration distance.

These results taken together with genetic and in vitro protein synthesis evidence suggest that it is likely that most if not all of the polypeptides seen on 2-D gels are gene products. Further since differences in amino acids of the same charge will not change the pI it is possible that the individual spots may represent the products of more than one structural gene.

Genetics

Genetic analysis using conventional crosses[14,15,16] and chromosome doubled monoploids[17] have shown that the B and C hordeins are controlled by two loci Hor-2 and Hor-1 respectively, approximately 10-15 recombination units apart and

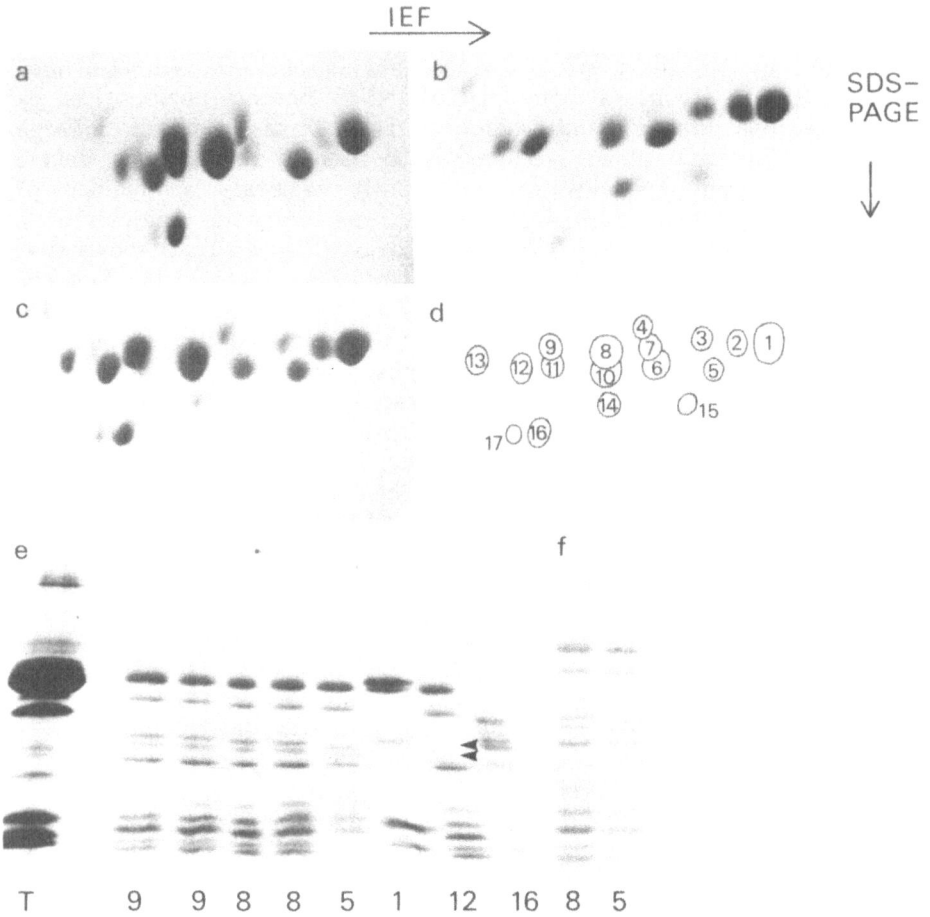

Fig. 1   Structural Homology of 'B' Hordein Polypeptides

a-c, 2-D analysis of PE 'B' hordein from a, Jupiter; b, Proctor; c, a mixture of fractions from Jupiter and Proctor.

d, Diagram of the major 'B' polypeptides present in Proctor and Jupiter. Polypeptides 2,3,6,10,11,13-15 occur only in Proctor, 4,5,7-9 only in Jupiter and 1,12,16,17 in both varieties.

e, Gradient SDS-PAGE of CNBr peptides of 6 major polypeptides of Jupiter. Arrows indicate minor peptides which are absent in the map of polypeptide 12, but are present in the maps of polypeptides 5, 8 and 9. The band of highest mol. wt. in each map is probably the uncleaved polypeptide. T = total hordein from Jupiter.

f, Gradient SDS-PAGE of Cyanogen Bromide peptides of polypeptides 5 and 8 of Jupiter. Arrows indicate peptides which differ in mobility. The separation was run for slightly longer than that in Fig.1e.

located either side of the Mla locus on chromosome 5 with the Hor-2 locus distal from the centromere and the Hor 1 locus close to the Mlk gene (Figure 2). Analysis of over 200 varieties of barley, including exotic lines, by SDS-PAGE[26] and of lesser numbers by IEF, Urea-PAGE and 2-D methods[6,27] have shown that a wide range of variation exists within each of the two groups. Analysis of crosses by 2-D methods have failed to indicate any crossing over within either locus but, given the numbers which can be analysed, this is not surprising. However 2-D analysis linked with CNBr peptide mapping as shown in Figure 1 indicates that many varieties have some polypeptides in common suggesting a degree of reassortment during the course of the development of modern varieties. Our current hypothesis is that the B and C loci are complex structural loci based upon two ancestral genes which have subsequently been duplicated many times.

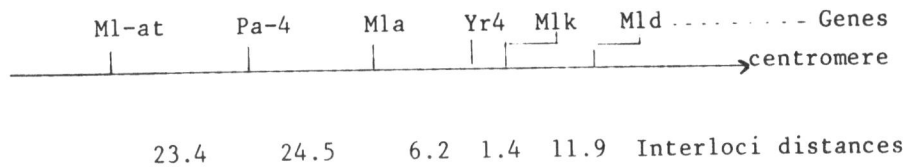

Figure 2   Genetic map of part of chromosome 5 of barley[18].

In vitro synthesis

Translation, in a wheat germ system of polysomes derived from a membranous pellet (sedimenting at 30,000 x g for 20 min) normally yields products that are similar to native hordeins in their solubility properties and mobility on SDS-PAGE, IEF and 2-D separation systems[19-22], (Figures 3 and 4). Recently we have found that certain wheat germ preparations produce a more complex pattern of products on SDS-PAGE (Figure 3); such preparations are also partially inhibited by MDMP (2-(4-methyl-2,6-dinitroanilino)-N-methyl propionamide) and 7meGMP consistent with the occurrence of reinitiation of the polysome message[23].

Isolation of poly-A containing mRNA from membrane bound polysomes has been shown in some instances (see Figure 6 of reference 24) but not others (see Figures 7 and 8 of reference 24) to form products similar to hordein but of lower mobility on SDS-PAGE. We have found that our poly-A preparation routinely produces polypeptides that are 75-85% soluble in 50% propan-2-ol and 2% 2-mercaptoethanol, have a

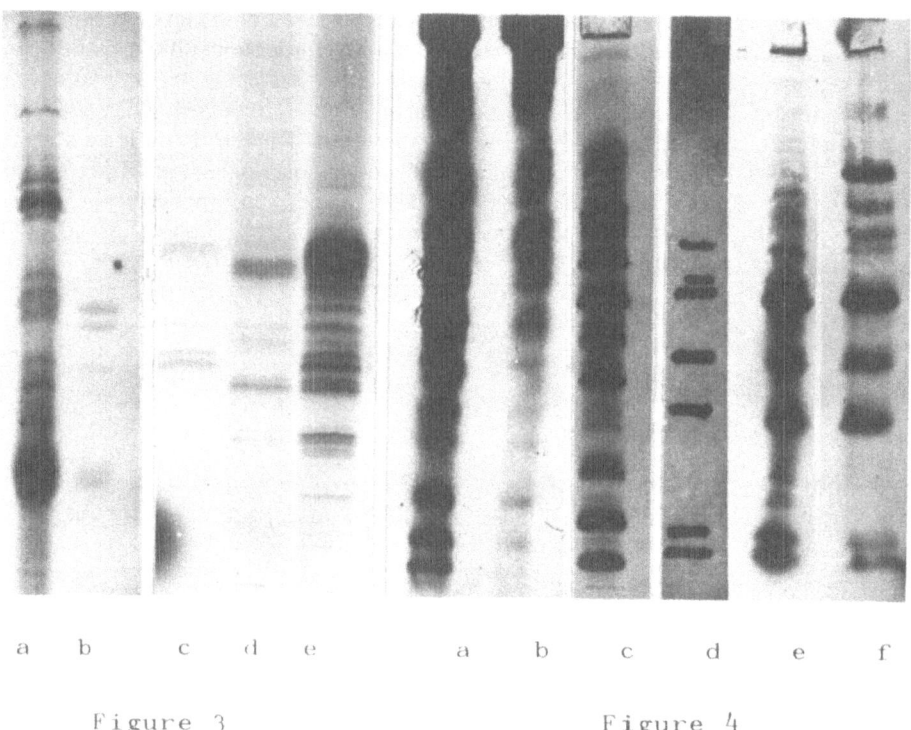

a   b   c   d   e   a   b   c   d   e   f

Figure 3   Figure 4

Figure 3 Autoradiographs of SDS-PAGE of $[^{35}S]$-methionine labelled polypeptides synthesised in a wheat germ translation system in the presence of.(a) and (c) membrane bound polysomes, (d) poly-A RNA derived from such polysomes. Tracks (b) and (e) contain hordein extracted from ears fed $[^{35}S]$ O_4^{2-}. Products in track (a) from a different wheat germ preparation from those in track (c).

Figure 4 Autoradiographs of IEF of products of $[^{35}S]$-methionine labelled polypeptides synthesized in vitro in the presence of (a) membrane bound polysomes, (b) free polysomes, and (e) poly-A RNA; (c) is a separately run track containing authentic hordein and stained with coomassie blue; (d) shows the position of stained proteins relating to the autoradiograph of (e); and (f) contains authentic hordein from $[^{35}S]O_4^{2-}$ fed ears.

high leucine to lysine incorporation ratio and have lower mobility on SDS-PAGE (Figure 3) but similar pI (Figure 4) to authentic hordeins[23]. These products have identical mobility to the new bands produced by reinitiating polysomes.

Based upon our present results we concluded that the most likely explanation is that the products synthesised in wheat germ systems in the presence of poly-A RNA contain an extra polypeptide sequence of uncharged (or net charge = 0) amino acids at the N terminus which is cleaved by an enzyme in the e.r. during synthesis. The products normally synthesized in the presence of polysomes are "run off" products of polypeptides from which the extra sequence had been cleaved before polysome isolation but that any products resulting from reinitiation are not processed. However it must be stated that the evidence from the existence of the extra sequence and its processing is largely circumstantial.

Based on the minimum mol. wts of the B and C hordein groups in Table 1 it is expected that the length of the mRNA coding for the polypeptides should be at a minimum of 1317 bases for the C and 813 for the B; to the coding sequence must be added a poly A tail of 6%[23] and an unknown length of leader sequence which of say 50 to 100 bases long, giving approximate minimal values in the region of 1000 to 1500 bases for the B and C groups respectively. Estimation of the sedimentation characteristics of the poly-A RNA isolated from membrane bound polysomes gives values of around 17S equivalent to about 1900 nucleotides[23]. This value is greater than that previously published by Brandt and Ingversen[24].

In vivo synthesis

The association of prolamin synthesis with the e.r. during endosperm development has been studied by homogenisation of the endosperms and subsequent fractionation on sucrose density gradients in the present and absence of Mg^{2+}. The distribution of NADH-cytochrome c reductase, RNA, total and alcohol-soluble proteins, and mitochondrial and plastid marker enzymes, and the appearance of fractions in the TEM have been followed[22,25]. A typical separation, in the presence of Mg^{2+}, of homogenates of developing endosperms of barley is shown in Figure 5. The most dense peak consists almost entirely of alcohol-soluble protein, has RNA and about 20% of the total NADH-cytochrome C reductase associated with it, and contains protein bodies which can be recognized under the TEM. The adjacent broad but less dense peak contains some alcohol-soluble proteins and the bulk of the NADH-cytochrome c reductase. When viewed under the TEM this peak chiefly consists

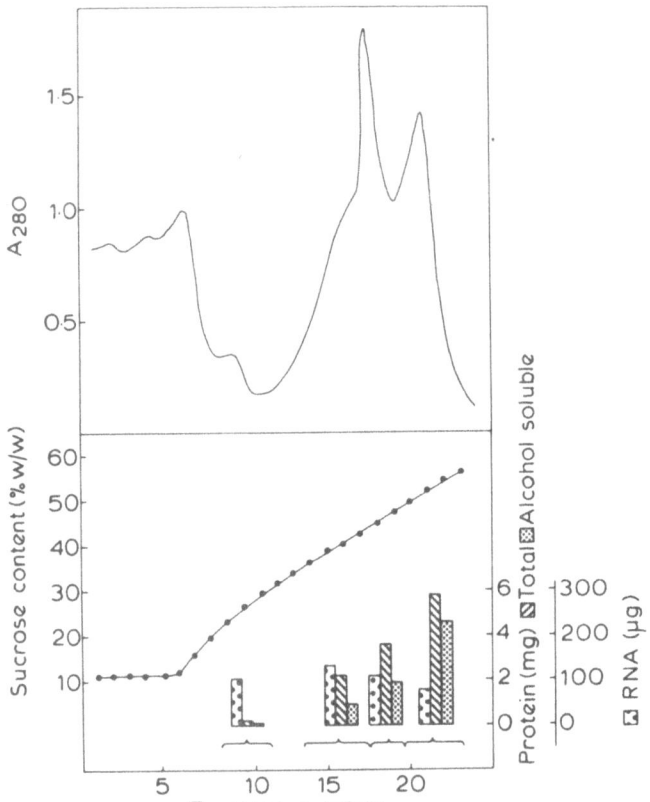

Figure 5 Separation of an homogenate of developing barley endosperms by sucrose density gradient centrifugation.

of rough e.r. with associated clumps of protein. Removal of Mg^{2+} decreases the density of this peak (Table 2). In comparison developing wheat endosperms have somewhat less (ca. 10%) of their NADH-cytochrome c reductase associated with the protein bodies but the major e.r. peak also has alcohol-soluble protein associated with it. In both wheat and barley the alcohol soluble protein has been identified by SDS-PAGE as the storage prolamin fraction. Homogenates of developing maize endosperms give two, about equal, peaks of NADH cytochrome c reductase activity, one associated with the

Table 2

Sucrose Content (Density g. cm -1) of Main Peak

Tissue	Endoplasmic Reticulum		Protein Bodies
	$-Mg^{2+}$	$+Mg^{2+}$	$-Mg^{2+}$
Barley Endosperm	41 (1.190)	45 (1.210)	53 (1.245)
Wheat Endosperm	29 (1.170)	42 (1.190)	57 (1.270)
Maize Endosperm	23 (1.100)	35 (1.155)	49 (1.230)
Castor bean Endosperm[28]	(1.12)	(1.16)	–

protein bodies and one at a ρ = 1.1 g cm^{-3} (this is of similar density to the e.r. from developing castor bean endosperm[28]). The latter peak does not appear to have alcohol-soluble protein associated with it. In contrast gradients of homogenates of developing pea cotyledons show only one peak of NADH-cytochrome c reductase associated with the e.r. Neither NADH-cytochrome c reductase nor RNA are associated with the protein bodies. Table 2 summarises the densities of the major e.r. and protein body peaks in the different tissues tested. Although not reported here there are minor variations in these values according to the age of the tissue.[25]

To test whether protein bodies are surrounded by an impermeable membrane homogenates were incubated with proteinase K (1.5 mg/10 ml) for 30 min at 10°C prior to centrifugation, and the amount of protein in the region of the protein body peak determined. A summary of the results in Table 3 shows that whereas the protein bodies from pea and maize were unaffected by this treatment those from wheat and barley were completely digested.

Table 3
Effects of Proteinase-K on the Protein Bodies of
Different Tissues

Tissue	Protein in protein body (mg)	
	Control	Proteinase-K
Barley Endosperm	1.95	0.04
Wheat Endosperm	6.05	0.13
Maize Endosperm	6.45	6.56
Pea Cotyledon	3.72	3.54

These results are interpreted as showing the close correlation between the synthesis of proteins on the rough e.r. and their deposition in cereals. Differences within the cereals exist in which the proteins of barley and wheat adhere to the e.r. and form clumps of protein (i.e. protein bodies) only partially surrounded by e.r. membranes and exposed to attack by proteases whereas zein does not adhere to the e.r. and forms spherical protein deposits completely surrounded by e.r.. Pea storage proteins also appear to be synthesized on the e.r. but are deposited in protein bodies bounded by membranes which lack NADH-cytochrome c reductase activity do not have RNA associated with them.

REFERENCES

1. P. R. Shewry, M. A. Kirkman, H. M. Pratt, and B. J. Miflin (1979) in "Carbohydrate and Protein Synthesis" B. J. Miflin and M. Zoschke ed. CEC Luxembourg EUR 6043 pp.155-171

2. B. J. Miflin and P. R. Shewry (1979) in "Seed Protein Improvement in Cereals and Grain Legumes" Vol. 1 IEAE Vienna STI/PUB/496, pp.137-157.

3. P. R. Shewry, J. M. Hill, H. M. Pratt, M. M. Leggatt, and B. J. Miflin (1978). J. exp. Bot. 29, 677-692.

4. P. R. Shewry, H. M. Pratt, and B. J. Miflin (1977) in Techniques for the Separation of Barley and Maize Proteins, B. J. Miflin and P. R. Shewry ed. CEC, Luxembourg EUR 5687, pp.37

5. P. R. Shewry, H. M. Pratt, M. J. Charlton, and B. J. Miflin (1977) J. exp. Bot. 104, 597-606.

6. P. R. Shewry, J. R. S. Ellis, H. M. Pratt and B. J. Miflin (1978). J. Sci. Fd. Agric. 29, 433-441.

7. E. Waldschmidt-Leitz and H. Kling (1966) Hoppe-Zeylers Zeitschrift f. Physiol. Chemie 346, 17-20.

8. B. Mesrob, M. Petrova, and Ch. Ivanov (1969). Biochim. Biophys. Acta 181, 482-484.

9. B. Koie, J. Ingversen, A. Andersen, H. Doll, and B. O.
       Eggum, 1976 in "Evaluation of Seed Protein Alterations by
       Mutation Breeding" IEAE Vienna, STI/PUB/426. pp.55-61.
10. B. J. Miflin and P. R. Shewry (eds.)(1977) "Techniques for
       the Separation of Barley and Maize Proteins" CEC,
       Luxembourg
11. P. R. Shewry, J. M. Field, M. A. Kirkman, A. J. Faulks,
       and B. J. Miflin (1979). manuscript submitted.
12. K. R. Sexson, Y. V. Wu, F. R. Huebner, and J. S. Wall
       (1978) Biochim. Biophys. Acta 532, 279-285.
13. B. Burr (1979) in "Seed Protein Improvement in Cereals
       and Grain Legumes" IEAE Vienna STI/PUB/496 pp.175-178.
14. P. R. Shewry, H. M. Pratt, R. A. Finch, and B. J. Miflin
       (1978). Heredity 40, 463-466.
15. H. Doll and A. H. Brown (1979) Canad. J. Genet. Cytol (in
       press).
16. J. Jensen and H. Doll (1979) manuscript in preparation.
17. P. R. Shewry, A. J. Faulks, R. J. Pickering, I. T. Jones,
       R. A. Finch, and B. J. Miflin (1979) manuscript in preparation.
18 J. Jensen (1978) Coordinators report: Chromosome 5 Barley
       Genetics Newsletter 8, 139-141.
19. J. E. Fox, H. M. Pratt, P. R. Shewry, and B. J. Miflin
       (1977) in "Nucleic Acid and Protein Synthesis in Plants"
       CNRS, Paris, pp.501-509.
20. A. Brandt and J. Ingversen (1976). Carlsberg Res. Commun.
       41, 312-320.
21. B. J. Miflin, J. E. Fox, and P. R. Shewry, (1977). Plant
       Physiol. 59, suppl. 580.
22. B. J. Miflin and P. R. Shewry (1979) in "Recent Advances in
       the Biochemistry of Cereals" D. Laidman and R. G. Wyn
       Jones eds., Academic Press, London. pp. 239-273
23. J. A. Matthews, B. J. Miflin, A. J. Faulks, and P. R. Shewry
       (1979), manuscript in preparation
24. A. Brandt and J. Ingversen (1978). Carlsberg Res. Commun.
       43, 451-469.
25. B. J. Miflin, S. R. Burgess, and P. R. Shewry (1979),
       manuscript in preparation.
26. P. R. Shewry, H. M. Pratt, A. J. Faulks, S. Parmar, and
       B. J. Miflin (1979) J. of Natnl. Institut. Agric. Bot. (in press)
27. P. R. Shewry, H. M. Pratt and B. J. Miflin (1978). J. Sci.
       Fd. Agric. 29, 587-596.
28. J. M. Lord, T. Kagawa, T. S. Moore, and H. Beevers (1973).
       J. Cell. Biol. 57, 659-667.
29. A. Brandt (1976) Cereal Chem. 53, 890-901.

BIOSYNTHESIS OF PEA SEED PROTEINS : EVIDENCE FOR PRECURSOR FORMS

FROM *IN VIVO* AND *IN VITRO* STUDIES

T.J.V. Higgins and Donald Spencer

CSIRO, Division of Plant Industry, P.O. Box 1600
Canberra City, A.C.T. 2601. AUSTRALIA

INTRODUCTION

In his review of seed formation, Dure (1975) pointed out that although relatively little is known about seed ontogeny, considerably more is known about the nature of seed storage proteins. This latter topic has frequently been reviewed, particularly for the legumes (Millerd 1975; Derbyshire et al. 1976) and for some of the cereals (Kasarda et al. 1976; Miflin and Shewry 1978). Until recently much less was known about the biosynthesis and deposition of storage proteins than about their size, amino acid composition and genetic variability (for review see Spencer and Higgins 1979; Müntz 1977). This state of affairs has now begun to change and a number of groups have tackled the problem of defining some of the factors regulating storage protein biosynthesis. The general approach so far has been to attempt to determine the role of mRNA level in controlling the rate of reserve protein synthesis. Little attention has been given so far to the role of post-transcriptional control (e.g., mRNA turnover), translational control (e.g., ribosome activity, tRNA levels, soluble enzymes of the translational apparatus) or post-translational controls (e.g., protein turnover, protein processing or protein sequestration).

Ultrastructural features of storage protein biosynthesis and deposition have also received some attention but considerably more work is required in order to unravel the complexities of the site of synthesis, sites of processing (if any occurs) and how storage protein is deposited in protein bodies (see Pernollet 1978 for some discussion). There is a need for a two-pronged approach to the study of seed storage protein biosynthesis in which ultrastructural investigations are carried out in conjunction with biochemical

245

studies. Examples of this are provided by the work of Burr and Burr (1976), Larkins and Hurkman (1978), Cameron-Mills and Ingversen (1978) and Miflin and Shewry (1979), in which some preliminary steps were taken to show localization of storage protein synthesis at both the ultrastructural and biochemical level. Higher resolution techniques will be required to detail the precise pathway of storage protein biosynthesis and deposition for example, by combining electron microscopy with an immunochemical probe (Bailey et al. 1970; Graham and Gunning 1970).

The biochemical approach to the study of storage protein biosynthesis has largely involved the establishment of cell-free systems capable of synthesis of the reserve proteins. Two types of *in vitro* system are in common use. The first involves the isolation of polysomes either attached to, or freed of, membranes and the subsequent completion of nascent chains in the presence of radioactive amino acids. This system is referred to as "run-off" and provides information on the current protein synthetic activity of a tissue. The second system involves the isolation of total or polysomal RNA (and is often followed by (poly A+) RNA isolation) which is then used to program a mRNA-dependent cell-free system derived from wheat-germ or rabbit reticulocytes. This is called an "initiating" system and provides a refinement of the "run-off" reaction by largely eliminating the translational controls (e.g., ribosome activity or soluble factors such as initiation factors or tRNAs) which may be contributing to polysome formation. The initiating system can provide a direct method of measuring storage protein mRNA.

The use of cell-free protein synthesizing systems has enabled a number of investigators to establish that storage protein synthesis is a major protein synthetic activity of developing seeds (Larkins and Dalby 1975; Sun et al. 1975; Brandt and Ingversen 1976; Higgins and Spencer 1977; Luthe and Peterson 1977). The availability of mutants in maize and barley with reduced levels of zein and hordein, respectively, has allowed Jones et al. (1976), Brandt and Ingversen (1976) Fox et al. (1977) and Miflin and Shewry (1979) to conclude from "run-off" experiments that the polysomes from mutant seeds have reduced capacity for storage protein synthesis. Further analysis of the "run-off" synthesis of hordein has revealed that the mutant may have reduced capacity for translation and processing of hordein polypeptides (Cameron-Mills and Ingversen 1978).

The cell-free systems developed to assay for the synthesis of storage proteins other than the cereal prolamins (which are alcohol-soluble), e.g., the globulins of cereals and legumes, have not been quite as successful largely because it is more difficult to identify and quantify the storage proteins among the translation products. The most useful technique is immuno-precipitation and this has been successfully applied to the storage proteins of bean

(Hall et al. 1978) and pea (Evans et al. 1979) as well as the prolamins of maize (Weinand and Feix 1978). Under our conditions of
translation of pea seed mRNA, we found that although storage protein
synthesis occured *in vitro* (monitored qualitatively and quantitatively
by tryptic peptide mapping), the translation products were inefficient-
ly precipitated by storage protein antibodies and did not coincide
with the authentic storage protein polypeptides when fractionated by
sodium dodecyl sulfate-polyacrylamide gel electrophoresis (SDS-PAGE)
(Higgins and Spencer 1977). We tentatively concluded that either
storage protein mRNAs constituted a small proportion of the total
mRNA population and that storage proteins must therefore accumulate
only by virtue of their different rate of turnover, or that storage
protein mRNAs were a quantitatively major proportion of the total
population (in agreement with the peptide mapping data) but that the
storage protein polypeptides synthesized *in vitro* were in the form
of precursors which lacked all the antigenic determinants needed
for immunoprecipitation.

In attempting to distinguish between these possibilities we have
found evidence that some pea storage protein polypeptides are
synthesized as precursors in our cell-free system, while evidence
for other precursor forms has come from *in vivo* labelling studies.
A summary of this evidence is included in the following discussion
on *in vitro* and *in vivo* systems for studying storage protein
synthesis.

Cell-Free Translation Studies

As we described earlier, polysomal RNA from developing pea seeds
programmes the synthesis of a wide size-range of polypeptides none
of which coincides precisely with any of the storage protein polypeptides found in mature protein bodies (Fig. 1a and Higgins and
Spencer 1977) nor with *in vivo* labelled polypeptides which are
known to be storage protein polypeptides (Fig. 1c and f). These
latter were synthesized by detached, intact cotyledons 11 days
after flowering (DAF) during a 2 h pulse with a mixture of $[^{14}C]$-
labelled amino acids (Spencer and Higgins, in preparation). However
if an EDTA-stripped microsomal membrane fraction (Blobel and
Dobberstein 1975) from dog pancreas (Fig. 1d) or developing pea
cotyledon (Fig. 1g) is added at the beginning of the cell-free
incubation, some of the translation products (one at 75,000 and two
of those at about 50,000 daltons) are modified so that they now
correspond in size to *in vivo*-synthesized major polypeptide components of vicilin (Fig. 1c and f). If 1% Triton X-100 is included in
the *in vitro* system it has no effect on the translation of the polysomal RNA alone (Fig. 1b) but largely abolishes the processing effect
of the membranes (Fig. 1e and h). The modified polypeptides and two
other unmodified bands (25,000 and 30,000 daltons) were shown to be
within membrane vesicles as judged by protection from trypsin added
at the end of the cell-free incubation (Higgins and Spencer, in

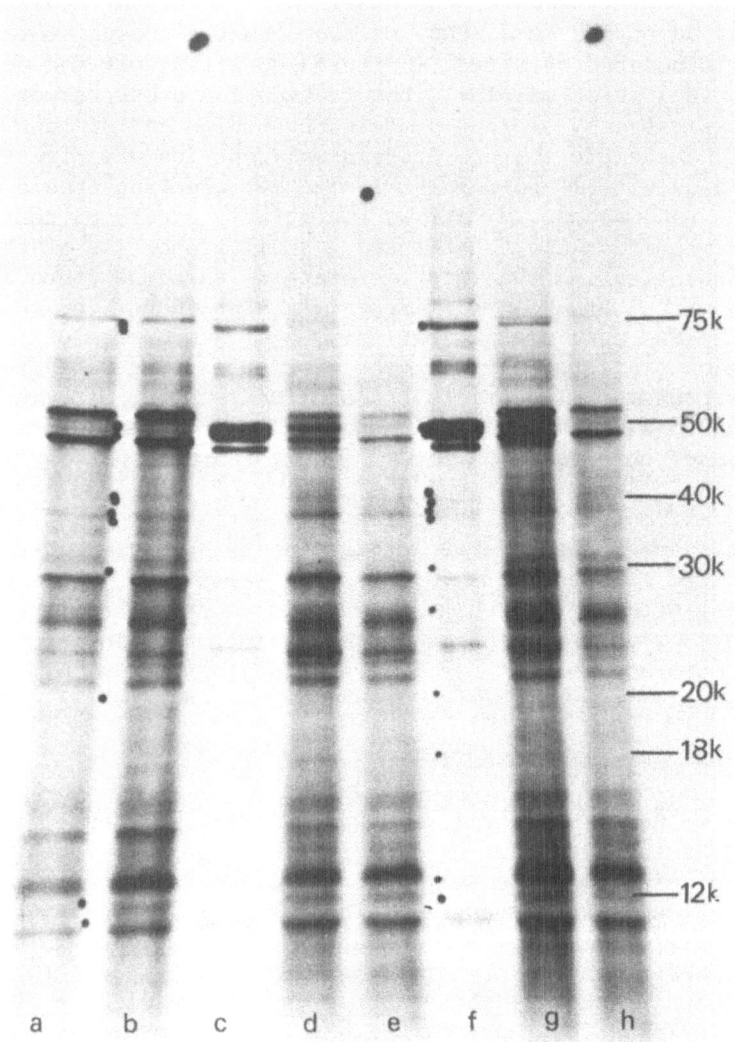

Fig. 1. *In vivo* and *in vitro* synthesized pea seed proteins fractionated by SDS–PAGE and fluorography. Cotyledons (11 DAF) were
labelled *in vivo* for 2 hr with a mixture of [^{14}C] amino·acids
and the salt-soluble proteins were extracted (c and f).
Polysomal RNA was isolated from cotyledons (11 DAF) and used
to program the wheat germ cell-free protein synthesizing
system (a, b, d, e, g, h). a, RNA alone; b, e, h, RNA in
the presence of Triton X-100; d, e, g, h, RNA in the presence of stripped microsomal membranes from dog pancreas (d,
e) or pea cotyledon (g, h). The numbers (75k etc.) refer
to the apparent molecular weight (k = kilodaltons) of some
of the major polypeptides of pea protein bodies.

preparation). The *in vitro* translation products were characterized
by immuno-affinity chromatography using antibodies to the total
storage protein fraction from mature seeds. The antibodies were
purified on antigen-affinity columns and coupled to UltroGel ACA 34
(Ternynck and Avrameas 1976). The affinity columns showed excellent
specificity for storage proteins isolated from mature seeds or
developing cotyledons (Spencer and Higgins, in preparation). When
cell-free translation products were chromatographed on these affinity
columns the specificity of binding was much higher than that obtained
by immunoprecipitation (either direct or indirect) but the efficiency
of binding was still fairly variable. When the translation products
programmed by polysomal RNA alone were chromatographed, polypeptides
in the 75,000, 50,000, 30,000 and 25,000 dalton region were bound
selectively (Fig. 2d) although sometimes a significant proportion
(10-50%) remained in the non-bound fraction (Fig. 2c). When membranes
were added during translation, however, specific binding was again
observed and the modified polypeptides were bound with high efficiency
(Fig. 2g). We conclude that at least some of the pea storage protein
polypeptides such as the 75,000 and 50,000 dalton components of
vicilin are synthesized as precursors which can be modified by the
addition of membranes during translation. The modified polypeptides,
as well as some bands which are not visibly modified by the membrane
treatment (in the 25,000 and 30,000 dalton regions) appear to be
taken up into the membrane vesicle during translation. All of these
polypeptides are bound selectively by antiserum to mature storage
proteins. Certain features of pea storage protein biosynthesis, as
studied at this level, are consistent with the "signal hypothesis"
developed for secretory proteins (Blobel and Dobberstein 1975) in
mammalian cells. An unexpected result was the fact that membrane
modification of the translation products greatly increased their
antigenic efficiency.

These results are in general agreement with those of Hall et
al. (1978) for two of the polypeptides of French bean Gl protein
which also are synthesized as polypeptides slightly different in
size to the authentic subunits. However, our data contrast somewhat with that obtained by Evans et al. (1979) for pea in which
they found that one mature-sized and antigenically active vicilin
polypeptide of 51,000 daltons was synthesized *in vitro* in response
to either added polysomes, fractionated polysomal RNA or poly(A+)
RNA. Possible explanations for this discrepancy include, (a) Evans
et al. (1979) used an SDS-PAGE system which may not have provided
sufficient resolution of precursor and mature polypeptides or (b)
it is possible that the cell-free system employed by Evans et al.
contained the necessary components to modify the cell-free translation products. In support of the latter, Walk and Hock (1978)
found that glyoxysomal malate dehydrogenase synthesized *in vitro*
by a wheat germ system probably underwent some co- or post-translational modification. Another feature of the data of Evans et al.
(1979) which might support the notion that their cell-free system

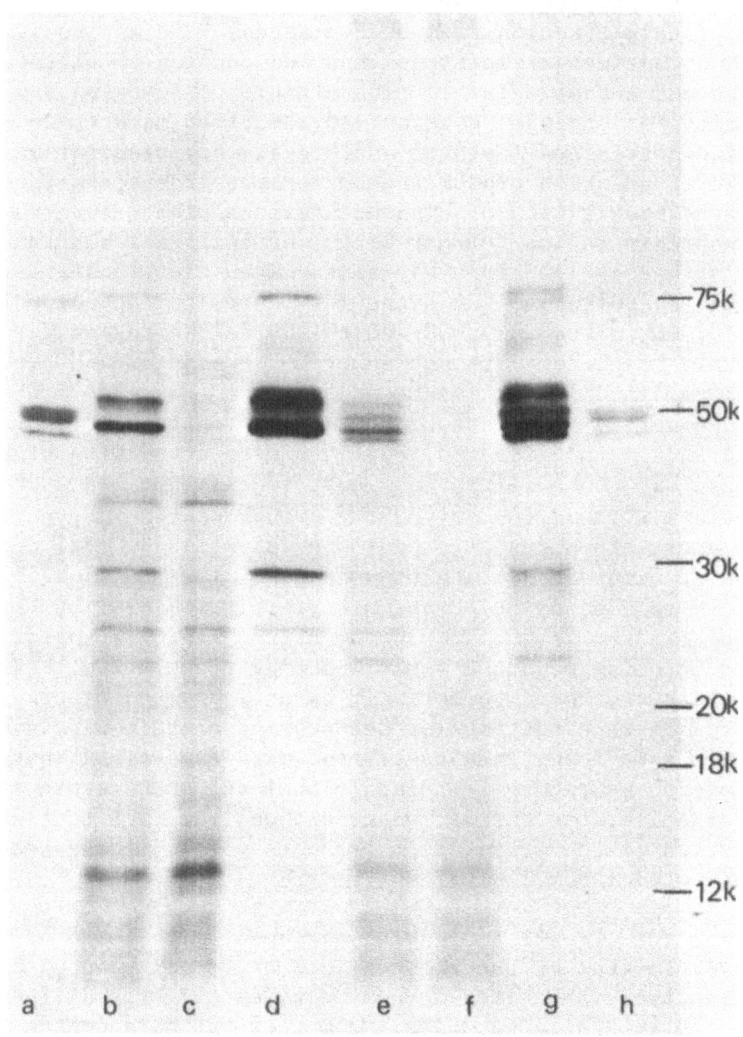

Fig. 2.   Immuno-chromatography and SDS-PAGE of *in vitro* synthesized
          pea seed proteins. Polysomal RNA (mRNA) isolated from pea
          cotyledons (11 DAF) was translated in the wheat germ system
          in the absence (b) and presence (e) of pea microsomal mem-
          branes. The translation products were chromatographed over
          columns containing immobilized antibody to pea storage
          proteins (legumin plus vicilin). The non-bound translation
          products of mRNA alone (c) and mRNA plus pea membranes (f)
          are compared with the bound fractions programmed by mRNA
          alone (d) and mRNA plus membranes (g). *In vivo* synthesized
          (11 DAF) polypeptides are shown (a and h) and the numbers
          refer to apparent molecular weights in kilodaltons.

contained processing activity is the fact that they found a similar-
ly-sized vicilin translation product whether they used a non-
initiating, polysomal "run-off" system or the "initiating" RNA-
dependent system. This also contrasts with our work (Higgins and
Spencer 1977) in which we found that the polysomal "run-off" products
were different to the mRNA-programmed polypeptides and to authentic
subunits. In our experience, the translation products of the "run-
off" system probably represent an intermediate form of processed
precursor storage protein (unpublished results). In agreement with
our earlier work, Evans et al. (1979) could not demonstrate the cell-
free synthesis of the other vicilin subunits nor legumin using
serological methods.

     As indicated earlier, a number of cell-free systems derived
from developing seeds have been studied for the synthesis of storage
protein. Storage protein products have usually been characterized
by solubility and/or size of the polypeptides. In a few cases it
has been shown that the cell-free "run-off" products are slightly
different in size to the products of the "initiating" system and
often co-migrate exactly with authentic subunits. This was shown
for zein (Larkins and Hurkman 1978) when membranes were also present
during the "run-off". However, Viotti et al. (1978) have used polysomes detached from membranes and "run-off" the nascent chains in
the absence of membranes and they found that the translation products
were identical in size and isoelectric point to authentic zein polypeptides. It is clear that the cell-free zein products of the
"initiating" system are synthesized as precursors (Burr et al. 1978;
Larkins and Hurkman 1978; Weinand and Feix 1978). A similar situation seems to exist in the case of barley hordeins in that the RNA-
programmed products are not identical to authentic hordeins but the
"run-off" products of detached polysomes are apparently processed
to the correct size (Brandt and Ingversen 1978). Processing activity
in the polysome fraction of French bean may also account for the
authentic size of G1 subunits detected in the "run-off" system by
Sun et al. (1975) whereas in the initiating system described by
Hall et al. precursors to two of the G1 polypeptides were found
(the third component of G1 was not detected by translation of RNA).

     It seems likely that microsomes of pea, maize and barley contain
processing activity which can convert those storage protein polypeptides which occur as precursors to their final, authentic size.
Further work is required to show that the polysome fractions thought
to contain processing activity are in fact free of membranes. Based
upon the available evidence, some of the storage protein polypeptides
of at least two legumes and two cereals exist as short-lived precursors which are processed co- or post-translationally.

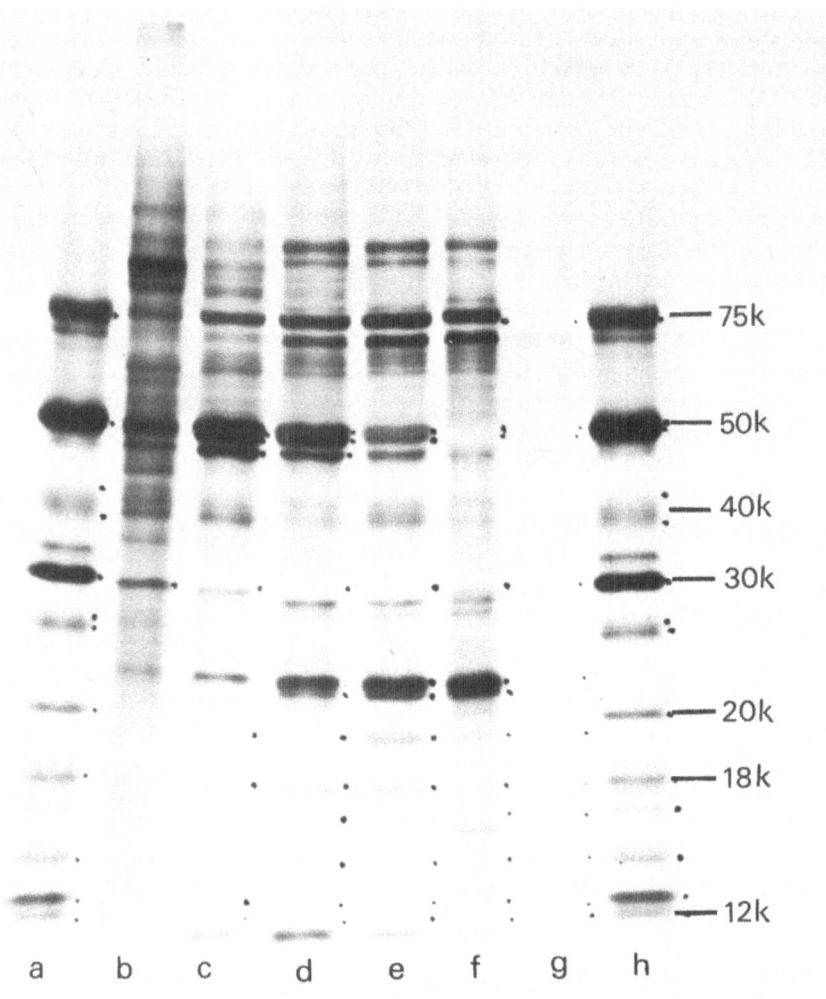

a   b   c   d   e   f   g   h

Fig. 3.   *In vivo* pulse-labelling of pea seed proteins at various
developmental stages and fractionation by SDS-PAGE and
fluorography.

Pea cotyledons were pulse-labelled for 2 hr with a mixture
of [14C] amino acids at various stages during development;
b, 8 DAF; c, 11 DAF; d, 14 DAF; e, 17 DAF; f, 20 DAF; g,
24 DAF. The total salt-soluble proteins were isolated and
compared with the salt-soluble proteins (labelled chemically
with [14C]-formaldehyde) from protein bodies of mature
seeds. The numbers refer to apparent molecular weights
in kilodaltons.

In vivo Pulse-Labelling Studies

In the earlier stages of this investigation we sought evidence for precursors of pea seed storage proteins by pulse-labelling cotyledons *in vivo* for times ranging from 2 min to 2 hr. It is now not surprising that we did not find precursors equivalent to those found *in vitro* because it seems likely that the precursors found in the cell-free system are co-translationally modified and have a lifetime less than that taken for translation of the complete chain *in vivo*.

In our *in vivo* labelling studies (for details see Spencer and Higgins, in preparation), we initially labelled for periods up to two hours and then extracted the salt-soluble proteins and separated the component polypeptides by SDS-PAGE followed by fluorography. We found that:

(a) Storage protein biosynthesis was a major activity of cotyledons during 10 to 15 days after flowering under constant environmental conditions. All components related to storage proteins were identified by immunoaffinity chromatography. Qualitative and quantitative analysis of the newly synthesized proteins showed that between 60 and 70% of the pulse-labelled protein was related to storage protein.

(b) The newly synthesized storage protein was quickly (less than 15 min) sequestered into a subcellular fraction sedimenting at 1000 g.

(c) The storage protein polypeptides were not all synthesized continuously during the 10-15 day period of protein accumulation, e.g., the 50,000 dalton complex of vicilin was mainly synthesized during the first half of the period (Fig. 3 c,d and e).

(d) There were complexities in some of the immature storage protein polypeptide classes (e.g., 50,000 dalton) which were not apparent in the mature proteins (Fig. 3 cf. c-e with a and h). We do not know the fate of these extra polypeptides but speculate that they may be precursors of mature storage protein polypeptides.

(e) Some storage protein polypeptides were not labelled by a 2 hr pulse at any stage of development, yet they were readily detectable by Coomassie staining. Longer-term labelling, (up to 72 hr), which is probably equivalent to a pulse-chase experiment in that all of the isotope was taken up and maximum incorporation was reached after 6-8 hours, suggested that some of the minor vicilin polypeptides were either synthesized very slowly or arose from precursors. For example, a major vicilin polypeptide of 30,000 daltons was labelled during the "chase" period. These data suggest the existence of a precursor and we are currently investigating this possibility.

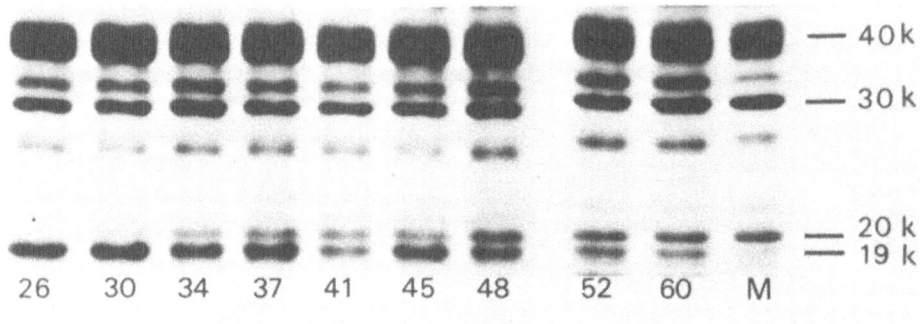

DAYS AFTER FLOWERING

Fig. 4.   Changes in the ratio of the large to the small subunit of
           legumin with maturation of pea seeds.   Protein bodies were
           isolated from pea seeds at various stages of maturation and
           the salt-soluble proteins were fractionated by SDS-PAGE.
           A portion of the stained gel containing the major legumin
           subunits (40 and 20 kilodaltons) is shown together with the
           polypeptide migrating with an apparent mobility of 19 kilo-
           daltons.   M refers to the extract of mature protein bodies.

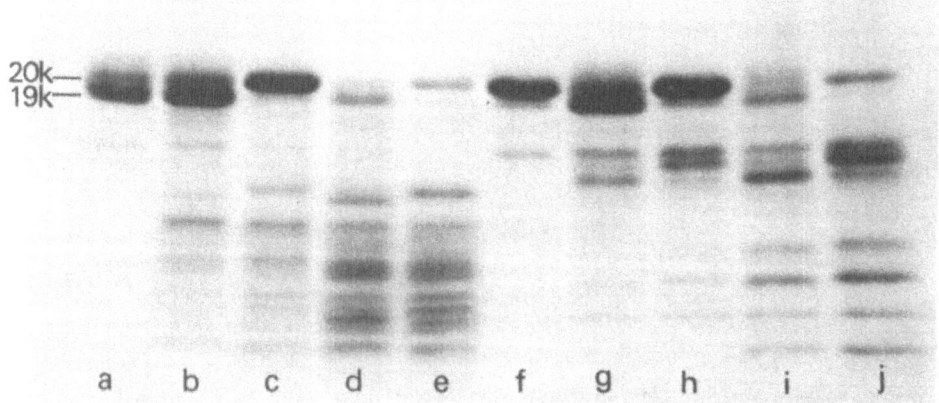

Fig. 5.   Partial proteolytic peptide mapping of the purified small
           subunits of immature and mature legumin by chymotrypsin
           and *Staphylococcus aureus* protease (S.A.P.).   Purified
           19 kilodaltons (a, b, d, g, i) and 20 kilodaltons (c, e,
           f, h, j) legumin subunits were digested at room temperature
           for 10 min with chymotrypsin (b-e) and S.A.P. (g-j), each
           at two concentrations. a, undigested 19 kilodaltons;
           f, undigested 20 kilodaltons;   b, c, chymotrypsin (5 µg/ml);
           d, e, chymotrypsin (20 µg/ml);   g, h, S.A.P. (30 µg/ml);
           i, j, S.A.P. (100 µg/ml).

(f)   One of the two major legumin subunits (20,000 daltons)
could not be detected as a major component either by Coomassie
staining (Spencer and Higgins, in preparation) or by radioactive
labelling (Fig. 3 cf. b-g with a and h) of cotyledons at any
developmental stage during the phase of protein deposition and
fresh weight increase i.e., up to 24 DAF. We could only detect
the 20,000 dalton polypeptide in seeds which had almost reached their
final moisture content (8-10%) (Fig. 4, 30-34 DAF). This suggested
the existence of a major component which was a long lived precursor
form of the 20,000 dalton subunit.

Another polypeptide, with a mobility equivalent to about 19,000
daltons, was detected as a major component by Coomassie staining and
by pulse labelling during the phase of protein accumulation (Fig. 3,
b-f). This polypeptide proved to be the precursor of the 20,000
dalton subunit found in legumin of mature seeds. The precursor
relationship was evidenced by the fact that the 19,000 dalton species
was the major small subunit of legumin purified from immature seeds.
In addition, the Coomassie-stained 19,000 dalton band was reduced
in color intensity as the 20,000 dalton subunit became increasingly
obvious with seed maturation (Fig. 4, 34-60 DAF). Purification of
both the 19,000 and 20,000 dalton polypeptides by DEAE-cellulose
chromatography in the presence of urea and reducing agent showed
that they were similarly charged (i.e., relatively basic). Partial
proteolytic digestion and peptide mapping, on SDS-PAGE, of the
purified polypeptides showed that there was extensive sequence homology between the two proteins (Fig. 5).

The nature of the modification which converts the precursor
to the product of 20,000 daltons is unknown. It is remarkable that
the modification apparently occurs after assembly of the legumin
holoprotein and during a stage of seed development when desiccation
seems complete.

CONCLUSIONS

Many of the pea storage protein polypeptides appear to be
synthesized as precursors. The lifetime of the precursors varies
over a wide range. Of those which we have studied in some detail,
the 75,000 and 50,000 dalton polypeptide precursors may exist only
during the translation phase of synthesis while the precursor of
the legumin 20,000 dalton subunit may have a lifetime of many days
or even weeks.

Conversion of the precursor to the product polypeptide in the
case of the 75,000 and 50,000 dalton bands probably occurs through
the removal of a short sequence (equivalent to the signal sequence
of mammalian secreted proteins) by an enzyme or enzymes associated
with the microsomal membranes. The function of the precursor may
be to provide the signal for recognition of the membrane which then

allows sequestration of the newly synthesized polypeptide. The
mechanism of conversion of the 20,000 dalton precursor is unknown,
as is the function of such a precursor. The precursor form is
apparently assembled into the holoprotein and only undergoes modification during the desiccation phase of seed maturation.

At least some of the pea storage protein precursors have
counterparts in other seed reserve proteins. Short-lived precursors
have been found in another legume, French bean and in some of the
cereals. However, it remains to be seen whether all seed reserve
proteins are synthesized via short-lived precursors and whether any
other seed proteins possess longer-lived precursor forms.

REFERENCES

Bailey, C.J., Cobb, A. and Boulter, D. (1970) A cotyledon slice
     system for the electron autoradigraphic studies of the
     synthesis and intracellular transport of seed storage protein
     of *Vicia faba*. Planta 95, 103-118.
Blobel, G. and Dobberstein, B. (1975) Transfer of proteins across
     membranes. I. Presence of proteolytically processed and
     unprocessed nascent immunoglobulin light chains. J. Cell.
     Biol. 67, 835-851.
Burr, B. and Burr, F.A. (1976) Zein synthesis in maize endosperm
     by polyribosomes attached to protein bodies. Proc. Natl. Acad.
     Sci. (U.S.A.) 73, 515-519.
Burr, B., Burr, F.A., Rubenstein, I. and Simon, M.N. (1978)
     Purification and translation of zein messenger RNA from maize
     endosperm protein bodies. Proc. Natl. Acad. Sci. (U.S.A.)
     75, 696-700.
Brandt, A. and Ingversen, J. (1976) *In vitro* synthesis of barley
     endosperm proteins on wild type and mutant templates. Carls-
     berg Res. Commun. 41, 312-320.
Brandt, A. and Ingversen, J. (1978) Isolation and translation of
     hordein messenger RNA from wild type and mutant endosperms in
     barley. Carlsberg Res. Commun. 43, 451-469.
Cameron-Mills, V. and Ingversen, J. (1978) Transfer of *in vitro*
     synthesized barley endosperm proteins into the lumen of the
     endoplasmic reticulum. Carlsberg Res. Commun. 43, 471-489.
Derbyshire, E., Wright, D.J. and Boulter, D. (1976) Legumin and
     vicilin, storage proteins of legume seeds. Phytochemistry.
     15, 3-24.
Dure, L.S. (1975) Seed formation. Annu. Rev. Plant Physiol.
     26, 259-278.
Evans, I.M., Croy, R.R.D., Hutchinson, P., Boulter, D., Payne, P.I.
     and Gordon, M.E. (1979) Cell free synthesis of some storage
     protein subunits by polyribosomes and RNA isolated from
     developing seeds of pea (*Pisum sativum* L.). Planta 144,
     455-462.

Fox, J.E., Pratt, H.M., Shewry, P.R. and Miflin, B.J. (1977)
    The *in vitro* synthesis of hordeins with polysomes from normal
    and high lysine varieties of barley. Acides nucleiques et
    synthese des proteines chez les vegetaux. No. 261. C.N.R.S.
    Paris pp. 501–509.
Graham, T.A. and Gunning, B.E.S. (1970) Localization of legumin
    and vicilin in bean cotyledons using fluorescent antibodies.
    Nature. 228, 81–82.
Hall, T.C., Ma, Y., Buchbinder, B.U., Pyne, J.W., Sun, S.M. and
    Bliss, F.A. (1978) Messenger RNA for G1 protein of French
    bean seeds: Cell-free translation and product characterization.
    Proc. Natl. Acad. Sci. (U.S.A.) 75, 3196–3200.
Higgins, T.J.V. and Spencer, D. (1977) Cell-free synthesis of pea
    seed proteins. Plant Physiol. 60, 655–661.
Jones, R.A., Larkins, B.A. and Tsai, C.Y. (1976) Reduced synthesis
    of zein *in vitro* by a high lysine mutant of maize. Biochem.
    Biophys. Res. Commun. 69, 404–410.
Kasarda, D.D., Bernardin, J.E. and Nimmo, C.C. (1976) Wheat
    proteins. In: "Advances in cereal science and technology"
    (ed. Y. Pomeranz) Publ. Amer. Assoc. Cereal Chem. Inc. St.
    Paul, Minn. pp. 158–236.
Larkins, B.A. and Dalby, A. (1975) *In vitro* synthesis of zein-
    like protein by maize polyribosomes. Biochem. Biophys. Res.
    Commun. 66, 1048–1054.
Larkins, B.A. and Hurkman, W.J. (1978) Synthesis and deposition
    of zein in protein bodies of maize endosperm. Plant Physiol.
    62, 256–263.
Luthe, D.S. and Peterson, D.M. (1977) Cell-free synthesis of
    globulin by developing oat (*Avena sativa* L.) seeds. Plant
    Physiol. 59, 836–841.
Miflin, B. and Shewry, P.R. (1978) The biology and biochemistry
    of cereal seed prolamins. In: "Int. Symp. Seed Protein
    Improvement in Cereals and Legumes" Int. Atomic Energy Agency,
    FAO. (in press).
Miflin, B.J. and Shewry, P.R. (1979) The synthesis of proteins
    in normal and high lysine barley seeds. In: "Cereals".
    (eds. D. Laidman and R.G. Wyn-Jones) Acad. Press. (in press).
Millerd, A. (1975) Biochemistry of legume seed proteins. Annu.
    Rev. Plant Physiol. 26, 53–72.
Müntz, K. (1977) Cell specialization processes during biosynthesis
    and storage of proteins in plant seeds. In: "Regulation of
    Developmental Processes in Plants". (ed. H.R. Shutte and
    D. Cross). VEB. G. Fischer Verlag, Jena. pp. 70–97.
Pernollet, J.C. (1978) Protein bodies of seeds: Ultrastructure,
    biochemistry, biosynthesis and degradation. Phytochemistry.
    17, 1473–1480.
Spencer, D. and Higgins, T.J.V. (1979) Molecular aspects of seed
    protein biosynthesis. Current Adv. Plant Science. 34, 1–13.
Ternynck, T., and Avrameas, S. (1976). Polymerization and immobil-
    ization of proteins using ethylchloroformate and glutaraldehyde.

Scand. J. Immun. Suppl. 3, 29-35.

Sun, S.M., Buchbinder, B.U. and Hall, T.C. (1975) Cell-free synthesis of the major storage protein of the bean, *Phaseolus vulgaris* L. Plant Physiol. 56, 780-785.

Viotti, A., Sala, E., Alberi, P. and Soave, C. (1978) Heterogeneity of zein synthesized *in vitro*. Plant Sci. Lett. 13, 365-375.

Walk, R-A. and Hock, B. (1978) Cell-free synthesis of glyoxysomal malate dehydrogenase. Biochem. Biophys. Res. Commun. 81, 636-643.

Weinand, U. and Feix, G. (1978) Electrophoretic fractionation and translation *in vitro* of poly (rA)-containing RNA from maize endosperm. Eur. J. Biochem. 92, 605-611.

BEAN SEED GLOBULIN mRNA: TRANSLATION, CHARACTERIZATION, AND ITS

USE AS A PROBE TOWARDS GENETIC ENGINEERING OF CROP PLANTS

Timothy C. Hall, Samuel M. Sun, Barry U. Buchbinder,
John W. Pyne, Fredrick A. Bliss and †John D. Kemp

Departments of Horticulture and †Plant Pathology,
University of Wisconsin, Madison, Wisconsin 53706, U.S.A.

INTRODUCTION

Single-celled organisms exhibit regulation of genome expression
through temporal control, some information being processed to yield
enzymes, RNA, and other substances used in the earliest stages of
growth while other regions of the DNA are utilized to provide the
late functions in cell maturation. Many other, more sophisticated
types of control over the expression of information that is encoded
in the nuclear DNA have evolved, making possible the development of
cells having specialized function and consequently the existence of
complex multicellular organisms. The beginning of scientific inquiry
into genome expression can be seen in the morphological studies on
the growth and differentiation of plants and animals that fascinated
the early biologists. Subsequently, the discovery and application
of genetic laws led to rapid improvements in the quality and produc-
tivity of crops and livestock. These improvements, which have
supported major changes in human populations and achievements, may
well prove to be as significant an evolutionary event as was the
discovery of agricultural principles that led to the dawn of civilization by permitting early man to cease his nomadic habits. It is
a happy coincidence that plant seeds, which provided this basis for
civilization, also provide excellent material for investigating the
molecular mechanisms of genome expression.

The Suitability of Phaseolus vulgaris for Genetic Engineering

Attributes for using the common bean, Phaseolus vulgaris L.,
in molecular studies include the widespread use of the seed as a
human protein source, its great genetic variability, and the
relative simplicity of the major seed storage protein. The flowers

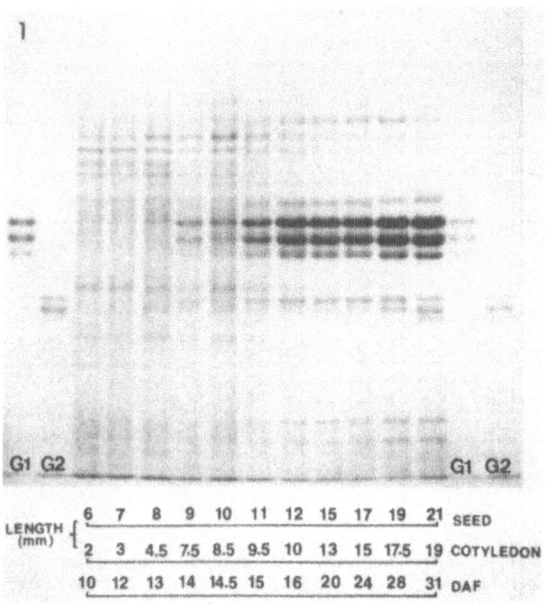

Fig. 1.   Electrophoretic separation of bean cotyledon proteins during
          growth.   Numbers under each lane of this 16% acrylamide gel
          denote the seed and cotyledon lengths and days after
          flowering (DAF).   Standard G1 and G2 proteins were run as
          markers in the outer lanes.

usually self-fertilize and the embryonic tissues include the cotyledons as well as the rudimentary stem axis and first leaves.   Hence,
the phenotype of the new plant is expressed in the cotyledons.   Part
of the dry seed can be analyzed to determine the phenotypic expression of the seed protein and the remainder germinated.   The testing
is, therefore, non-destructive and seed having useful characteristics
can be propagated for use in heritability studies (Romero et al.,
1975).

     Analysis of the proteins present in the ripening cotyledons of
the bean cultivar 'Tendergreen' by electrophoresis in acrylamide
gels under denaturing conditions (Sun et al., 1978) clearly shows
that three polypeptides are dominant in older seeds (Fig. 1).
Sensitive immunoelectrophoretic techniques have failed to detect
these polypeptides in any bean tissues other than the cotyledons
and even in the cotyledons prior to 10 days after flowering (Sun
et al., 1978; M. A. Mutschler, unpublished observations).   Expression of the genetical information for these peptides is, therefore,
repressed throughout development of the plant except in the ripening
cotyledons where its expression is intense.   How is this repression
achieved?   What events take place to remove its influence?   What

contributes to the intense expression during cotyledon development?
We expect to answer these questions and also important practical
questions such as whether the quantity and nutritional quality of
the storage protein is amenable to significant modification through
the application of biochemical techniques.

STRUCTURE OF THE MAJOR BEAN STORAGE PROTEIN

     A detailed knowledge of the structure and biosynthesis of the
major seed protein is a prerequisite for understanding genomic
controls. We have shown that the three predominant polypeptides
present in the cotyledons (Fig. 1) reversibly associate according to
pH conditions to a monomeric form at neutral pH and a tetramer at
pH 4 (Sun et al., 1974). The tetrameric form is readily precipitated
from acidic saline solutions by dialysis or dilution with distilled
water. This solubility in saline, but insolubility in aqueous
solutions, identifies this storage protein as a globulin. Because
it is the first globulin to precipitate on addition of water, we have
called it globulin-1 (G1) protein (McLeester et al., 1973). Further
reduction in the ionic strength of saline solutions of bean seed

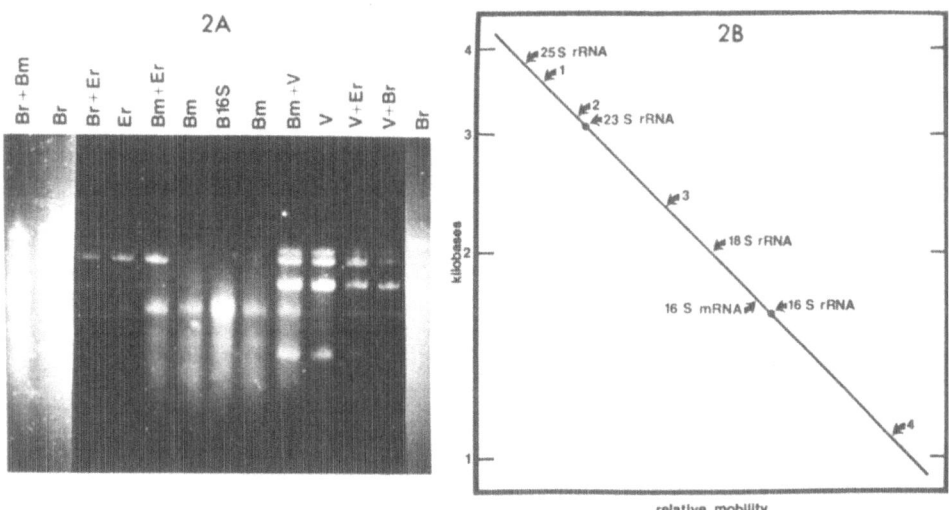

Fig. 2A. Electrophoretic separation of RNAs. Agarose gels (2%)
          were run in the presence of 2.2 M formaldehyde. Br is
          bean rRNA; Bm is bean mRNA; B16S is bean 16S mRNA enriched
          by one cycle of sucrose gradient sedimentation; Er is
          E. coli rRNA; and V is bromegrass mosaic virus (BMV) RNA.

Fig. 2B. Size determination of RNAs. 18S and 25S rRNAs are from
          beans; 16S and 23S rRNA are from E. coli; 1, 2, 3, and 4
          denote RNAs from BMV. The 16S mRNA migrated with an
          apparent length of 1.7 kilobases.

protein yields a second globulin fraction (G2) which contains several polypeptides which appear to be chemically unrelated. Each of the three polypeptides of Gl protein from Tendergreen seeds (α, 53 kilodaltons; β, 47 kilodaltons; and γ, 43 kilodaltons) is glycosylated (Hall et al., 1977); indeed, Gl is the same as glycoprotein II isolated by Pusztai & Watt (1970) from bean seeds. It comprises the majority of the phaseolin fraction isolated by Osborne (1894). A detailed review of bean storage protein nomenclature, which has caused much confusion, is in preparation (Buchbinder & Hall, 1980); in this article, Gl will be used to denote the major reserve protein of the seed.

PREPARATION OF mRNA FROM BEAN COTYLEDONS

The rapid accumulation of Gl protein during seed ripening clearly reflects the high activity of protein biosynthesis in the cotyledons. Addition of polysomes isolated from cotyledons 12-18 mm in length to cell-free translation reactions containing wheat germ extracts results in the incorporation of radioactive amino acid substrates into polypeptides that co-migrate on electrophoresis with the Gl polypeptides and which are precipitated by antibody to Gl protein (Sun et al., 1975; 1978). The isolation of free mRNA from polysomes was found to be possible, but the procedure was tedious, the yields low, and cell-free translation products contained many peptides in addition to those tentatively identified as being Gl subunits (Hall et al., 1977a). Because of these difficulties, we explored the possibility of extracting mRNA directly from bean cotyledons without first extracting polysomes (Buchbinder et al., 1980). Homogenization of ripening cotyledons in sodium borate buffer (0.2 M, pH 9.0, 65°C) containing 1% SDS followed by deproteinization with proteinase K (0.5 mg/ml) from Beckman at 37°C for 60 min yielded a total RNA fraction that was rich in poly(A)-containing RNA (about 10-20 µg/g fresh wt cotyledons) that could be isolated by modification of the Aviv & Leder (1972) technique for chromatography on oligo(dT)-cellulose columns (Fig. 2). This RNA was further separated by sucrose gradient centrifugation and the material sedimenting at 16S found to be highly enriched in mRNA that could be translated in vitro to yield polypeptides that co-migrated with Gl subunits (Fig. 3).

The translation products of 16S mRNA from bean cotyledons have been shown by immunoprecipitation and partial proteolytic hydrolysis to consist mainly of Gl polypeptides (Hall et al., 1978), but only two of the three Gl subunits could be discerned. The apparent absence of the α subunit of Gl from mRNA translation products was disquieting. However, it was apparent that all three Gl polypeptides were present in products translated from membrane-bound polysomes (Sun et al., 1978), and the knowledge that each of the native subunits was glycosylated (Hall et al., 1977) suggested that an inability of the wheat germ or rabbit reticulocyte lysate

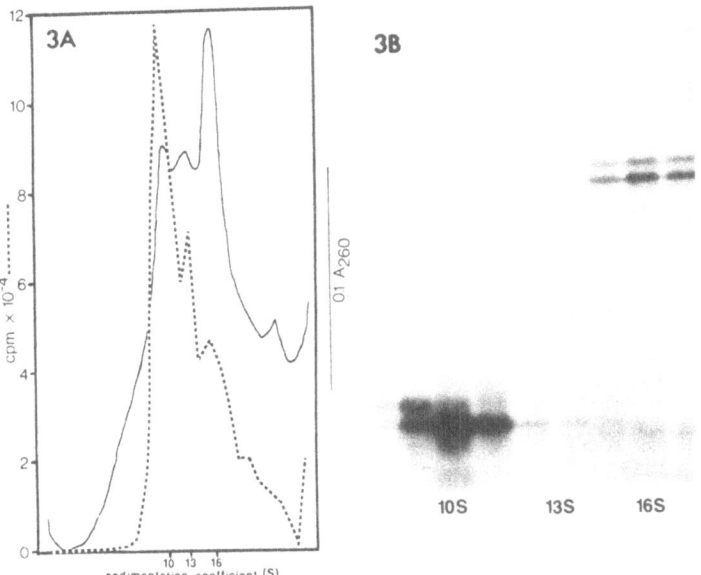

Fig. 3A.   Sucrose gradient sedimentation of cotyledon poly(A) RNA.
           Fractions were taken from a linear log sucrose gradient
           (———— A_{260}) and translated in the wheat germ cell-free
           system (- - -).

Fig. 3B.   Gel electrophoretic analysis of translation products of
           RNA fractions.   The 16S peak directed synthesis of Gl
           polypeptides.

cell-free translation systems to glycosylate the mRNA-derived polypeptides might account for the disparity between the Gl peptides
synthesized in vitro and in vivo.   Strong support for this hypothesis
has been adduced by inhibition of Gl glycosylation in vivo by
tunicamycin.   Gl protein isolated from cotyledon slices incubated in
a medium (Bailey et al., 1970) containing [^{35}S]met shows the typical
electrophoretic profile of three subunits, but only two subunits are
apparent in profiles of Gl protein isolated from incubations that
differed only by the addition of tunicamycin (2 µg/ml) to the medium
(Fig. 4).   These two subunits co-migrate on electrophoresis with the
in vitro translation products.   Taken together with additional
evidence from peptide mapping studies (Ma et al., 1980), it is
becoming clear that the glycosylated forms of the α and β subunits
migrate on electrophoresis as if they were about six kilodaltons
heavier than the non-glycosylated molecules.   The γ subunit appears
to carry a relatively smaller proportion of sugar residues, and the
glycosylated and non-glycosylated forms differ little in electrophoretic mobility.   As a consequence, the non-glycosylated β and
γ subunits translated in vitro co-migrate as a single band with a
mobility similar to that of the native (glycosylated) γ subunit;

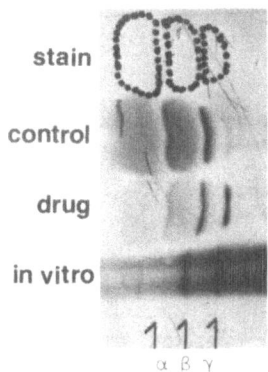

Fig. 4.   Incorporation of [^{35}S]methionine into Gl in the cotyledon
          slices in the presence and absence of tunicamycin.

the in vitro-synthesized α subunit has a mobility similar to that of
the native β subunit.

Heritable variation of the Gl polypeptides has been clearly
documented for the α subunit (Romero et al., 1975; Hall et al.,
1977). Comparison of the translational activity of mRNA from
different bean lines is greated facilitated by the direct extraction
technique because cotyledons (excised from ripening seeds under
essentially aseptic conditions, then initially frozen in liquid
nitrogen) can be stored at -80°C and the mRNA extracted when
convenient. A comparison of the native Gl polypeptides and Gl polypeptides translated in vitro from polysomes or mRNA isolated from
six different bean lines revealed that the variation in α subunit
is encoded in the mRNA (Fig. 5) and, hence, is not due to post-
translation modification (processing or glycosylation).

CLONING OF cDNA

The availability of microgram to milligram amounts of relatively
pure Gl mRNA made it realistic to undertake the synthesis of cDNA
(using reverse transcriptase) and subsequently to use both this cDNA
and cloned Gl cDNA in screening a library (Maniatis et al., 1978;
Blattner et al., 1978) of cloned bean genome DNA for Gl sequences.
Radioactive cDNA is a useful probe in itself, and if it were possible
to be sure that the mRNA sample used as a template for cDNA synthesis
contained a single sequence (or, as in the case of Gl polypeptides,
closely related sequences), it could be used directly for detecting
specific mRNA sequences in total DNA extracts and in the estimation
of reiteration frequency of specific genes. Despite the facts that
cDNA of discrete lengths can be synthesized using 16S mRNA as a

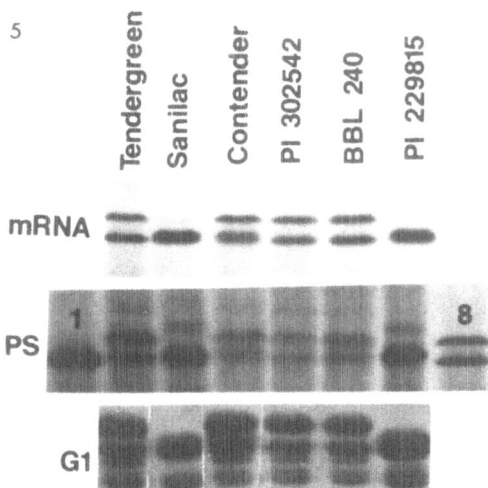

Fig. 5.   Electrophoretic analysis of products from translations
          directed by templates isolated from different lines of
          beans. mRNA are products synthesized by isolated mRNA in
          wheat germ extract; PS are incorporations directed by
          polysomes; G1 are authentic proteins isolated from cotyle-
          dons. Lanes 1 and 8 in the PS row are products whose
          synthesis was directed by mRNAs isolated from P.I. 229815
          and Tendergreen, respectively.

template (Fig. 6A), and that G1 peptides are essentially the only
protein products synthesized in vitro from 16S mRNA preparations
(Fig. 6B), it is difficult to be sure that low levels of contaminating mRNAs (which might possibly be preferentially transcribed
by reverse transcriptase and, hence, cDNAs) are absent. For this
reason, a cloned cDNA probe is preferable since it can be defini-
tively characterized (e.g. by hybrid-arrested translation: Paterson
et al., 1977) as containing the desired sequence. In collaboration
with Dr. Harvey Faber, W. B. Gurley, and others, we have cloned G1
mRNA-cDNA hybrids in bacteriophage λ by techniques similar to those
of Kinniburgh et al. (1978). The Charon series of λ phages (Blattner
et al., 1977) were used together with E. coli strain DP50supF to
provide an EK2 host-vector system permitting the work to be carried
out under P2 conditions. Radioactive cDNA prepared from a G1 mRNA
template was used to screen plaques by the Benton and Davis (1977)
technique; the DNA from some plaques (e.g. Ch16A·G1·9-5; Fig. 7)
was found to hybridize strongly to the cDNA probe. Hybrid-arrested
translation (HART) was used to confirm the identity of the inserted
DNA. Since substantial amounts of DNA greatly inhibit translation,
the HART technique was attempted with a fragment of the phage DNA
presumed to contain a G1 cDNA insert. Digestion of Ch16A·G1·9-5

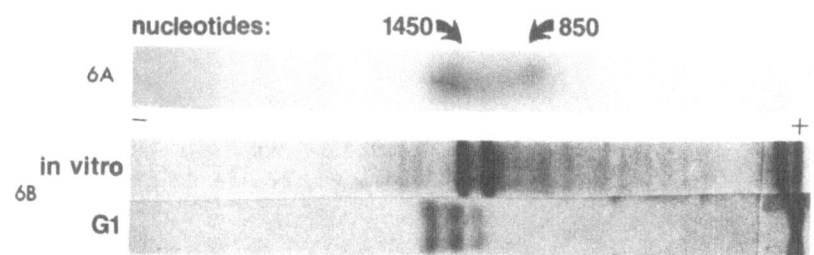

Fig. 6A.   Electrophoretic analysis of cDNA reverse transcribed from
           16S mRNA.

Fig. 6B.   Electrophoretic analysis of products synthesized in a 16S
           mRNA directed reticulocyte lysate reaction. Products were
           not immunoprecipitated.

DNA with the restriction endonuclease HpaI provided a set of DNA
fragments for which the electrophoretic mobility of one fragment in
an agarose gel (fragment 2) was decreased by the amount expected for
the insertion of a DNA sequence 1200 base pairs long (Fig. 8A).
cDNA prepared from a G1 mRNA template hybridized preferentially to
this fragment (Fig. 8B), indicating that it contains a G1 cDNA
insert. This fragment was isolated from a preparative agarose gel
by the method of Jeffreys & Flavell (1977). The isolated DNA was
used in the HART technique along with total bean cotyledon poly(A)+

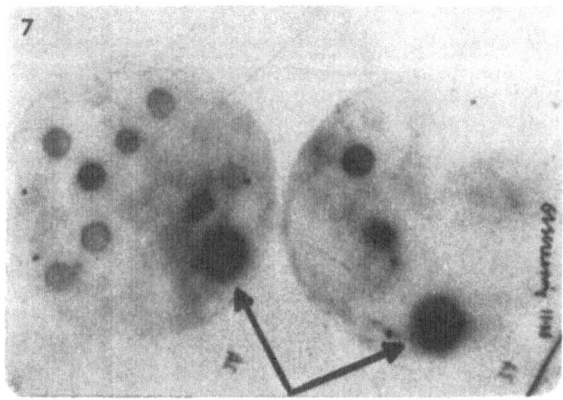

Clone Ch16A.G1.9-5

Fig. 7.   Hybridization of cDNA to DNA from phage plaques of cloned
          G1 cDNA. The arrows indicate the position of the strongly
          hybridizing clone, Charon 16A·G1·9-5.

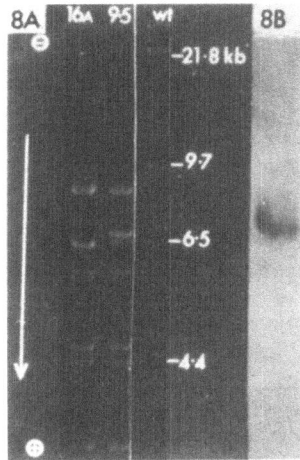

Fig. 8A. Electrophoretic separation of restriction digests of
          isolated DNAs. Lane 16A is Charon 16A DNA digested by
          HpaI; 9·5 is Charon 16A·G1·9-5 DNA digested by HpaI; and
          wt is wild type λ DNA digested Hind III.

Fig. 8B. Southern blot hybridization of bean cDNA to the restriction
          digest of Ch16A·G1·9-5.

RNA. The in vitro translation of G1 peptides was specifically
arrested by the added DNA (Fig. 9). The evidence presented confirms
that Ch16A·G1·9-5 contains a sequence corresponding to a section of
G1 cDNA. Interestingly, translation of all the G1 peptides is
arrested, providing support for evidence obtained by partial peptide
mapping (Ma et al., 1980) that there is considerable sequence homology among the G1 subunit sequences.

    These cDNA probes will be valuable for detecting the earliest
stage of cytoledon development at which G1 mRNA is transcribed and
for determining if G1 mRNA is transcribed, besides not being
translated, in tissues other than those of the cotyledon. The amount
of G1 mRNA present can also be measured; this will be valuable in
determining if the large variation in G1 content (from less than
20% to over 70% of the total seed protein: Mutschler & Bliss, 1978)
results from a differential efficiency for translation of similar
levels of G1 mRNA or because these lines contain different absolute
levels of G1 mRNA.

CLONING OF GENOMIC SEQUENCES CODING FOR STORAGE PROTEINS

    Despite the value of DNA sequences complementary to G1 mRNA,
the isolation of the G1 gene itself is a more exciting goal. It has

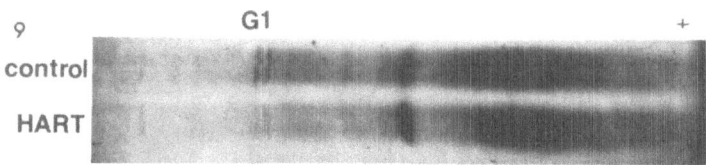

Fig. 9.   Hybrid-arrested translation (HART) of total cotyledon mRNA
          with DNA isolated from Ch16A·G1·9-5.

long been appreciated that the genomic DNA is likely to contain
regulatory sequences (e.g. those promoting transcription) not present
in the mRNA itself.   The discovery that many, if not all, eukaryotic
genes contain non-coding sequences (introns) within the coding
sequences (Jeffreys & Flavell, 1977a) makes the isolation of genomic
DNA all the more cogent.

    An estimate of the number of G1 genes can be obtained through
hybridization kinetic analysis using cDNA (Rosen & Barker, 1976).
Such information will be useful (although recently reservations
regarding the accuracy of the results have appeared: Kiper, 1979),
and it is possible to use radioactive DNA probes to determine the
chromosomal location of specific sequences (e.g. highly reiterated
sequences: Flavell et al., 1978).   If multiple copies of G1 genes
exist, it is by no means certain that they would be arranged conti-
guously.   Indeed, the genes for the α and β subunits of globin lie
on different chromosomes (Deisseroth et al., 1977; 1978), and
classical genetical analysis has shown that genes coding for zein
(corn storage protein) genes are distributed on three chromosomes
(Valentini et al., 1979).   As noted above, it will be important to
locate sequences regulating expression (transcription and translation) of G1 polypeptides.   Genomic clones are more likely to contain
these regions, but at present, we cannot be certain that the position
of the control sequences bears any correlation with the position of
the coding sequences.   Such information can only be obtained
experimentally.

    Recently, we have isolated DNA of excellent purity from bean
embryos.   This DNA is >60 kilobases long and can be completely
cleaved by Eco RI (Fig. 10A).   Fragments obtained by Eco RI cleavage
under restrictive conditions have been size-separated on salt (NaCl)
gradients (Fig. 10B) and fragments 8-20 kilobases long isolated.
In collaboration with Drs. Jerry Slighton and Oliver Smithies, these
fragments have been cloned in Charon 24A.   At present, screening of
these genomic clones using G1 cDNA probes is in progress, and
several showing strong hybridization have been identified.

    If a clone containing an apparently complete G1 sequence can
be isolated, the ability of the DNA to be transcribed (with RNA
polymerase II) and of the resulting mRNA transcripts to direct

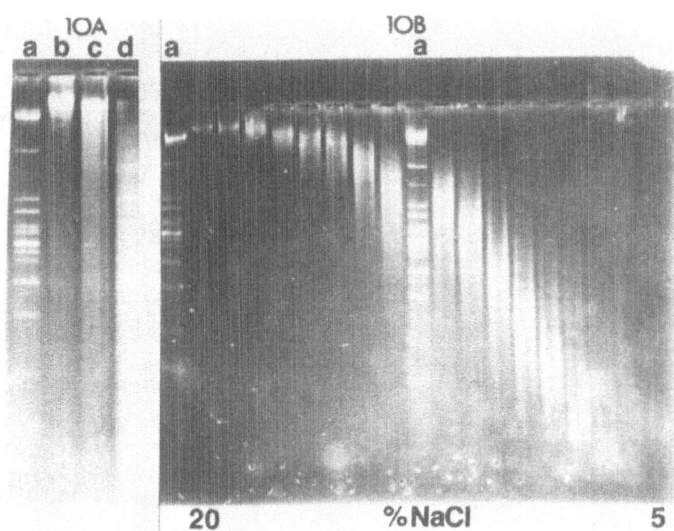

Fig. 10A.   Electrophoretic analysis of DNA isolated from bean
            embryonic axes. a, molecular weight standards; b,
            undigested high molecular weight bean DNA; c and d,
            bean DNA digested with increasing amounts of Eco RI
            restriction enzyme.

Fig. 10B.   Electrophoretic analysis of bean DNA Eco RI restriction
            fragments separated by salt gradient centrifugation.
            a, molecular weight standards.

cell-free protein synthesis will be assayed. These are important
steps in understanding the arrangement and function of economically
important genes and the potential for chemical modification (e.g.
the chemical insertion of additional methionine sequences into the
sulfur-poor G1 protein) evaluated. If a suitable vector for
insertion of functional genes into plant genomes is developed, the
exciting potential for genetic modification through the use of
chemical recombination may indeed be realized.

Acknowledgments

     We are indebted to many for their help and advice in this work,
especially J. Bedbrook, F. Blattner, J. W. S. Brown, J. Callis,
H. Faber, R. B. Flavell, W. B. Gurley, J. M. Lamb, Y. Ma, R. C.
McLeester, M. A. Mutschler, J. Romero, J. Ross, J. L. Slightom,
and O. Smithies. We thank Dr. J. W. Beard of Life Sciences, Inc.
for reverse transcriptase which was supplied through the Viral
Cancer Program of the National Cancer Institute. This study was
supported by grants from NSF (PCM 78-11804), USDA/SEA (5901-0410-8-
0138-0) and the Herman Frasch Foundation.

REFERENCES

Aviv, H., and Leder, P., 1972, Purification of biologicably active
    globin messenger RNA by chromatography on oligothymidylic
    acid-cellulose, Proc. Natl. Acad. Sci. USA, 60:1408.
Bailey, C. J., Cobb, A., and Boulter, D., 1970, A cotyledon slice
    system for electron autoradiographic study of the synthesis and
    intracellular transport of the seed storage protein of Vicia
    faba, Planta (Berl.), 95:103.
Benton, W. D., and Davis, R. W., 1977, Screening λgt recombinant
    clones by hybridization to single plaques in situ, Science,
    196:180-182.
Blattner, F. R., Blechl, A. E., Denniston-Thompson, K., Faber, H. E.,
    Richards, J. E., Slightom, J. L., Tucker, P. W., and Smithies,
    O., 1978, Cloning human fetal γ globin and mouse α-type globin
    DNA: preparation and screening of shotgun collections, Science,
    202:1279.
Blattner, F. R., Williams, B. G., Blechl, A. E., Denniston-Thompson,
    K., Faber, H. E., Furlong, L.-A., Grunwald, D. J., Kiefer,
    D. O., Moore, D. D., Schumm, J. W., Sheldon, E. L., and
    Smithies, O., 1977, Charon phages: safer derivatives of
    bacteriophage lambda for DNA cloning, Science, 196:161.
Buchbinder, B. U., and Hall, T. C., 1980, A critical reexamination
    of Phaseolus storage protein nomenclature, manuscript in
    preparation.
Buchbinder, B. U., Ma, Y., and Hall, T. C., 1980, Messenger RNA
    for Gl protein of French bean seeds: mRNA characterization
    and phenotypic variation, manuscript in preparation.
Deisseroth, A., ·Nienhuis, A., Lawrence, J., Giles, R., Turner, P.,
    and Ruddle, F., 1978, Chromosomal location of human β globin
    gene on human chromosome 11 in somatic cell hybrids, Proc.
    Natl. Acad. Sci. USA, 75:1456.
Deisseroth, A., Nienhuis, A., Turner, P., Velez, R., Anderson, W. F.,
    Ruddle, F., Lawrence, J., Creagen, R., and Kucherlapati, R.,
    1977, Localization of the human α-globin structural gene to
    chromosome 16 in somatic cell hybrids by molecular hybridiza-
    tion assay, Cell, 12:205.
Flavell, R., O'Dell, M., Rinpau, J., and Smith, D., 1978, Biochemi-
    cal detection of alien DNA incorporated into wheat by
    chromosome engineering, Heredity, 40:439.
Hall, T. C., Ma, Y., Buchbinder, B. U., Pyne, J. W., Sun, S. M.,
    and Bliss, F. A., 1978, Messenger RNA for Gl protein of French
    bean seeds: Cell-free translation and product characteriza-
    tion, Proc. Natl. Acad. Sci. USA, 75:3196.
Hall, T. C., McLeester, R. C., and Bliss, F. A., 1977, Equal
    expression of the maternal and paternal loci for the polypep-
    tide subunits of the major storage protein of the bean,
    Phaseolus vulgaris L., Plant Physiol., 57:1122.

Hall, T. C., Sun, S. M., Buchbinder, B. U., and Belozerskii, M. A.,
    1977a, The translation of mRNA for storage globulin of the bean,
    Phaseolus vulgaris, in: "Translation of Natural and Synthetic
    Polynucleotides," A. B. Legocki, ed., University of Agriculture
    in Poznan (Poland).
Jeffreys, J. A., and Flavell, R. A., 1977, A physical map of the DNA
    region flanking the rabbit β-globin gene, Cell, 12:429.
Jeffreys, A. J., and Flavell, R. A., 1977a, The rabbit β-globin gene
    contains a large insert in the coding sequence, Cell, 12:1097.
Kinniburgh, A. J., Mertz, J. E., and Ross, J., 1978, The precursor
    of mouse β-globin messenger RNA contains two intervening RNA
    sequences, Cell, 14:681.
Kiper, M., 1979, Gene numbers as measured by single-copy DNA satura-
    tion with mRNA are routinely overestimated, Nature, 278:279.
Ma, Y., Bliss, F. A., and Hall, T. C., 1980, Peptide mapping analysis
    of Gl, the major storage protein of Phaseolus vulgaris L.,
    manuscript in preparation.
Maniatis, T., Hardison, R. C., Lacy, E., Lauer, J., O'Connell, C.,
    Quon, D., Sim, G. K., and Efstratiadis, A., 1978, The isolation
    of structural genes from libraries of eucaryotic DNA, Cell, 15:
    687.
McLeester, R. C., Hall, T. C., Sun, S. M., and Bliss, F. A., 1973,
    Comparison of globulin proteins from Phaseolus vulgaris with
    those from Vicia faba, Phytochemistry, 12:85.
Mutschler, M. A., and Bliss, F. A., 1978, Inheritance of total pro-
    tein and globulin-1 protein in the seed of common bean, Agron.
    Abs., 70th Meeting, p. 57.
Osborne, T. B., 1894, The proteids of the kidney bean, J. Amer.
    Chem. Soc., 16:633, 703, 757.
Paterson, B. M., Roberts, B. E., and Kuff, E. L., 1977, Structural
    gene identification and mapping by DNA·mRNA hybrid-arrested
    cell-free translation, Proc. Natl. Acad. Sci. USA, 74:4370.
Pusztai, A., and Watt, W. B., 1970, Glycoprotein II: the isolation
    and characterization of a major antigenic and non-haemaggluti-
    nating glycoprotein from Phaseolus vulgaris, Biochim. Biophys.
    Acta, 207:413.
Romero, J., Sun, S. M., McLeester, R. C., Bliss, F. A., and Hall,
    T. C., 1975, Heritable variation in a polypeptide subunit of
    the major storage protein of the bean Phaseolus vulgaris L.,
    Plant Physiol., 56:776.
Rosen, J. M., and Barker, S. W., 1976, Quantitation of casein
    messenger ribonucleic acid sequences using a specific comple-
    mentary DNA hybridization probe, Biochemistry, 15:5272.
Sun, S. M., Buchbinder, B. U., and Hall, T. C., 1975, Cell-free
    synthesis of the major storage protein of the bean Phaseolus
    vulgaris L., Plant Physiol., 56:780.
Sun, S. M., McLeester, R. C., Bliss, F. A., and Hall, T. C., 1974,
    Reversible and irreversible dissociation of globulins from
    Phaseolus vulgaris seed, J. Biol. Chem., 249:2118.

Sun, S. M., Mutschler, M. A., Bliss, F. A., and Hall, T. C., 1978,
    Protein synthesis and accumulation in bean cotyledons during
    growth, Plant Physiol. 61:918.
Valentini, G., Soave, C., and Ottaviano, E., 1979, Chromosomal
    location of zein genes in Zea mays, Heredity, 42:33.

THE mRNAs THAT CODE FOR SOYBEAN SEED PROTEINS[1]

R.N. Beachy[*], K.A. Barton[+], J.T. Madison[+], J.F. Thompson[+], and N. Jarvis[*]

[*]Dept. of Biology, Washington University, St. Louis, MO 63130
[+]U.S. Department of Agriculture, Science and Education Administration - Agricultural Research, U.S. Plant, Soil, and Nutrition Laboratory, Ithaca, NY 14853

SUMMARY

Soybean (Glycine max L., Merr.) seeds produce large amounts of protein during the period of rapid seed expansion. Storage proteins comprise 60 to 80% of the accumulated proteins. Messenger RNAs (mRNAs) isolated from developing seeds ranged from approximately 8S to 25S, and *in vitro* translation reactions containing the RNAs produced two subunits of the 7S storage protein (conglycinin) and polypeptides which may be some or all of the acidic and basic subunits of the 11S protein (glycinin). Electrophoresis of fractionated mRNAs in several gel systems separated several species of mRNA (M_r = 1.1, 0.84, and possibly 0.75 X 10^6) in fractions of mRNA that code for the *in vitro* synthesis of subunits of the 7S protein. No other species of RNA was prominent. Complementary DNA produced using the high M_r mRNAs as templates included nearly full-length transcripts.

INTRODUCTION

Mature soybean seeds contain from 40 to 55% protein, most of which is produced over an 18 to 24 day period of seed growth

---

[1]This research was partially supported by a research grant to RNB from the U.S. Plant, Soil and Nutrition Laboratory, Ithaca, NY. KAB was supported by Fellowship Funds from W.S. and E.M. Meyers.

(Hill and Breidenbach, 1974). Depending upon the cultivar and the growth conditions, there may be from 1.5 to 4 mg of protein produced per seed per day during periods of maximum protein biosynthesis, making the soybean seed an interesting model for studies of gene regulation. Our interest in seed proteins has arisen from several questions: (1) What factors control the variable amounts of proteins produced in different legume species, i.e., from the approximately 20% protein in seeds of Pisum sativum to the greater than 40% found in soybeans? (2) What is the molecular basis for the "turning on" and "turning off" of the synthesis of seed storage proteins during the period of time from anthesis to seed maturity, and what effect does the nutritional status of the plant, or other factors (pathogen infection, for example) have on the synthesis of these proteins? These and other questions may be approached by first identifying the mRNAs which code for the synthesis of the seed storage proteins.

Most of the seed proteins that accumulate are the storage proteins which have no known enzymatic function, and which are normally broken down for use during germination of the seed (Bewley and Black, 1978). They are soluble in high salt solutions (0.4 $\underline{M}$ NaCl) and are characterized by their sedimentation rates in sucrose density gradients of approximately 7S and 11S. The 11S protein (glycinin or legumin) has been well characterized in terms of its subunit composition and the requirements for dissocia- tion and re-assembly of the protein (Kitamura et al., 1977). The 11S protein consists of four basic subunits of about 20,000 M_r, and four or six acidic subunits of 30 to 45,000 M_r (Draper and Catsimpoolas, 1977: Moreira et al., 1978, 1979; Kitamura and Shibasaki, 1975). The 7S protein is glycosylated and contains three major subunits, α', α, and β with M_r = 0.91, 0.84, and 0.44 x 10^5, respectively (according to our measurements in several PAGE systems), although Thanh and Shibasaki (1977) reported different M_r values. We have described an approximate timetable for the production of these subunits in developing seeds (Beachy et al., 1979). In short, all subunits are produced during periods of rapid protein accumulation; however, early in seed development the subunits of the 7S protein apparently accumulate during different time periods.

When polyribosomes isolated from immature seeds were used as templates in a wheat germ cell-free translation reaction it appeared that most of the 11S subunit components, as well as one or two subunits of the 7S protein, were being formed (Beachy et al., 1978). As expected, the translation of the progressively larger polypeptides occurred on polyribosomes with progressively higher rates of sedimentation. Our next step was to isolate and characterize the mRNAs involved.

METHODS AND RESULTS

Isolation and Chacterization of the mRNAs.

        We first began to isolate large quantitites of mRNAs from
immature seeds by isolating polyribosomes from which poly(A)+ RNA
was selected by oligo(dT)-cellulose chromatography. These experiments yielded RNA which was translated efficiently in *in vitro*
reactions (Beachy *et al.*, 1979). However, the yields of mRNA
were relatively low and the RNAs differed qualitatively between
extractions. For example, we did not always extract the mRNAs
which coded for the *in vitro* synthesis of the α'- and α-like
polypeptides (Beachy *et al.*, 1979). We subsequently began to
isolate total seed nucleic acids in the absence of phenol, employing
the method described by Hall *et al.* (1978), using SDS and proteinase
K. The method yielded large amounts of nucleic acids, but the
relative activity of the RNA in *in vitro* translation reactions
was variable. In all cases poor translation of RNA was correlated
with the presence of a gelatinous material in the RNA presumed to
be polysaccharides, not unlike the material common in phenol
extracts of plant tissues. Most of the material has been removed
by slightly modifying the procedure of Hall *et al.* We found it
necessary to suspend the LiCl-precipitated nucleic acids (NA)
with a Dounce homogenizer, and to wash the precipitate several
times with 2 $\underline{M}$ LiCl. After the final LiCl pellet was dissolved
in water, the RNA solution was clarified by centrifugation. The
NAs were precipitated several times with ethanol, each time taken
up in a minimal amount of sterile water and clarified by centrifugation. These steps resulted in the loss of most of the gelatinous
material in the extracts, as well as a small percentage of the
nucleic acids, which may be partially recovered by re-extracting
the undissolved residue with smaller amounts of water. After the
final alcohol precipitation, NAs were prepared for affinity
chromatography on oligo(dT)-cellulose. The contaminating material
co-purified with the poly(A)+ RNA, and was removed by binding the
mRNA to hydroxylapatite (HAP) since the contaminant could be
washed from the column with 10 m$\underline{M}$ phosphate buffer. RNA was
removed from HAP with high molarity (0.4 $\underline{M}$) phosphate buffer, the
PO_4^- was removed by G-25 Sephadex, and the RNA was rechromatographed
on oligo(dT)-cellulose. Bound RNA was removed in a no-salt
buffer and prepared for DMSO-sucrose gradient centrifugation
(Bishop *et al.*, 1968). RNA prepared by this method was highly
active both in *in vitro* translation and cDNA synthesis.

        Most of the mRNA had sedimentation values of 12 to 18S in
sucrose gradients containing DMSO. mRNA activity, measured by
the incorporation of labeled amino acids in *in vitro* protein
synthesis reactions, closely paralleled the optical density
(A_{280}) profiles of the sucrose gradients (Beachy *et al.*, submitted).
The polypeptides synthesized in translation reactions containing

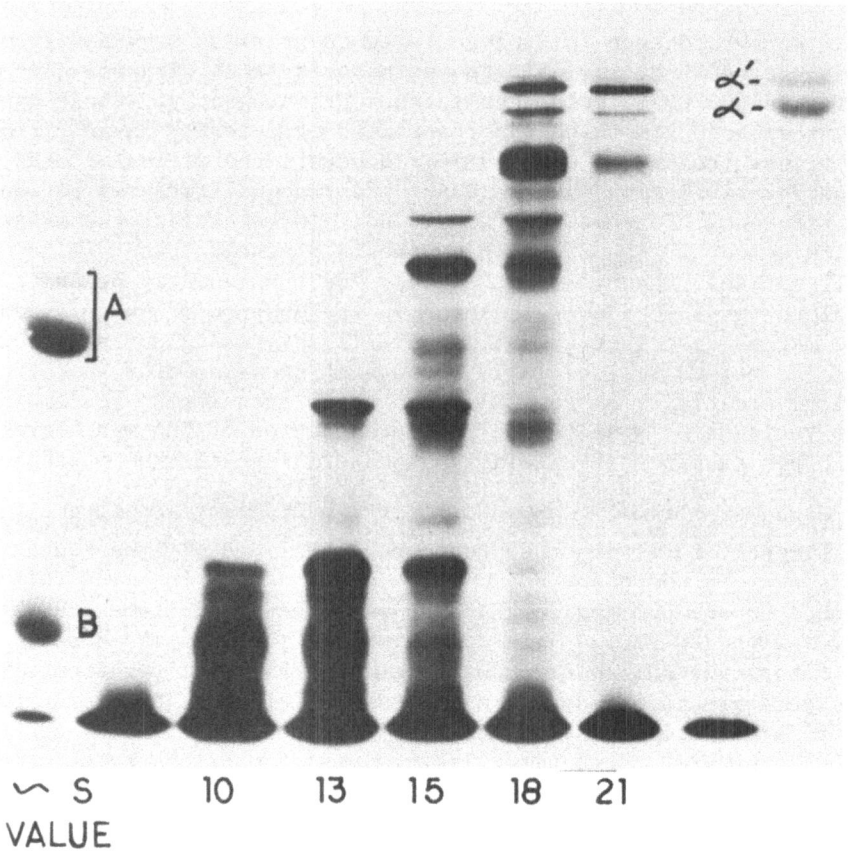

Fig. 1.   Polyacrylamide gel electrophoresis of polypeptides that
           were produced in cell-free protein synthesis reactions
           to which soybean mRNAs were added. The mRNAs were iso-
           lated from developing seeds and were subjected to centri-
           fugation in sucrose density gradients prepared in di-
           methyl sulfoxide prior to translation *in vitro*: approx-
           imate S values for the mRNAs that were translated are
           indicated. Radioactivity was detected by fluorography.

RNAs that sedimented at different rates are shown in Figure 1,
which demonstrates that some of the polypeptides produced *in
vitro* had electrophoretic mobilities similar to the mobilities of
some of the authentic protein subunits. We have completed tryptic
fragment mapping of the α'- and α-like subunits of the 7S proteins
(using high pressure liquid chromatographic analyses according to
Vogt *et al.*, 1975) and have verified that the α' and α subunits

are synthesized *in vitro* (Beachy *et al*., submitted). Similar
experiments were done to verify that some or all of the basic
subunits of the 11S protein were synthesized *in vitro*, but with
equivocal success: even though prominent bands of radioactivity
produced *in vitro* co-migrated with authentic 11S basic subunits,
only a portion of the trypsin-induced peptides eluted from the
ion exchange resin with peptides of authentic subunits. Similar
results were obtained with the acidic subunits (Barton, unpublished). We have not done comparable analyses using peptides produced
by "read-off" of polyribosomes *in vitro*, which might produce
polypeptides more nearly identical to the authentic subunits.

Analysis of the RNAs found in different fractions of the
sucrose gradients by polyacrylamide gel electrophoresis in the
presence of 8 $\underline{M}$ urea gave results as shown in Figure 2. Note
that no outstanding bands of RNA are observed in fractions 7
through 13, and that RNA bands (other than rRNA components) are
seen in fractions 14 through 20. The molecular weights of the
RNAs migrating between the 18S and 25S RNAs range in molecular
weight from 1.1 to 0.75 x 10^6 when analyzed in several gel
systems (Beachy *et al*., submitted), large enough to contain the
genetic information to code for the α' and α subunits of the 7S
storage protein. These subunits are produced by the RNAs found
in fractions 15 through 20. On the other hand, the fractions of
RNA coding for the presumed 11S basic subunits (fraction 10) and
acidic subunits (fraction 13) contain only broadly diffuse bands
of RNA.

Synthesis of cDNA

We used the RNAs in fractions 16 and 17 as templates in
reactions containing reverse transcriptase (using the methods of
Meyers and Spiegelmann, 1978), and produced cDNA of high molecular
weight. The fluorograph shown in Figure 3 demonstrates that our
reactions produced a mixture of cDNAs containing two distinct
species with electrophoretic mobilities between that of the 18S
and 25S rRNAs. What relationship these cDNAs have to the bands
of presumed mRNAs described above remains to be determined, but
it is tempting to think that they represent full-length transcripts
of several species of mRNA.

DISCUSSION

Isolation of mRNA from immature, developing soybean seeds in
a form that is highly active in *in vitro* translation reactions
can be accomplished using either of several methodologies. We
routinely use the proteinase K ~ SDS method (Hall *et al*., 1978)
and include the HAP and oligo(dT)-cellulose chromatographic steps

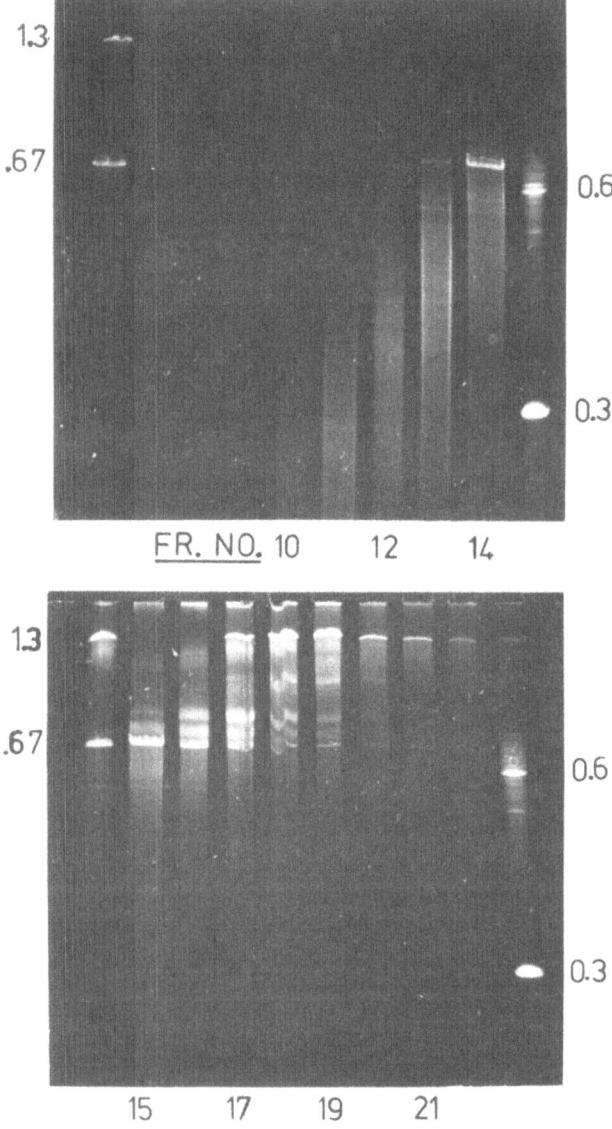

Fig. 2.   Electrophoresis of mRNAs (taken from fractions of a
           DMSO-containing sucrose density gradient) in a 4%
           polyacrylamide gel containing 8 $\underline{M}$ urea and 0.2% SDS.
           Ethidium bromide staining was used to visualize the
           RNAs. Molecular weight markers used were 18S and 25S
           soybean rRNA and 0.6 and 0.3 x 10^6 M_r RNAs isolated
           from subgenomic fragments of tobacco mosaic virus.

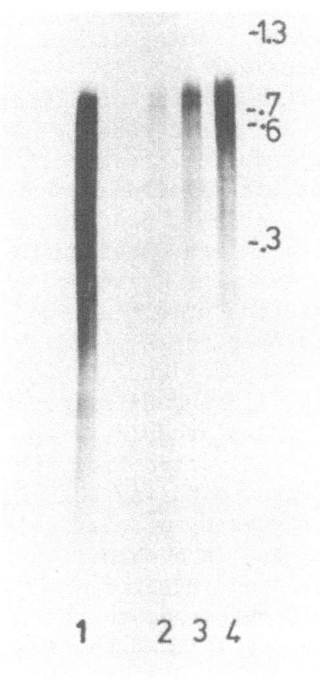

Fig. 3. Fluorograph of complementary DNA transcribed from the
mRNA in fraction 17 shown in
Figure 2. The cDNA was electrophoresed in gels contain-
ing 4% polyacrylamide, 8 $\underline{M}$
urea and 0.2% SDS. Lane $\underline{1}$,
unfractionated cDNA; lanes 2,
3, 4 size-fractionated cDNAs.
M_r x 10^{-6} markers are indicated on the right.

as outlined above.

The activity of the mRNAs in *in vitro* translation reactions,
both in mRNA-dependent reticuloycte lysates (Pelham and Jackson,
1976) and in wheat germ extracts, was comparable (Beachy *et al.*,
submitted), and the products synthesized in the two systems were
similar. There were minor differences in the synthesis of several
polypeptides, but not in the area of the gel to which subunits of
storage proteins migrated. Tryptic fragment mapping conclusively
demonstrated that the α' and α subunits were synthesized *in
vitro*. We have also shown that the α' and α subunits have nearly
identical tryptic peptide elution profiles (Beachy *et al.*, submitted). They also have very similar amino acid compositions (Thanh
and Shibasaki, 1977).

Since there appears to be a greater amount of 11S than 7S
protein (two or three times more) in soybean seeds, we expected
that there would be large amounts of the 11S protein subunits
synthesized *in vitro*. We have attempted numerous peptide mapping
experiments with polypeptides synthesized *in vitro* having electrophoretic mobilities equal to, faster than, and slower than authentic
acidic and basic subunits. To date we have not identified polypeptides sufficiently similar to the authentic subunits to be

convinced that they are indeed the storage protein subunits
(Barton, unpublished). Several possible reasons, among others,
for our results are: (1) The peptides synthesized in vitro may
react differently to the trypsin digestion thus leading to different
peptide elution patterns. Authentic acidic and basic subunits
are highly sensitive to trypsin and give reproducible fragment
elution patterns. The in vitro polypeptides with similar electrophoretic mobilities, on the other hand, give variable peptide
elution patterns. (2) The mRNAs may not be efficient messengers
in wheat germ extracts or reticulocyte lysates. For example,
there may be requirements for specific initiation or elongation
factors (from soybeans) to be efficiently translated in vitro.
This reasoning requires, however, that the mRNAs for the 7S
subunits have different requirements than the 11S subunits, since
the α' and α subunits of the 7S protein are translated in vitro.
(3) The mRNAs for the 11S subunits may not be extracted from the
tissues by our procedures, or they may be degraded during the
extraction. (4) The mRNAs in question may represent a relatively
low percentage of the total mRNA population, and in combination
with suggestion (2) may be very poorly translated in vitro.
Since we can see no band of mRNA which might correspond to the
mRNA for the 11S basic subunits, the easy answer is that these
are not abundant mRNAs in our extracts. Obviously more experiments
are needed to resolve these questions.

cDNAs made to fractions of mRNAs encoding the α' and α
subunits of the 7S protein apparently included full- or nearly
full-length transcripts of the mRNAs. Since the predominant
mRNAs in the transcription reactions represent approximately 60%
of the non-ribosomal RNAs in the fraction (see Figure 2), selecting
cDNAs, or cDNA clones, containing sequences complimentary to
those RNAs should be feasible. Presuming that they represent the
mRNAs for the α' and α subunits we will be in a position to
conduct experiments relating the biosynthesis of the 7S storage
protein to the development of the soybean seed.

REFERENCES

Beachy, R.N., Barton, K.A., Thompson, J.F., and Madison, J.T., In
    vitro synthesis of the α' and α subunits of the 7S storage
    protein of soybean (submitted for publication).
Beachy, R.N., Thompson, J.F., and Madison, J.T., 1978, Isolation
    of polyribosomes and messenger RNA active in in vitro synthesis
    of soybean seed proteins, Plant Physiol. 61: 139-144.
Beachy, R.N., Thompson, J.F., and Madison, J.T., 1979, Isolation
    and characterization of messenger RNAs which code for the
    subunits of soybean seed proteins, in: "The Plant Seed:
    Development, Preservation, and Germination", I. Rubenstein,
    E. Green, R. Phillips and B. Gengenbach, eds., Academic

NY (in press).

Bewley, J.D., and Black, M., 1978, "Physiology and Biochemistry of Seeds in Relation to Germination" Vol. 1, Development, Germination and Growth", Springer Verlag, NY.

Bishop, J.M., Koch, G., Evans, B., and Merriman, M., 1969, Polio-virus replicative intermediate: structural basis of infectivity, J. Mol. Biol. 46: 235-249.

Draper, M., and Catsimpoolas, N., 1977, Isolation of the acidic and basic subunits of glycinin, Phytochemistry 16: 25-27.

Hall, T.C., Ma, Y., Buchbinder, B.U., Pyne, J.W., Sun, S.M., and Bliss, F.A., 1978, Messenger RNA for G1 protein of French bean seeds: cell-free translation and product characterization, Proc. Natl. Acad. Sci. USA 75:3196-3200.

Hill, J.E., and Breidenbach, R.W., 1974, Proteins of soybean seeds. II. Accumulation of the major protein components during seed development and maturation, Plant Physiol. 53: 747-751.

Kitamura, K., and Shibasaki, K., 1975, Isolation and some physico-chemical properties of the acidic subunits of soybean 11S globulin, Agr. Biol. Chem. 39: 945-951.

Kitamura, K., Takagi, T., and Shibasaki, K., 1977, Renaturation of soybean 11S globulin, Agr. Biol. Chem. 41: 833-840.

Meyers, J.C., and Spiegelman, S., 1978, Sodium pyrophosphate inhibition of RNA·DNA hybrid degradation by reverse transcriptase, Proc. Natl. Acad. Sci. USA 75: 5329-5333.

Moreira, M.A., Larkins, B.A., Hermondson, M.A. and Nielsen, W.C., 1978, Soybean storage proteins: isolation and characterization of the largest acidic subunit of glycinin, Plant Physiol. 61: supplement (abstract No. 213).

Moreira, M.A., Hermondson, M.A., Larkin, B.A., and Nielson, N.C., 1979, Soybean storage proteins: Further characterization of the acidic subunits of glycinin, Plant Physiol. 63: supplement (abstract No. 526).

Pelham, H.R.B., and Jackson, R.J., 1976, An efficient mRNA-dependent translation system from reticulocyte lysates, Eur. J. Biochem 67: 247-256.

Thanh, V.H., and Shibasaki, K., 1977, Beta-conglycinin from soybean proteins. Isolation and immunological and physico-chemical properties of the monomeric forms, Biochem. Biophys. Acta 490: 370-384.

Vogt, V.M., Eisenman, R., and Diggelmann, H., 1975, Generation of avian myeloblastosis virus structural protein by proteolytic cleavage of a precursor polypeptide, J. Mol. Biol. 96: 471-493.

DEVELOPMENTAL REGULATION OF STORAGE PROTEIN SYNTHESIS IN SEEDS

I. M. Sussex, R. M. K. Dale, and M. L. Crouch[1]

Biology Department
Yale University
New Haven, CT   06520, U. S. A.

INTRODUCTION

Biosynthesis of storage proteins in angiosperm seeds represents a massive, one time, embryo-specific expression of a few genes during a short period of the plant's developmental cycle. Storage protein begins to accumulate rapidly about one third to one half of the way through seed development at a time when cell divisions are completed or nearly so (Millerd, 1975; Dure, 1975). Thus there is a numerically stabilized embryo cell population in which protein composition is changing due to the preferential synthesis of one or a few storage proteins.

This fact is important when considering ways, other than the selection of plants with altered amino composition of storage proteins, by which the nutritional quality of seed proteins may be improved. By altering harvest time in relation to seed maturity, or by selecting regulatory mutants in which the timing or the rate of storage protein accumulation is altered, or by treating plants so as to enhance or inhibit storage protein accumulation it should be possible to modify the seed protein content in predictable ways.

In this paper we review studies that we have carried out to examine ways in which the accumulation and synthesis of seed storage proteins can be modified during embryogeny in the bean, Phaseolus vulgaris, and in rape seed, Brassica napus.

---

[1] Present address: Biology Department, Indiana Univeristy, Bloomington, IN   87401.

EXPERIMENTAL MATERIALS

     All the plants used in this study were grown in controlled
environment rooms. Phaseolus plants were grown in 16/8 hr light/
dark photoperiods, at 25°C, and Brassica plants were grown in
10/14 hr light/dark photoperiods for the first two weeks, then in
16/8 hr photoperiods, at 18°C/14°C day/night. All plants were
supplied with Hyponex nutrient solution. Sterile cultured embryos
were grown in 16/8 hr light/dark photoperiods at 25°C on nutrient
media described in the text.

EMBRYOGENY IN BEAN

     Phaseolus vulgaris cv. Taylor's Horticultural seeds mature
36 days after fertilization (Walbot et al., 1972). During the
first 20 days of development, water and chlorophyll content of the
embryo are high, there is a high rate of RNA synthesis (Walbot,
1972) and dry weight, protein content, and concentration of abscisic
acid (ABA) (Hsu, 1979) are low. After day 20 water content decreases,
chlorophyll is degraded, RNA synthesis declines, and dry weight,
protein content, and ABA concentration all increase rapidly. In
this latter period the concentration of protein increases progressively (Table 1). Sun et al. (1978) have shown that in P. vulgaris
cv. Tendergreen 60% of the protein accumulated during this period
is a single storage protein. We shall examine how the change from
a low to a high rate of storage protein accumulation is controlled.

STORAGE PROTEIN ACCUMULATION AND SYNTHESIS IN BEAN

     The principal storage protein accumulated in the seed of P.
vulgaris is the vicilin, Glyoprotein II (Gl II) (Pusztai and Watt,
1970). We purified this protein from seeds of P. vulgaris cv.

Embryo age (days)	Protein/wet wt. (μg/mg)	Embryo age (days)	Protein/wet wt. (μg/mg)
10	17.0	22	87.3
12	18.6	26	133.0
14	25.3	30	148.0
16	34.5	34	191.0
18	37.0	36	180.0

     Table 1.   Total protein content of developing bean embryos.
                Protein was determined from tissue homogenized in
                0.1N NaOH by a modified Lowry procedure (Hartree,
                1972).

Taylor's Horticultural and prepared an antibody against it (Sussex and Dale, 1979) which was used in rocket immunoelectrophoresis to examine the accumulation and synthesis of Gl II during embryogeny. In embryos younger than 20 days old the concentration of Gl II was very low, and from day 20 on there was a large and increasing amount of this storage protein present in the embryo (Sussex and Dale, 1979). The basis of this change in the pattern of storage protein accumulation was examined in 9 and 10 day old embryos removed aseptically from the seed and grown in sterile culture on agar-solidified media containing hormone supplements.

It was first necessary to show that embryos in culture continue to develop as embryos, and do not undergo precocious germination. To do this 10 day embryos were grown on a basal medium consisting of Mante and Boll (1975) inorganic salts and organic nitrogen compounds, and 3% sucrose. These continued to develop embryonically for a further 10 days as indicated by the continued accumulation of Gl II. They then began to germinate as indicated by elongation of the radicle and the emergence of root hairs, by the development of geotropic sensitivity of the root, and by expansion of the primary leaves.

Embryos were cultured for 6 days on basal medium or on this supplemented with sterile filtered indole acetic acid, kinetin, gibberellic acid, or ABA as growth regulators. Only ABA had a significant effect on the accumulation of Gl II, 10^{-5}M ABA causing an approximate 3 fold increase in Gl II content of the embryo (Table 2). The increased accumulation of Gl II caused by ABA results from an increased rate of synthesis as indicated by the approximate 3 fold increase in ^{14}C amino acid counts incorporated into Gl II in embryos cultured for 3 days on 10^{-5}M ABA medium as

Medium addition	Gl II/embryo (μg)
Initial	3.2
basal medium	47.0
ABA 10^{-8}M	39.0
ABA 10^{-7}M	31.0
ABA 10^{-6}M	66.0
ABA 10^{-5}M	117.0
ABA 10^{-4}M	88.0
ABA 5×10^{-4}M	31.0

Table 2. Glycoprotein II content of cultured embryos. Embryos cultured for 6 days were homogenized in buffer and Gl II content of the homogenate was determined by rocket immunoelectrophoresis.

compared with those cultured on basal medium (Table 3).

The next question we examined was whether ABA is required continuously for stimulation of Gl II synthesis, or whether brief exposure to ABA results in a persistent increase in Gl II synthesis. Ten day embryos were cultured for 3 days either on basal medium, or with 10^{-5}M ABA. Some embryos from each treatment were then processed for Gl II determination; others were cultured for a further 3 days on fresh culture medium of the same composition as they had been on previously, or were transferred from basal medium to 10^{-5}M ABA medium or vice versa. On day 6 all embryos were processed for Gl II determination. In the culture period day 3-6 embryos transferred from ABA to basal medium synthesized less Gl II than those that were continuously on ABA, and those transferred from basal to ABA medium synthesized more Gl II than those continuously on basal medium (Table 3). These results indicate that ABA is continuously required to maintain the high rate of Gl II synthesis , and that embryos initially cultured in the absence of exogenously supplied ABA can subsequently respond to it by increasing their rate of Gl II synthesis.

Culture medium day 0-3 day 3-6	Gl II concentration (μg/5ul)	Gl II/embryo (μg)	^{14}C cpm
initial	0.023	0.09	–
basal	0.175	1.66	307
ABA	0.535	4.14	845
basal basal	1.9	30.23	1,400
basal ABA	3.3	41.46	1,800
ABA basal	1.75	18.14	1,430
ABA ABA	4.6	45.72	2,822

Table 3. Glycoprotein II accumulation and synthesis in cultured bean embryos. Embryos were cultured either on basal medium or on this supplemented with 10^{-5}M ABA for 3 days, then were assayed for Gl II or were transferred to basal or ABA supplemented medium for a further 3 days and then assayed. Radioactivity was determined from embryos pulsed for 2 hours in ^{14}C mixed amino acids, homogenized, and run in rocket immunoelectrophoresis. Rockets were cut out of the gel, solubilized in Aquasol, and the radioactivity in Gl II determined by liquid scintillation counting.

STORAGE PROTEIN ACCUMULATION AND SYNTHESIS IN RAPE SEED EMBRYOS

The major storage protein accumulated in the embryo of Brassica napus is a 12S legumin-like protein. This protein was purified, and an antibody prepared against it was used in rocket immunoelectrophoresis to study its accumulation and synthesis (Crouch, 1979). To investigate whether ABA also stimulated storage protein synthesis in this species 27 day embryos, that had just begun to accumulate 12S protein, were cultured on Monnier's medium (1976) with supplements of sterile filtered ABA. The amount of 12S storage protein and its rate of synthesis were determined after embryos had been in culture for 0.25, 2 and 5 days. The concentration of 12S storage protein accumulated by cultured embryos is shown in Table 4. The highest concentration occurred in embryos cultured on the highest level of ABA, and there was progressively less 12S protein as the level of ABA in the medium was reduced. That the increased accumulation of 12S protein resulted from an increased rate of synthesis was shown by pulsing embryos cultured for 2 or 5 days on different media with ^{14}C mixed amino acids and determining incorporation of radioactive counts into 12S protein (Table 5).

	Days in culture		
Medium addition	0	2	5
0	0.65	1.04	0.78
10^{-7}M ABA		1.43	3.64
10^{-6}M ABA		3.12	11.70
10^{-5}M ABA		4.16	15.47

Table 4.   Concentration (μg/mg F.W.) of 12S storage protein in embryos cultured for 2 or 5 days on media with or without ABA. Assay procedure as in Table 3.

Medium addition	Days in Culture		
	0.25	2	5
0	410	234	250
10^{-7}M ABA	324	459	1,945
10^{-6}M ABA	416	1,582	5,400
10^{-5}M ABA	250	1,945	5,144

Table 5.   Radioactivity (cpm) incorporated into 12S storage protein in embryos cultured 0.25, 2 or 5 days on various media. Embryos were cultured for the time period indicated, and then incubated in ^{14}C-labelled amino acids for 5 hours. The precipitate formed during immunoelectrophoresis of sample extracts was cut out of the agarose gel, melted in water, and mixed with Aquasol for liquid scintillation counting.

Research supported by NIH grant GM 24775, and USDA grant 5901-0410-8-0174-0, R. M. K. Dale was supported on a NIH Post-doctoral Fellowship, and M. L. Crouch was supported on a NIH Predoctoral Traineeship. We thank P. Naples, N. Gaedke and A. Zuckerman for their expert technical assistance.

REFERENCES

Crouch, M. L., 1979. Storage proteins as embryo-specific
    developmental markers in zygotic, microsporic, and somatic
    embryos of Brassica napus L., Ph. D. thesis, Yale
    University, New Haven, CT., USA.

Dure, L. S., 1975. Seed formation. Annu. Rev. Plant Physiol.,
    26:259.

Hartree, E. F., 1972. Determination of protein: a modifica-
    tion of the Lowry method that gives a linear photometric
    response. Anal. Biochem., 48:422.

Hsu, F. C., 1979.   Abscisic acid accumulation in developing
    seeds of Phaseolus vulgaris. Plant Physio., 63:552.

Mante, S. and Boll, W. G., 1975. Comparison of growth and
    extracellular polysaccharide of cotyledon cell suspension
    cultures of bush bean (Phaseolus vulgaris cv. Contender)
    grown in coconut milk medium and synthetic medium. Cana-
    dian J. Bot., 53:1542.

Millerd, A., 1975. Biochemistry of legume seed proteins. Annu.
    Rev. Plant Physiol., 26:53.

Monnier, M., 1976. Culture in vitro d'l'embryon immature de
    Capsella bursa-pastoris Moench (L.). Rev. Cytol. Biol.
    Veget., 39:1.

Pusztai, A. and Watt, W. B., 1970. Glycoprotein II. The
    isolation and characterization of a major antigenic and
    non-haemagglutinating glycoprotein from Phaseolus vulgaris.
    Biochim. Biophys. Acta., 207:413.

Sun, S. M., Mutschler, M. A., Bliss, F. A. and Hall, T. C.,
    1978. Protein synthesis and accumulation in bean cotyle-
    dons during growth. Plant Physiol., 61:918

Sussex, I. M., and Dale, R. M. K., 1979. Hormonal control of
    storage protein synthesis in Phaseolus vulgaris, in:
    "The Plant Seed: Development, Preservation, and Germina-
    tion," I. Rubenstein, ed., Academic Press, NY.

Walbot, V., 1972. Rate of RNA synthesis and tRNA end-labeling
    during early development of Phaseolus. Planta, 108:161.

Walbot, V., Clutter, M. and Sussex, I., 1972. Reproductive
    development and embryogeny in Phaseolus. Phytomorph.,
    22:59.

ORGANIZATION AND TRANSCRIPTION OF

MAIZE CHLOROPLAST GENES

Lawrence Bogorad, Setsuko O. Jolly, George Kidd,
Gerhard Link and Lee McIntosh

The Biological Laboratories, Harvard University
Cambridge, MA 02138, U.S.A.

INTRODUCTION

The Zea mays chloroplast chromosome is an 91,100,000 dalton calculated to be compromised of about 135 kilobase pairs (KBP) (Kolodner and Tewari, 1975; Bedbrook and Bogorad, 1976). All the recognition sites for the restriction endonuclease Sal I as well as some recognition sites for Eco RI and Bam HI have been mapped on it (Bedbrook and Bogorad, 1976) and the locations and arrangements of two sets of genes for 16S, 23S and 5S chloroplast rRNAs of maize were established by Bedbrook, Kolodner and Bogorad (1977). In addition, structural genes for two proteins have been identified and assigned to mapped and cloned restriction fragments (Coen et al.; 1977; Bedbrook McIntosh, Link and Bogorad, 1979).

The structural gene for the large subunit (LS) of ribulose bisphosphate carboxylase (RuBPCase) is in Bam fragment 9 (Coen et al., 1977). Maize chloroplast DNA (cpDNA) sequence Bam 9 cloned in E. coli (as part of pZmc 37 for which the vehicle is RSF 1030) was shown to direct the synthesis in vitro of LS RuBPCase in an E. coli RNA polymerase-rabbit reticulocyte lysate linked transcription-translation system. LS labeled with radioactive methionine was identified by its size, by precipitability with antibody against RuBPCase and by the identity of proteolytic fragments produced by digestion of the in vitro product and authentic LS isolated from maize seedlings supplied with radioactive methionine.

Photogene 32 has been located on maize cpDNA Bam fragment 8 (Bedbrook et al., 1978). A 32 kilodalton (kd) polypeptide is added to chloroplast photosynthetic membranes during photoinduced development. Isolated maize chloroplasts produce a 34.5 kd polypeptide which is processed to the 32 kd form (Grebanier et al., 1978; Grebanier, Steinback and Bogorad, 1979). Maize chloroplasts and greening plastids in illuminated dark-grown maize seedlings contain an RNA which directs the synthesis in vitro of the 34.5 kd polypeptide and which hybridizes Bam fragment 8. Etioplasts lack this RNA. A portion of Bam fragment 8 which contains Photogene 32 has now been cloned and the precise location of the DNA complementary to Photogene 32 mRNA has been established (McIntosh, Link and Bogorad,1979).

We will describe experiments which have permitted the LS gene to be mapped precisely on Bam 9 and will outline the discovery, also on Bam 9, of part of a gene which is transcribed into a 2.2 KB RNA (Bogorad et al., 1979) The directions of transcription of these two genes and their differential expression in mesophyll and bundle sheath cells of maize leaves will also be discussed.

Understanding the regulation of gene expression in chloroplasts requires knowing not only the organization of genes on cpDNA but also the role of maize cpRNA polymerase. A small polypeptide which affects selective transcription of maize cpDNA sequences has been purified and some of its effects on the in vitro transcription of cpDNA sequences cloned in F. coli will be described (Jolly and Bogorad, 1979).

GENES ON CHLOROPLAST DNA SEQUENCE BAM 9.

The LS of maize RuBPCase is about 52 kd in size. The minimum length of DNA required to code for such a polypeptide is about 1.3-1.6 KB (assuming the average molecular weight of its amino acids as 100 or 120 daltons respectively) but Bam 9,which directs the synthesis of LS,is 4.35 KPB in length.

The first problem was to determine whether Bam 9 contains one or more copies of the LS gene and, if a single copy is present, to determine its approximate position on this larger DNA fragment. This was accomplished by first mapping recognition sites for the restriction enzymes Pst I, Bgl II, Eco RI and Sma I on Bam 9. Then, cloned Bam 9, as part of the chimeric plasmid pZmc 37, was digested with Bam HI plus one of the restriction enzymes whose recognition sites had been mapped on Bam 9. Digestion with Bgl II or Sma I did not interfere with the ability of Bam 9 to direct the synthesis of full-sized LS, thus sites recognized by these enzymes were judged to be outside of the structural gene sequence. On the other hand, digestion with Eco RI or Pst I eliminated the capacity of this chloroplast DNA sequence to direct the synthesis of the full sized polypeptide. In this way it was judged that the LS gene lies within a 2.5 KB portion of Bam 9 delimited by one recognition site for Sma I and one for Bgl II (Bedbrook et al., 1979). A diagram of these results is shown in Fig. 1.

Bam 9 from pZmc 37

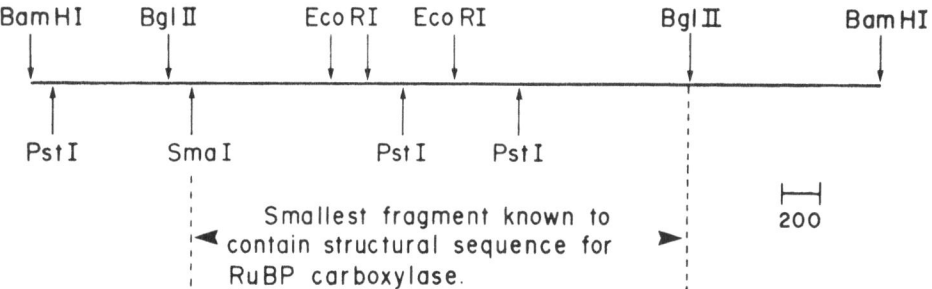

Figure 1. The location of recognition sites for restriction enzymes Bam HI, Bgl II, Fco RI, Pst I and Sma I on maize cpDNA sequence Bam 9. The smallest fragment that directed the synthesis of LS RuBPCase in a linked transcription-translation system (Bedbrook et al.,1979) is also indicated.

The availability of LS DNA in a clone permitted probing RNAs from bundle sheath and mesophyll cells of Zea mays for LS mRNA. The two cells were separated enzymatically and their RNAs extracted. RNA from bundle sheath cells directed the synthesis of LS but RNA from mesophyll cells did not. Furthermore, RNA from the former but not from the latter cell type hybridized to the 2.5 KB fragment within Bam 9 which was known to contain the LS gene (Link, Coen and Bogorad, 1978). In the course of this work it was found that RNA from both cell types hybridizes to cloned Bam 9 to some extent; this suggested the possibility that although mesophyll cells did not contain LS mRNA they might contain transcripts of another gene which had not yet been identified on Bam 9. Still, knowing that the entire gene for LS was located on Bam 9 and knowledge of its approximate position on this DNA segment permitted using this cloned cpDNA sequence to "fish" in an unfractionated cpRNA preparation to determine the size of the functional LS mRNA, the size of the gene which codes for it, and the precise location of the DNA complementary to LS mRNA.

IDENTIFYING CHLOROPLAST DNA COMPLEMENTARY TO LS mRNA.

In previous work (Bedbrook et al.,1979) the gene for LS RuBPCase was determined to lie between restriction sites for Bgl II and Sma I within a 2.5 KB region of the maize cpDNA sequence Bam 9 by in vitro transcription-translation analyses of cloned DNA. This indicated that there would be space for little non-coding DNA with 1.3-1.6 KB of DNA required 'for LS and,in view of the more-or-less faithful transcription-translation, that any additional DNA was not likely to be interspersed among coding regions. We sought to identify directly which portions of cpDNA sequence Bam 9 are complementary to LS mRNA.

Berk and Sharp (1977) have shown that nuclease S1, which digests single stranded nucleic acids, can be used to identify DNA sequences complementary to RNA. Single-stranded, i.e. denatured, DNA is hybridized with RNA and then the entire mixture is treated with nuclease S1. All unhybridized RNA and DNA strands are digested away; hybrids from which protruding single nucleic acid strands have been removed remain. In the original method, the radioactive DNA which was employed was analyzed by electrophoresis in alkaline agarose gels (thus destroying the RNA) and located on the gels by autoradiography.

We started with pZmc 37 DNA or cpDNA sequence Bam 9 (Fig. 1) prepared from the plasmid. This DNA was denatured and hybridized with total, i.e. unfractionated, cpRNA. After treatment with nuclease S1, surviving DNA was subjected to agarose gel electrophoresis and the DNA fragments were transferred to nitrocellulose filter sheets (Southern, 1975). To identify the DNA, the Bgl-Bgl portion of Bam 9 (Fig. 1) was used as a template to prepare copy RNA (cRNA) labeled with radioactive phosphorus. The labeled cRNA was incubated with the nitrocellulose filter sheet to permit hybridization to DNA complementary to it. The position of the DNA was then revealed by autoradiography. From these experiments it was determined that maize LS mRNA is colinear with a 1.6 KB cpDNA sequence. Thus, the functional message for LS is 1.6 KB long and the gene is not interrupted by non-coding sequences (Bogorad et al., 1979; Link and Bogorad, 1979).

The size of LS mRNA was also determined to be 1.6 KB by another method. Chloroplast RNA was denatured by glyoxylation and separated according to size electrophoretically on 1.5% agarose gel slabs. The cpRNA was transferred to diazobenzyloxymethyl (DBM) paper and hybridized with nick-translated DNA essentially according to the method of Alwine et al. (1977).

The method of Berk and Sharp (1977) was further modified to determine the precise location of the LS gene in the 2.5 KB region of Bam 9 in which it was known to reside. Instead of incubating cpRNA with pZmc 37 or Bam 9 DNA, the DNA was first digested with restriction enzymes known to cleave within the 2.5 KB region of interest. Chloroplast RNA was incubated with this fragmented DNA, hybridized and the hybridization mixture was subjected to digestion with Sl nuclease. Again, surviving DNA fragments were separated by gel electrophoresis, transferred to nitrocellulose paper, and identified by hybridization with radioactive cRNA. Using cRNA transcribed from the 2.7 KBP Bgl-Bgl portion of Bam 9, those restriction fragments which were fully protected by cpRNA could be identified but the sources in Bam 9 of partially protected sequences could not be ascertained. By probing with cRNA transcribed from, e.g., F 2.16, the Eco RI-Bam fragment at the right-hand end of Bam 9 (Fig. 2), restriction fragments from the right-hand portion of Bam 9 could be identified and their reduced sizes could be determined by including appropriate markers in the gel. Similarly, by using cRNA transcribed from a fragment in the left-hand portion of the Bam 9 sequence (e.g. F 1.55), DNA sequences only partially protected from Sl digestion by hybridization to RNA could be located and sized. The results of such experiments (Link and Bogorad, 1979) are shown diagramatically in Fig. 3.

Once it was known which restriction fragments of the Bgl-Bgl sequence were wholly or partially included within the 1.6 KB sequence that codes for LS mRNA, a further modification of the S1 nuclease digestion method was made to determine the direction of transcription of the gene. In these experiments, Bgl-Bgl DNA was digested with the restriction endonuclease Pst I and the resulting fragments were labeled with radio-phosphorus either at their 5' ends using polynucleotide kinase (Maxam and Gilbert, 1977) or, alternatively, at their 3' ends using terminal deoxynucleotidyl transferase (Roychoudhury et al., 1976). Terminally labeled fragments were then hybridized with cpRNA, treated with nuclease Sl, and surviving DNA fragments were separated by gel electrophoresis to identify them by size. Fragments retaining terminal label were identified by autoradiography of dried gels. The direction of transcription (from the 3' to the 5' end of the gene) could be determined by establishing which of the Pst cuts retained their label. As is shown in the diagram in Fig. 3, one of the Pst fragments is completely protected by LS mRNA while two of the other Pst fragments are partially digested. The protection of the label on the 3' end of two Pst fragments, the fully protected one and the one to its right, together with the loss of this label from the partially cut Pst fragment to the left of the fully protected one indicated that, as represented in Fig. 3, the direction of transcription is from left to right. Data from 5' labeled strands supported this conclusion. (Link and Bogorad, 1979).

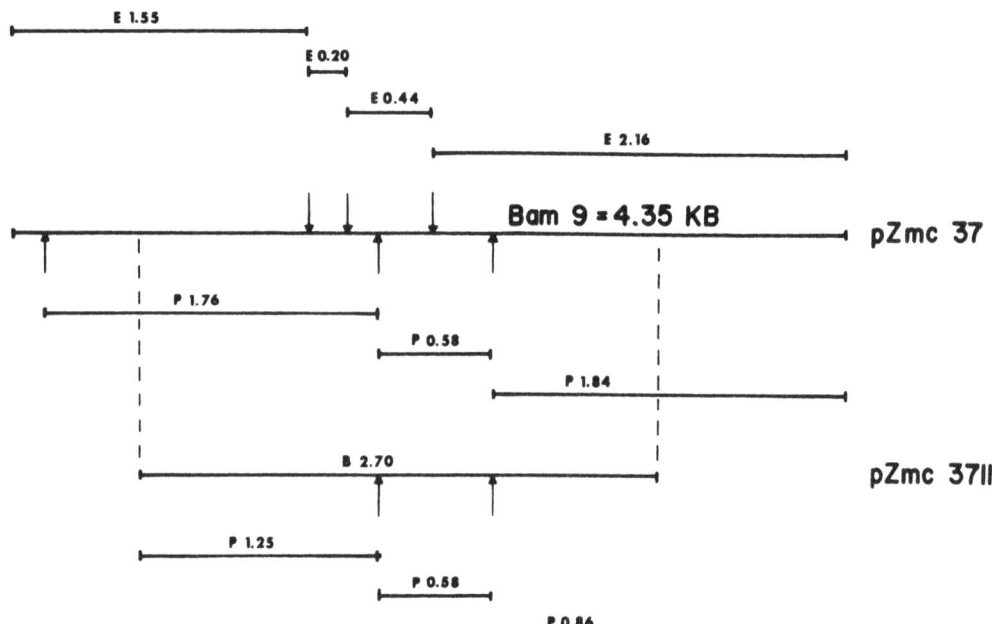

Figure 2.   A diagram showing recognition sites for endonucleases
             Eco RI, Pst I, Bgl I and Bam HI on cpDNA fragment Bam 9
             and the sizes of DNA fragments generated by cleavage
             with each of these enzymes. Bam 9 has been cloned in
             E. coli as part of pZmc 37 (Coen et al., 1977). The
             2.70 KBP fragment produced from Bam 9 by Bgl II (see
             Fig. 1), designated B 2.70, has been cloned separately
             as part of pZmc 3711. Fragments produced by digestion
             of Bam 9 or B 2.70 with Eco RI are designated "F";
             products of digestion with Pst I are labeled "P".
             Numerals following these designations indicate the size
             of each fragment in KBP.

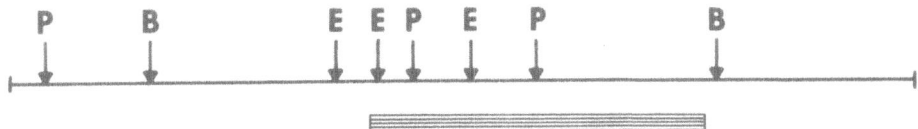

Figure 3. The location of DNA complementary to LS mRNA (box with
           horizontal   lines)   in   relation   to   endonuclease
           recognition sites on cpDNA sequence Bam 9 (upper line).
           P=Pst I. B= Bgl II. F=Eco RI.

1.4 KB OF A 2.2 KB GENE IS ON BAM 9.

In the course of studying the size of the DNA sequence
complementary to LS mRNA by the S1 nuclease protection method,
another fragment to which cpRNA hybridizes was discovered on Bam
9.   In these experiments cRNA transcribed from Bgl-Bgl DNA was
used as a probe to detect DNA that survived S1 nuclease digestion
because of hybridization to cpRNA. When Bam 9 DNA was used for
hybridization with cpRNA before S1 nuclease digestion, a 1.4 KB
fragment of DNA was protected from digestion. However, when the
2.7 KBP Bgl-Bgl DNA was used for the hybridization,only a 0.7 KB
fragment was protected. (In both cases, as described above, the
1.6 KB sequence complementary to LS mRNA was also protected.) The
region from the left-hand Bgl II recognition site (Fig. 4) to the
Bam restriction site is 0.7 KBP. Thus, cpRNA includes a species
complementary to this 0.7 KBP region from the Bgl to the Bam
recognition sites on the left-hand portion of Bam 9.   From this
and another S1 experiment using Bgl-Bgl DNA digested with Pst for
hybridization with cpRNA before nuclease treatment, 0.7 KB to the
right of that Bgl recognition site was also found to be
complementary to this RNA species. The gene was determined to be
2.2 KBP long (of which 1.4 KBP are on Bam 9) by the method of
Alwine et al.   (1977) in which total cpRNA was separated
electrophoreticaly according to size and probed with
nick-translated DNA taken from an appropriate portion in the
left-hand side of Bam 9.

The space between the LS and the 2.2 KBP genes is 0.33 KBP.
By the method outlined earlier, we have found that transcription
of the 2.2 KB gene commences from within the 0.33 KBP sequence
intervening between the 2.2 KB and the LS genes.   Since
transcription of the LS gene is also initiated here,these

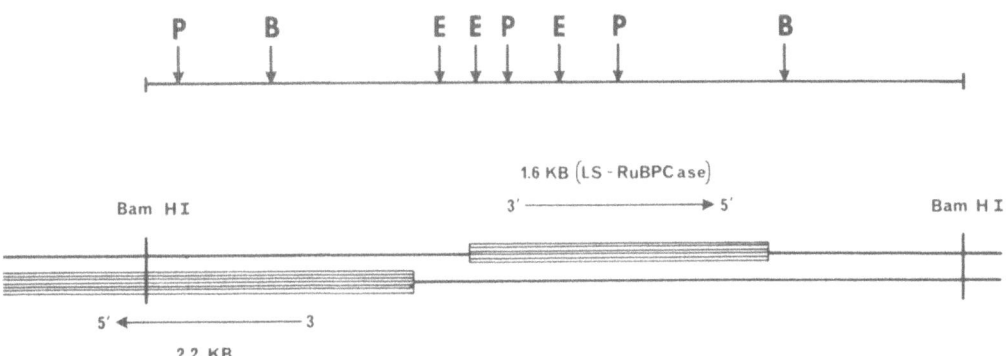

Figure 4.   The locations and directions of transcription of the LS
            RuBPCase gene and 1.4 KB of the 2.2 KB gene on cpDNA
            Bam 9 sequence.

two genes must be on opposite strands of DNA in Bam 9 (Link and
Bogorad, 1979).

RNA TRANSCRIBED FROM THE 2.2 KB GENE IS EXPRESSED IN BOTH
MESOPHYLL AND BUNDLE SHEATH CELLS.   LS mRNA IS FOUND IN BUNDLE
SHEATH CELLS BUT NOT MESOPHYLL CELLS.

        As discussed in the Introduction, LS mRNA is found in bundle
sheath and not mesophyll cells of maize (Link et al., 1978).   In
the present series of experiments, Eco RI fragment E 0.44 (Fig.
2) from within the LS gene was found to be hybridized by bundle
sheath but not by mesophyll cell RNA. On the other hand, an Eco
RI fragment (E 1.55) which contains the 1.4 KB portion of the 2.2
KB gene is hybridized by RNA from both bundle sheath and mesophyll
cells with about equal intensity.   Signals involved in the
differential expression of these two genes can be presumed to lie
within the 0.33 KB region between them. Detailed analysis of the
sequence of bases in this region is in progress.

PHOTOGENE 32 IS 1.2 KBP LONG

        Photogene 32 codes for a 34.5 kd polypeptide which is
processed to 32 kd in maize plastids in vivo, but not in vitro
(Grebanier et al., 1978). The 32 kd polypeptide is absent from
membranes of etioplasts; it is synthesized during light-induced
plastid development (Grebanier, Steinback and Bogorad, 1979).

Photogene 32 has been traced to maize cpDNA fragment 8 (Bedbrook et al., 1978). A 2.2 KBP portion of Bam 8 limited by recognition sites for Bam HI and Fco RI, has been cloned in pBR 322. It directs the synthesis of a 34.5 kd polypeptide in a linked E. coli RNA polymerase-rabbit reticulocyte lysate transcription-translation system in vitro. This product and the similar-sized product of translation of cpRNA yield similar proteolytic fragments. (McIntosh, Link and Bogorad, 1979.)

The cpDNA insert derived from Bam 8 and cloned in pZmc 427 is complementary to an uninterrupted 1.2 KBP long cpRNA species. The total message has been shown to be of this length by the RNA blotting-DNA/RNA hybridization method of Alwine et al. (1977) and the precise location of DNA complementary to cpDNA has been mapped on the sequence inserted into pZmc 427. The kinds of S1 nuclease procedures described above in the analysis of LS RuBPCase and the 2.2 KB genes were also employed here. The results are shown diagramatically in Fig. 5.

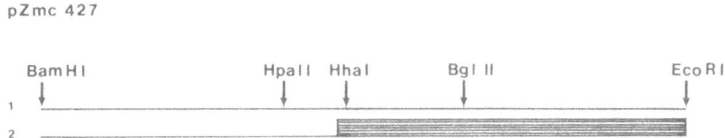

pZmc 427

Figure 5. The location of maize cpDNA complementary to Photogene 32 mRNA on the Bam-Fco RI cpDNA sequence cloned as part of pZmc 427.

About 0.9-1.0 KBP of DNA is the minimum required to code for a 34.5 kd polypeptide. Our data show that the functional message is complementary to an uninterrupted region 1.2 KBP in length. LS RuBPCase mRNA, for comparison, is probably no more that 100 ribonucleotides longer than the minimum length required to code for a polypeptide of 52 kd but Photogene 32 mRNA may be 200-300 nucleotides larger than the minimum needed. It seems unlikely that the 34.5 kd form is an already highly processed intermediate because of the in vitro transcription-translation data. The 1.2 kb mRNA may contain an extended leader sequence with some processing signal. The identification and study of genes for additional thylakoid polypeptides and other photogenes will be required before the possible significance of the extra length of this message will be understood.

MAIZE CHLOROPLAST RNA POLYMERASE

Maize chloroplasts contain a multimeric RNA polymerase with unique properties (Smith and Bogorad, 1974). However, this enzyme and the type II nuclear polymerase of maize have four pairs of similar sized polypeptides: each has a 180 kd polypeptide; the chloroplast enzyme has a 140 kd polypeptide while the second largest polypeptide of the nuclear enzyme is 160 kd; both enzymes have polypeptides in the 42-43 and 27-28 kd size ranges. Analyses of tryptic peptides generated from the subunits in these size classes have now been compared and shown to be distinct (Kidd and Bogorad, 1979). If there are any polypeptides which affect both enzymes they must be only loosely associated.

To understand the control of transcriptional programs of the sort which occur during mesophyll-bundle sheath cell differentiation (Link, Coen and Bogorad, 1978), and photoregulated plastid development (Bogorad,1967; Bedbrook et al., 1978), we must understand the action of the chloroplast's RNA polymerase as well as the organization of the genes on the chloroplast chromosome (and, eventually, mechanisms for integrating nuclear and organelle gene expression). Prior work has shown that the RNA synthesizing activity of maize chloroplast RNA polymerase rises during photo-induced plastid development (Bogorad, 1967) unaccompanied by a commensurate increase in the major polypeptides of the enzyme (Apel and Bogorad, 1976).

A major impediment to the in vitro study of RNA polymerase under conditions approximating those in vivo is the unavailability of the enzyme's template in its native form. The 85×10^6 dalton maize chloroplast chromosome can be isolated in its supercoiled form but it has a high probability of breaking. We have instead used DNA of chimeric plasmids, i.e. cpDNA sequences inserted in plasmid vehicles, as templates as well as probes in the study of maize chloroplast RNA polymerase (Jolly and Bogorad, 1977).

The RNA polymerase of maize chloroplasts is solubilized readily when the thylakoid membrane fraction of plastids purified by sucrose density gradient centrifugation is incubated at 37° in the presence of EDTA or in lower yield without this chelating agent (Bottomley, Smith and Bogorad, 1971). In the experiments to be described here, solubilization was without EDTA. The solubilized enzyme was further purified by chromatography on DEAE cellulose from which it elutes with about 0.2 M KCl. This enzyme preparation is about 50 times more active with calf thymus DNA than with supercoiled bacterial plasmid DNA as a template. However, a fraction which elutes from the column with 0.5 M KCl greatly stimulates the transcription of supercoiled plasmid DNA but does not alter the activity of the enzyme with calf thymus DNA as a template. Active material in this fraction has been designated the "stimulator." It is inactivated by pronase or by boiling for 5 min., but it is stable against heating to 50 for 5 min. Further purification of the stimulator by phosphocellulose chromatography and filtration through Sephadex G-150 yields a 99% pure polypeptide of molecular weight 26,000-29,000. The stimulator is not replaced by nor does it replace the sigma factor of F. coli RNA polymerase.

The influence of the stimulator on the specificity of transcription by the plastid RNA polymerase has been studied by analyzing the transcripts of cpDNA sequence-containing chimeric bacterial plasmids. An example is some work with pZmc 134 DNA. This chimeric plasmid is comprised of the vehicle pMB 9 into which has been inserted maize cpDNA fragments Fco RI a and 1 (Bedbrook and Bogorad, 1976; Bedbrook, Kolodner and Bogorad, 1977). The organization of maize chloroplast genes for 5S, 16S and 23S rRNAs on Fco RI fragment a has been described in detail (Bedbrook, Kolodner and Bogorad, 1977).

pZmc 134 DNA was incubated with solubilized maize chloroplast RNA polymerase in a transcription mixture containing ^{32}P-UTP as one of the four ribonucleotide substrates. To analyze the products, pZmc 134 DNA was digested with one or more restriction endonucleases and the fragments so generated were separated by gel electrophoresis and then transferred to nitrocellulose filter sheets by the method of Southern (1975). The radioactive products of the in vitro RNA polymerase reactions were hybridized against the DNA fragments. RNA produced in the absence of added stimulator hybridized somewhat more to chloroplast than to vehicle DNA sequences. In the presence of stimulator, the activity was enhanced 5-15 fold and transcription of the cpDNA inserts was very highly favored. Furthermore, some portions of the cloned cpDNA sequence Fco RI a were transcribed more than others.

The demonstration that transcription of various portions of Fco R1 fragment a is differentially influenced by stimulator indicated the desirability of using a smaller cpDNA sequence for quantitative analyses of the effects of the stimulator. We employed pZmc 150 DNA as a template, and subsequently as a probe. This plasmid is comprised of the approximately 1.8 KBP maize cpDNA sequence 1 incorporated into the plasmid vehicle pMB9. A constant amount of radioactively labeled transcripts of pZmc 150 produced by the maize chloroplast enzyme in the abence or presence of stimulator were hybridized against varying amounts of pMB9 or pZmc 150 DNA bound to nitrocellulose filters. Taking into account the relative proportions of pZmc 150 comprised of its two components, Fco RI fragment 1 and pMB9, in the absence of stimulator chloroplast and vehicle DNA sequences were transcribed about equally well. On the other hand, in the presence of stimulator, cpDNA sequences were transcribed about 8-fold more than vehicle DNA sequences. Thus, the quantitative data support the conclusions drawn from the more qualitative evidence provided by hybridization followed by autoradiography.

Similar analyses have been conducted with pZmc 37-11. This chimeric plasmid is comprised of the vehicle RSF 1030 into which has been inserted the 2.7 KBP cpDNA sequence described in an earlier part of this paper. This 2.7 KBP sequence includes the entire structural gene for LS RuBPCase. The stimulator promotes synthesis of the cpDNA sequence but not as much as of Eco RI fragment 1 included in pMB 9.

It remains to be seen whether this stimulating protein itself associates with cpDNA and, if so, whether it is at the RNA polymerase bonding site. It will also be interesting to learn whether, as suggested by differences between our results with cpDNA sequences in pZmc, 134, 150 and 37-11, this stimulator does indeed affect transcription of some chloroplast genes differently from others and thus plays a role in plastid differentiation.

SUMMARY AND DISCUSSION

Several features of maize cpDNA organization have been revealed in the course of the studies described here but our total knowledge of this chromosome is minute compared with the total problem and almost all functional problems remain unanswered. On the other hand, these new data have made it possible to frame numerous new questions.

The entire gene for LS RuPBCase and part of the 2.2 KBP gene are on cpDNA Bam fragment 9. One general question about chloroplast gene organization is whether genes expressed during a particular development program are physically close to one another? We have learned that two adjacent genes are expressed differently in the differentiation of mesophyll and bundle sheath chloroplasts in maize. The gene for LS RuBPCase is transcribed in bundle sheath chloroplasts but the 2.2 KB gene, which is only 330 base pairs away, is transcribed in both mesophyll and bundle sheath plastids. We have not detected transcripts which hybridize to the right of LS on Bam 9 (Fig. 1) in preparations of chloroplast RNAs. Further study of DNA in that region and beyond should help resolve this question of whether other genes that are silent in mesophyll cells of maize are near the LS gene. Sequencing DNA beyond the right-hand end of LS should help in the search for promoters or sequences which can code for polypeptides. At the moment our attention is focused on the 330 base pair long region between the 2.2 KBP and the LS genes. Further work may reveal whether other genes expressed in bundle sheath but not mesophyll cells are on the same strand as the LS gene.

Questions about the possible clustering of transcriptionally related genes are also applicable to Photogene 32. In this case, we have less information about adjacent genes than on Bam 9. Other genes nearby are being sought.Sequencing of the DNA flanking Photogene 32 will also be instructive.

We cloned segments of the maize chloroplast chromosome in part because of our interest in the action of the chloroplast RNA polymerase. Work conducted to date indicates that this should continue to be a fruitful approach. Additional genes need to be identified on the chloroplast chromosome and to be studied as templates for maize chloroplast RNA polymerase in vitro as parts of chimeric plasmids. One immediately obvious question that requires further study is whether the stimulator described in this report affects the transcription of various genes in maize cpDNA differently.

ACKNOWLEDGEMENTS

We are indebted to Susan F. Adam, Diane Headen, Brigitte Link and Linda Scrafford-Wolff for technical assistance in various aspects of this work.

This research was supported in part by grants from the National Institute of General Medical Sciences from the National Science Foundation and from the Competitive Research Grants Program of the United States Department of Agriculture-SEA. It was also supported in part by the Maria Moors Cabot Foundation of Harvard University. During the course of this work Gerhard Link was the recipient of fellowships from the Deutsche Forschungsgemeinschaft and from EMBO. His present address is: Biologisches Institut II, der Universitaet Freiburg, Schanzlestrasse 1, D-7800 Freiburg i. Br., FRG. Lee McIntosh is a Maria Moors Cabot Foundation Fellow. George Kidd was a trainee of the National Institutes of Health on a Multibiological Sciences Training Grant. His present address is: Dr. George H. Kidd, Battelle Columbus Laboratories, 505 King Avenue, Columbus, Ohio 43201.

REFERENCES

Alwine, J.C., Kemp, D. J. and Stark, G. R., 1977, Proc. Natl. Acad. Sci. U.S.A. 74, 5350-5354.
Apel, K. and Bogorad, L., 1976, Eur. J. Biochem 67, 615-620.
Bedbrook, J. and Bogorad, L., 1976, Proc. Natl. Acad. Sci. U.S.A. 73, 4309-4313.

Bedbrook, J. R., Kolodner, R. and Bogorad, L., 1977, Cell 11, 739-749.

Bedbrook, J. R., Link, G., Coen, D. M., Bogorad, L. and Rich, A., 1978, Proc. Natl. Acad. Sci. U.S.A. 75, 3060-3064.

Bedbrook, J. R., Coen, D. M., Beaton, A. R., Bogorad, L. and Rich, A. 1979, J. Biol. Chem. 254, 905-910.

Berk, A. J. and Sharp, P. A., 1977, Cell 12, 721-732.

Bogorad, L., 1967, Devel. Biol. Supplement 1 pp. 1-31.

Bogorad, L., Link, G., McIntosh, L. and Jolly, S., 1979, Genes on the maize chloroplast chromosome. in: "ICN-UCLA Symposium on Fxtrachromosomal DNA," D.J.Cummings, P.Borst, I. Dawid and S. Weissman, eds., Academic Press, N. Y. In press.

Bottomley, W., Smith, H. J. and Bogorad, L., 1971, Proc. Natl. Acad. Sci. U.S.A. 68, 2412-2416.

Coen, D. M., Bedbrook, J. R., Bogorad, L. and Rich, A., 1977, Proc. Natl. Acad. Sci. U.S.A. 74, 5487-5491.

Grebanier, A., Coen, D. M., Rich, A. and Bogorad, L., 1978, J. Cell. Biol. 78, 734-746.

Grebanier, A., Steinback, K. and Bogorad, L., 1979, Plant Physiology 63, 436-439.

Jolly, S. O. and Bogorad, L., 1977, J. Cell Biol. 75: 338a

Kidd, G.H. and Bogorad, L,, 1979, Proc. Natl. Acad. Sci. U.S.A. In press.

Kolodner, R. and Tewari, K., 1975, Biochim. Biophys. Acta 402, 372-390.

Link, G. and Bogorad, L., 1979, In preparation.

Link, G., Coen, D. M., and Bogorad, L., 1978, Cell 15, 725-731.

Maxam, A. and Gilbert, W., 1977, Proc. Natl. Acad. Sci. U.S.A. 74,560-564.

McIntosh, L., Link, G. and Bogorad, L., 1979, In preparation.

Roychoudhury, R., Jay, F. and Wu, R., 1976, Nucl. Acids. Res. 3, 863-878.

Southern, F. M., 1975, J. Molec. Biol. 98, 503-517.

THE ORGANISATION IN HIGHER PLANTS OF THE GENES CODING FOR

CHLOROPLAST RIBOSOMAL RNA

Tristan A. Dyer and John R. Bedbrook

Plant Breeding Institute
Maris Lane
Cambridge CB2 2LQ, U.K.

INTRODUCTION

The formation of ribosomes is a complicated process as not only
do they contain several species of RNA but also numerous different
types of protein. How then is the formation of these components
co-ordinated ? Are there specific groups of genes which are transcribed en bloc so as to achieve the appropriate proportion of each
or does the control reside in the regulation of individual genes
or even in the destruction of components produced in excess ?   In
the study which we describe here, the primary concern was the determination of the organisation of genes coding for chloroplast
ribosomal RNAs of flowering plants (in particular maize) with a
view to giving some insight into how transcription and processing
of ribosomal RNA (rRNA) occurs during ribosome formation.

Types of chloroplast rRNA

The large subunit of chloroplast ribosomes from higher plants
contains three RNA components. The largest of these (23S RNA) has
a molecular weight of about 1.05×10^6 (Table 1) while the other
two types of RNA (5S and 4.5S RNA) are much smaller. The small subunit contains just one RNA component (16S RNA) with molecular weight
0.56×10^6. The high-molecular-weight chloroplast components are
very similar in size to those found in bacteria and smaller than
those of the cytosol ribosomes of the same plants, a fact which has
contributed to the speculation concerning the evolutionary origin
of chloroplasts (Phillips and Carr, 1977; Woese and Fox, 1977). A
5S RNA component like that of chloroplasts (Dyer and Bowman, 1979) is
found in all ribosomes except those of the mitochondria of fungi and

Table 1.   Estimated size of chloroplast rRNAs of maize

| Source of rRNA | Size parameters | | |
	S coefficient	Mol. wt.	No. of nucleotides
Large subunit	23S	1.05×10^6*	$2,890^+$
	5S	3.94×10^3	122
	4.5S	3.04×10^3	95
Small subunit	18S	0.56×10^6*	1,490

* Determined from their elecrophoretic mobility relative to
  the high-molecular-weight rRNAs of E. coli which were assumed
  to have molecular weights of 1.1 and 0.56 $\times 10^6$ (Kurland,
  1960). These values are presumably for the salts of these
  polymers (Van Holde & Hall, 1974); based on a mean molecular
  weight of 320 for each residue, these values would be
  9.25×10^5 and 4.77×10^5 respectively.

+ Calculated from its molecular weight relative to that of the
  16S rRNA.

animals (Borst, 1972). However, the chloroplast ribosomes of
higher plants are unusual in having a further low-molecular-weight
RNA component (Whitfeld et al., 1978; Bowman and Dyer, 1979).
Several variants of this 4.5S rRNA molecule may occur in one
type of plant although one form usually predominates. The predominant forms in maize and duckweed, of which the complete
sequences are now known (Dyer, Bedbrook and Bowman, unpublished
results), differ appreciably in size, duckweed 4.5S rRNA containing
103 nucleotides and maize 4.5S rRNA 95 nucleotides. Other unusual
low-molecular-weight rRNA species (7S and 3S RNA) have been
found in the chloroplast ribosomes of the green alga Chlamydomonas
reinhardii (Rochaix and Malnoë, 1978) but these do not appear to
be homologues of the 4.5S rRNA of higher plants.

Distribution of rRNA genes

    Two sets of ribosomal genes per DNA molecule occur in most
chloroplasts. The sets of ribosomal genes are in segments of DNA
which are inverted repeats of one another and hence the coding
sequences are on opposite strands. In maize the sets of genes
are 18,500 base pairs (bp) apart in one direction and 106,100 bp

apart in the other in a circular molecule which contains a total
of 135,000 bp (Bedbrook et al., 1977). Not all clusters of ribosomal genes are arranged in this way; in Euglena (Gray and
Hallick, 1978; Rawson et al., 1978; Jenni and Stutz, 1978) and
pea (Chu and Tewari, 1979; Kolodner and Tewari, 1978) they are
in tandem.

SEQUENCE ANALYSIS OF CHLOROPLAST rRNA GENES

     In order to determine the precise relative position of the
genes which code for the chloroplast rRNAs, the DNA containing
the coding regions for these RNAs is being studied by sequence
analysis. Schwartz and Kössel (1979) have analysed, in particular,
the region containing the 16S rRNA coding sequence while we have
been most interested in the sequence of the DNA coding for the
4.5S and 5S rRNA and for the 3' end of the 23S rRNA. These studies
have been greatly helped by the cloning of the rRNA genes as an
insert in a plasmid pZmc134 (Bedbrook et al., 1977). This plasmid
contains a 12 kbp Eco RI fragment with a complete set of rRNA genes
from maize. As we were specifically interested in a relatively
small part of the insert, DNA from plasmid pZmc134 was used in the
construction of another plasmid pJB822 which contains an insert of
about 970 bp. This was digested with restriction enzymes Hpa II,
Mbo II and Taq I and the resulting fragments hybridised to purified

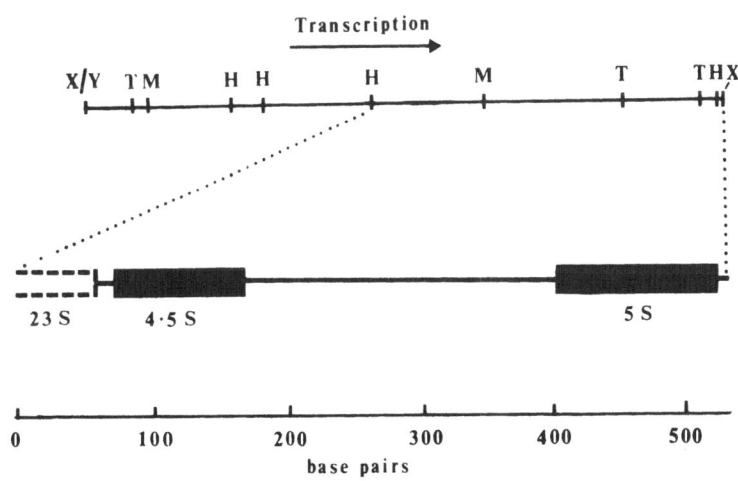

Fig. 1. Cleavage map of the insert in plasmid pJB822 and distribu-
         tion of rRNA coding sequences in the part of it which was
         sequenced. Cleavage sites for the enzymes Hind III,
         Hae III, Taq I, Mbo II and Hpa II are denoted by X, Y,
         T, M and H respectively.

rRNA species. From this it was established that just under half
of the insert in pJB822 comprises the 23S rRNA coding sequence and
the remaining part contains the complete coding sequences of 4.5
and 5S rRNA. The 529 bp segment of the DNA extending from an
Hpa II restriction site and coding for the low-molecular-weight
rRNAs and the 3' end of 23S rRNA was then sequenced by the method
of Maxam and Gilbert (1977).

    The segment of DNA coding for the 5S rRNA was readily
identified as the sequence for this RNA was already known for
duckweed (Dyer and Bowman, 1979) and the maize 5S rRNA apparently
differs only in a few positions from this as indicated by the DNA
sequence. These differences were confirmed using a rapid RNA
sequencing technique (Stanley and Vassilenco, 1978). The sequence
of the 4.5S rRNA was established by a combination of methods. A
gel "read off" method (Simoncsits et al., 1977) was used to establish the positions of G, A and pyrimidine residues and C and U
distinguished from one another by reference to the DNA sequence and
by alignment with the oligonucleotides of known sequence derived
from duckweed 4.5S rRNA by digestion with T1 and pancreatic A ribonuclease (Bowman and Dyer, 1979). As yet we have not been able to
determine exactly where the coding sequence of the 23S rRNA ends.
Whereas the sequence coding for the 3' end of 16S rRNA of chloroplasts could be identified by alignment with the sequence at the
3' end of the 16S rRNA of E. coli (Schwartz and Kössel, 1979), the
sequence coding for the comparable region of chloroplast 23S rRNA
could not be identified in the same way because the 3' end of
E. coli 23S rRNA (Branlant, personal communication) does not appear
to be very similar to that of chloroplast 23S rRNA. However, at
least some of the 72 nucleotides which precede the 4.5S rRNA coding
sequence in the DNA segment which was sequenced (Fig. 1) must code
for 23S rRNA as shown by both Northern and Southern blot hybridisations. Therefore the 23S rRNA and 4.5S rRNA coding sequences are
probably close to one another.

    The sequence shows that the 4.5S and 5S coding regions are
separated by 231 nucleotides. Seventy nucleotides out from the 3'
end of the 4.5S rRNA coding sequence there is a region which is
particularly high in A.T base pairs (23 out of 28) which indicates
that transcription could be terminated there, as sequences of this
type are found at the known ends of genes (Federoff, 1979).
Furthermore, a sequence of nucleotides -GGTATTAA- which resembles
the promoters of E. coli (Scherer et al., 1978) occurs 29 nucleotides before the region coding for the 5S rRNA.

    These results suggest the 5S rRNA of chloroplasts is not part
of a large primary transcript containing the sequences of all the
rRNAs as it is in E. coli (Ginsberg and Steitz, 1975) but may be
separately transcribed. Certainly the mature 5S rRNA does not

appear to be a primary transcription product as it only has a single phosphate residue at the 5' end whereas primary transcripts usually have di- or triphosphates in this position. If the chloroplast 5S rRNA is separately transcribed, one would predict that the DNA which codes for 5S rRNA would not hybridise to a large ribosomal precursor of the type which has been suggested may occur in spinach (Hartley and Ellis, 1973; Hartley and Head, 1979) and that there should be a precursor to the 5S rRNA which is a primary transcription product and is slightly larger than the mature molecule. These two propositions are at present being investigated.

The close proximity of the 4.5S and 23S rRNA coding sequences agrees well with the suggestion that these two molecules are part of a common precursor. A molecule of 1.28×10^6 molecular weight has been found in spinach which shows sequence homology with both as judged by DNA/RNA competition hybridisation. However, the sequencing results indicate that there is only one copy each of the 5S and 4.5S rRNA coding genes not 4 - 5 as suggested by DNA/RNA hybridisation (Hartley, 1979).

A detailed physical description may now be made of the organisation of each of the two sets of ribosomal genes in maize chloroplast DNA. Their relative order is 16S, 23S, 4.5S and 5S rRNA and the first three at least are probably transcribed together as parts of a common precursor. The 1,490 bp of the 16S coding sequence is followed by a spacer region of about 2,100 bp (Bedbrook et al., 1977) and the 2,890 bp of the 23 rRNA gene. The 95 bp coding sequence of the 4.5S rRNA is close to that of the 23S rRNA and separated by 231 bp from that of the 122 bp segment which codes for the 5S rRNA. Thus, in maize, about 13,850 bp (10%) of the chloroplast DNA is occupied by the ribosomal genes and the spacers between them. The ribosomal operons are probably somewhat larger than this for, as in E. coli (Glaser and Cashel, 1979), the promoter for the 16S rRNA gene may precede it by some distance. Transcription probably stops abruptly, however, at the end of the 5S rRNA gene as there is an A.T-rich segment in this region.

REFERENCES

Bedbrook, J. R., Kolodner, R., and Bogorad, L., 1977, Zea mays chloroplast ribosomal RNA genes are part of a 22,000 base pair inverted repeat, Cell, 11: 739.

Borst, P., 1972, Mitochondrial nucleic acids, Ann. Rev. Biochem., 41: 333.

Bowman, C. M., and Dyer, T.A., 1979, 4.5S RNA, a novel ribosome component in the chloroplasts of flowering plants, Biochem. J., in press.

Chu, N., and Tewari, K.K., 1979, Arrangement of the ribosomal RNA
    genes in the restriction endonuclease map of pea chloroplast
    DNA, submitted.
Dyer, T.A., and Bowman, C.M., 1979, Nucleotide sequence of chloro-
    plast 5S rRNA of flowering plants, Biochem J., in press.
Federoff, N.V., 1979, On Spacers, Cell, 16: 697.
Ginsberg, D., and Steitz, J.A., 1975, The 30S ribosomal precursor
    RNA from Escherichia coli, J. Biol. Chem., 250: 5647.
Glaser, G., and Cashel, M., 1979, In vitro transcripts from the
    rrn B ribosomal cistron originate from two tandem promoters,
    Cell: 16: 111.
Gray, P.W., and Hallick, R.B., 1978, Restriction endonuclease map
    of Euglena gracilis chloroplast DNA, Biochemistry, 17: 284.
Hartley, M.R., and Ellis, R.J., 1973, Ribonucleic acid synthesis
    in chloroplasts, Biochem. J., 134: 249.
Hartley, M.R., 1979, The synthesis and origin of chloroplast low-
    molecular-weight ribonucleic acid in spinach, Eur. J. Biochem.,
    96: 311.
Hartley, M.R., and Head, C., 1979, The synthesis of chloroplast
    high-molecular-weight ribosomal ribonucleic acid in spinach,
    Eur. J. Biochem., 96: 301.
Jenni, B., and Stutz, E., 1978, Physical mapping of the ribosomal
    DNA region in Euglena gracilis, Eur. J. Biochem., 88: 127.
Kolodner, R., and Tewari, K.K., 1979, Inverted repeats in chloro-
    plast DNA from higher plants, Proc. Nat. Acad. Sci. U.S.,
    76: 41.
Kurland, C.G., 1960, Molecular characterisation of ribonucleic
    acid from Escherichia coli ribosomes. I. Isolation and
    molecular weight, J. Mol. Biol., 2: 83.
Maxam, A.M., and Gilbert, W., 1977, A new method for sequencing
    DNA, Proc. Nat. Acad. Sci. U.S., 74: 560.
Phillips, D.O., and Carr, N.G., 1977, Nucleic acid analysis and
    the endosymbiotic hypothesis, Taxon, 26: 3.
Rawson, J. R. Y., Kushner, S.R., Vapnek, D., Alton, N.K., and
    Boerma, C.L., 1978, Chloroplast ribosomal RNA genes in
    Euglena gracilis exist as three clustered tandem repeats,
    Gene, 3: 211.
Rochaix, J.D., and Malnoë, P., 1978, Anatomy of the chloroplast
    ribosomal DNA of Chlamydomonas reinhardii, Cell, 15: 661.
Scherer, G.E.F., Walkinshaw, M.D., and Arnott, S., 1978, A
    computer aided oligonucleotide analysis provides a model
    sequence for RNA polymerase-promoter recognition in E. coli,
    Nucleic Acid Res., 5: 3759.
Schwartz, Zs., and Kössel, H., 1979, Sequencing of the 3' terminal
    region of a 16S rRNA gene from Zea mays chloroplast reveals
    homology with E. coli 16S rRNA, Nature, 279: 520.

Simoncsits, A., Brownlee, G.G., Brown, R.S., Rubin, J.R., and
     Guilley, H., 1977, New rapid gel sequencing method for RNA,
     Nature, 269: 833.
Stanley, J., and Vassilenco, S., 1978, A different approach to
     RNA sequencing, Nature, 274: 87.
Van Holde, K.E., and Hill, W.E., 1974, General properties of ribo-
     somes, in: "Ribosomes", M. Nomura, A. Tissiéres and P. Lengyel,
     eds, Cold Spring Harbor Laboratory.
Whitfeld, P.R., Leaver, C.J., Bottomley, W., and Atchison, B.A.,
     1978, On 4.5S rRNA in higher plant chloroplasts, Biochem. J.,
     175: 1103.
Woese, C.R., and Fox, G.E., 1977, Phylogenetic structure of the
     prokaryotic domain: the primary kingdoms, Proc. Nat. Acad.
     Sci. U.S., 74: 5088.

TRANSFER RNAs AND AMINOACYL-tRNA SYNTHETASES IN PLANT ORGANELLES

G. Burkard, J. Canaday, E. Crouse, P. Guillemaut,
P. Imbault, G. Keith, M. Keller, M. Mubumbila
L. Osorio, V. Sarantoglou, A. Steinmetz and J.H. Weil

Institut de Biologie Moléculaire et Cellulaire
Université Louis Pasteur
15 rue Descartes, 67084 Strasbourg, France

INTRODUCTION

Chloroplasts contain their own DNA and have their own systems
for replication, transcription and translation. As far as the
translation machinery is concerned, chloroplasts contain tRNAs and
aminoacyl-tRNA synthetases which are different from their cytoplasmic counterparts[1,2].

This report presents recent results on the primary structure
of chloroplast tRNAs, on the localization of their genes on
chloroplast DNA and describes comparative studies on cytoplasmic
and chloroplast aminoacyl-tRNA synthetases.

MAPPING OF tRNA GENES ON CHLOROPLAST DNA

Chloroplast specific tRNAs have been shown to hybridize to
chloroplast DNA, which codes for about 30 tRNA species, that is,
probably a full set of tRNAs necessary for protein synthesis. We
have studied the mapping of tRNA genes on chloroplast DNA from
spinach (in collaboration with H.J. Bohnert, A.J. Driesel,
K. Gordon and R.G. Herrmann, University of Düsseldorf, Germany),
maize (in collaboration with L. Bogorad, M. McIntosh and R. Selden,
Harvard University, USA) and Euglena gracilis.

Fractionation of chloroplast tRNAs by two-dimensional polyacrylamide gel electrophoresis yielded about 35 spots in the case
of spinach[3] and Euglena, about 30 in the case of maize. Most of

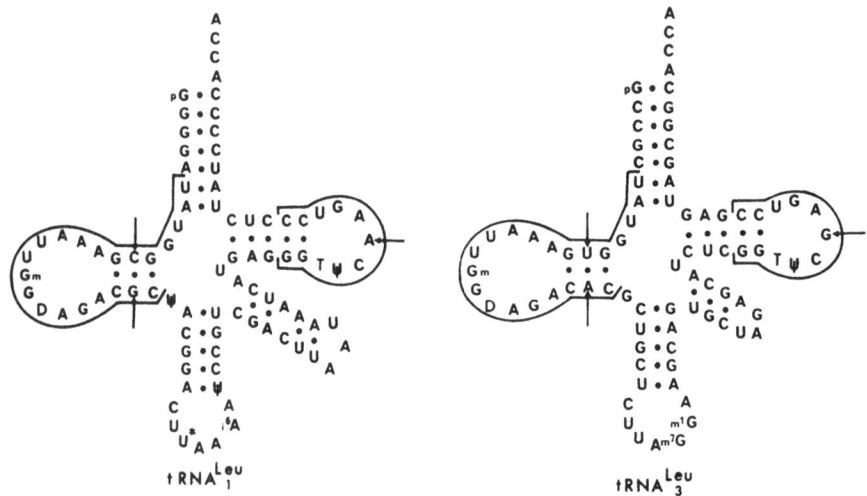

Fig. 2.   Structure of bean chloroplast tRNA$_1^{Leu}$ and tRNA$_3^{Leu}$. Nucleotide sequences showing a large extent of homology are
underlined (arrows designate substitutions within these
sequences).

tRNA SEQUENCES

Only two chloroplast tRNA sequences have been published so
far, those of Euglena[7] and Phaseolus[8] chloroplast tRNAs^{Phe}. The
primary structures of four more bean chloroplast tRNAs have been
studied, namely the three isoaccepting tRNAs^{Leu} and the initiator
tRNA$_F^{Met}$; the latter tRNA sequence has been compared to that of
the bean cytoplasmic initiator tRNA$_i^{Met}$.

Bean chloroplasts contain three tRNAs^{Leu} isoacceptors as
revealed by RPC-5 chromatography[9] or by two-dimensional gel electrophoresis[3]. The bean chloroplast tRNA gel pattern is very similar to that of the spinach chloroplast tRNA presented in the last
reference ; the three tRNA^{Leu} isoacceptors are in the chloroplasts
of both species among the slowest migrating tRNAs because of their
large extra loop (see fig. 2). All three chloroplast isoacceptors
recognize UUG[10]. Bean chloroplast tRNA$_1^{Leu}$ and tRNA$_3^{Leu}$ have been
sequenced and were found to differ quite extensively from one
another, except for the D loop and the Tψ loop. The two Tψ loops
are identical, except for a A - G substitution, but it should be
pointed out that the 11 nucleotide-long sequence (underlined in

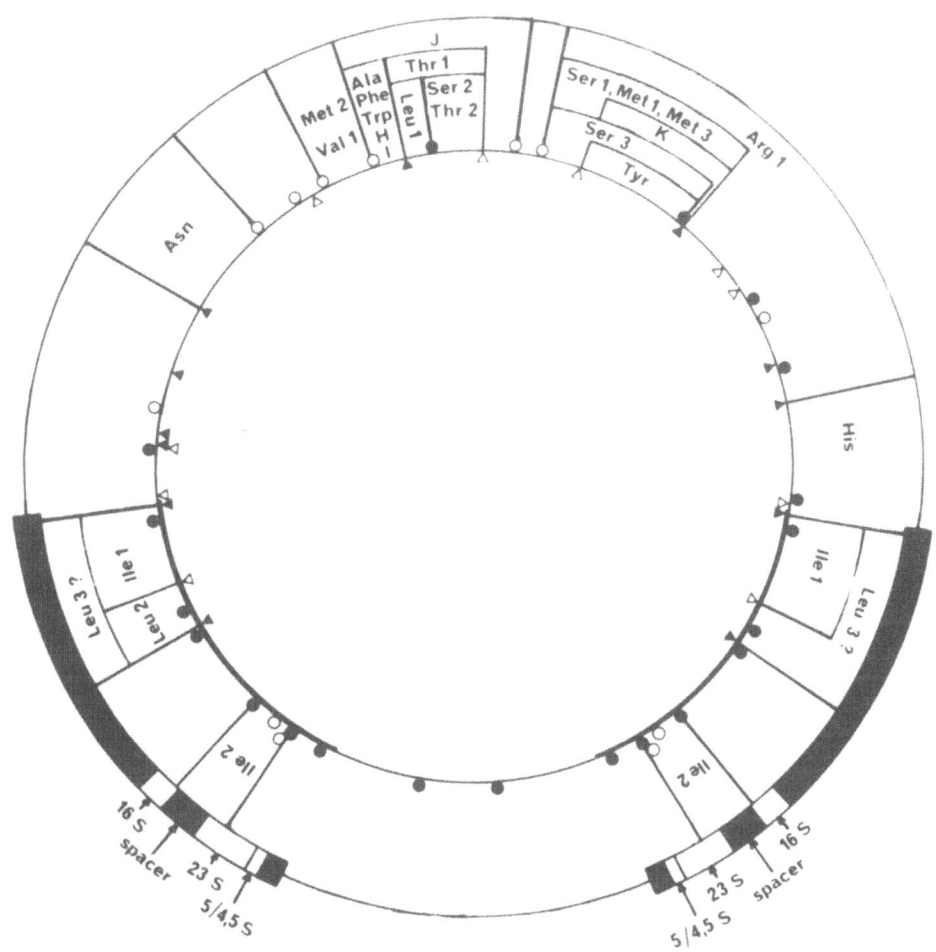

Fig. 1. Localization of tRNA genes on the restriction endonuclease cleavage site map of spinach chloroplast DNA. The
smallest DNA segment to which hybridization was observed
is indicated for each chloroplast tRNA tested. The thick
line corresponds to the two copies of the inverted repeat
region, containing the two polycistronic ribosomal DNA
units ; the arrangement of the rRNA genes in each unit
is : 16S-spacer-23S-4,5/5S, with the transcription polarity in that order (as in E.coli). ▼ Sal I cleavage
site ; ▽ Pst I cleavage site ; O Kpn I cleavage site ;
● Xma I cleavage site.

these species could be identified by aminoacylation using E.coli
or chloroplast aminoacyl-tRNA synthetases : 27 (corresponding to
16 amino acids) in the case of spinach[3], 23 (corresponding to 16
amino acids) in the case of maize, and 23 (corresponding to 18
amino acids) in the case of Euglena. Hybridization of individual
tRNAs (re-extracted from the gel and labeled in-vitro with either
[125]I or ^{32}P) to DNA fragments obtained by digestion of chloroplast
DNA with various restriction endonucleases, was used to localize
tRNA genes on the circular map of the chloroplast DNA molecule
from these 3 species. All three DNA molecules have a molecular
weight of about 90×10^6 daltons. Spinach chloroplast DNA contain
a 15×10^6 dalton region which is present twice but in inverted
orientation ("inverted repeat") ; these 2 regions are separated
from each other by a large and a small single-copy region (which
have a molecular weight of 52×10^6 and 12×10^6 daltons respectively). Each inverted repeat contains a set of ribosomal RNA
genes. A similar organization exists in maize chloroplast DNA,
whereas Euglena chloroplast DNA contains 3 tandemly repeated sets
of ribosomal RNA genes[4].

At present a minimum of 21 genes, corresponding to tRNAs for
14 different amino acids,have been localized on the map of the
spinach chloroplast DNA molecule (fig. 1). Of these, 15 genes,
corresponding to tRNAs for 12 amino acids, are located in the
large single-copy region. Each copy of the "inverted repeat" contains a gene for $tRNA_2^{Ile}$ in the "spacer" sequence between the 16S
and the 23S rRNA genes. The genes for $tRNA_1^{Ile}$, $tRNA_2^{Leu}$ and $tRNA_3^{Leu}$
also map in the inverted repeat, but outside the rDNA unit.

In maize, most of the tRNA genes located so far also map in
the large single-copy region and a $tRNA_2^{Leu}$ gene is located in each
of the two inverted repeats. In Euglena, a $tRNA^{Ala}$ and a $tRNA^{Ile}$
genes have been found in the region containing the 3 sets of rRNA
genes.

The presence of tRNA genes close to rRNA genes has also been
reported in E.coli, where tRNA genes have been found both in the
"spacer" between the 16S and the 23S rRNA genes ($tRNA^{Ile}$, $tRNA^{Ala}$,
$tRNA^{Glu}$) and at the distal end of some rRNA operons[5,6].

Of the several chloroplast isoacceptors for the same amino
acid, some are coded for by different genes, as shown for instance
in the case of the leucine, isoleucine, methionine and serine isoacceptors (fig. 1). In the case of the leucine isoacceptors, that
they are coded for by different genes is confirmed by the fact
that they differ in their primary structure (fig. 2).

fig. 2), has been highly conserved in procaryotic and eucaryotic tRNAs^{Leu} (for a review, see 11). The two D loops are also identical, and there is a common sequence of 21 nucleotides (which is underlined), except for the replacement of a AU by a GC base pair ; this D loop is very similar to that found in E.coli tRNA^{Leu} 11. The third chloroplast isoacceptor, tRNA$_2^{Leu}$ is presently being sequenced.

Bean chloroplast initiator tRNA^{Met}, like procaryotic initiators, is formylated in-vivo and has been characterized [12-14]. Its sequence shows two main features common to procaryotic tRNA$_F^{Met}$: Its 5' terminal nucleotide (which is a C) is not base-paired, and it has a TψCA sequence, whereas eucaryotic initiators have their 5' terminal nucleotide base-paired and contain a AUCG or a AψCG sequence. We have found that these last two features are also present in bean cytoplasmic initiator tRNA$_i^{Met}$.

Euglena and bean chloroplast tRNAs^{Phe} only differ by the nature of 5 nucleotides, in addition to small differences in post-transcriptional modifications[7,8] and this could suggest that tRNAs have not changed very much during evolution. However, heterologous hybridization reactions performed between bean chloroplast DNA and chloroplast tRNAs from various organisms show that tRNA^{Phe} is an exceptional case. Bean and spinach total chloroplast tRNAs hybridize to the same extent with bean chloroplast DNA, while maize total chloroplast tRNAs hybridize to a lower level (about 65%) and Euglena total chloroplast tRNAs hybridize only to a level of about 15%. When individual Euglena chloroplast tRNAs were studied, only tRNA^{Phe} was found to hybridize to bean chloroplast DNA (to the same extent as bean chloroplast tRNA^{Phe}), while all other individual tRNAs tested (about 10) did not hybridize at all.

COMPARATIVE STUDIES ON EUGLENA CYTOPLASMIC AND CHLOROPLAST VALYL-tRNA SYNTHETASES

Whereas chloroplast tRNAs are coded for by chloroplast DNA, chloroplast aminoacyl-tRNA synthetases appear to be coded for by the nucleus and synthesized in the cytoplasm before being imported into the chloroplasts[15,16]. The cytoplasmic and the chloroplast enzymes,catalyzing the attachment of a given amino acid to the cognate tRNA,differ not only in their intracellular localization, but also in their chromatographic behaviour and in their substrate (tRNA) specificity. However it is not known how different the two enzymes are, and whether they are coded for by different genes. In order to approach these problems, we have purified Euglena cytoplasmic and chloroplast valyl-tRNA synthetases, to be able to compare their structural and catalytic properties.

Purification of the chloroplast enzyme was achieved by ammonium sulfate precipitation and chromatography on Sephadex G-50,

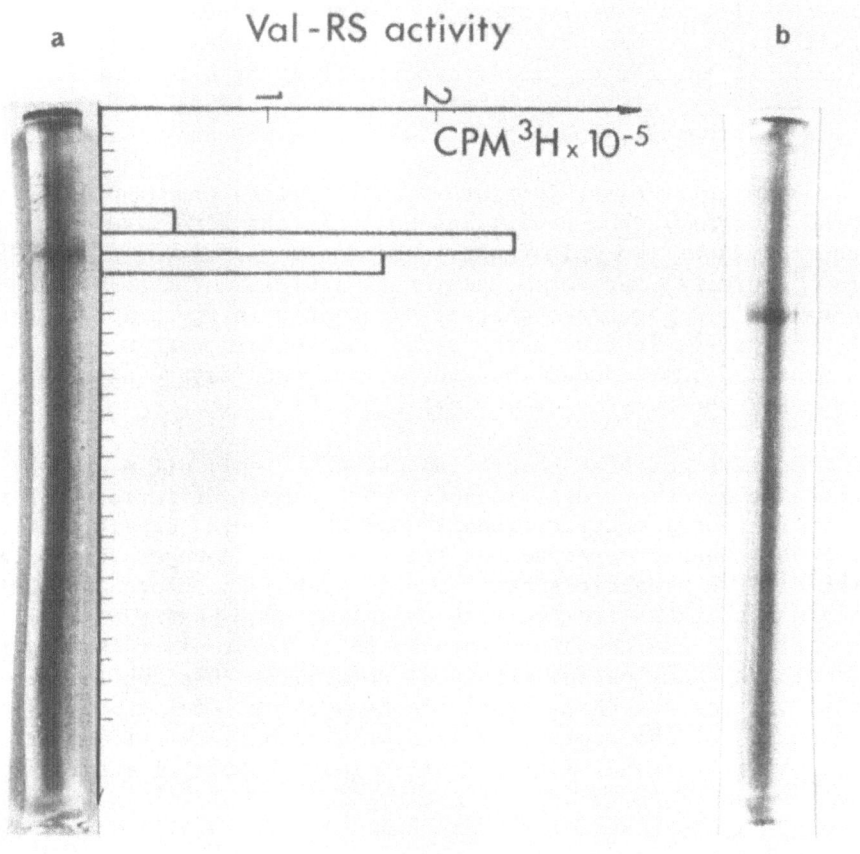

Fig. 3.   Polyacrylamide gel electrophoresis of the purified
          chloroplast valyl-tRNA synthetase.

          a) Under non-denaturating conditions (simultaneously on
          two 7.5% gels). One gel was stained with Coomassie
          Brillant Blue, the other was cut into fragments which
          were tested for enzymatic activity.

          b) Under denaturing conditions (on a 7.5% gel containing
          0.1% SDS, 4 M urea, 5% β-mercaptoethanol) after denatur-
          ation of the enzyme at 100° for 4 min in the presence of
          the same concentrations of the above-mentioned denaturing
          agents.

hydroxyapatite (this step separates the cytoplasmic and the chloroplast enzymes), DEAE-cellulose, Blue-Dextran Sepharose (which is
the most efficient step in the purification procedure) and
Sephadex G-200 [17]. After separation from the chloroplast enzyme by
hydroxyapatite chromatography, the cytoplasmic enzyme was further
purified by chromatography on phosphocellulose, DEAE-cellulose and
Blue-Dextran Sepharose.

As shown on fig. 3, polyacrylamide gel electrophoresis in
non-denaturating conditions of the purified chloroplast enzyme
yields a single protein band which coincides with valyl-tRNA synthetase activity. Upon polyacrylamide gel electrophoresis in urea-
SDS, the denatured enzyme also yields one single band corresponding to a M.W. of 126,000 daltons. As a similar M.W. value was
found for the non-denatured enzyme upon Sephadex G-200 filtration,
the chloroplast enzyme appears to have a monomeric structure.

The chloroplast valyl-tRNA synthetase has been purified about
1000 fold, to apparent homogeneity. It has a very high specific
activity of about 1100 units/mg (one unit catalyses the aminoacylation of 1 nmole tRNA in 1 min at 30°), comparable to that of the
enzymes purified from E.coli and yeast. Km values have been deter-
mined for L-valine (1.5×10^{-5} M), ATP (5×10^{-5} M) and tRNA^{Val}
(6×10^{-8} M). The structural features of the cytoplasmic and chloroplast enzymes are presently being studied.

Acknowledgments : The excellent technical assistance of Mrs
Annelyse Klein and Miss Claire Arnold is gratefully acknowledged.

REFERENCES

1. W.E. Barnett, S.D. Schwartzbach and L.I. Hecker. The transfer
   RNAs of eukaryotic organelles, in Progr. Nucl. Ac. Res. Mol.
   Biol., W.E. Cohn ed., Academic Press, New York, 143, 1978.
2. J.H. Weil. Cytoplasmic and organellar tRNAs in plants in
   Nucleic Acids in Plants, T.C. Hall and J. Davies eds, CRC
   Press, West Palm Beach, 1979 (in press).
3. A.J. Driesel, E.J. Crouse, K. Gordon, H.J. Bohnert, R.G.
   Herrmann, A. Steinmetz, M. Mubumbila, M. Keller, G. Burkard
   and J.H. Weil. Fractionation and identification of spinach
   chloroplast tRNAs and mapping of their genes on the restriction map of chloroplast DNA, Gene, 6, 285-306, 1979.
4. J.R. Bedbrook and R. Kolodner. The structure of chloroplast
   DNA, Ann. Rev. Plant Physiol. 30, 593-620, 1979.
5. E.A. Morgan, T. Ikemura and M. Nomura. Identification of spacer
   tRNA genes in individual ribosomal RNA transcription units
   of E.coli. Proc. Natl. Acad. Sci. US 74, 2710-2714, 1977.

6. E.A. Morgan, T. Ikemura, L. Lindahl, A.M. Fallon and M. Nomura. Some rRNA operons in E.coli have tRNA genes at their distal ends, Cell 13, 335-344, 1978.

7. S.H. Chang, C.K. Brum, M. Silberklang, U.L. RajBhandary, L.I. Hecker and W.E. Barnett. The first nucleotide sequence of an organelle transfer RNA : chloroplast tRNA^{Phe}, Cell 9, 717-724, 1976.

8. P. Guillemaut and G. Keith. Primary structure of bean chloroplastic tRNA^{Phe}, FEBS Lett. 84, 351-356, 1977.

9. P. Guillemaut, A. Steinmetz, G. Burkard and J.H. Weil. Aminoacylation of tRNA^{Leu} species from E.coli and from the cytoplasm, chloroplasts and mitochondria of Phaseolus vulgaris by homologous and heterologous enzymes, Biochim. Biophys. Acta 378, 64-72, 1975.

10. J. Ramiasa, P. Guillemaut and J.H. Weil. Codon recognition pattern of Phaseolus vulgaris cytoplasmic and chloroplastic tRNAs, FEBS Lett. 75, 128-132, 1977.

11. D.H. Gauss, F. Grüter and M. Sprinzl. Compilation of tRNA sequences, Nucl. Ac. Res. 6, r1-r19, 1979.

12. G. Burkard, B. Eclancher and J.H. Weil. Presence of N-formyl-methionyl-transfer RNA in bean chloroplasts, FEBS Lett. 4, 285-287, 1969.

13. P. Guillemaut, C. Burkard, A. Steinmetz and J.H. Weil. Comparative studies on the tRNAs^{Met} from the cytoplasm, chloroplasts and mitochondria of Phaseolus vulgaris, Plant Science Lett. 1, 141-149, 1973.

14. P. Guillemaut and J.H. Weil. Aminoacylation of Phaseolus vulgaris cytoplasmic, chloroplastic and mitochondrial tRNAs^{Met} by homologous and heterologous enzymes, Biochim. Biophys. Acta 407, 240-248, 1975.

15. B. Parthier. Cytoplasmic site of synthesis of chloroplast aminoacyl-tRNA synthetases in Euglena gracilis, FEBS Lett. 38, 70-74, 1973.

16. L.I. Hecker, J. Egan, R.J. Reynolds, C.E. Nix, J.A. Schiff and W.E. Barnett. The sites of transcription and translation for Euglena chloroplastic aminoacyl-tRNA synthetases, Proc. Natl. Acad. Sci. US 71, 1910-1914, 1974.

17. P. Imbault, V. Sarantoglou and J.H. Weil. Purification of the chloroplastic valyl-tRNA synthetase from Euglena gracilis, Biochem. Biophys. Res. Comm. 88, 75-84, 1979.

SYNTHESIS, TRANSPORT AND ASSEMBLY OF

CHLOROPLAST PROTEINS

R.John Ellis, Steven M.Smith and
Roger Barraclough

Department of Biological Sciences
University of Warwick
Coventry, West Midlands
U.K. CV4 7AL

INTRODUCTION

Why should we be interested in chloroplast protein synthesis? There are two answers to this question. The major conceptual challenge in biology at the present time is to unravel the molecular basis of differentiation. The leaf is a highly differentiated tissue because of the presence of chloroplasts. Moreover, chloroplasts are easy to isolate, and contain massive amounts of ribulose bisphosphate carboxylase (or Fraction I protein), which catalyses the initial steps in both photosynthesis and photorespiration. The sheer abundance of this protein makes it ideal for studies on the control of protein synthesis, and it is no accident that the first reported in vitro translation of a specific messenger RNA for a plant enzyme produced the large subunit of Fraction I protein[1]. The second reason for being interested in chloroplast protein synthesis derives from the fact that chloroplasts represent an extranuclear genetic system. When it is realised that most, if not all, eukaryotic cells possess extranuclear genetic systems, the significance of this aspect of chloroplasts is seen to extend beyond photosynthesis and differentiation. The hope is that the study of the genetic aspect of chloroplasts, as expressed in protein synthesis, will provide insights not just into chloroplast differentiation, but also into the modus operandi of extranuclear genetic systems in general.

Most research has concentrated on answering the basic question as to where in the cell the more tractable

321

chloroplast polypeptides are synthesised. It is clear
from this work that a division of labour exists in the
photosynthetic cells of eukaryotes; some chloroplast
polypeptides are synthesised inside the developing organelle, while others are made in the cytoplasmic compartment.
It follows that a traffic of polypeptides must flow across
the chloroplast envelope, and recent advances have allowed
this process to be studied in cell-free extracts. The
incoming polypeptides are assembled into their functional
configuration inside the organelle. The assembly process
can also now be demonstrated in cell-free extracts, and
involves interactions with chloroplast-synthesised
polypeptides. This review will discuss selected aspects
of these synthetic, transport, and assembly processes.

SYNTHESIS
      The most direct way to establish which polypeptides
are made on chloroplast ribosomes and which are made on
cytoplasmic ribosomes is to identify the products of
protein synthesis by isolated subcellular systems. Current
methods allow discrete polypeptides to be synthesised by
intact chloroplasts isolated from several higher plant
species and Euglena, as well as lysed chloroplasts, free
and bound chloroplast ribosomes, etioplasts and proplastids,
and heterologous systems programmed with either chloroplast
messenger RNA, chloroplast DNA, or cytoplasmic messenger
RNA. Table 1 lists the methods now available for studying
the in vitro synthesis of chloroplast proteins.

      The first in vitro system which produced discrete
polypeptides used light energy to drive protein synthesis
in intact isolated chloroplasts[2]. Since only intact
chloroplasts can generate ATP from light in the absence of
added cofactors, this method allows the use of crude
chloroplast preparations which can be made very rapidly.
With practice, the time taken to extract chloroplasts
from their normal environment, and place them in an
illuminated tube with labelled amino acids can be as short
as three minutes. Since the envelope around the chloroplasts is intact, the microenvironment around the polysomes
is likely to be more normal than in lysed systems. The
light-driven system thus provides a baseline against
which to judge the results of further fractionating the
system.

      The products of light-driven protein synthesis were
first analysed on cylindrical sodium dodecylsulphate (SDS)
polyacrylamide electrophoretic gels, which were cut into
slices for counting[2,3]. These old studies have been

Table 1. In Vitro Chloroplast Protein
Synthetic Systems

System	Species & Reference
1. Intact chloroplasts	Pisum[2-4], Spinacia[5,6] Hordeum[4], Zea[4,7,8] Euglena[9]
2. Lysed chloroplasts	Pisum[4]
3. Free chloroplast ribosomes	Pisum[4]
4. Bound chloroplast ribosomes	Pisum[4,10] Chlamydomonas[11]
5. Etioplasts & proplastids	Pisum[12], Euglena[13]
6. Chloroplast mRNA plus cell-free extracts of E.coli, wheat-germ or reticulocyte lysate	Spinacia[1,14,15] Zea[16], Spirodela[17] Cucumis[18], Euglena[20] Chlamydomonas[19]
7. Chloroplast DNA plus cell-free extracts of E.coli, and reticulocyte lysate	Spinacia[21], Zea,[22] Chlamydomonas[23]
8. Cytoplasmic mRNA plus cell-free extracts of wheat-germ	Pisum[24-27], Spinacia[27] Lemna[28], Chlamydomonas[29]

repeated using slab gels and autoradiography, which gives much higher resolution. Fig.1 shows an autoradiograph of the products of light-driven protein synthesis by isolated pea chloroplasts incubated with [S-35]methionine. In terms of the amount of incorporated methionine, there are two major products, and many minor ones. One major product (mol.wt.55,000) is soluble, and it has been identified as the large subunit of Fraction I protein[2]. The abundance of this protein could explain why chloroplast ribosomes account for up to 50% of the total complement

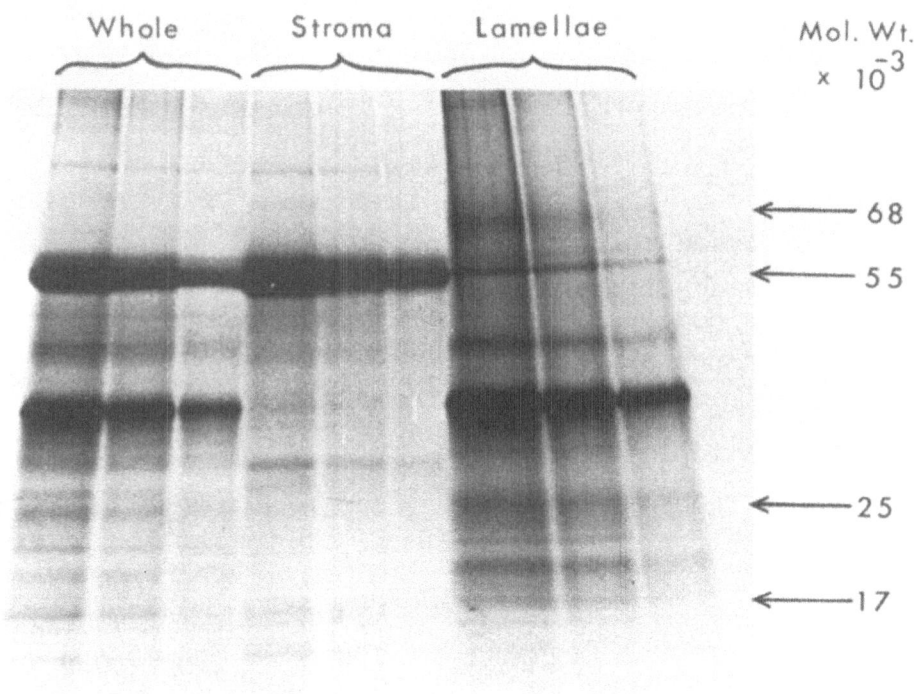

Fig.1. Autoradiograph of the products of light-driven
        protein synthesis by isolated Pisum chloroplasts
        incubated with [S-35]methionine. An aliquot of
        chloroplasts was lysed, and separated into stromal
        and lamellar fractions by centrifugation at
        38,000 x g for 10 min. The fractions were
        electrophoresised on an SDS polyacrylamide gel.

of leaf ribosomes. The other major product is attached to
the chloroplast lamellae, and has an apparent molecular
weight of 32,000. We refer to this product as peak D,
since it was the fourth labelled peak seen on cylindrical
gels in the early studies[3]. The identity of peak D
still eludes us. Some of the label in peak D can be
immunoprecipitated by antiserum to the chloroplast
coupling factor, suggesting that it may be a membrane

component of the ATP synthase complex[30]. However work in the authors' laboratory has found that peak D is lost early in the purification of this complex from Pisum chloroplast lamellae by a published method[31].
Although peak D is a major product of chloroplast protein synthesis in terms of incorporation, it does not accumulate as does the large subunit of Fraction I protein. It can be separated from closely running staining polypeptides by extraction in chloroform/methanol[32]. Peak D thus appears to be a minor hydrophobic component of thylakoids which turns over rapidly. Its synthesis can even be detected in plastids old enough to have ceased the manufacture of Fraction I protein[12]. · A polypeptide with similar properties has been described from Zea[7] and Spirodela[17]; the Zea polypeptide has been shown to be encoded in the chloroplast genome[16]. The major unresolved problem in this field is to identify the function of this polypeptide, and to determine why chloroplasts spend so much time in continually making it.

What can be said about the minor products of chloroplast protein synthesis? The number of resolvable products has increased steadily with the improvement in techniques. About 37 labelled bands can be routinely observed on an autoradiograph of the type shown in Fig.1. This number rises dramatically if two-dimensional gels are employed; about 80 labelled spots are seen if the stromal fraction from chloroplasts is analysed on an O'Farrell gel[30]. The identities of these soluble products are unknown for the most part. The published data from in vivo inhibitor experiments provide no clues as to what they might be. On the other hand, it is largely the enzymes of the Calvin cycle that have been examined in such experiments; enzymes of the many other metabolic pathways found in chloroplasts have not been examined in this regard. Thus a large discrepancy presently exists between the results of in vivo inhibitor experiments and those of in vitro incorporation experiments as to the function of chloroplast ribosomes.

An interesting calculation can be made on the basis of two enormous and unproven assumptions viz. that each labelled spot from in vitro chloroplast incubations represents a unique polypeptide, and that each such polypeptide is encoded in the chloroplast genome. The total molecular weight of polypeptides being synthesised is then about 3×10^6 which would account for about 50% of the total potential coding capacity of the unique · sequences in this genome.These estimates are almost certainly too low, because not all polypeptides contain

Table 2.   Identified Products of In Vitro
            Chloroplast Protein Synthesis

Polypeptide	Reference
1. Large subunit of Fraction I protein	2,5
2. Three subunits of chloroplast coupling factor	4,7,33
3. Elongation factors T and G of chloroplast protein synthesis	34
4. Cytochrome f	35
5. Dicyclohexylcarbodiimide-binding protein	36
6. Cytochrome b_{559}	37
7. Apoprotein of chlorophyll protein complex I	37

methionine, and because valid two-dimensional resolution
of the products of chloroplast protein synthesis has so
proved possible for the stromal fraction only. So we
may be within striking distance of being able to account
for the information content of chloroplast DNA in terms
of resolvable polypeptides, and the task for the future
is to identify these, and to establish their site of
encoding.

A number of laboratories in the pastfew years have
entered the game of identifying some of the minor products of in vitro chloroplast protein synthesis. Table 2
summarises the present position. All the identified
products are components of the chloroplast lamellae with
the exception of the large subunit of Fraction I protein
and the elongation factors. It is disappointing that
there is no discernable pattern in this list of chloroplast-synthesised polypeptides. If there are rules
which dictate that a given polypeptide must be synthesised inside the chloroplast rather than outside it, these
are not obviously related to the nature, location or
function of the polypeptides so far identified.

TRANSPORT
     Two polypeptides that are not labelled in isolated
chloroplasts are the small subunit of Fraction I protein
and the chlorophyll a/b binding protein. _In vivo_
inhibitor experiments clearly suggest that these are
products of cytoplasmic protein synthesis[38,39]. When
cytoplasmic poly (A)-containing RNA from greening
Pisum leaves is translated in a wheat-germ extract,
these polypeptides are produced as higher molecular
weight precursors (Fig.2). One of the more interesting
advances in the field of chloroplast development in the
last two years has been the demonstration that both
these precursors will enter isolated chloroplasts with
processing to the mature size, and will then assemble
correctly inside the organelle. The data on the processing and uptake of the small subunit of Fraction I protein
have been published from two laboratories[24,25,27]; elsewhere in this volume G.W.Schmidt discusses the data on
the uptake, processing and assembly of the chlorophyll
a/b binding polypeptides. In the remainder of this
section, some general ideas about the uptake of polypeptides into chloroplasts will be discussed, and
questions that need tackling will be posed.

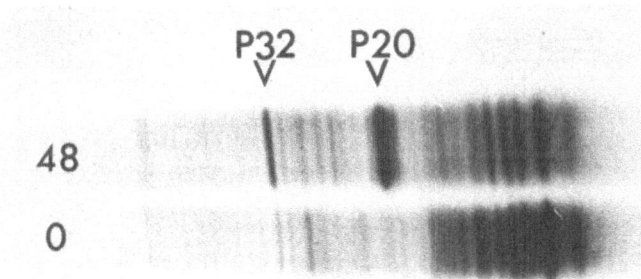

Fig.2. Autoradiograph of the products of protein synthesis
       by polysomal poly (A)-containing RNA isolated from
       either etiolated _Pisum_ shoots (0) or from shoots
       greened for two days (48). The RNA was translated
       in a wheat-germ extract. Electrophoresis on an
       SDS gel is from left to right. Use of specific
       antisera shows that P32 is a higher mol.wt. precurs-
       or (mol.wt.32,000) of a polypeptide of the chloro-
       phyll a/b binding protein (mol.wt.26,000), while
       P20 is a precursor (mol.wt.20,000) of the small
       subunit (mol.wt.14,000) of Fraction I protein.
       Unpublished work of S.M.Smith and A.Cuming.

There are two main types of hypothesis to explain
how polypeptides traverse membranes. These are
illustrated in Fig.3. The signal hypothesis relies on
a leader sequence or signal at the N-terminus of the
polypeptide chain which binds the polysome to the membrane to be traversed by a specific interaction with
membrane proteins. The polypeptide chain then goes
through a pore created from these membrane proteins as
it is being lengthened during protein synthesis. The
essential feature of this model is that transport is
co-translational, that is, protein synthesis must occur
at the same time as transport. It is not an essential
feature of this model that the signal sequence should
subsequently be removed, although this often happens.
This model is well supported by evidence from animal
and bacterial systems, but it is important to note
that some variants are being discovered. For example,
ovalbumen does not have a cleaved signal sequence at
its N-terminus; nevertheless it traverses the endoplasmic reticulum in a cotranslational manner. The signal in
ovalbumen appears to be some 200 amino acid residues
from the N-terminus,which poses an interesting problem
in topology[46].

Extension of the signal hypothesis to chloroplasts
predicts that bound ribosomes should be seen on the
chloroplast envelope during development. There is little
evidence for this. A modified version of the hypothesis
supposes that the chloroplast polypeptides are first
inserted by bound ribosomes into vesicles of endoplasmic reticulum, and it is these vesicles which fuse with
chloroplasts[4]. This model requires specific recognition
sites to be present on these vesicles; transport into
chloroplasts would require vesicles to be present, but
would be a post-translational event instead of a co-
translational event.

An alternative model we term the envelope carrier
hypothesis (Fig.3B). The polypeptide is made by free
instead of by bound ribosomes, and it is released from
these ribosomes prior to transport. The released chain
folds up, and some aspect of its tertiary structure,
analogous to the active site of an enzyme, binds to
specific carriers in the chloroplast envelope. It was
originally suggested that all those proteins destined
to enter chloroplasts possessed a common site[2]. This
site could involve amino acid residues at the N-terminus,
C-terminus, or in between, or any combination of these
(the zig-zag lines in Fig.3B represent such possible
sites). Binding to the envelope leads to transport;

IDEAS   ABOUT   PROTEIN   TRANSPORT   ACROSS   MEMBRANES

A. THE   SIGNAL   MECHANISM   B. THE   CARRIER   MECHANISM

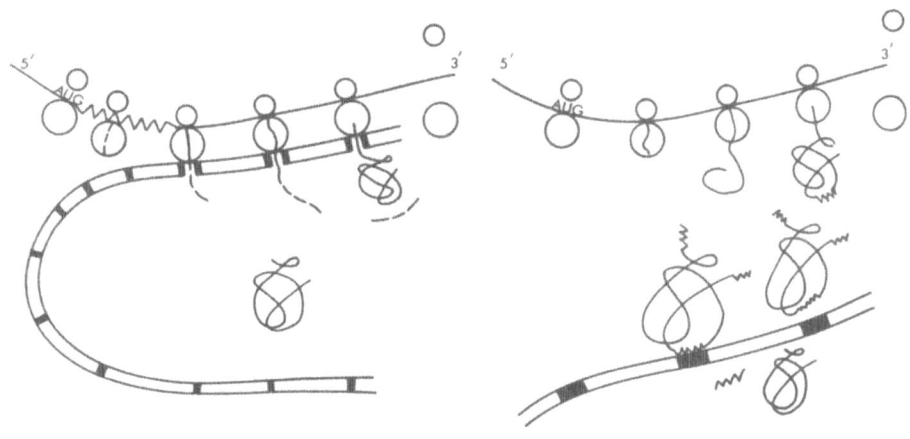

Fig.3.   Ideas on the transport of polypeptides across
         membranes.
         A: the signal hypothesis[41],
         B: the envelope carrier hypothesis[2,4,24]

involved in this process is a cleavage event which
removes sequences from either or both termini. In this
model, transport is post-translational; the completed
polypeptide chain enters chloroplasts in the absence of
both protein synthesis, and of any membranous vesicles
other than the chloroplasts themselves,

    All the evidence from both N-H.Chua's laboratory at
the Rockefeller University and from R.J.Ellis' laboratory
at Warwick University shows that it is the envelope
carrier type of mechanism that operates for chloroplasts.
Recently some evidence that this is also true for
mitochondria has appeared[42]. Table 3 summarises the data
which indicates that the small subunit of Fraction I
protein enters isolated chloroplasts by a post-translational mechanism.

    What are the oustanding questions that need to be
answered in the field of protein transport into chloroplasts? The most important question asks about the mechanism by which a completed polypeptide chain crosses the
two membranes of the chloroplast envelope. It is a

Table 3.   Evidence for The Envelope Carrier
                    Hypothesis

Prediction from the signal hypothesis	Data for the small subunit of Fraction I protein
1. Nascent N-terminal hydrophobic signal triggers ribosome binding to membranes	Precursor polypeptide made on free cytoplasmic ribosomes[26,43]
2. Processing and transport are concomitant with translation	Processing and transport are post-translational events; no requirement for ribosomes[24,27,45]
3. Processing activity is membrane-bound	Processing activity is soluble inside the chloroplast[25]

weakness of the carrier hypothesis that it does not
provide a mechanism of transport as does the signal
hypothesis. The best suggestion we can make is to fall
back on the philosopher's stone of molecular biologists
and invoke the magic of conformational change. If the
amino acid sequence removed is both sufficiently large
and sufficiently charged,its loss could trigger a
conformational change that brings the molecule through
the envelope. In the case of the small subunit of
Fraction I protein, the extra sequence is about 40
amino acids in length, and it is basically charged
(our previous report[32] that it was acidic is in error).
Is the extra sequence required for transport? A key
experiment that needs to be done is to remove the extra
sequence from the precursor, and then to determine
whether the residual polypeptide will enter isolated
chloroplasts. This may seem an easy experiment to
perform but it is not, because the processing enzyme
is stromal in location. If small subunit precursor is

added to stromal extracts, not only does processing
occur, but the released small subunit assembles into
holoenzyme. Is processing required for transport, or for
assembly, or for both events? It is clear that the
purification of the processing enzyme should be given
priority so that these questions can be answered. A more
distant question asks how the processing enzyme for the
small subunit precursor itself enters the chloroplast,
since it is known to be a product of cytoplasmic
ribosomes[44]. Perhaps there is a master processing
enzyme which is a product of chloroplast protein
synthesis, and which serves to process the precursor
forms of thoseprocessing enzymes made in the cytoplasm.

ASSEMBLY

    Three multisubunit complexes have been shown to
assemble in isolated chloroplasts,viz. the chloroplast
coupling factor[4,33], the chlorophyll a/b binding
protein, and Fraction I protein[25,46]. Recent work in the
authors' laboratory on the assembly of Fraction I
protein has led to the discovery of a chloroplast
protein which binds the large, but not the small,
subunit. The large subunit binding protein may be an
intermediate in the assembly process[46].

    The first report of the synthesis of the large
subunit in isolated chloroplasts showed that, under the
conditions used, the large subunit did not enter pre-
existing Fraction I protein[2]. Instead the in vitro-
synthesised large subunit behaved as an aggregate, and
analysis on non-denaturing polyacrylamide gels
revealed an apparent molecular weight of $6-7 \times 10^5$ for
this aggregate[4]. Such analysis also revealed that at a
gel concentration of about 5% acrylamide the in vitro-
synthesised large subunit migrates fortuitously with
the pre-existing Fraction I protein; at other gel
concentrations the large subunit aggregate can be
clearly resolved from the holoenzyme. Thus we dismissed
two subsequent reports[5,33] that in vitro-synthesised
large subunit does assemble into the holoenzyme, because
the analyses in both these reports were carried out at
a gel concentration of 5% only. However our recent
demonstration that assembly will occur readily from
small subunits in stromal extracts[25] led us to reinvest-
igate why we failed to observe assembly from large
subunit in the early studies. The answer is that whether
or not assembly occurs in isolated intact chloroplasts
depends on the incubation conditions. Assembly occurs
in a medium where sorbitol provides the osmoticum, but

not in a medium where KCl provides the osmoticum; the other laboratories used the sorbitol medium, so it is likely that their findings were not due to a fortuitous choice of gel concentration, but resulted from true assembly.

In the course of this work, we noticed that the aggregated large subunit, seen after incubation of chloroplasts in both sorbitol and KCl media, invariably and exactly coincided with a staining band of protein on non-denaturing gels. A time course analysis of incubation under assembly conditions,i.e. in sorbitol medium, showed that the labelled large subunit appeared first in this band, and subsequently appeared in the holoenzyme. Analysis of this band on SDS polyacrylamide gels showed that all its radioactivity migrated with large subunit, but that the staining material did not. Instead an unlabelled polypeptide of molecular weight about 60,000 was stained. This polypeptide we have named the large subunit binding protein[46].

What do these results mean? There are two possible interpretations:-
1. Large subunit accumulates in isolated chloroplasts because of a low concentration of small subunits with which to combine. Large subunit happens to combine with another chloroplast protein. This does not happen in vivo because assembly proceeds as fast as synthesis. We term this the pessimistic hypothesis.
2. Chloroplasts contain a protein which specifically functions to bind the large subunit. The resulting aggregate acts as a store for large subunit prior to assembly of Fraction I protein. Isolated large subunit is insoluble in aqueous solvents, so the binding protein could serve to stop it precipitating from solution or dissolving in the thylakoids. The lack of label in the binding protein when isolated chloroplasts incorporate [S-35]methionine implies that it is a product of cytoplasmic ribosomes. We term this the optimistic hypothesis (Fig.4). Our current research aims to decide between these hypotheses.

PRINCIPLES OF ORGANELLAR PROTEIN SYNTHESIS?
It is the authors' view that in a rapidly advancing field it is important to try to formulate some general principles, imperfect and transient though they may turn out to be, to provide a mental guide through the jungle of new data. Such principles have been proposed before for organellar protein synthesis[4]. They did not survive for long[32],but it is clear that their formulation

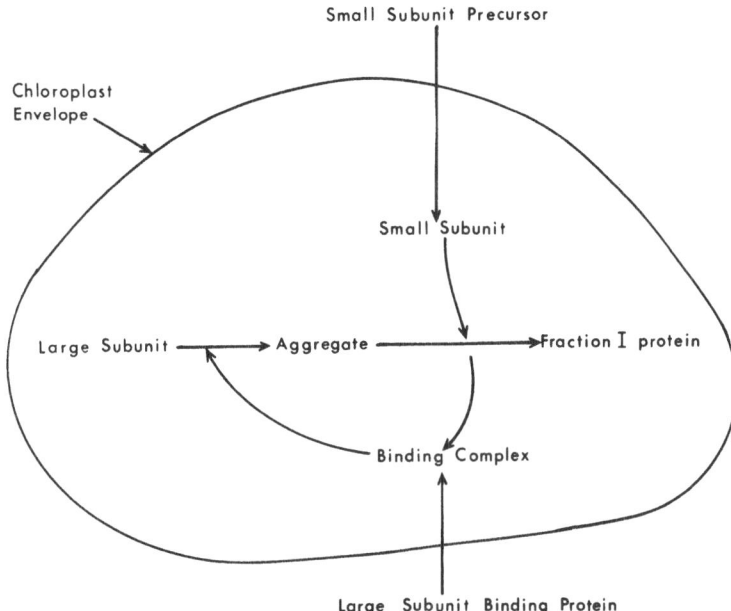

Fig.4.   The optimistic hypothesis for the function of
         the large subunit binding protein. For details
         see ref.46.

provoked experimentation to disprove them. It is in this
provocative spirit that we suggest another set of
principles:-
1.   Any multisubunit protein located in an organelle
has some subunits encoded in the organellar genome,
and other subunits encoded in the nuclear genome.
2.   Polypeptides encoded in the organellar genome are
synthesised by the organellar ribosomes, whereas those
encoded in the nucleus are synthesised by cytoplasmic
ribosomes.
3.   Transport of polypeptides into organelles involves
a post-translational mechanism.
4.   Assembly of organellar proteins occurs inside the
organelle.

     Principle no.3 suggests a new line of research.
The operation of a post-translational mechanism for the
import of polypeptides implies that the absence of
chloroplast ribosomes on the <u>inner</u> side of the chloroplast envelope cannot be used to rule out the possibility
that polypeptides are also exported from chloroplasts
into the cytoplasm. Such exported polypeptides could
include enzymes that function in the cytoplasm in

pathways interacting with chloroplast metabolism, and
regulatory molecules which inform the nucleo-cytoplasmic
genetic system of the current state of affairs in the
chloroplast compartment. It is to be hoped that attempts
will be made, both by in vitro techniques and by the use
of mutants in suitable species, to investigate the
possibility that polypeptides are exported from
chloroplasts.

REFERENCES

1. M.R.Hartley, A.M.Wheeler and R.J.Ellis, J.Mol.Biol.
   91:67(1975).
2. G.E.Blair and R.J.Ellis, Biochim.Biophys.Acta 319:
   223 (1973).
3. A.R.J.Eaglesham and R.J.Ellis, Biochim.Biophys.Acta
   335:396 (1974).
4. R.J.Ellis, Biochim.Biophys.Acta 463:185 (1977).
5. W.Bottomley, D.Spencer and P.R.Whitfeld, Arch. Bio-
   chem.Biophys. 164:106 (1974).
6. J.J.Morgenthaler and L.Mendiola-Morgenthaler, Arch.
   Biochem.Biophys.172:51 (1976).
7. A.E.Grebanier, D.M.Coen, A.Rich and L.Bogorad, J.Cell
   Biol.78:734 (1978).
8. A.E.Grebanier, K.E.Steinback and L.Bogorad, Plant
   Physiol. 63:436 (1979).
9. A.C.Vasconcelos, Plant Physiol.58:719 (1976).
10.R.Alscher,R.Patterson and A.T.Jagendorf, Plant Physiol.
   62:88 (1978).
11.A.Michaels and M.M.Margulies, Biochim.Biophys.Acta
   390:352 (1975).
12.S.G.Siddell and R.J.Ellis, Biochem.J. 146:675 (1975).
13.A.Dockerty and M.J.Merrett, Plant Physiol.63:468
   (1979).
14.A.M.Wheeler and M.R.Hartley, Nature 257:66 (1975).
15.J.Silverthorne and R.J.Ellis, Biochim.Biophys.Acta
   in press (1979).
16.J.Bedbrook,G.Link, D.M.Coen, L.Bogorad and A.Rich,
   Proc.Natl.Acad.Sci.75:3060 (1978).
17.A.Reisfeld, J.Gressel, K.M.Jakob and M.Edelman,
   Photochem. and Photobiol.27:161 (1978).
18.R.Walden and C.J.Leaver in "Chloroplast Development"
   G.Akoyunoglou et al eds.,p.251, Elsevier/North-holland
   Biomedical Press (1978).
19.S.H.Howell, P.Heizmann,S.Gelvin and L.L.Walker, Plant
   Physiol.59:464 (1977).
20.D.Sagher, H.Grosfeld and M.Edelman, Proc.Natl.Acad.
   Sci. 73:722 (1976).

21. W.Bottomley and P.R.Whitfeld, Eur.J.Biochem. 93: 31 (1979).
22. D.M.Coen, J.R.Bedbrook, L.Bogorad and A.Rich, Proc.Natl.Acad.Sci.74:5487 (1977).
23. J.D.Rochaix and P.Malnoe in "Chloroplast Development" G.Akoyunoglou et al eds.,p.581 Elsevier/North-Holland Biomedical Press (1978).
24. P.E.Highfield and R.J.Ellis, Nature 271:420 (1978).
25. S.M.Smith and R.J.Ellis,Nature 278:662 (1979).
26. A.R.Cashmore, M.K.Broadhurst and R.E.Gray, Proc. Natl.Acad.Sci.75:655 (1978).
27. N-H.Chua and G.W.Schmidt, Proc.Natl.Acad.Sci.75: 6110 (1978).
28. E.M.Tobin, Proc.Natl.Acad.Sci.75:4749 (1978).
29. B.Dobberstein, G.Blobel and N-H.Chua, Proc.Natl. Acad.Sci.74:1082 (1977).
30. R.J.Ellis, P.E.Highfield and J.Silverthorne in Proc.Fourth Int.Cong.Photosynthesis, p.497 Biochemical Society Press (1977).
31. G.D.Winget, N.Kanner and E.Racker, Biochim.Biophys. Acta 460:490 (1977).
32. R.J.Ellis and R.Barraclough in "Chloroplast Development", G.Akoyunoglou et al eds.,p.185, Elsevier/ North-Holland Biomedical Press (1978).
33. J.J.Morgenthaler, M.P.F. Marsden and C.A.Price, FEBS Lett.62:96 (1976).
34. O.Tiboni,G.Pasquale and O.Ciferri in "Chloroplast Development", G.Akoyunoglou et al eds.,p.675, Elsevier/North-Holland Biomedical Press (1978).
35. A.Doherty and J.G.Gray, Eur.J.Biochem.98:87 (1979).
36. A.Doherty and J.G.Gray, unpublished.
37. R.E.Zielinski and C.A.Price, submitted to J.Cell.Biol.
38. R.Barraclough and R.J.Ellis, Eur.J.Biochem. 94:165 (1979).
39. R.J.Ellis, Phytochem. 14:89 (1975).
40. D.F.Steiner, Nature 279: 674 (1979).
41. G.Blobel and B.Dobberstein, J.Cell Biol.67:835 (1975).
42. M-L.Maccecchini,Y.Rudin, G.Blobel and G.Schatz, Proc. Natl.Acad.Sci.76:343 (1979).
43. H.Roy, B.Terenna and L.C.Cheong, Plant Physiol.60: 532 (1977).
44. J.Feierabend and G.Wildner, Arch.Biochem.Biophys. 186:283 (1978).
45. N-H.Chua and G.W.Schmidt,J.Cell Biol. 81:461 (1979).
46. R.Barraclough and R.J.Ellis, submitted to Biochim. Biophys.Acta (1979).

IN VITRO SYNTHESIS, TRANSPORT, AND ASSEMBLY OF THE CONSTITUENT

POLYPEPTIDES OF THE LIGHT-HARVESTING CHLOROPHYLL a/b PROTEIN COMPLEX

Gregory W. Schmidt*, Sue Bartlett, Arthur R. Grossman
Anthony R. Cashmore, and Nam-Hai Chua
The Rockefeller University
New York, New York 10021

INTRODUCTION

Recent studies have established that transport across chloro-plast envelopes of proteins which are synthesized by cytoplasmic ribosomes can occur by a post-translational mechanism (1,2). Dobberstein et al. (3) first discovered that a major chloroplast stromal protein, the small subunit (S) of ribulose 1,5-bisphosphate carboxylase (RuBPCase) is synthesized by free polysomes in the green alga, Chlamydomonas reinhardtii. Moreover, they found that translation of the small subunit messenger RNA in vitro yields a precursor (pS) 4000-5000 daltons larger than the mature protein. Upon incubation with a cell-free Chlamydomonas extract pS can be processed to the mature form and a small peptide fragment desig-nated the transit peptide (4,5). Dobberstein et al. (3) proposed that transport of the RuBPCase small subunit in vivo occurs after it is completely synthesized and that the transit sequence on pS facilitates its post-translational interaction with the chloroplast envelope. This proposed mechanism is fundamentally distinct from the co-translational transport across endoplasmic reticulum membranes of proteins which are synthesized by membrane-bound ribosomes (6,7). Precursor forms of the RuBPCase small subunit also have been found among the translation products of spinach (1), pea (1,2,8) and duckweed (9) mRNA in cell-free systems. In vitro reconstitution experiments have established that uptake of pS into purified intact chloroplasts does not require concomitant protein synthesis (1, 10). During or immediately following transport through the chloroplast

*Present· address: Botany Department, University of
Georgia, Athens, Georgia 30602

envelope pS is processed to S (1,2,10). Moreover, the newly-transported S is localized in the chloroplast stroma since 80% of the
protein assembles with the chloroplast-synthesized RuBPCase large
subunit to form the 18S holoenzyme (1,10).

Since S is a soluble protein the question arises whether insoluble chloroplast proteins, e.g., integral membrane proteins of
the thylakoids, also follow a similar biosynthetic pathway and
mechanism of transport into the chloroplasts. The major integral
protein of the thylakoid membrane is the apoprotein of the light-
harvesting chlorophyll a/b protein complex (CP II) (cf. 11). In
this paper, we present evidence that the two constituent polypeptides of pea CP II are synthesized as soluble, larger precursors in
cell-free translation systems. After synthesis, these precursors
can be transported into intact chloroplasts, converted to their
mature forms, and become inserted into thylakoids as insoluble,
integral membrane polypeptides. The disposition of the newly-assem-
bled CP II polypeptides within the membrane is indistinguishable
from those synthesized in vivo.

RESULTS

Purification of Pea CP II and Identification of its Constituent
Polypeptides

Various studies of the polypeptide composition of the light-
harvesting chlorophyll a/b protein complex (CP II) have been published but, as yet, no consensus has been reached as to the number
and identity of these thylakoid membrane proteins (cf. 12). A novel
approach toward this problem exploits the observation that thylakoid
membranes can be solubilized with the lithium salt of dodecyl sulfate ($LiDoDSO_4$) at low temperatures and subsequently resolved by gel
electrophoresis at 4° (13). This treatment reduces dissociation of
chlorophyll-protein complexes which occurs during migration in polyacrylamide gels (13). Aro and Valanne (14) reported that the
electrophoretic mobility of CP II is altered by the inclusion
Mg^{2+} in $NaDoDSO_4$ gels and anodal buffer. We have confirmed this
phenomenon for $LiDoDSO_4$ gels. Electrophoretic mobility of CP II in
polyacrylamide gels is reduced as a function of increasing Mg^{2+} concentration (Fig. 1). Therefore, Mg^{2+} can shift CP II to a gel region where few other thylakoid membrane polypeptides are found,
greatly facilitating the purification of the complex. Since this
shift is also obtained with CP II derived from trypsinized membranes
(Fig. 7A), this effect must be unrelated to the Mg^{2+}-induced stacking of thylakoid membranes mediated by CP II (15). Rather, the
effect on mobility is probably due to stabilizing the chlorophyll
protein interaction and, consequently, altering the amount of de-
tergent which binds to the complex. In contrast, electrophoretic
mobility of polypeptides from samples heated at 100^oC is not altered by Mg^{2+}.

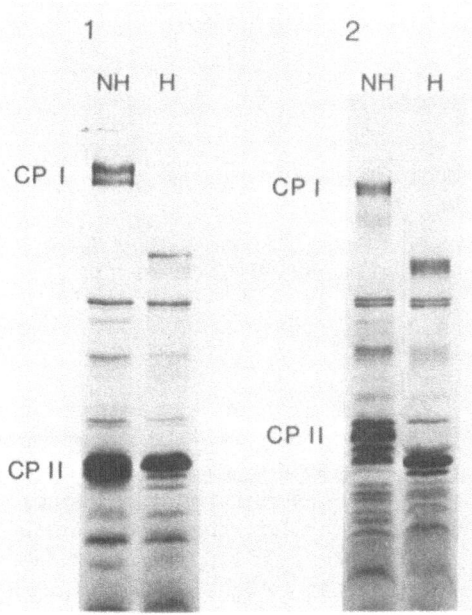

Figure 1. Effect of Mg^{2+} on the electrophoretic mobility of
CP II. Pea thylakoid membranes were purified by
flotation through sucrose gradients as described by
Chua and Bennoun (16) and solubilized in 60 mM
Na$_2$CO$_3$, 60 mM dithiothreitol (DTT), 2% LiDoDSO$_4$,
12% sucrose. An aliquot (H) was heated at 100° for
45 seconds before loading onto 9% polyacrylamide
gels with (1) and without (2) 0.7 mM MgCl$_2$ in the
stacking gel, resolving gel and anodal reservoir
buffers of Laemmli (17). The positions of CP I,
the chlorophyll a-P700-protein complex, and CP II
are indicated in the stained gel patterns of non-
heated samples (NH).

    Pea CP II excised from gels (Fig. 1) and then dissociated by
heating, contains a single polypeptide with a molecular weight of
28,000 when analyzed in a number of gel systems (data not shown).
Süss et al. (18) reported that urea profoundly affects the resolution of thylakoid membrane polypeptides on polyacrylamide gels.
When CP II excised from a 0.7 mM Mg^{2+}, 9% polyacrylamide gel is
denatured and run on a 10% gel containing a transverse gradient of
0-8 M urea, the 28,000-dalton polypeptide found in gels without
urea is resolved as two molecular weight species designated polypeptides 15 and 16 (Fig. 2). We do not know whether each polypeptide binds both chlorophyll a and b.

0————————————————Urea————————————————→ 8 M

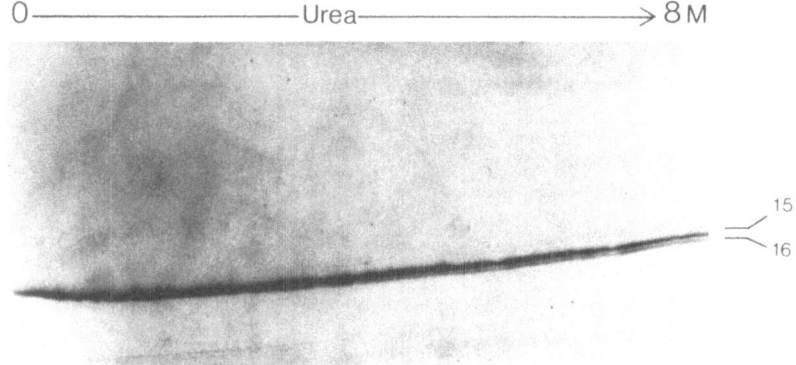

Figure 2.   Effect of urea on the electrophoretic mobility of
            CP II apoproteins. CP II was purified from a 9%
            polyacrylamide gel containing 0.7 mM MgCl$_2$ as in
            Fig. 1. The chlorophyll-protein complex was dis-
            sociated by boiling in 2% NaDoDSO$_4$, 50 mM DTT and
            applied to a 10% polyacrylamide gel containing,
            from left to right, a 0-8 M urea concentration
            gradient.

Site of Synthesis of CP II Polypeptides

     Machold and Aurich (19) and Cashmore (20) used specific inhi-
bitors of cytoplasmic and chloroplast protein synthesis to show
that the polypeptide which co-migrates with CP II in polyacrylamide gels is a product of cytoplasmic protein synthesis. We re-
examined the site of synthesis of the two CP II polypeptides using
the protocol of Cashmore (20), except that the pulse-labeled membrane polypeptides were analyzed by high-resolution NaDoDSO$_4$ polyacrylamide gels followed by autoradiography. Both CP II polypeptides are labeled in the control and chloramphenicol-treated samples but are not labeled in the cycloheximide-treated sample (Fig.
3). These results confirm previous findings (19, 20) that the
constituent polypeptides of CP II are synthesized in the cytosol
and therefore must be imported into the chloroplast before integration into the thylakoid membrane.

Chemical Properties of CP II Polypeptides

     We have examined some of the chemical properties of the CP II
polypeptides. These polypeptides must be integral components of
thylakoid membranes since they are not extracted with 0.1 N NaOH
(Fig. 6) or 6 M urea (not shown), reagents which dislodge polypeptides that are associated with the membranes by ionic interactions (24). Consistent with the notion that polypeptides 15 and

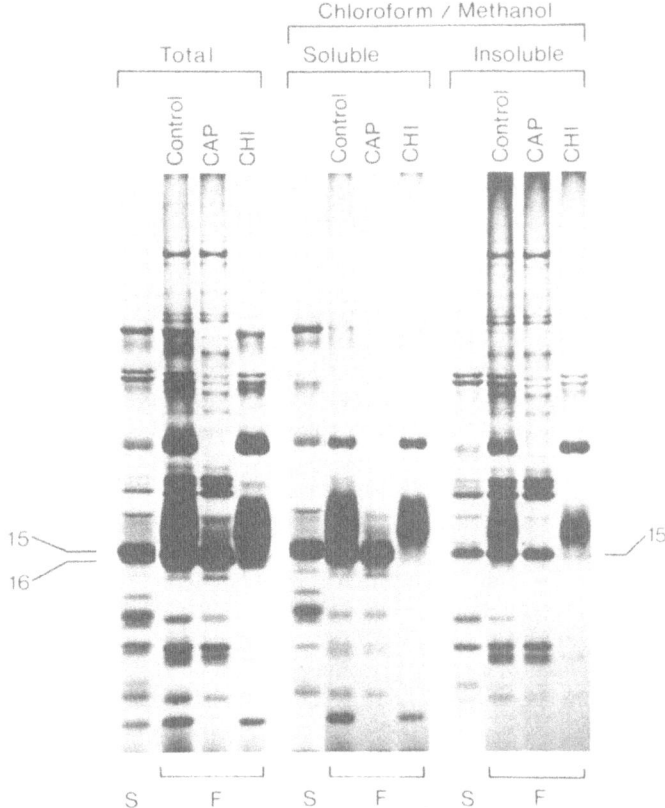

Figure 3. Sites of synthesis of thylakoid membrane polypeptides from pea. 7-day old pea seedlings were pre-
treated for 30 minutes with 20 µg/ml cycloheximide
or 200 µg/ml chloramphenicol and then pulse-labeled
with 10 mCi of ^{35}S-SO$_4$ per seedling. After 3
hours of incubation in the presence and absence of
the inhibitors, intact chloroplasts were purified
by silica sol gradient centrifugation (21). Thylakoid membranes were purified and their polypeptides were resolved in NaDoDSO$_4$ gels containing
4 M urea, 9% polyacrylamide and the Laemmli (17)
buffer system. After photographing, the stained
gels (S) were prepared for fluorography (F) (22)
and exposed to X-ray film for 18 hours. Extraction of membranes with 2:1 (vol/vol) mixtures of
CHCl$_3$ and CH$_3$OH was performed as described by Chua
et al. (23). The positions of CHCl$_3$/CH$_3$OH-soluble
CP II polypeptides 15 and 16 and a CHCl$_3$/CH$_3$OH-
insoluble polypeptide, 15', which comigrates with
polypeptide 15 are indicated.

16 are embedded in the thylakoid membrane lipid bilayer, they represent the most abundant polypeptides which are soluble in a 2:1 (vol/vol) mixture of chloroform:methanol (Fig. 3) and therefore are hydrophobic. A fraction of polypeptide 15, designated 15', is not extracted with chloroform:methanol. Immunological studies indicate that the chloroform:methanol soluble and insoluble forms of polypeptide 15 share antigenic determinants (see below) and therefore are probably the same polypeptide which becomes partitioned in the two fractions.

We purified polypeptides 15, 15' and 16 by subjecting chloroform:methanol extracts and residues of pea thylakoids to preparative gel electrophoresis. Antibody against the major consituent apoprotein of CP II in the green alga Chlamydomonas reinhardtii was employed to assess the antigenic properties of the pea polypeptides. The antibody reacts with both polypeptides 15 and 15' as well as with polypeptide 16. Most importantly, the precipitin lines obtained with all three CP II constituent polypeptides completely fuse with one another. Thus, all of the CP II polypeptides in pea are immunologically and structurally related.

A trivial explanation for the immunochemical finding is that polypeptide 16 is a proteolytic product of polypeptide 15. Such spurious proteolysis is unlikely since all steps in thylakoid purification included a battery of protease inhibitors (1 mM phenyl-

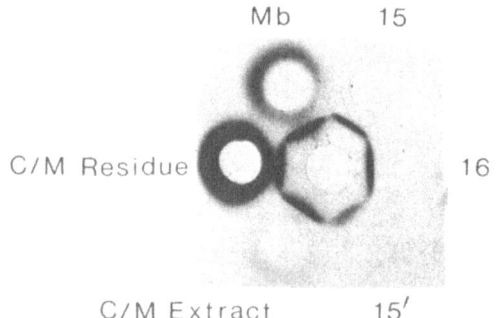

Figure 4.   Immunological characterization of CP II polypeptides. 25 μg of pea thylakoid membrane protein (Mb), an equivalent amount of the polypeptides which are soluble and insoluble in $CHCl_3/CH_3OH$, and 5 μg each of polypeptides 15, 16 and 15' purified from $CHCl_3/CH_3OH$ extracts or residues resolved on 4 M urea, 9% polyacrylamide gels were applied to 1% agarose gels as described by Chua and Blomberg (15). The center well contained IgG from antiserum against the major CP II polypeptide (polypeptide 11) of Chlamydomonas reinhardtii (25).

methylsulfonyl fluoride (PMSF), 5 mM aminocaproic acid and 1 mM benzamidine). However, polypeptide 16 could result from the phys- iological processing of polypeptide 15 in vivo. If this were the case, the primary translation product should be a single polypep- tide which reacts with antibody directed against CP II polypeptides.

In Vitro Synthesis of CP II Polypeptide Precursors

To characterize the primary translation product(s) of mRNA encoding the CP II polypeptides, pea polyadenylated RNA was trans- lated in the wheat germ system. Immunoprecipitation of the trans- lation products with antibody against C. reinhardtii polypeptide 11, a constituent polypeptide of CP II, yielded specifically two major polypeptides (MW 33,000 and 32,000) (Fig. 5). However, the possibility remained that the wheat germ extract, which is derived from a plant source, contains specific proteases which process primary translation products of plant mRNAs. Therefore, we trans- lated pea mRNA in the reticulocyte lysate system (26) and immuno- precipitated the products as above. Again, two major polypeptides of 33,000 and 32,000-daltons were synthesized in vitro and both were immunoprecipitated by antibody against CP II apoproteins (Fig. 5). Addition of wheat germ tRNAs to the reticulocyte lysate system did not alter the ratio of label in these two polypeptides indicating that the 32,000-dalton polypeptide is not a product of premature chain termination. Since both the wheat germ and retic- ulocyte lysate systems synthesize the 33,000 and 32,000-dalton products, we conclude that these polypeptides are primary trans- lation products of distinct mRNA species. We identify these poly- peptides as precursors (p15, p16) of pea thylakoid membrane poly- peptides 15 and 16, the CP II apoproteins. A precursor of similar size (MW 29,500) has been reported for the CP II apoproteins (MW 25,000) from barley (29). We propose that the precursor chain extension is a transit sequence (cf. 4,5) required for transport of a cytoplasmically-synthesized polypeptide into the chloroplast.

Since p15 and p16 are recovered in post-ribosomal supernates of the translation mixtures, we consider them to be soluble pro- teins. However, these precursors may be associated with membrane vesicles or lipid micelles endogenous to the translation system. To test this possibility we treated post-ribosomal supernatants from translation mixtures with trypsin. If p15 and p16 were asso- ciated with such vesicles they would be completely or partially protected against proteolysis. Analysis by $NaDoDSO_4$ gel electro- phoresis followed by autoradiography revealed that all proteins synthesized in the wheat germ system are susceptible to complete proteolytic degradation (data not shown). Thus, we conclude that p15 and p16 are soluble precursors to integral membrane polypep- tides.

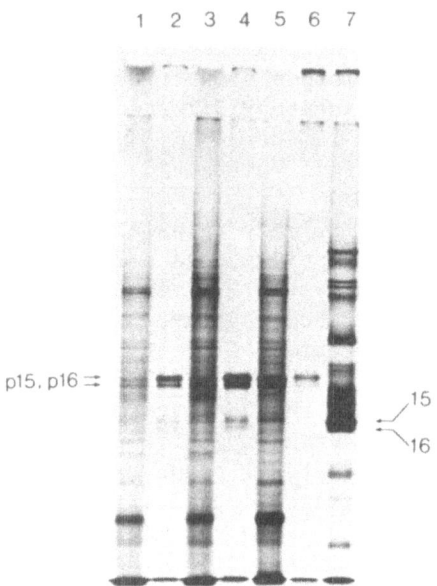

p15, p16 ⇒

Figure 5.　Immunological identification of higher molecular
weight precursors of polypeptides 15 and 16.　Polyadenylated RNA from pea was purified (8) and translated in either the reticulocyte lysate (26) or
wheat germ (27) cell-free systems.　Wheat germ tRNA
was purified by the method of Rogg et al (28).
Both systems contained 0.4 volumes of wheat germ
extract or reticulocyte lysate, 110 mM KAc, 0.75
mM MgAc$_2$, 20 mM HEPES-KOH, pH 7.5, 2 mM DTT, 200
µM spermine, 24 µM of amino acids except methionine,
12 µM GTP, 0.5 mM ATP, 8.4 mM phosphocreatine, 40
µg/ml creatine phosphokinase, 450 µCi/ml ^{35}S-methi-
onine (>600 Ci/mM)　and 50 µg/ml pea polyadenylated
RNA.　Translation in the wheat germ system was
carried out at 26° whereas reticulocyte lysates
were incubated at 37°.　Immunoprecipitation of
post-ribosomal supernates was with antibody to
C. reinhardtii polypeptide 11 and formalin-fixed
Staphylococcus aureus as described previously (1).
Pea mRNA products and immunoprecipitates from the
reticulocyte lysates (1,2), reticulocyte lysates
supplemented with 100 µg/ml wheat germ tRNA (3,4)
and the wheat germ (5,6) cell-free systems are
compared with ^{35}SO$_4$-labelled pea thylakoids (7)
in fluorographs of 4 M urea, 9% polyacrylamide
NaDoDSO$_4$ gels.

Post-Translational Transport of the CP II Polypeptide Precursor into Intact Chloroplasts

To investigate the physiological importance of p15 and p16 we constructed reconstitution experiments for transport of these precursors into intact chloroplasts in vitro. Post ribosomal supernates from ^{35}S-labeled translation mixtures were incubated with intact pea chloroplasts using conditions optimal for post-translational transport (30). Polypeptides not imported into the chloroplasts during the incubation were degraded by trypsin and chymotrypsin. Intact chloroplasts were reisolated by silica sol gradient centrifugation (21). Purified thylakoids contained several labeled polypeptides. prominent among these were polypeptides which co-migrate with the pea CP II polypeptides in polyacrylamide gels (Fig. 6). Thus, we have established that precursors to the pea CP II polypeptides are imported post-translationally into intact pea chloroplasts in our in vitro reconstitution system. Pea thylakoid membranes do not contain p15 and p16 and we have not detected any forms of the CP II polypeptides in the stroma or envelope fractions of chloroplasts. Thus, the precursors must be processed to their mature forms during or immediately following transport across the chloroplast envelopes or upon association with thylakoids.

We considered the possibility that developing chloroplasts may be more active in the import of cytoplasmically-synthesized membrane polypeptides than mature chloroplasts. Consequently ^{35}S-translation products from pea were incubated with intact chloroplasts from the inner leaves of romaine lettuce (31). As was found with the reconstitution experiments using homologous components, many ^{35}S-labeled pea polypeptides, including polypeptides 15 and 16, were recovered in purified lettuce thylakoids (Fig. 6). Since the imported CP II polypeptides co-migrate with those of pea rather than lettuce in 9% polyacrylamide gels containing 4 M urea, we conclude that the pea precursors, p15 and p16, are processed correctly by lettuce chloroplasts.

In the experiments presented in Fig. 6, equal numbers of chloroplasts from pea and lettuce were employed for in vitro transport. Comparison of the fluorograms (Fig. 6) reveals that, on a chloroplast basis, more ^{35}S-labeled polypeptides were associated with lettuce thylakoids than pea thylakoids. This finding suggests that developing chloroplasts from the inner leaves of romaine lettuce possess a substantially higher transport activity than mature pea chloroplasts.

Characterization of the Newly-Transported CP II Polypeptides

In order to establish that the association of newly-transported CP II polypeptides with thylakoids represents correct assembly we

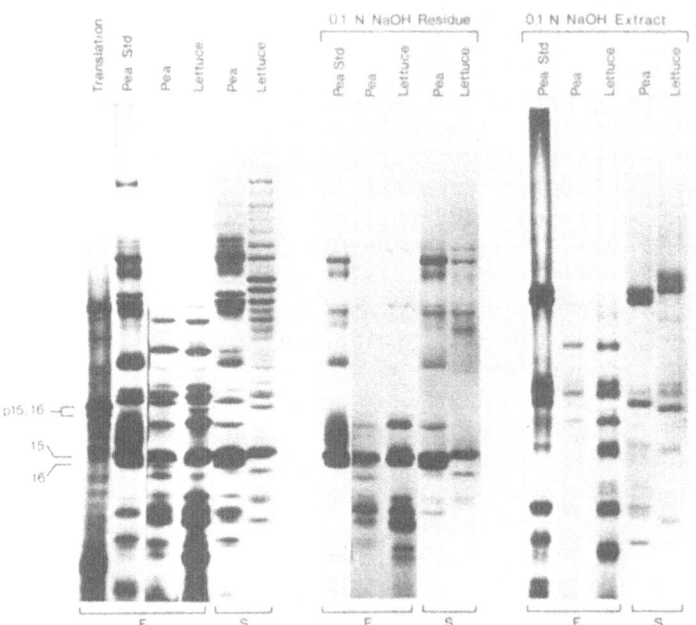

Figure 6. In vitro transport into chloroplasts, processing and
           assembly of CP II polypeptides into thylakoid mem-
           branes. Translation of polyadenylated RNA from pea,
           isolation of chloroplasts, incubation for uptake and
           digestion of polypeptides not transported into
           chloroplasts were as described previously (30). In-
           cubation mixtures contained 2 x 10^9 chloroplasts
           equivalent to 7 x 10^7 and 2 x 10^7 µg chlorophyll for
           pea and romaine lettuce, respectively. After diges-
           tion with trypsin and chymotrypsin, PMSF was added
           to 2 mM and intact chloroplasts were repurified by
           silica sol gradient centrifugation. Chloroplasts
           were lysed in 50 mM HEPES-NaOH, pH 7.5, 1 mM PMSF
           and thylakoids were purified as in Fig. 1. Aliquots
           of thylakoid pellets were resuspended in 0.1 N NaOH
           and incubated at 5^0 for 1 hr. before centrifugation
           at 140,000 x g for 1 hr. The extracts were neutral-
           ized with HCl and precipitated with 10% TCA. The
           membrane pellet was washed with 50 mM HEPES-NaOH,
           pH 7.5, 10 mM EDTA. Samples from the transport
           experiment, together with identically-treated thyla-
           koids from pea seedlings labeled with $^{35}SO_4$ (Pea Std)
           were electrophoresed in 4 M urea, 9% polyacrylamide
           gels. Fluorograms (F) correspond to the stained gel
           profiles (S) of pea and lettuce thylakoid polypep-
           tides.

must demonstrate that (1) these polypeptides are embedded in the thylakoid lipid bilayer, and (2) they have the same disposition as native CP II polypeptides. Thus, we compared the characteristics of newly imported CP II polypeptides in pea and lettuce with the CP II polypeptides from pea thylakoids.

Thylakoids from pea and lettuce chloroplasts were extracted with 0.1 N NaOH which removes peripheral membrane proteins (24) (Fig. 6). In both cases, labeled polypeptides 15 and 16 were recovered in the NaOH-insoluble membrane fraction, indicating that the newly-transported polypeptides are integral components of the thylakoids.

To probe the dispositions of the newly-inserted CP II polypeptides, we subjected thylakoids to limited proteolysis by 40 µg/ml trypsin for 30 minutes at 25°. Since proteolysis under these conditions does not affect the electrophoretic mobility of CP II in 9% polyacrylamide gels containing Mg^{2+}, we could purify CP II from trypsin-treated membranes and determine the fate of the constituent polypeptides by subsequent electrophoresis of the dissociated complex. Polypeptides 15 and 16 are proteolyzed only partially, presumably because the bulk of these proteins is protected by the lipid phase of the membrane (Fig. 7a). The extent of proteolysis of polypeptides 15 and 16 newly imported into either pea or lettuce chloroplasts is the same as that of pea CP II polypeptides synthesized in vivo (Fig. 7b). Again, this demonstrates that newly-transported CP II polypeptides are correctly assembled into the thylakoids. Assembly of polypeptides 15 and 16 in lettuce thylakoids is correct with respect to pea CP II polypeptides by this criterion, suggesting that the extent to which these polypeptides are embedded in the thylakoid lipid bilayer is largely an inherent property of the proteins.

In Vitro Transport of Other Thylakoid Membrane Proteins

Thus far we have restricted our discussion to the in vitro transport and assembly of the CP II polypeptides. Numerous other pea polypeptides are synthesized in vitro and are transported into intact chloroplasts by a post-translational mechanism (Figs. 6 and 7). In virtually every case, the polypeptides imported into either pea or lettuce chloroplasts which are recovered with purified thylakoids display identical electrophoretic mobility with one another. Moreover, the imported polypeptides correspond to those in stained gels of pea thylakoids, not to those of lettuce. Since the gel pattern of imported membrane polypeptides is distinct from that of the pea translation products (Fig. 6), most polypeptides probably are synthesized as precursors which are modified upon their import into the chloroplasts.

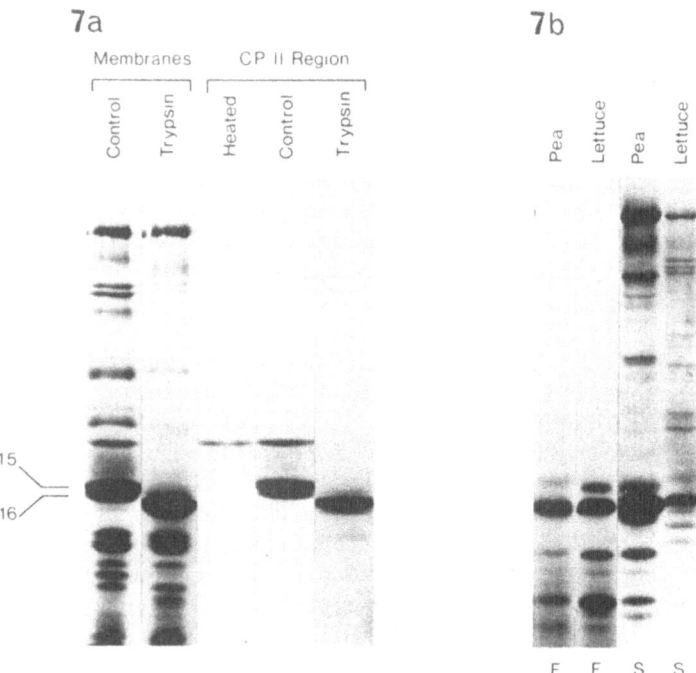

Figure 7.   Trypsin digestion of CP II polypeptides in the thyla-
             koid membrane. Thylakoids from pea and from chloro-
             plasts of pea and lettuce which had transported pea
             mRNA translation products were treated with 40 μg/
             ml trypsin for 30 minutes at 25°. The membranes
             were harvested by centrifugation in the presence of
             2 mM PMSF and processed for electrophoresis in 4 M
             urea, 9% polyacrylamide gels. A: Trypsin-digestion
             products of CP II polypeptides of pea are identified
             by comparison of the total thylakoid membrane poly-
             peptide profiles with those of CP II purified from
             0.7 mM $MgCl_2$, 9% acrylamide gels. The polypeptide
             which migrates behind the CP II polypeptides of
             undigested membranes is shown to be a contaminant
             which migrates in the CP II region in Mg^{2+}-gels
             even when chlorophyll-protein complexes are disso-
             ciated by heating before electrophoresis. B: Fluro-
             grams (F) are compared with the stained gel profiles
             (S) of polypeptides from trypsin-treated thylakoids
             of pea and romaine lettuce chloroplasts employed in
             protein transport.

CONCLUSIONS

In this study, we have reconstructed in vitro the biosynthesis of the most abundant integral proteins of the thylakoid, the subunits of the light-harvesting chlorophyll-protein complex (CP II). We have shown that the CP II polypeptides are synthesized as precursors which possess an apparent molecular weight 4-5000 daltons greater than the mature polypeptides. A striking characteristic of the CP II protein precursors is that they are soluble proteins even though they give rise to polypeptides which can be solubilized from membranes only with detergents or organic solvents. When post-ribosomal supernates of in vitro translation mixtures are incubated with intact chloroplasts, the completely synthesized precursors are transported across the envelopes by a post-translational mechanism. We have shown that during or following transport the precursors are processed to their mature size and are assembled into the thylakoids as integral membrane polypeptides. The newly assembled CP II polypeptides assume a disposition in the thylakoid lipid bilayer which is identical to that of the in vivo synthesized proteins inasmuch as they are degraded by trypsin to the same extent. Thus, the post-translational transport, processing and assembly of the CP II proteins in vitro appears to reconstruct with fidelity the in vivo biosynthetic pathway for these proteins.

By analogy to the precursor of the small subunit of RuBPCase, we propose that the CP II polypeptide precursors possess an extension of amino acids designated the transit sequence (4, 5). We believe the transit sequence facilitates post-translational transport of cytoplasmically-synthesized proteins into chloroplasts and, in the case of p15 and p16, may be required for the solubility of these proteins in aqueous solutions. Because p15 and p16 are not detected in either the stromal or envelope fractions of chloroplasts, processing of these precursors must occur either during transport or immediately thereafter. At present, we do not know whether processing precedes assembly into the thylakoid membranes or whether processing intermediates (pro-forms) are involved in the maturation of the CP II polypeptides. If such intermediates are present in the chloroplast stroma their concentrations must be so low that they escape detection by immunological methods. Finally, since processing of p15 and p16, as well as other imported thylakoid polypeptides, is identical when either pea or lettuce chloroplasts are employed for in vitro transport, the information for the processing of presumptive precursors has been evolutionarily conserved.

Preliminary studies indicate that the in vitro transport and assembly of the CP II polypeptides culminates in chlorophyll binding. Thus, polypeptides 15 and 16 synthesized in vitro are found in the CP II complex purified from Mg^{2+}-containing polyacrylamide gels. These results, combined with those presented here, constitute a stringent proof of the relevance of the in vitro system for studies of chloroplast development and thylakoid biogenesis.

Acknowledgements

We thank Michael Becker-Fluegel for excellent technical assistance. This work was supported by NIH grants GM-21060, GM-25114 and S07 RR 07065. N.-H.C. is the recipient of NIH Research Career Development Award GM-00223. A.R.G. and S.B. are supported by NIH National Service Award Postdoctoral Fellowships GM-06444 and GM-06678, respectively.

REFERENCES

1. Chua, N.-H. and Schmidt, G.W. , 1978, Proc. Natl. Acad. Sci. USA 75:6110.
2. Highfield, P.E. and Ellis, R.J., 1978, Nature (London) 271:420.
3. Dobberstein, B., Blobel, G., and Chua, N.-H., 1977, Proc. Natl. Acad. Sci. USA 74:1082.
4. Chua, N.-H. and Schmidt, G.W., 1979, J. Cell Biol. 81:461.
5. Schmidt, G.W., Devillers-Thiery, A., Desruisseaux, H., Blobel, G., and Chua, N.-H. submitted to J. Cell Biol.
6. Blobel, G. and Dobberstein, B., 1975, J. Cell Biol. 67:835.
7. Blobel, G. and Dobberstein, B., 1975, J. Cell Biol. 67:852.
8. Cashmore, A.R., Broadhurst, M.K., and Gray, R.E., 1978, Proc. Natl. Acad. Sci. USA 75:655.
9. Tobin, E.M., 1978, Proc. Natl. Acad. Sci. USA 75:4749.
10. Smith, S.M. and Ellis, R.J., 1979, Nature (London) 278:662.
11. Anderson, J.M., 1975, Biochim. Biophys. Acta 238:324.
12. Machold, O. and Meister, A., 1979, Biochim. Biophys. Acta. 546:472.
13. Delepelaire, P. and Chua, N.-H., 1979, Proc. Natl. Acad. Sci. USA 75:655.
14. Aro, E.-M. and Valanne, N., 1978, Physiol. Plant. 43:261.
15. Carter, D.P. and Staehelin, L.A., 1978, J. Cell Biol. 79:313a.
16. Chua, N.-H. and Bennoun, P., 1975, Proc. Natl. Acad. Sci. USA 72:2175.
17. Laemmli, U.K., 1970, Nature (London) 227:680.
18. Süss, K.-H., Schmidt, O., and Machold, O., 1976, Biochim. Biophys. Acta 448:103.
19. Machold, O. and Aruich, O., 1972, Biochim. Biophys. Acta 281:103.
20. Cashmore, A.K., 1976, J. Biol. Chem. 251:2848.
21. Morgenthaler, J.-J., Marden, M.P.F., and Price, C.A., 1975, Arch. Biochem. Biophys. 168:289.
22. Bonner, W.M. and Laskey, R.A., 1974, Eur. J. Biochem. 46:83.
23. Chua, N.-H., Matlin, K., and Bennoun, P., 1975, J. Cell Biol. 67:361.
24. Yu, J. and Steck, T.L., 1975, J. Biol. Chem. 250:9170.
25. Chua, N.-H. and Blomberg, F., 1979, J. Biol. Chem. 254:215.

26. Pelham, H.R.B. and Jackson, R.J., 1976, Eur. J. Biochem. 67: 247.
27. Roman, R., Brooker, J.D., Seal, S.N., and Marcus, A., 1976, Nature (London) 260:359.
28. Rogg, H., Wehrli, W., and Staehelin, M., 1969, Biochim. Biophys. Acta 195:13.
29. Apel, K. and Kloppstech, K., 1978, Eur. J. Biochem. 85:581.
30. Grossman, A.R., Bartlett, S.G., Schmidt, G.W., and Chua, N.-H., 1979, Annals N.Y. Acad. Sci. In Press.
31. Henriques, F. and Park, R., 1976, Proc. Natl. Acad. Sci. USA 73:4560.

SYNTHESIS, PROCESSING AND FUNCTIONAL PROBING OF P-32000, THE

MAJOR MEMBRANE PROTEIN TRANSLATED WITHIN THE CHLOROPLAST

Marvin Edelman and Avi Reisfeld

Department of Plant Genetics
Weizmann Institute of Science
Rehovot, Israel

INTRODUCTION

Recently expanded investigations of plastid membranes has indicated the complex nature of the protein fraction. A multi-component ATPase system (Nelson, 1976), and several pigment-protein complexes with their light harvesting (Thornber, 1975) and reaction center (Nelson and Notsani, 1977) proteins, cytochromes and structural components (Arntzen, 1978), have been identified and their sites of synthesis determined (Chua and Gillham, 1977). In addition to these relatively abundant proteins, Eaglesham and Ellis (1974) uncovered a group of quantitatively minor membrane components synthesized by intact, isolated chloroplasts. Most prominently labeled among these was a 32000 d polypeptide (P-32000).

We have investigated synthesis and metabolism of P-32000 in Spirodela with emphasis on its photo and molecular control in vivo (Rosner et al., 1977; Reisfeld et al., 1978a,b; Edelman and Reisfeld, 1978; Weinbaum et al., 1979). A picture emerges of a light-induced thylakoid precursor protein, rapidly synthesized, processed and metabolized within the organelle. P-32000-like molecules have also been identified in maize (Grebanier et al., 1978).

PROPERTIES OF P-32000 IN SPIRODELA

Spirodela oligorrhiza is a particularly suitable organism for investigating physiological synthesis and molecular control of chloroplast proteins. The entire plant is only a few mm long, consisting mainly of frond-like leaves with several rhizoids (Figure 1). Spirodela can be cultured axenically on defined mineral media in liquid or agar (Posner, 1967). Due to its natural

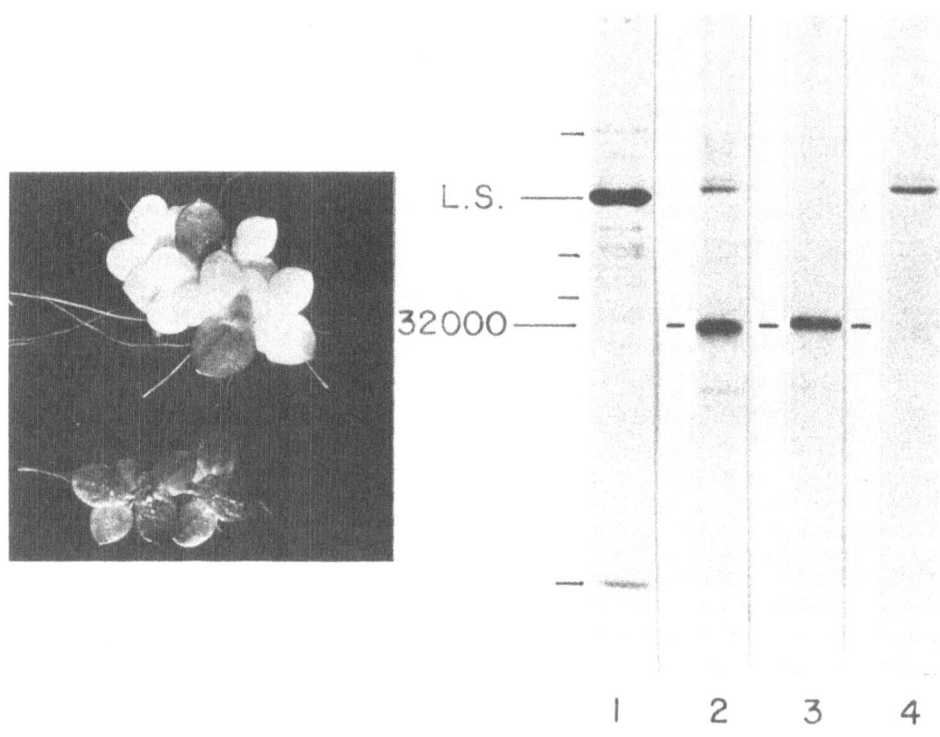

L.S. ——

32000 ——

1   2   3   4

Figure 1. Colonies of Spirodela oligorrhiza. Left - light grown
culture. Right - dark grown culture (3 months). The green fronds
already present when the colony was placed in the dark retained
their chloroplasts (courtesy of D. Porath).

Figure 2. Rapidly labeled chloroplast polypeptides. Light grown
Spirodela were labeled with ^{35}S-met for 2 h, chloroplasts isolated
(Blair and Ellis, 1973) and the soluble and membrane fractions
prepared. Samples were applied to an SDS-polyacrylamide slab-gel
which was electrophoresed (Laemnili, 1970) and fluorographed
(Bonner and Laskey, 1974). (1) Soluble polypeptides. (2) Membrane
polypeptides. (3) Membrane polypeptides not released by 1 M LiCl.
(4) Membrane polypeptides released by 1 M LiCl. Molecular weight
markers: 68,000d, 43,000d, 36,000d and 12,000d.

habitat, uptake of biochemicals by the entire underside of the
frond is rapid. This plant can grow phototrophically with a doubling time of 2-3 days or indefinitely as a colorless heterotroph
in the dark upon the addition of a carbon source (doubling time
∿10 days). Light induced study of chloroplast development starting
from steady state heterotrophy thus becomes possible.

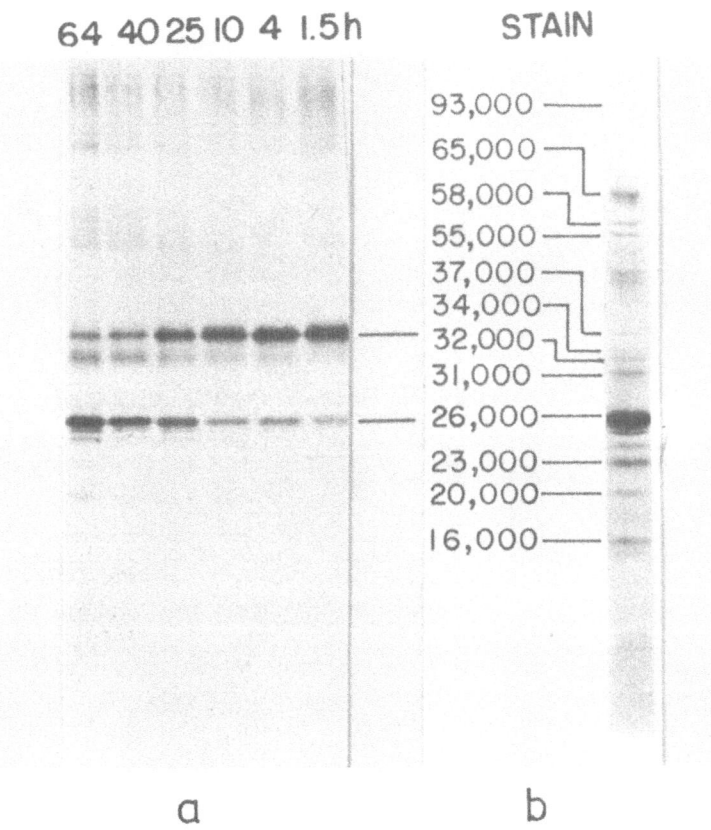

Figure 3. (a)- Time course labeling of membrane proteins. Light grown Spirodela were labeled with ^{35}S-met for various times as indicated. Membranes from whole cells were isolated and aliquots electrophoresed and autoradiographed. (b)- Stained membrane proteins from Spirodela chloroplasts. Chloroplast membranes from Spirodela, were isolated, analyzed on SDS polyacrylamide slab-gels and stained with Coomassie brilliant blue.

In Spirodela, P-32000 is located in the chloroplast fraction. It forms an integral part of the photosynthetic membrane (Figure 2), remaining insoluble after treatments with salts (such as LiCl), chelating agents, or other procedures which remove more loosely-bound proteins.

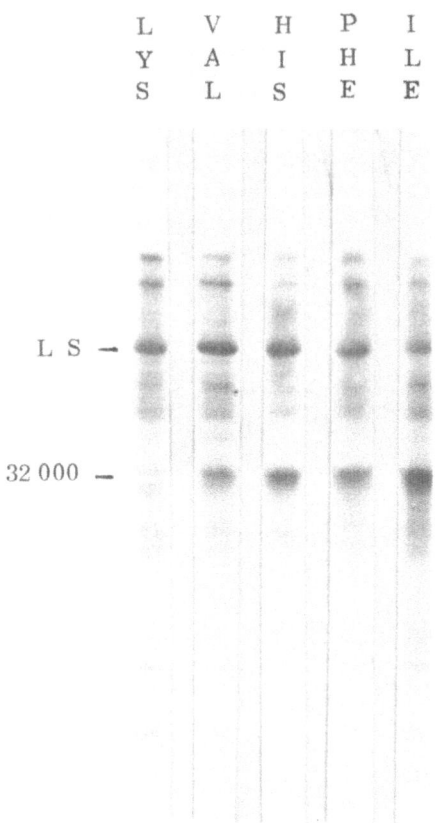

Figure 4.   Soluble proteins labeled with ^{3}H-amino acids. Cultures
of light grown Spirodela were labeled, individually, with the indicated ^{3}H-amino acid for 3 h. Whole cell homogenates   were prepared and equal amounts of counts applied to each well. Analysis
was by SDS-polyacrylamide slab-gel electrophoresis and fluorography.

P-32000 is a rapidly synthesized but relatively short-lived
protein (Figure 3a). Its half-life, estimated from pulse-chase
experiments with ^{3}H-leucine labeled fronds, shows it to possess
at least an order of magnitude more rapid decay than that of other
major plastid polypeptides (H. Hoffman, unpublished). As a result,
this protein does not accumulate and represents, at best, a minor
band in stained preparations (Figure 3b). Nonetheless, production
of P-32000 represents a major activity of the chloroplast protein-

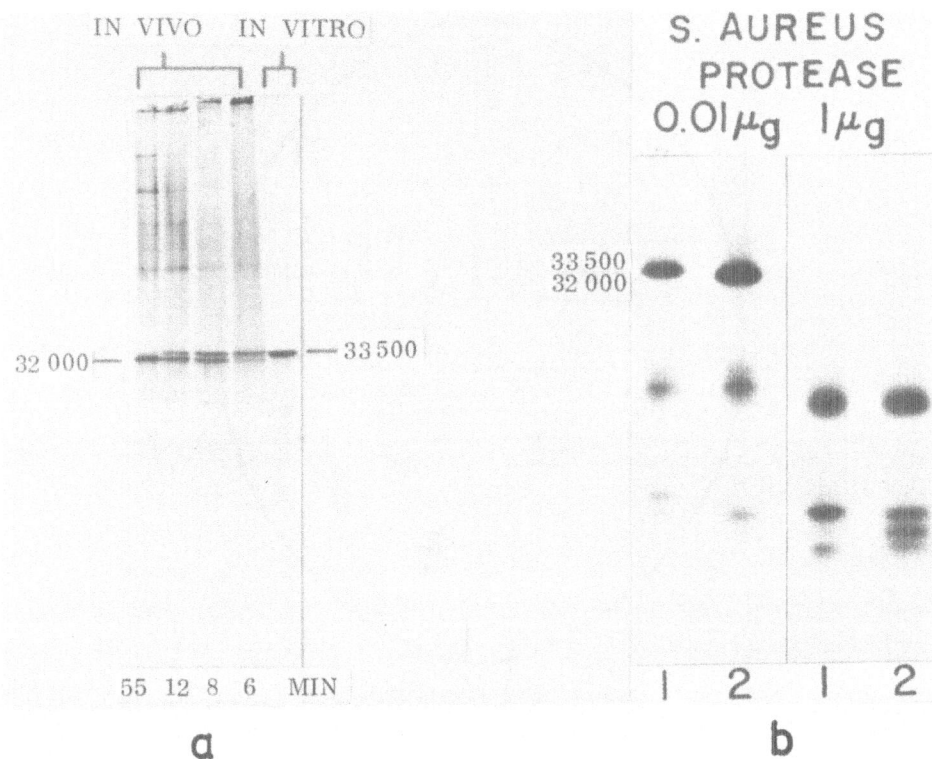

Figure 5. (a)- Appearance and processing of P-33500. Chloroplast
membranes were isolated from light grown Spirodela and labeled in
vivo with ^{35}S-met for the times indicated. These samples were
electrophoresed on SDS-polyacrylamide gels alongside of ^{35}S-met-
labeled polypeptide products, induced by Spirodela chloroplast
RNA in vitro, in a wheat germ system. The gel was autoradiographed.
(b)- Limited proteolysis products of P-32000 and P-33500. The
P-32000 and P-33500 bands from the 8 min sample of Figure 5(a) were
separately excised, electrophoresed on a second SDS slab-gel in
the presence of two concentrations of S.aureus prolease and fluorographed as described by Cleveland et al. (1977). Frames (1)-
digestion patterns of P-32000. Frames (2)- digestion patterns of
P-33500.

synthesizing apparatus. Indeed, in steady state light grown
Spirodela it is the main membrane protein translated within the
organelle, and on a molecule to molecule basis, probably rivals
synthesis of the large subunit (LS) of ribulose bisphosphate carboxylase (Figure 4).

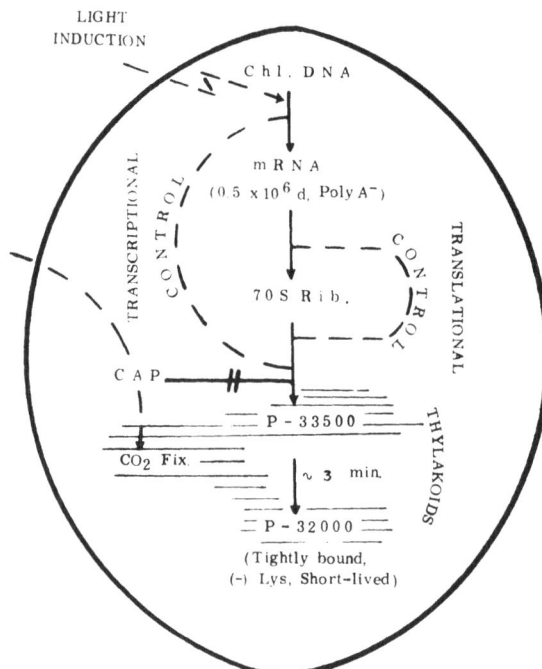

Figure 6. Flow diagram of P-32000 synthesis in Spirodela. The mRNA for P-32000 has a molecular weight of $\sim0.5\times10^6$ d, lacks poly(A), and is light induced. It is translated on 70S chloroplast ribosomes to yield a precursor polypeptide, P-33500, which is processed to P-32000 within a few minutes. Translation of P-32000 in vivo is inhibited by chloramphenicol (CAP) under conditions where CO_2 fixation continues. The processed P-32000 product is tightly bound to the thylakoid membranes, lacks lysine residues and is itself relatively short-lived. Control of P-32000 synthesis appears mainly at the transcriptional level; however, translational control also exists under certain physiological conditions (from Edelman and Reisfeld, 1978).

PRECURSOR SYNTHESIS IN VIVO AND IN VITRO

P-32000 first appears in vivo as a discrete 33500 d precursor (P-33500) associated with the membrane fraction (Figure 5a). The precursor is shown to be related to P-32000 by partial proteolytic digestion (Figure 5b) and the mutual lack of lysine residues. In the light, P-33500 is rapidly processed to P-32000; in the dark, processing is inhibited. Based on experiments involving heterotrophic and regreening fronds, it appears that chloroplast structure and light are required for conversion of P-33500 to P-32000

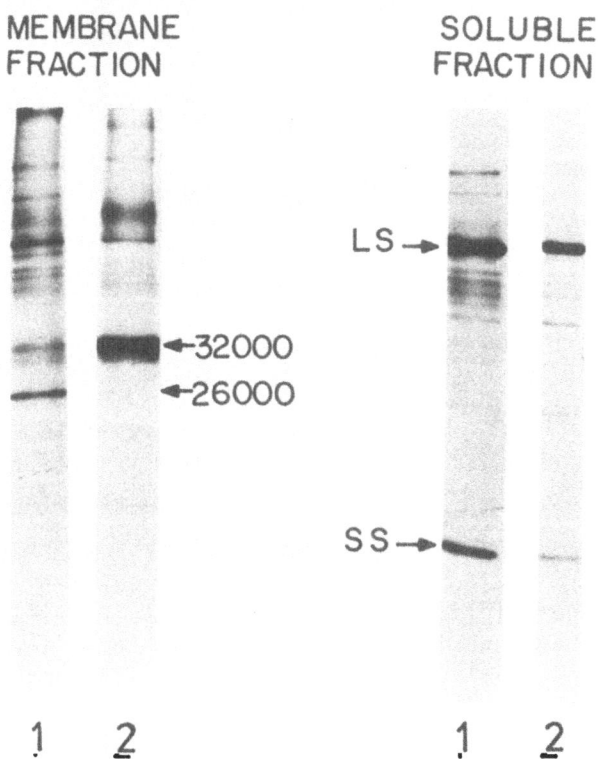

MEMBRANE
FRACTION

SOLUBLE
FRACTION

LS→

←32000
←26000

SS→

1   2   1   2

Figure 7. Rapidly labeled proteins of juvenile and mature fronds. Cultures were labeled for 1 h with ^{35}S-met, washed, frozen, and microdissected into expanding and mature tissues. Membrane and soluble fractions were prepared, fractionated on SDS-polyacryl-amide slabs and fluorographed. LS and SS refer to the large and small subunits of RUBPCase. Frames (1)- juvenile fronds. Frames (2)- mature fronds (from Weinbaum et al, 1979).

(A. Mattoo, unpublished). In steady-state, light-grown fronds this conversion occurs in the presence of both plastid and cyto-plasmic protein synthesis inhibitors.

    In vitro, Spirodela chloroplast RNA induces translation of a polypeptide of 33500 d as a major product in wheat germ (Figure 5a) and reticulocyte systems. The mRNA activity for this polypeptide is found in the poly(A)$^{-}$ fraction, at a molecular weight of ∿500,000 (Reisfeld et al., 1978b). It has several properties in

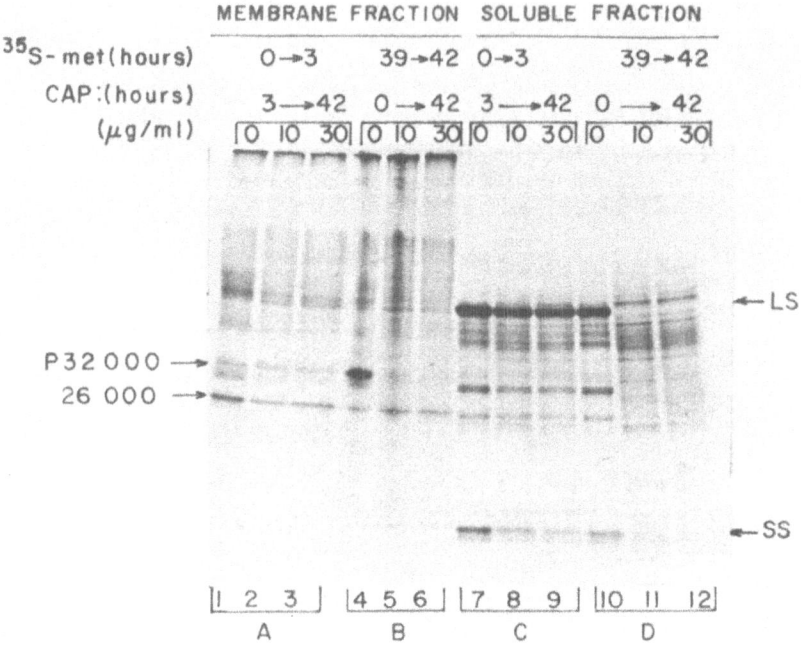

^{35}S-met(hours)   0→3   39→42 0→3   39→42
CAP:(hours)   3——→42 0——→42 3———→42 0———→42
(µg/ml)

Figure 8. Specific depletion of P-32000 by chloramphenicol (CAP). D-threo-chloramphenicol (as indicated) was added to light-grown cultures of Spirodela during a 39 h chase period after a 3 h pulse of ^{35}S-met (A and C); or as a 39 h preincubation before a 3 h pulse of ^{35}S-met (B and D). In A and C the ^{35}S-met was removed and the plants were rinsed several times in unlabeled methionine. Membrane and soluble fractions were prepared, electrophoresed on a SDS slab-gel and fluorographed. Equal numbers of counts (80,000 cpm) were applied to all slots (from Weinbaum et al, 1979).

common with a rapidly labeled, non-ribosomal, poly(A)⁻ RNA of this same size (Rosner et al., 1975) which can be isolated from Spiro-dela chloroplasts and which is induced by low-energy blue light (Gressel, 1978). Several lines of evidence were obtained showing that the in vitro translated 33500 d product is similar, if not identical, to the P-33500 chloroplast membrane precursor found in vivo (Edelman and Reisfeld, 1978).

PHYSIOLOGICAL CONTROLS AND FUNCTION

     Both P-32000 and its RNA template activity undergo major induction in the light. Under several physiological conditions there is a stoichiometric relation between the levels of P-32000 and its mRNA activity, suggesting transcriptional control (Reisfeld et al., 1978a; Edelman and Reisfeld, 1978). However, under certain stress conditions, this coordination becomes uncoupled with control being post-transcriptional. A diagramatic representation of these controls is shown in Figure 6.

Two hypotheses were tested concerning the physiological function of P-32000 (Weinbaum et al., 1979). 1. P-32000 synthesis occurs in fully mature, expanded tissues to a greater extent than in juvenile, expanding fronds. It is, in fact, the main synthesis product in mature membranes. In juvenile fronds, on the other hand, the main membrane polypeptide synthesized is the 26000d component of the light-harvesting chlorophyll a/b protein complex (Figure 7). Thus, a transient, developmental function for P-32000 during thylakoid biogenesis seems unlikely.

2. Thylakoids can be selectively depleted of a large percentage of P-32000 by treatment, in vivo, with low concentrations of D-threo-chloramphenicol, an inhibitor of chloroplast protein synthesis. A fair degree of selectivity is possible in this case because of the polypeptide's relatively rapid rate of decay (Figure 8). When photoassimilation of CO_2 by such fronds was compared with that of non-depleted controls, near normal levels of CO_2 fixation were obtained. This suggests that P-32000 is not rate limiting in this process and argues against its being a direct, integral part of the photosynthetic pathway. The function of P-32000, one of the most abundantly synthesized polypeptides in the biosphere, remains an enigma and a challenge.

ACKNOWLEDGEMENTS

Research supported in part by a grant from The United States-Israel Binational Science Foundation.

REFERENCES

Arntzen, 1978, in: "Current Topics in Bioenergetics" Vol.8. L. Vernon and R. Sanadi, eds., Academic Press, New York, pp.111-160.
Blair, G.E. and Ellis, R.J., 1974, Biochim.Biophys.Acta, 319:223-234.
Bonner, W.M. and Laskey, R.A., 1974, Eur.J.Biochem., 46:83-88.
Chua, N.-H. and Gillham, N.W., 1977, J.Cell Biol. 74:441-452.
Cleveland, D.W., Fischer, S.G., Kirschner, M.W. and Laemmli, U.K., 1977, J.Biol.Chem., 252:1102-1106.
Eaglesham, A.R.J. and Ellis, R.J., 1974, Biochim.Biophys.Acta, 335:396-407.
Edelman, M. and Reisfeld, A., 1978, in: "Chloroplast Development", G. Aklyunoglon et al., eds., Elsevier/North Holland, Amsterdam, pp. 641-652.

Grebanier, A.E., Coen, D.M., Rich, A. and Bogorad, L., 1978, J.Cell Biol. 78:734-746.

Gressel, J., 1978, Photochem.Photobiol., 27:167-169.

Laemmli, U.K., 1970, Nature,New Biol., 227:680-685.

Nelson, N., 1976, Biochim.Biophys.Acta, 456:314-338.

Nelson, N. and Notsani, B.C., 1977, in: "Bioenergetics of Membranes", L. Packer, et al., eds., Elsevier/North Holland, Amsterdam, pp.233-243.

Posner, H.B., 1967, in: "Methods in Developmental Biology", F.H. Witt and N.K. Wellels, eds., T.Y. Crowel Co., New York, p.301.

Reisfeld, A., Gressel, J., Jakob, K.M. and Edelman, M., 1978a, Photochem.Photobiol., 27:161-165.

Reisfeld, A., Jakob, K.M. and Edelman, M., 1978b, in: "Chloroplast Development", G. Akoyunoglon et al., eds., Elsevier/North Holland, Amsterdam, pp.669-674.

Rosner, A., Jakob, K.M., Gressel, J. and Sagher, D., 1975, Biochem. Biophys.Res.Commun., 67:383-391.

Rosner, A., Reisfeld, A., Jakob, K.M., Gressel, J. and Edelman, M. 1977, Colloques Internationaux C.N.R.S., 261:561-568.

Thornber, J.P., 1975, Ann.Rev.Plant Physiol., 26:127-158.

Weinbaum, S.A., Gressel, J., Reisfeld, A. and Edelman, M., 1979, Plant Physiol. (in press).

THE CHARACTERISATION OF LEAF MESSENGER RNAs AND THEIR USE

IN THE SYNTHESIS OF COMPLEMENTARY DNAs

Anthony R. Cashmore and Nam-Hai Chua

The Rockefeller University

New York, N.Y. 10021

The major products of cytoplasmic protein synthesis in pea leaves correspond to polypeptide components of two chloroplast proteins. These polypeptides are the small subunit of ribulose-1,5-bisphosphate (RuBP) carboxylase and the constituent polypeptides of the chlorophyll-protein complex II (CP_{II}) or light-harvesting chlorophyll a/b protein (Cashmore, 1976 and Schmidt et al., this volume). Both the small subunit and the CP_{II} polypeptides are synthesized, on free cytoplasmic polyribosomes, as soluble precursors which function in the post-translational transport of the polypeptides from their site of synthesis into the chloroplast (Dobberstein et al., 1977; Cashmore et al., 1978; Highfield and Ellis, 1978; Apel and Kloppstech, 1978; Chua and Schmidt, 1978; Schmidt et al., this volume). From studies on the mode of inheritance of peptide variants it appears that these cytoplasmically synthesized polypeptides are encoded by nuclear genes (Kawashima and Wildman, 1972; Kung et al., 1972). In contrast, the large subunit of RuBP carboxylase is translated on chloroplast ribosomes (Blair and Ellis, 1973) and is encoded by the chloroplast genome (Coen et al., 1977).

Of major interest in studies concerned with the biosynthesis of organelle proteins is the means by which the nuclear and organelle genomes are coordinately regulated. Polynucleotide hybridization studies have proven to be a particularly useful means of examining various aspects of genetic regulation. We have recently reported studies concerned with the nuclear DNA sequence encoding the small subunit of RuBP carboxylase. Using purified mRNA in hybridization studies we have shown that the gene encoding the small subunit is present as single, or close to single, copies per haploid nuclear genome (Cashmore, 1979). This observation raises an additional aspect of the interactions between nuclear and organ-

elle genomes.   The circular DNA molecules encoding the large subunit of RuBP carboxylase are present in 15-30 copies per chloroplast (Whitfeld et al., 1973) and leaf cells may contain several
hundred chloroplasts (Possingham and Saurer, 1969).   Consequently,
the gene for the large subunit of RuBP carboxylase may be reiterated
several thousand fold per cell.   What are the features of these two
distinct genetic and biosynthetic systems which have resulted in
such vastly different reiteration frequencies for the nuclear and
chloroplast genes?   We report here our initial experiments aimed
at developing hybridization probes which should prove useful in
studying the regulation of the nuclear genes encoding the major
cytoplasmically synthesized polypeptides.

FRACTIONATION AND CHARACTERIZATION OF LEAF mRNA

When pea leaf cytoplasmic poly(A)$^{+}$ mRNA is fractionated by
sucrose gradient centrifugation, two partially resolved mRNA peaks

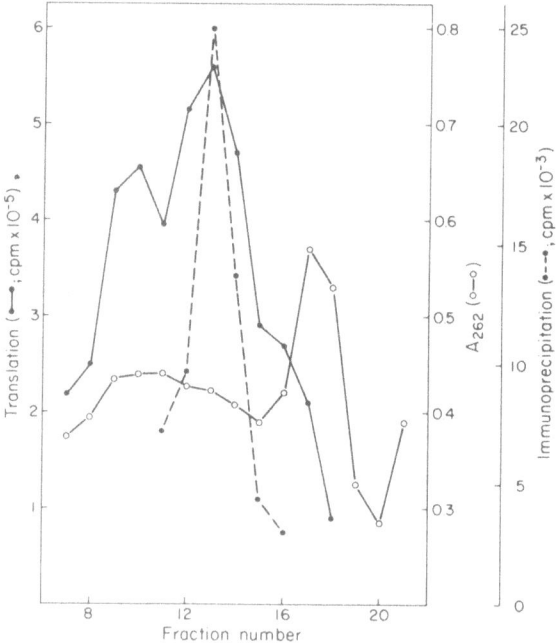

Fig. 1   Sucrose gradient fraction of pea leaf mRNA.   Conditions
         for sucrose gradient centrifugation, cell-free translation
         and the preparation of poly(A)$^{+}$ mRNA have been described
         (Cashmore, 1979 and Cashmore et al., 1978).   Immunoprecip-
         itation was performed according to Chua and Schmidt (1978).
         Sedimentation was from left to right (18S rRNA is the RNA
         peak corresponding to fraction 17).

are obtained with sedimentation coefficients of 11.5S and 14S
(Cashmore, 1979). The slower sedimenting species has been purified
and shown to encode the precursor for the small subunit of RuBP
carboxylase (Cashmore, 1979). When mRNA, fractionated by sucrose
gradient centrifugation, is assayed by cell-free translation, again
two peaks are observed (Figure 1). These translation products have
been resolved by SDS-polyacrylamide gel electrophoresis and ^{35}S-
methionine labeled polypeptides detected by autoradiography
(Figure 2). The sucrose gradient centrifugation conditions have
clearly resulted in a successful fractionation of the various cytoplasmic mRNA species.

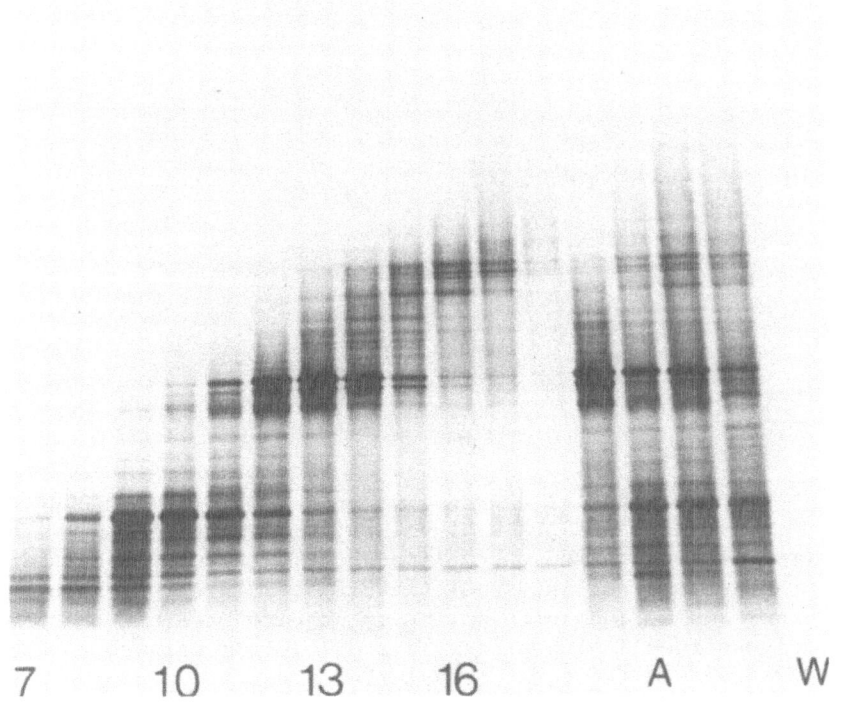

Fig. 2.   Cell-free translation of fractionated pea leaf mRNA. Cell-
          free translation was carried out with mRNA recovered from
          fractions 7-18 (Fig. 1), poly(A)$^+$ mRNA (A) and no RNA (W).
          SDS-polyacrylamide gel electrophoresis was as described by
          Chua and Bennoun (1975) and the ^{35}S-methionine labeled
          translation products were detected by autoradiography.

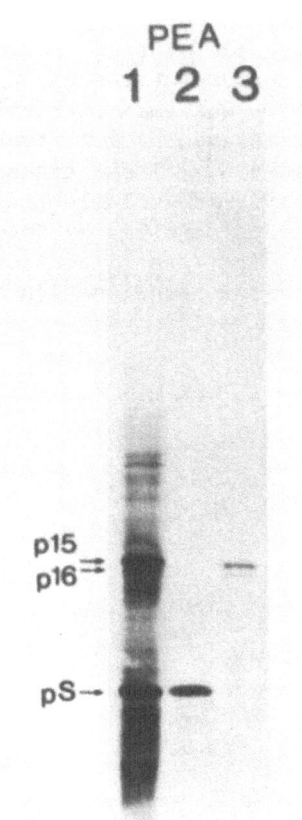

PEA
1 2 3

p15
p16

pS

Fig. 3   Immunoprecipitation of pea leaf mRNA translation products.
Poly(A)$^+$ mRNA was translated in a wheat germ extract.   The
translation products were immunoprecipitated with antiserum
prepared against the small subunit of RuBP carboxylase from
pea (sample 2) or with antiserum prepared against the CP_{II}
polypeptides (polypeptides 15 and 16) from spinach.   Sample
1 represents the total translation products.   Samples were
fractionated by SDS-polyacrylamide gel electrophoresis and
the ^{35}S-methionine labeled polypeptides were detected by
autoradiography.

Using antiserum prepared against polypeptides 15 and 16 (CP_{II}
polypeptides) from spinach chloroplast membranes, we have shown
that the translation products corresponding to the 14S mRNA contain polypeptides which are immunologically related to the CP_{II}
polypeptides.   In fact, 29% of the ^{35}S-methionine labeled translation products which are encoded by fraction 13 mRNA are immunoprecipitated with the CP_{II} antiserum (Figure 1).   These polypeptides are characterized in the experiment shown in Figure 3.   It
is observed that a major and minor polypeptide, with respective

molecular weights of approximately 33,000 and 32,000, are immunoprecipitated from cell-free translation products which are encoded
by unfractionated poly(A)$^+$ mRNA. We have identified these products
as biosynthetic precursors for the CP_{II} polypeptides (Schmidt et
al., this volume). From the experiments shown in Figures 1-3 it
is seen that the precursors for the CP_{II} polypeptides represent
the major products of translation of the 14S mRNA from pea leaves.

SYNTHESIS OF SINGLE-STRANDED cDNA

The partially purified mRNAs, encoding the small subunit of
RuBP carboxylase and the CP_{II} polypeptides, have been used as templates for transcription with reverse transcriptase from avian
myeloblastosis virus. The conditions of synthesis were similar to
those used by a number of laboratories (Efstratiadis et al., 1976;
Friedman and Rosbash, 1977; Buell et al., 1978). One important
variable examined was the amount of reverse transcriptase in the
reaction mixture relative to the amount of mRNA template. Maximum
incorporation of α^{32}P-dCTP was achieved with 84.5 units of reverse
transcriptase per µg of mRNA and this corresponded to approximately
0.5 mole equivalents of cDNA synthesized per mRNA template (data
not shown). However, when the products of cDNA synthesis were
examined by alkaline agarose gel electrophoresis (Figure 4; lanes
3 and 7) we obtained a rather broad size range of products with
this relatively high level of reverse transcriptase. The highest
percentage of long transcripts were attained with a much lower in-
put of reverse transcriptase. These results presumably reflect
contamination of the reverse transcriptase with ribonuclease activity. At 16.9 units of reverse transcriptase per µg of mRNA,
relatively high yields of apparently full-length transcripts were
obtained (Figure 4; lanes 1 and 5). For the small subunit mRNA
the reverse transcripts were estimated to be approximately 850
nucleotides long. We have previously estimated, by polyacrylamide
gel electrophoresis in the presence of formamide, the molecular
length of purified small subunit mRNA to be 800 nucleotides
(Cashmore, 1979). These two estimates agree within experimental
error and are therefore consistent with the notion that full-length
cDNA copies are being synthesized for the small subunit mRNA. For
the mRNAs encoding the CP_{II} polypeptides the transcripts were
estimated to be approximately 1250 nucleotides in length.

SYNTHESIS OF DOUBLE-STRANDED cDNA

Under standard conditions of reverse transcription of mRNA
to produce cDNA, the resulting mRNA/DNA hybrid is stable and little
synthesis of DNA duplexes occurs. However, the mRNA can be readily
removed, either by treatment with alkali or by heat-denaturation,
and the cDNA can then be used as a template for the synthesis of
a second, complementary strand. The self-priming characteristic

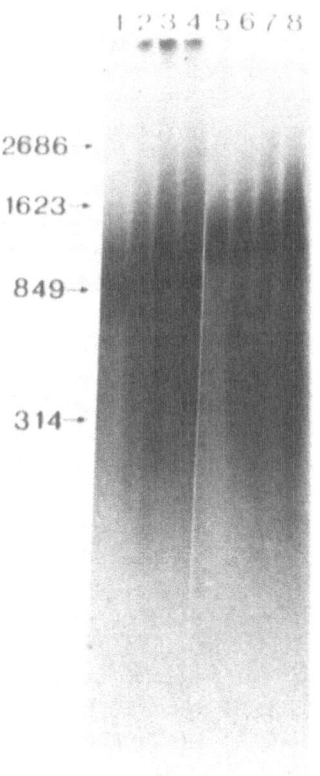

Fig. 4. Synthesis of single-stranded cDNAs from mRNAs encoding the small subunit of RuBP carboxylase (S) and the CP_{II} polypeptides (a/b). Reaction mixtures were 92 mM KCl, 50 mM Tris-HCl (pH 8.5 at 25°), 10 mM Mg acetate, 10 mM DTT, 20 µg/ml oligo dT_{12-18}, 0.5 mM deoxynucleotide triphosphates (including ^{32}P-dCTP) and 15 µg/ml mRNA. Reverse transcriptase (lot # G-1378, kindly supplied by Dr. J.W. Beard, Life Sciences, Inc., St. Petersburg, Florida) was included at the following units per µg mRNA: 16.9 (samples 1 and 5) 42.3 (samples 2 and 6), 84.5 (samples 3 and 7), and 169 (samples 4 and 8). Incubation was for 60 minutes at 42° and then the samples were fractionated by electrophoresis on alkaline 2% agarose gels (McDonell et al., 1977). After washing the gels in 5% TCA the products were detected by autoradiography.

of this second-strand synthesis results from double-stranded hairpin loops which exist at the 3'-termini of the cDNA. This synthesis of the second DNA strand can be achieved with DNA polymerase I from Escherichia coli, and detailed studies of this reaction have been made (Efstratiadis et al., 1976; Wickens et al., 1978). DNA polymerase I is an extremely complicated enzyme and the synthesis of full-length second strands is critically dependent on several variables.

An alternate approach to the synthesis of double-stranded cDNA is to use the DNA dependent DNA polymerizing ability of reverse transcriptase (Rougeon and Mach, 1976; Ullrich et al., 1977). Here we have examined, as in the case of first-strand synthesis, the effect of various levels of reverse transcriptase on the

Table 1   Resistance of cDNAs to S1 nuclease

Experiment	Sample	RNA	Reverse Transcriptase (Units/µg mRNA)	S1 Resistance (%)
1	1	A	150	77.7
	2	A	30	66.1
	3	A	30	63.9
	4	A	7.5	54.8
	5	A	0	13.1
2	1	S	152	82.8
	2	CP_{II}	152	91.7
	3	A	75	68.8
	4	A	0	8.6

The samples in experiment 1 represent those described in Figure 5. The samples in experiment 2 represent second-strand cDNA synthesis (42°/2h) for small subunit (S), CP_{II} and mixed poly(A)$^+$ (A) mRNAs. First-strand synthesis in experiment 2 was, for A, as described in Figure 5 and for S and CP_{II} as described for samples 1 and 5 in Figure 4 (except that mRNA was 100 µg/ml and oligo dT was 80 µg/ml). The reverse transcriptase units in the table are for second-strand synthesis and are expressed per µg mRNA present for first-strand synthesis. The S1 digestion was similar to that described for Figure 5 and is expressed as % acid-insoluble counts resistant to S1 nuclease. No incorporation of ^{32}P-dCTP occurs during second-strand synthesis (dCTP is apparently destroyed during the 100° incubation following first-strand synthesis) and thus the S1 figures refer to the conversion of the ^{32}P-labeled first-strand into a nuclease-resistant form.

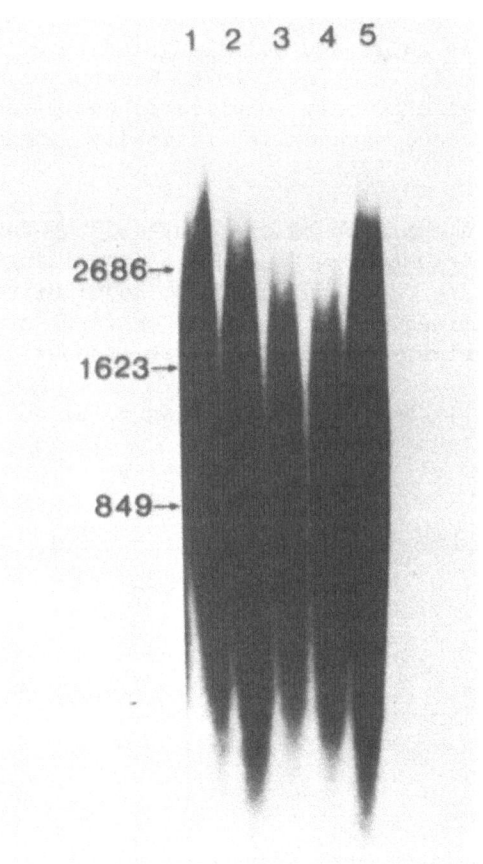

Fig. 5.   Synthesis of double-stranded cDNAs from mixed poly(A)$^+$
          mRNA. Synthesis of single-stranded cDNA was as for
          Figure 4 except that mixed poly(A)$^+$ mRNA was used at 340
          μg/ml, oligo dT was 100 μg/ml and reverse transcriptase
          was 7.5 units/μg mRNA. For synthesis of double-stranded
          cDNA, the single-stranded cDNA reaction mixture was in-
          cubated for 2 minutes at 100° (to denature the DNA/RNA
          hybrid), deoxynucleotide triphosphates were added to 0.5
          mM and reverse transcriptase was added at 150 (sample 1),
          30 (samples 2 and 3), 7.5 (sample 4), and 0 (sample 5)
          units per μg of original mRNA template. Synthesis was at
          42° for 2 h (samples 1, 3, 4 and 5) or 3 h (sample 2).
          Samples 1-4 were diluted with 4 volumes of S1 buffer
          (0.3 M NaCl, 30 mM Na acetate, 3 mM Zn acetate; pH 4.4)
          and incubated for 30 minutes at 37° with 150 units per ml
          of S1 nuclease (Sigma). Samples were fractionated accord-
          ing to the legend in Figure 4.

synthesis of double-stranded cDNA. We have used the development
of resistance to S1 nuclease as an assay for the synthesis of
double-stranded cDNA (Table 1). In direct contrast to the results
with single-stranded cDNA synthesis, we found that relatively high
levels of reverse transcriptase were required for efficient synthesis of the second strand. This conclusion is corroborated by
an examination of the products by alkaline agarose gel electrophoresis. In Figure 5 it is seen that cDNA prepared from unfractionated poly(A)$^+$ mRNA is converted, by high levels of reverse
transcriptase, into double-stranded cDNA which is similar in length
to the original single-stranded cDNA template. The level of reverse transcriptase used in this case (150 units per μg of initial
RNA), represents 20 times that used for the first-strand cDNA synthesis. It should be noted that in Figure 5 the products of second-strand synthesis, but not the products of first-strand synthesis, have been treated with S1 nuclease prior to assaying by
gel electrophoresis. This treatment degrades single-stranded DNA,
including the hairpin loop which exists at one end of the DNA
product resulting from second-strand synthesis. When the cDNA
product of first-strand synthesis is exposed to the reaction conditions used for second-strand synthesis, but omitting the polymerase, minimal development of S1 nuclease resistance occurs
(Table 1 and unpublished gel electrophoresis results). Using conditions similar to those described above we have also successfully
synthesized double-stranded cDNAs from the partially purified mRNAs
encoding the small subunit of RuBP carboxylase and the CP_{II} polypeptides.

CONCLUSION

     We have recently reported the purification from pea leaves of
the mRNA encoding the precursor for the small subunit of RuBP carboxylase (Cashmore, 1979). In this article we have reported the
partial purification of the mRNAs encoding the precursors for the
CP_{II} polypeptides. We have examined the reaction of reverse transcriptase with both unfractionated poly(A)$^+$ mRNA and with the mRNAs
encoding the small subunit and the CP_{II} polypeptides. The synthesis of apparently full-length single and double-stranded cDNA
has been achieved for both the unfractionated and the purified
mRNAs. These cDNAs should prove useful as hybridization probes
for analyzing the factors regulating the expression of plant
nuclear genes.

REFERENCES

Apel, K., and Kloppstech, K., 1978, Eur. J. Biochem., 85:581.
Blair, G.E., and Ellis, R.J., 1973, Biochim., Biophys. Acta, 319: 223.
Buell, G.N., Wickens, M.P., Payvar, F., and Schimke, R.T., 1978, J. Biol. Chem., 253:2471.
Cashmore, A.R., 1976, J. Biol. Chem. 251:2848.
Cashmore, A.R., 1979, Cell, 17:383.
Cashmore, A.R., Broadhurst, M.K., and Gray, R.E., 1978, Proc. Natl. Acad. Sci. USA, 75:655.
Chua, N.-H., and Bennoun, P., 1975, Proc. Natl. Acad. Sci. USA, 72:2175.
Chua, N.-H., and Schmidt, G.W., 1978, Proc. Natl. Acad. Sci. USA, 75:6110.
Coen, D.M., Bedbrook, J.R., Bogorad, L., and Rich, A., 1977, Proc. Natl. Acad. Sci. USA, 74:5487.
Dobberstein, B., Blobel, G., and Chua, N.-H., 1977, Proc. Natl. Acad. Sci. USA, 74:1082.
Efstratiadis, A., Kafatos, F.C., Maxam, A.M., and Maniatis, T., 1976, Cell, 7:279.
Friedman, E.Y., and Rosbash, M., 1977, Nucleic Acids Res., 4:3455.
Highfield, P.E., and Ellis, R.J., 1978, Nature, 271:420.
Kawashima, N., and Wildman, S.G., 1972, Biochim. Biophys. Acta, 262:42.
Kung, S.D., Thornber, J.P., and Wildman, S.G., 1972, FEBS Lett 24:185.
McDonell, M.W., Simon, M.N., and Studier, F.W., 1977, J. Mol. Biol., 110:119.
Possingham, J.V., and Saurer, W., 1969, Planta, 86:186.
Rougeon, F., and Mach, B., 1976, Proc. Natl. Acad. Sci. USA, 73: 3418.
Schmidt, G.W., Bartlett, S.G., Grossman, A.R., Cashmore, A.R., and Chua, N.-H., This volume.
Ullrich, A., Shine, J., Chirgwin, J., Pictet, R., Tischer, E., Rutter, W.J., and Goodman, H.M., 1977, Science, 196:1313.
Whitfeld, P.R., Spencer, D., and Bottomley, W., 1973, In The Biochemistry of Gene Expression in Higher Organisms. Pollak, J.K., and Lee, J.W., eds. (Sydney: Australia and New Zealand Book Co.) pp.504.
Wickens, M.P., Buell, G.N., and Schimke, R.T., 1978, J. Biol. Chem., 253:2483.

SITES OF SYNTHESIS AND CODIFICATION OF CHLOROPLAST ELONGATION FACTORS

Orio Ciferri, Orsola Tiboni, Giuseppe Di Pasquale and Daniela Carbonera

Istituto di Microbiologia e Fisiologia Vegetale, Università di Pavia, Pavia, Italia

INTRODUCTION

It is now well established that chloroplasts and mitochondria are endowed with different elongation factors[1]. Utilizing selective inhibitors of cytoplasmic or chloroplastic protein synthesis, we have found that in the alga Chlorella vulgaris, the chloroplast- -specific elongation factor G (EF-G_{chl}), unlike the mitochondrion- -specific one[2], is synthesized in the chloroplast itself[3]. Experi - ments on protein synthesis by isolated spinach chloroplasts have indicated that, also in this organism, EF-G_{chl} and another elon- gation factor (EF-$T_{u\ chl}$) appear to be produced in the organelle[4,5]. We present now more rigorous evidence confirming this conclusion. In addition, preliminary experiments strongly indicate that the two elongation factors are among the proteins synthesized when spin- ach chloroplast DNA is transcribed and translated in vitro by a cell-free extract from E. coli.

CHLOROPLAST ELONGATION FACTORS ARE SYNTHESIZED IN THE CHLOROPLAST

Availability of highly purified EF-G_{chl}, and EF-$T_{u\ chl}$ from spinach[6] has allowed the unambiguous localization of these pro- teins in two-dimensional gels of the soluble proteins present in spinach chloroplast (Fig. 1, left). Therefore it became possible to establish if the two proteins became labelled when isolated spin-

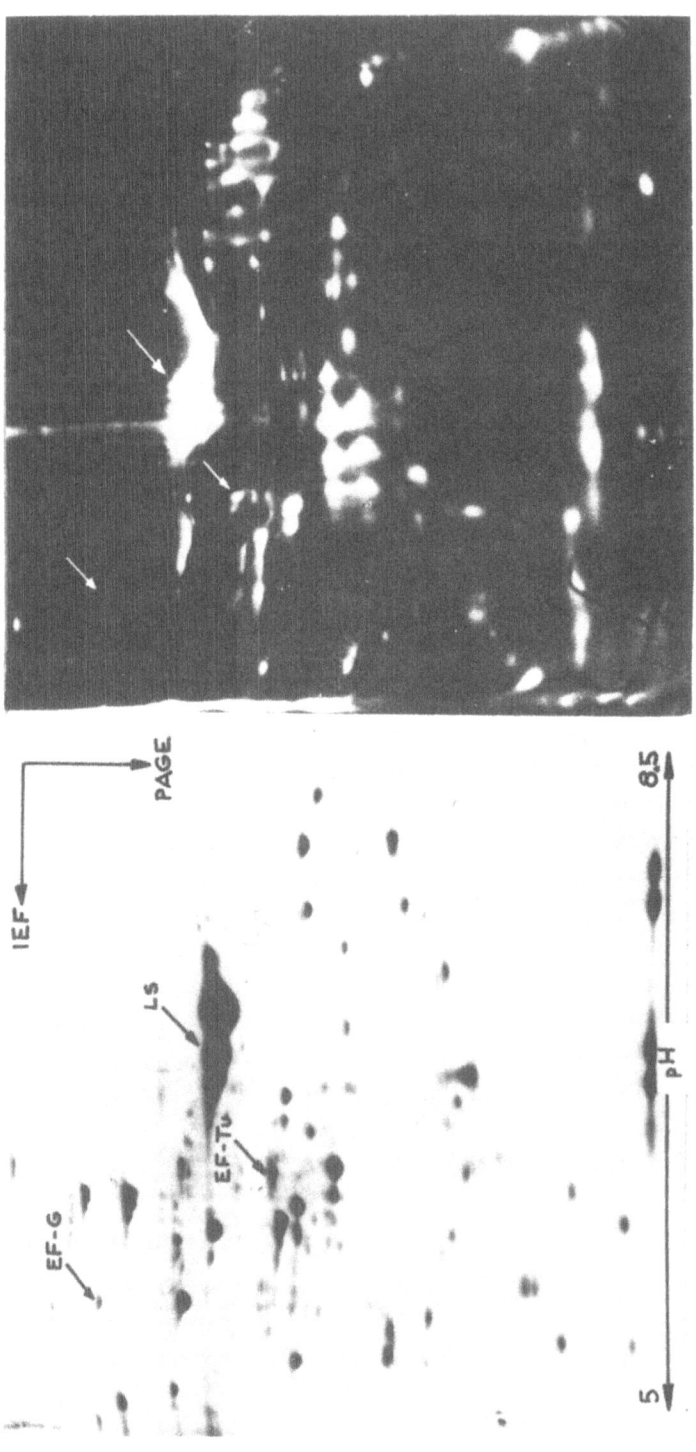

Fig. 1 Two-dimensional separation of chloroplast soluble protein.

Left, stain; right, fluorography. Approximately 0.14 mg of protein (left) or 0.35 mg of protein and 45,000 cpm (right) were separated by IEF-SDS-polyacrylamide gel electrophoresis.

ach chloroplasts were incubated in the light in the presence of
radioactive amino acids. As shown in Fig. 1, right besides the
large subunit of ribulose-bisphosphate carboxylase (LS), many
other proteins became labelied in isolated chloroplasts. In all experiments, a radioactive spot was always found in the position corresponding to $EF-T_{u\ chl}$ while, occasionally, a faint radioactive
spot was evident also in the position of $EF-G_{chl}$. Identification
of radioactive chloroplast elongation factors, synthesized in these
experiments, was confirmed by immunoautoradiography using monospecific antisera[7] (Fig. 2). From these results it appears quite
clear that $EF-G_{chl}$ and $EF-T_{u\ chl}$ are synthesized in isolated
spinach chloroplasts thus confirming the conclusion reached in
previous experiments performed on intact cells of C. vulgaris[3].

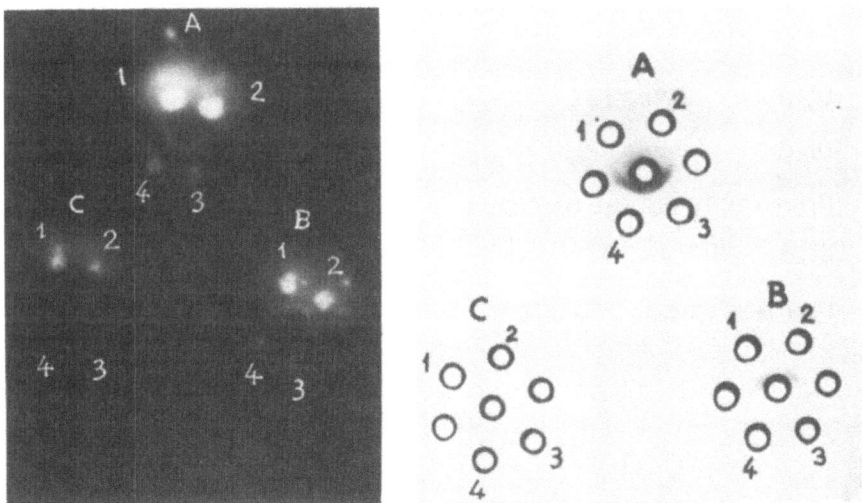

Fig. 2 Immunoprecipitation and radioimmunoprecipitation of partially purified $EF-G_{chl}$ and $EF-T_{u\ chl}$.

Left, stain; right, autoradiography. A, antiserum to
$EF-T_{u\ chl}$; B, antiserum to $EF-G_{chl}$; C, non-immune serum. 1-2, partially purified $EF-G_{chl}$; 3-4, partially purified $EF-T_{u\ chl}$.

CHLOROPLAST ELONGATION FACTORS ARE (PRESUMABLY) CODED IN CHLOROPLAST DNA

Bottomley and Whitfeld have demonstrated that DNA isolated from higher plant chloroplasts is transcribed and translated in vitro with good fidelity by an E. coli extract[8]. As shown in Table I, when spinach chloroplast DNA is added to the reaction mixture incorporation of labelled methionine is stimulated ca. 20-fold. Such a stimulation is abolished by adding to the reaction mixtures either DNase or rifampycin or by omitting CTP and UTP. When the reaction products were analysed by SDS-gel electrophoresis, three radioactive protein bands were identified with the same mobility of the LS, $EF-G_{chl}$ and $EF-T_{u \ chl}$ (Fig. 3).

Table I. Transcription and translation of spinach chloroplast DNA

	^{35}S methionine incorporated, cpm x 10^{-4}/assay
− DNA	2.7
+ DNA	38.2
+ DNA + rifampicin (0.1 μg)	2.3
+ DNA − UTP, CTP	2.8
+ DNA + DNase (3 μg)	1.9
− RNA	2.7
+ RNA	140.8
+ RNA + rifampicin (0.1 μg)	130.8
+ RNA − UTP, CTP	136.1

These three proteins were also evident when the products were analysed by two-dimensional electrophoresis (Fig. 4).

In conclusion, it appears quite likely that $EF-G_{chl}$ and $EF-T_{u \ chl}$ are produced when spinach chloroplast DNA is tran-

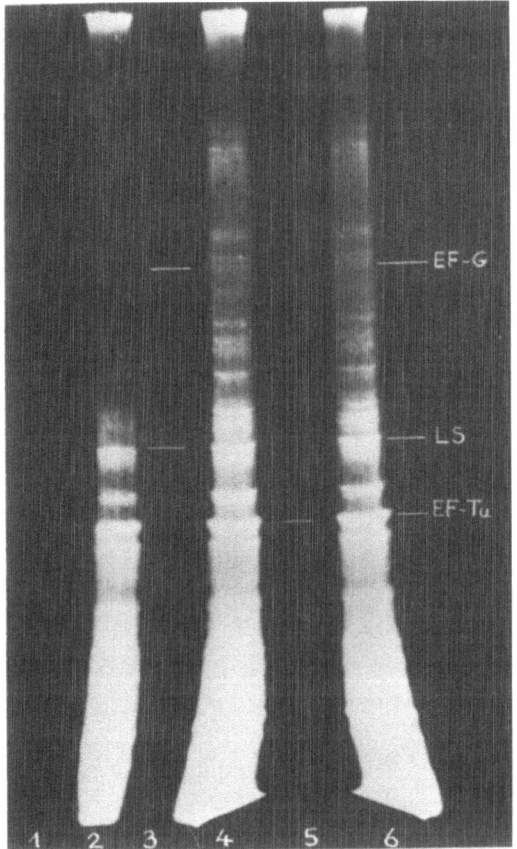

Fig. 3 Fluorography of the products of coupled transcription and translation.

DNA was isolated from spinach chloroplasts and assayed in a E. coli cell-free extract according to the reported procedures[8,9]. The reaction products were separated by SDS-10% polyacrylamide gel electrophoresis. 1, no DNA; 2, 2 μg of DNA; 3, $EF\text{-}G_{chl}$ and LS markers; 4, 3 μg of DNA; 5, $EF\text{-}T_{u\ chl}$ marker; 6, 4 μg of DNA.

scribed and translated in vitro. Thus the two elongation factors are coded most probably in the organellar DNA. Assuming the presence of one gene for each elongation factor per chloroplast genome and the absence of intervening sequences, leader sequences, etc., it may be calculated that the genetic information for the two proteins accounts for approximately 2.5% of the chloroplast ge-

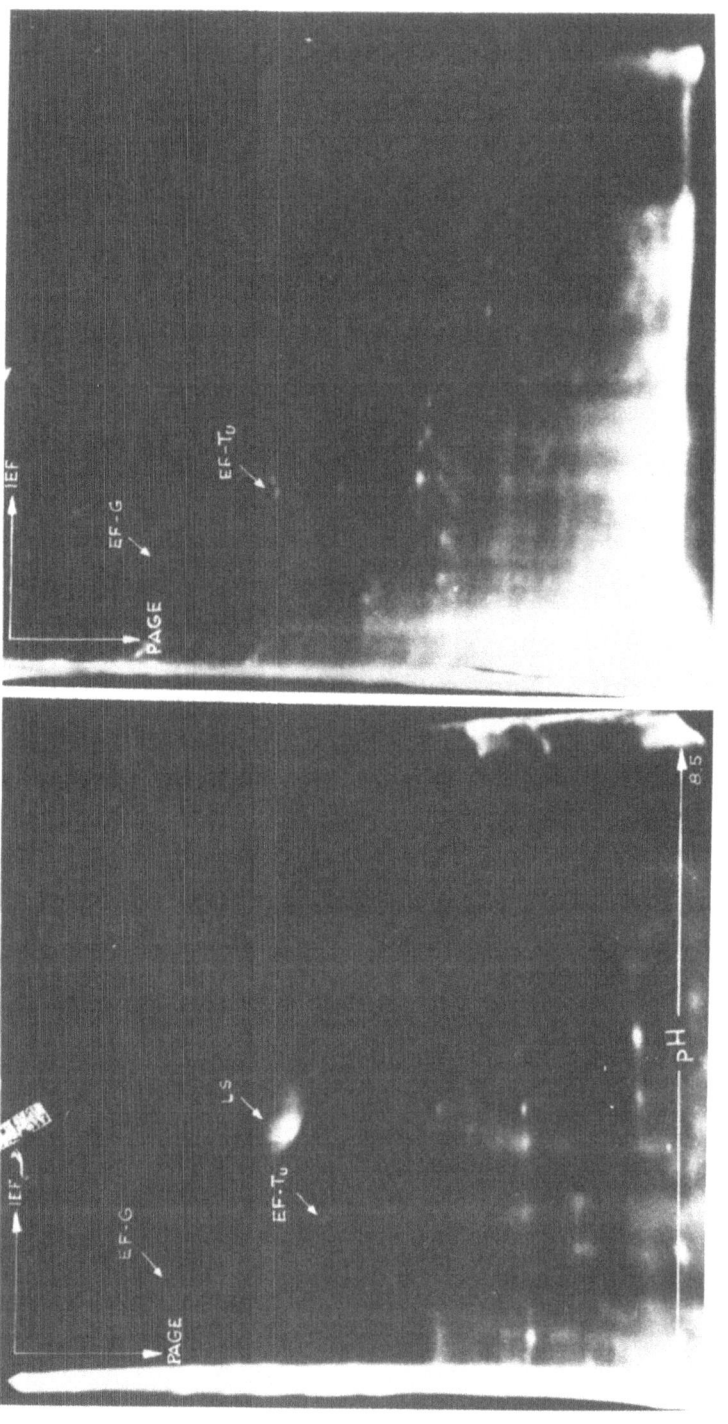

Fig. 4 Fluorography of crude or partially purified reaction products.

Aliquots of the crude reaction products (left) or the partially purified EF-G_{chl} + EF-T_u $_{chl}$ fraction (right) were separated by IEF-SDS polyacrylamide gel electrophoresis

nome.

As far as we know this is the first demonstration that complete, soluble proteins are coded and synthesized in any kind of cell organelle.

ACKNOWLEDGEMENTS

This work was supported by grants from the Consiglic Nazionale delle Ricerche to O.T. and to O.C. (the latter through the Laboratorio di Genetica Biochimica ed Evoluzionistica). D.C. is a postdoctoral fellow of the Programma Finalizzato "Nuove Fonti Proteiche" (Subprogetto 3) of the Consiglio Nazionale delle Ricerche.

REFERENCES

1. O. Ciferri, O. Tiboni, M.L. Munoz-Calvo, and G. Camerino, in "Nucleic Acids and Protein Synthesis in Plants" L. Bogorad and J.H. Weil, eds., Plenum Press, p. 155-166, (1977).

2. O. Ciferri, O. Tiboni, G. Lazar, and J. Van Etten, in "The Biogenesis of Mitochondria", A.M. Kroon and C. Saccone, eds., Academic Press, p. 107-115, (1974).

3. O. Ciferri, and O. Tiboni, Plant Sci. Lett. 7: 455 (1976).

4. O. Tiboni, B. Parisi, and O. Ciferri, in "Acides nucléiques et synthèse des protéines chez les végétaux" L. Bogorad and J.H. Weil, eds., Coll. Int. C.N.R.S., p. 345-349, (1977).

5. O. Ciferri, and O. Tiboni, in "Translation of Natural and Synthetic Polynucleotides", A.B. Legocki, ed., Poznan Agricultural University, p. 173-177, (1977).

6. O. Tiboni, G. Di Pasquale, and O. Ciferri, Eur. J. Biochem. 92: 471 (1978).

7. O. Tiboni, C.J. Leaver, and O. Ciferri, submitted for publication.

8. W. Bottomley, and P.R. Whitfeld, Eur. J. Biochem. 93: 31 (1979).

9. P.R. Whitfeld, R.G. Herrmann, and W. Bottomley, Nucleic Acid Res. 5: 1741 (1978).

NUCLEAR GENES CONTROLLING CHLOROPLAST DEVELOPMENT

Virginia Walbot, Deborah Thompson, G.M. Veith

Department of Biology
Washington University
St. Louis, MO 63130

E. H. Coe, Jr.

U.S. Department of Agriculture -- Science and Education
  Administration
University of Missouri
Columbia, MO 65201

I.   INTRODUCTION

We are interested in defining the controls of plastid differentiation in _Zea mays_ L., a C_4 plant containing dimorphic mesophyll and bundle sheath plastids as well as chromatoplasts, amyloplasts and proplastids in meristematic zones. One approach is to examine normal plastid developmental pathways at the light and electron microscope level and by description of the protein composition of plastids of various stages and terminal phenotypes. This study is complemented by examining the time of action and phenotypic result of nuclear genes which act to prevent normal, green plastid differentiation. So far we have utilized the nuclear genes _iojap_, _chloroplast mutator_, and _japonica 1ℓ2_. From a study of these mutants we hope to gain some insight into the types of nuclear genes required for normal development and to determine what aspects of plastid function can be autonomously expressed by either the plastid or nuclear genome. In this report, we describe the theoretical background to our ongoing study and some results on the terminal phenotypes and time of action of the nuclear genes _ij_, _cm_ and _j_   1ℓ2.

A.   Nuclear Controls of Plastid Development

        Although the plastid contains a genome and protein
synthetic machinery, the ontogeny and maintenance of the
organelle are dependent on nuclear-coded products. The
cooperation between the plastid and nuclear genomes occurs at
several levels:

    1.   The two genomes can specify subunits of the same protein
          complex: RuBPCase, for example, is composed of 55,000
          Dalton chloroplast-coded and 12-14,000 Dalton nuclear-
          coded subunits (1);

    2.   The nuclear genome-coded enzyme systems responsible for
          nucleic acid precursor, lipid, and amino acid synthesis
          contribute to organelle maintenance (2);

    3.   The state of differentiation of the cell, primarily a
          reflection of nuclear gene activity, correlates with the
          ultrastructural and functional capabilities of the
          plastid compartment: root cells contain amyloplasts in
          which starch and protein are stored, chromatoplasts are
          found in petal tissue, and photosynthetically competent
          chloroplasts are found in leaves and other green plant
          parts (3). These differentiated states of the plastid are
          not terminal since amyloplasts, for example, can be induced
          to form chloroplasts (4). In fact, it is instructive to
          view the plastid as capable of many alternative pathways
          of differentiation, dedifferentiation and redifferentia-
          tion (5).

This last point is the subject of our current research. That is,
we would like to understand how nuclear gene activity regulates
the ability of the plastid to differentiate into an organelle
appropriate for that particular cell type and to elucidate the
biochemical and ultrastructural steps the plastid traverses to
reach a particular phenotype.

    In maize a variety of nuclear genes have been described which
influence the phenotype of the plastid. Several categories of
nuclear effects can be recognized:

    1.   Specific genes coding for photosynthetic functions, such
          as genes of pigment pathways, cytochromes, and accessory
          proteins; these genes are recognized by scoring for
          photosynthetically incompetent (6) or pigment-deficient
          (7,8) plastids.

    2.   Nuclear background regulating the timing or extent of
          expression of photosynthetic functions. For example,

zebra striping in which low temperature suppresses
greening, virescence mutants, and albescence mutants may
result from deficiencies in particular nuclear-coded
products (e.g. a temperature-sensitive allele of a
specific nuclear gene); alternatively, such mutants may
present cases of a more fundamental regulatory nature.
Examples of putative variants in underlying developmental
programs regulating plastid differentiation were uncovered
in a survey of maize seedlings for RuBPCase phenotype
(9). Using antibody prepared against the holoenzyme,
Simpson demonstrated that in most seedlings chlorophyll
and carboxylase content increased in parallel as part of
a "greening" program of the leaf, however, instances of
green leaves lacking enzyme and yellow stripes containing
carboxylase were detected. Examples of virescent
seedlings in which greening is delayed several days
after leaf emergence were detected in which carboxylase
was present before, in parallel with, or after greening.
These results suggest that the nuclear gene contributions
resulting in a photosynthetically competent plastid can
be divided into at least two components or programs:
the program for greening and a program for acquiring
carboxylase and perhaps other enzymes as well. These
two programs may or may not be coordinate in time. The
basis for regulating the time of appearance may reside
at the individual loci of each program. Alternatively,
we may invoke a regulatory locus for each program, the
activity of which determines when and to what extent
greening or carboxylase synthesis will occur.

3. Specific nuclear genes which cause a permanent change in
    the ability of the plastid to differentiate; iojap (ij)
    and chloroplast mutator (cm) are genes of this class,
    recognized by the Mendelian inheritance of the gene but
    cytoplasmic (maternal) inheritance of the affected
    plastid. Homozygous recessive plants (ij/ij) will
    elaborate green and white striped leaves and, when used
    as the maternal parent in a cross to a normal individual
    (♀ ij/ij x ♂ +/+), will produce albino, green, and (rarely)
    striped individuals. If a homozygous ij/ij plant is
    used as the pollen parent, however, only green progeny
    are produced (10). Albinism can persist even in a
    subsequent generation in which the nuclear background is
    restored to wild type (+/+). This class of nuclear gene
    mutations, therefore, brings about a more fundamental
    disruption of plastid capability than simple pigment or
    enzyme deficiency of a particular developmental pathway
    since wild type nuclear background cannot restore normal
    plastid function.

B.   Ontogeny of the Maize Plant

     At the time of fertilization the maize egg sac contains
approximately thirty monoploid antipodals, a triploid
endosperm cell and a diploid zygote. Rapid mitoses along
fixed planes of division in the endosperm result in a
syncitial nuclear endosperm; subsequently the egg sac cytoplasm
is partitioned into several hundred cells which proliferate
to form the nutritive endosperm. The zygote also undergoes a
defined set of divisions in which the plane of cell division
is regulated. The first division of the zygote defines the
suspensor, a terminal embryonic organ, and the embryo.
Subsequent cleavage of the embryo results in a cell mass of
approximately 10,000 cells (2^{13}) within 12-14 days (11;
Thompson, unpublished). At this stage the apical meristem is
organized on the distal face of the embryo [relative to the
main plant axis] at a position such that cells derived from
the first cleavage division cells are equally represented.
The meristem, a single layer of epidermal tissue ultimately
formed into a flattened cone, gives rise to the leaves, stem,
ear shoot and tassel of the plant body. A fate diagram
describing the sequential use of the meristem cells is presented
in Figure 1 (adapted from 12).

     During embryonic growth 4-5 leaves are differentiated,
but the bulk of the plant body is formed after germination.
Organs are initiated 180° apart on the meristem at a point
that represents the line of contact between the two cells
founding the meristem [and hence the two cleavage cells]
(13). The two cells lying on the 180° axis undergo alternate
divisions to initiate the midribs of successive leaves which
are inserted at progressively higher positions on the meristem.
A subsequent wave of division around the meristem recruits
cells to form the leaf (14), as illustrated in Figure 2.
More precisely, meristem cells divide periclinally to produce
an initial cell, which subsequently divides anticlinally to
produce leaf and stem progenitors at that position. Therefore,
the leaf and its subtending node are simultaneously derived
from the activity of a single meristem region. The 16-32
stem and leaf progenitors proliferate to keep pace with
lateral stem expansion. The result is the formation of zones
of cells, called intercalary meristems, from which the bulk
of stem and leaf tissue will be derived. Specifically, the
stem intercalary meristem lies just below the node, while the
leaf intercalary meristem (often called the zone of cell
proliferation) lies just above the node and at the base of
the leaf.

     The pattern of leaf emergence is such that the oldest
tissue of the leaf is located at the tip, while a meristematic

zone occupies the leaf base near the node. The leaf is
pointed because the tip is composed of cells derived from one
or a few apical meristem cells. Derivatives of each meristem
cell are progressively recruited until about one-third down
from the tip, progeny of all 16-32 cells participate in leaf
formation.

II. APPEARANCE OF ALBINO SECTORS ON THE PLANT

We have previously reported (15) that iojap events often
occupy entire leaf or stem clones, that is 1/16 - 1/32 of the
organ area, but that stripes are not concordant between successive
leaves nor between the stem and leaf derived from the same meristematic activity. These observations are interpreted as evidence
that the iojap event occurs after separation of the leaf and stem
progenitors of a node and that the earliest time of iojap activity
is in the individual progenitor cells. The macroscopic iojap
phenotype is illustrated in Figure 3 in which it is clear that
clonal-size stripes as well as narrow stripes and subclonal areas
of switching from green to white are evident. At the light microscopic level we find evidence for abrupt changes of cell phenotype
(Figure 3) as well as examples of pale green and mosaics of green
and white cells in leaf tissue. At the electron microscopic level
we have found examples of individual chloroplasts with normal
ultrastructure in what appear to the eye to be pure albino clones
(Figure 4); similarly, we find "pale" proplastids of the iojap
type in what appear macroscopically to be normal green areas.

III. IOJAP, A CHLOROPLAST RIBOSOME DEFICIENT MUTANT OF MAIZE

The terminal phenotype of iojap-affected plastids is a
ribosome-less, rRNA-less plastid containing an apparently normal
plastid genome (15 and Table 1). The iojap-affected plastids
replicate as a compartment and the genome also replicates for many
cell generations and through successive plant generations. Presumably, therefore, nuclear-coded gene products maintain these plastid
activities in the absence of plastid protein synthesis. A similar
conclusion has been reached by other investigators utilizing
chloroplast ribosome-deficient mutants of Pelargonium and barley
to delineate the contribution of organellar genes to chloroplast
structure and function, Hagemann & Börner (16) also find that the
maintenance of a plastid compartment containing DNA does not
require organellar protein synthesis. A similar conclusion has
been drawn for algae plastid maintenance (17).

The plastid does, on the other hand, contribute substantially
to other organellar functions. Plastid protein synthesis plays a
major role in the greening process, and plastid-coded gene products

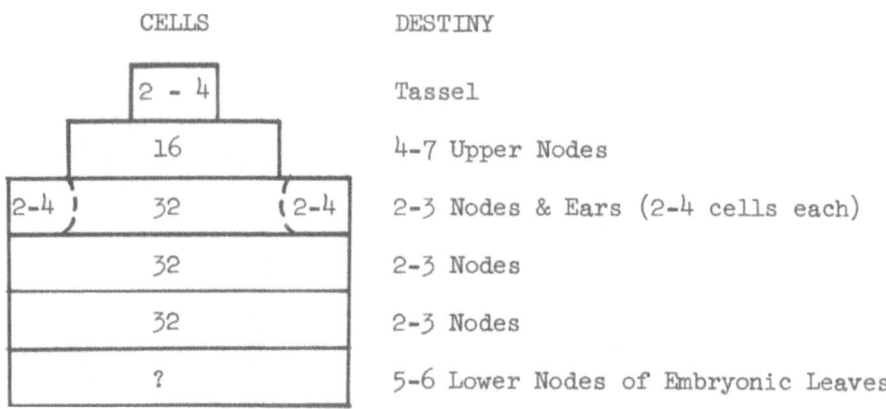

CELLS   DESTINY

CELLS	DESTINY
2 - 4	Tassel
16	4-7 Upper Nodes
2-4) 32 (2-4	2-3 Nodes & Ears (2-4 cells each)
32	2-3 Nodes
32	2-3 Nodes
?	5-6 Lower Nodes of Embryonic Leaves

Figure 1.   Ontogeny of the Maize Plant.

Cell marking experiments indicate that the apical meristem functions in a regular manner in which rings of cells of the meristem give rise to a discrete number of nodes (12). The ear tissue is derived from a subset of 2 - 4 cells whose progeny also form somatic tissue. Tassel tissue arises from a group of 2 - 4 cells which do not contribute to the somatic body. In tillers, however, all somatic and reproductive tissue is derived from the 2 - 4 cells which found the tiller apical meristem; therefore, there is not a "germ line" of determined cells.

Successive leaves are initiated 180° apart on the meristem axis as shown in the cross section through an apical meristem and leaf initials below. The ears arise 180° apart on successive nodes at the approximate position of the midrib.

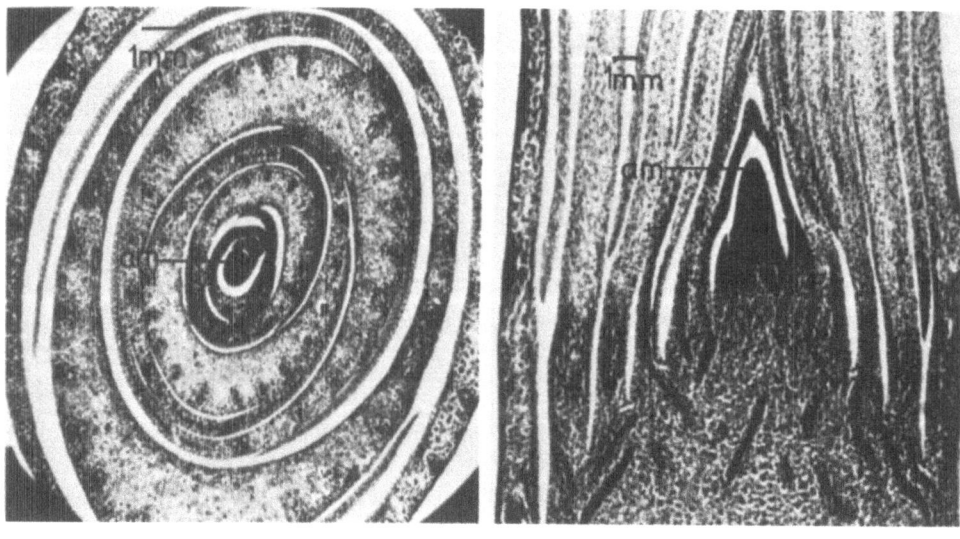

Figure 2.   Model of Maize Leaf Formation

Meristem cells divide outward to give rise to a ring of derivatives (cells depicted in Figure 5B). The meristem derivative cell subsequently divides anticlinally to give rise to an upper cell which is the leaf progenitor cell and a lower cell which is the stem progenitor cell. The node forms at the junction between leaf and stem progenitors (see ref.13).

   The leaf progenitor cells are the founders of the clones which occupy the leaf. Individual leaf progenitor cells divide in what appears to be a regulated order of plane of division to give rise to the three dimensional intercalary meristem from which the leaf will arise. The order and plane of cell divisions within the individual leaf clones can be deduced from examination of the width and length of stripes on the leaf.

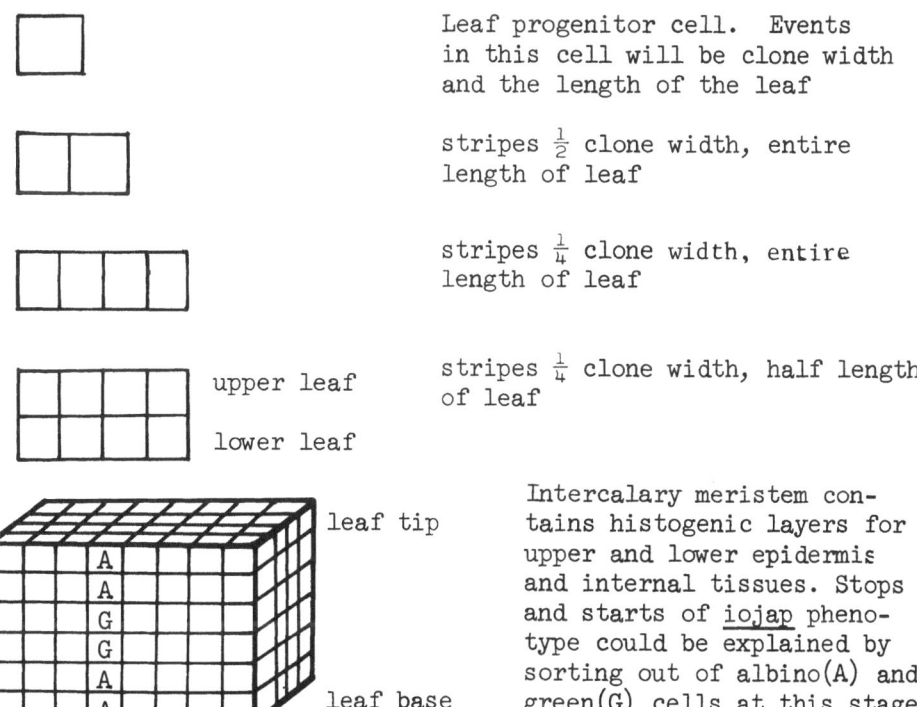

Leaf progenitor cell. Events in this cell will be clone width and the length of the leaf

stripes $\frac{1}{2}$ clone width, entire length of leaf

stripes $\frac{1}{4}$ clone width, entire length of leaf

stripes $\frac{1}{4}$ clone width, half length of leaf

Intercalary meristem contains histogenic layers for upper and lower epidermis and internal tissues. Stops and starts of _iojap_ phenotype could be explained by sorting out of albino(A) and green(G) cells at this stage.

Figure 3.   Description of Normal and Iojap-affected Tissue and
             Plastids

A.   A field-grown Zea mays (ij/ij) leaf expressing the iojap
      phenotype of white sector variegation.

B. - F.   Tissue from greenhouse-grown macroscopically albino
      (ij/ij) sectors of seedlings of K55 background. No green
      tissue is observed in the albino sectors; color ranges from
      white to creamy in appearance.
      B.   a cross section of a leaf at a secondary vascular
            bundle in which a number of cells contain a "normal
            looking" population of plastids=(n.p.) X30.
      C.   Heterogeneous populations of "normal looking" and
            aberrant plastids=(m.p.) within the same cell   X80.
      D.   Electron micrograph of the "normal looking" plastids
            observed in albino mesophyll tissue. Note the presence
            of ribosomes and grana=(g) in this plastid. Bar = 1
            micron.
      E.,F. iojap-affected plastids from an expanded leaf; note
            the absence of ribosomes and poorly developed lamella.
            Bar = 1 micron.

Albino tissue in a homozygous recessive plant (ij/ij) results from
the propagation of aberrant plastids typically produced in the
leaf clone progenitor cells and in the intercalary meristem. The
iojap trait does not always affect all of the plastids of a cell,
however, so there are also plastids with the ability to green. In
the examples above, such plastids are rare. However, if an intercalary meristem cell contains only unaffected plastids, a small
green streak will be produced within an otherwise albino region or
clone on the leaf. The extent of albino tissue, therefore, depends
on both the number and timing of aberrant plastid production and
subsequent sorting out during leaf development of plastids from
cells with a mixed plastid population.

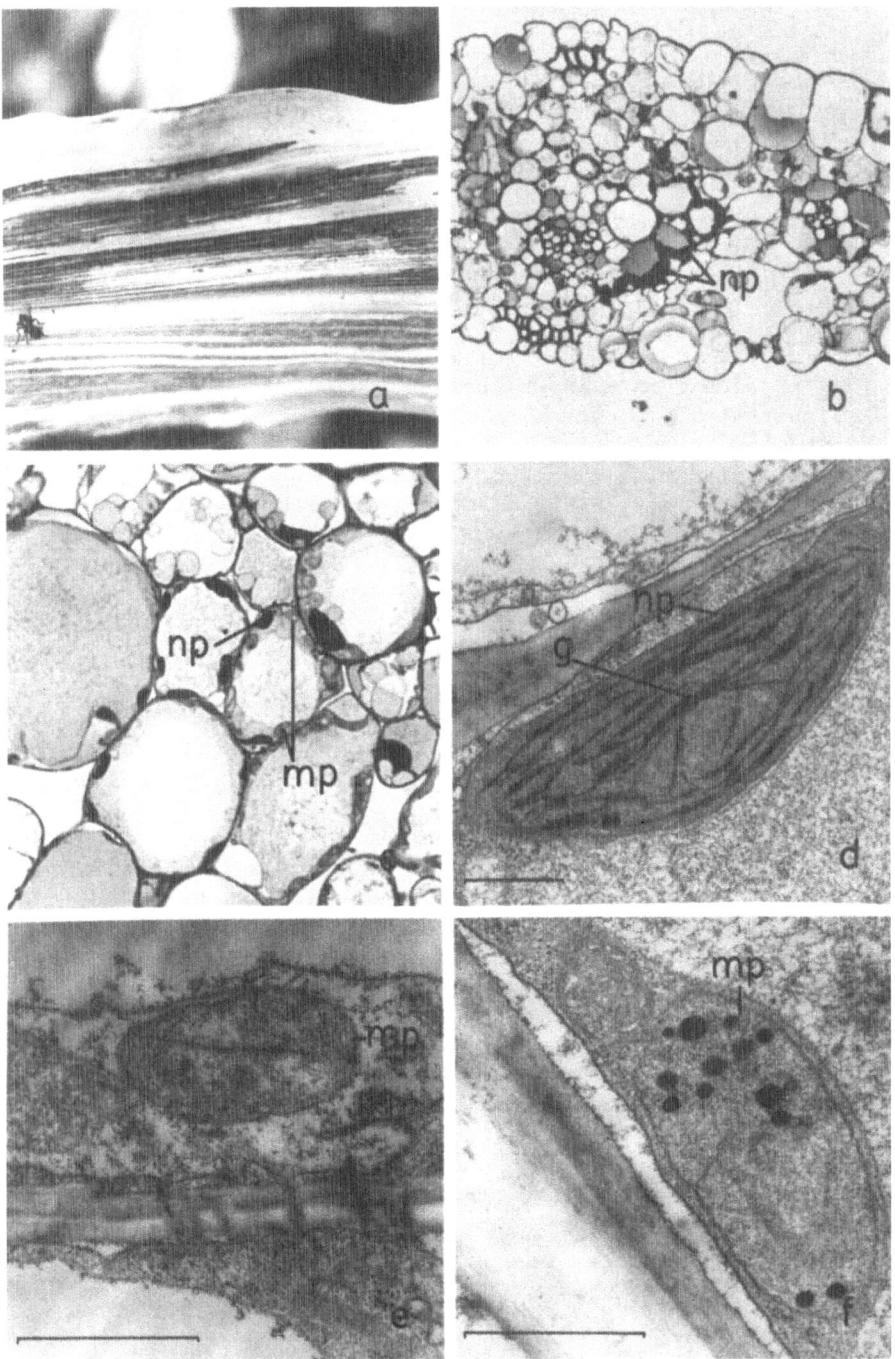

Figure 4.   Description of Iojap-affected Plastids

A.   Iojap-affected plastid from white, mesophyllic tissue of a
      sectored (ij/ij) leaf, K55 background.   M = mitochondrion.
      Bar = 1 micron.
B. - F.   Aberrant plastids from mesophyllic tissue of albino,
      ij/ij plants in Oh51a background.   This set of micrographs
      illustrates a progression in the ultrastructural complexity
      achieved by iojap-affected plastids.   There is no predictable
      progression within a given leaf, however, as the plastids of
      individual cells often reach different stages of development
      independent of the position of the cell within the leaf.   The
      extent of whole plant albinism varies considerably with
      nuclear background; moderate striping is typical of K55 back-
      ground (A. above) whereas striping affecting 95% of leaf
      tissue is common in Oh51a background.   Bar = 1 micron.
      B., C.   Aberrant plastids show progression from a circular
            body to a more ameboid form; note the cytoplasmic inclu-
            sion surrounded by double membranes (arrow).
      D., E.   Mutant iojap-affected plastids increase in size but
            have not been observed to develop beyond this ameboidal
            proplastid phase.   n = nucleus.
      F.   Vesicles (v) are commonly observed in iojap-affected
            plastids of mature or senescing leaf tissue.

The complexity of affected plastid organization may reflect the
length of time some plastid protein synthesis was possible.   To
investigate whether there is a correlation between plastid ribosome
content and the extent of plastid differentiation, we have developed
an RNA-specific stain which can differentiate affected from unaffected proplastids in thin sections.   We plan to use this staining
procedure to follow the dilution of plastid ribosomes, the plastid
phenotype in proplastids with a deficiency in ribosomes, and the
sorting out of unaffected from affected plastids during the construction of the intercalary meristem.

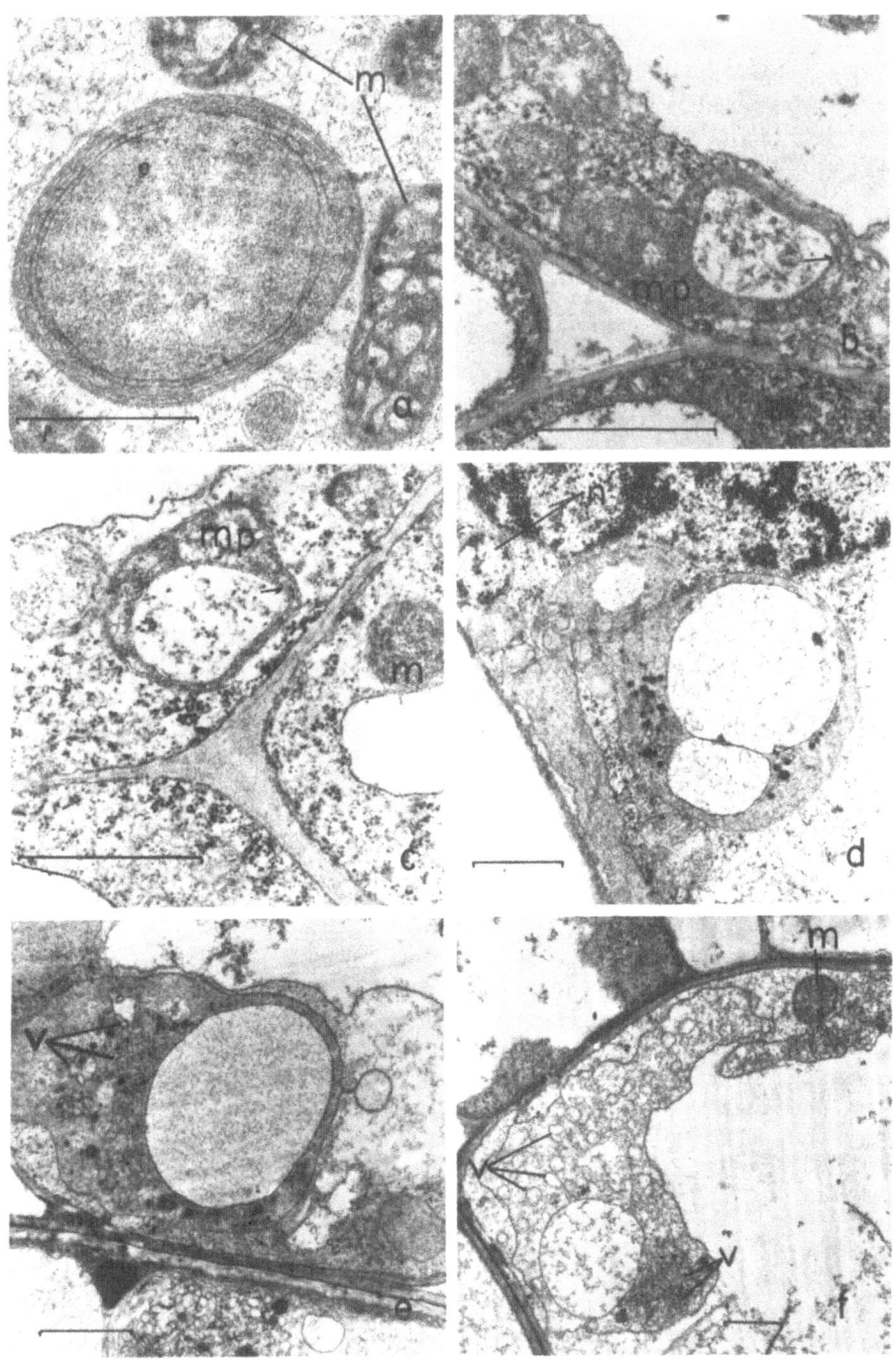

may also be required for aspects of proplastid, amyloplast or other non-green plastid differentiated states. To date, however, plastid ribosome-deficient plants are found to suffer from pleio-tropic photosynthetic deficiencies, and such phenotypes mask any possible effects on non-green plastid functions (18,19).

The clonal pattern of corn development allows an analysis of the timing of iojap activity; that is, to ask when, and possibly how, the plastid ribosome-less state is reached. Our hypothesis is that iojap causes a deficiency in a nuclear-coded product required to maintain plastid ribosomes, perhaps a ribosomal protein. In rapidly proliferating cells, the iojap mutation results in the progressive dilution of plastid ribosomes as the proplastid popula-tion expands. That rapid changes occur in plastid number is shown in Figure 5. The top plate depicts an inactive area of the apical meristem, a single, large subnuclear mitochondrion and a sparse population of small, dark-staining proplastids can be identified. The middle plate shows cells of the active meristem zone; there are multiple, small mitochondria (derived by fragmentation?) and increased numbers of proplastids. The bottom plate illustrates the zone of initials just derived from the meristem, in which enlargement and membrane disposition in the proplastids is clearly evident. In the leaf proper, normal plastid ontogeny proceeds rapidly through development of the prolamellar body, acquisition of an ameboid shape, and subsequent thylakoid proliferation [(5) and Figure 3]. Over a few cell generations, therefore, plastids must not only differentiate but also rapidly proliferate to keep pace with cell division and to expand their number on a per cell basis. The iojap-affected plastids illustrated in Figure 3 have accomplished proliferation and some aspects of differentiation, but they have failed to proceed past the ameboid stage.

Since normal plastids and tissue occur in ij/ij plants, it follows that either the frequency of the iojap-mediated deficiency varies [i.e. an all-or-none effect in certain cells or plastids], or else that the iojap defect affects all plastids, but to a variable extent. As a working hypothesis, we prefer the second alternative, and propose that all plastids of the proliferating stem and leaf meristematic zones are affected to some extent by iojap, but that variable numbers of individual plastids will be able to recover a normal phenotype. This restitution could occur in plastids which, by chance or as a result of a more favorable environment or slower division rate or other factors, retained a sufficient number of plastid ribosomes to manufacture plastid-encoded products required for maintenance of the plastid protein synthetic machinery, including the ribosomes. This hypothesis allows explanation of the macro- and microscopic phenotypes. The presence of iojap establishes a population of proplastids undergoing dilution of ribosome content in intercalary meristem cells; cell lineages of pure albino would be established whenever all plastids

of a cell lost the possibility of restitution; green lineages
would be established whenever restitution occurred, although such
areas would likely contain non-green iojap-affected plastids in a
tissue mosaic. The pattern of the green-white mosaic, its length
and width, would mark points during development when pure white
lineages were established by sorting out or continuous or renewed
iojap-mediated deficiency. It should be possible, therefore, to
utilize the pattern of albinism to deduce the three-dimensional
configuration, number and pattern of use of cells in zones of cell
division.

IV. OBSERVATIONS ON CHLOROPLAST MUTATOR AND JAPONICA-AFFECTED
     PLASTIDS

The phenotype of chloroplast mutator-affected tissue is
summarized in Table I. The salient features of cm are that it
can occur during embryo development prior to or just at the time of
meristem establishment. Thus, a chimeral meristem is formed with
a sector of cells containing cm-affected plastids from which only
albino-yellow tissue will arise. Events mediated by cm also occur
in individual meristem cells, thus marking a clone over several
nodes, as well as later in development (as found with iojap).
Although cm action occurs earlier than ij, the terminal phenotype
of cm-affected plastids is less deficient in enzyme complement and
ultrastructure. Albino cm-mediated plastids are CRM+ for RuBPCase,
suggesting the presence of plastid ribosomes. Despite the ability
of cm-affected plastids to express a characteristic chloroplast
phenotype, the cm-affected plastids occupy a permanent, heritable
non-green compartment. The basis for the lesion in the ability of
these plastids to differentiate is not known. However, as with
iojap, plants affected by cm also contain normal appearing plastids
and tissues; cm therefore has either variable expressivity or
results in a phenotype from which restitution is possible.

Our preliminary observations on j   1ζ2 suggest that these
genes act early in development resulting in clone size or large
sectors rather than fine striping. Most japonica events could be
explained by action in the meristem or leaf and stem progenitors
followed by very rapid sorting out or by a complete change in all
plastids of these cells.

V. SPECULATION ON THE NATURE OF CONTROL OF PLASTID DEVELOPMENT

Aberrant plastid differentiation can be used to probe the
relationship between various controls of plastid development. For
example, green plastids may be devoid of RuBPCase or otherwise
photosynthetically incompetent (6,9), suggesting that the decision
to produce pigments does not depend on the photosynthetic capacity

Figure 5.   Electron Micrographs of Meristematic Maize Tissue.

Bar represents 1 micron.   Magnification of plates X 9,975.

M = mitochondrion   P = proplastid
N = nucleus   V = vacuole

A. Uppermost cells of the apical meristem of an immature tiller.
    Nuclei are prominent as are large, subnuclear mitochondria;
    there may be a single mitochondrion per cell in these quiescent
    cells destined to be tassel tissue. Few proplastids are seen;
    they are small, dark-staining bodies.

B. Meristem cell derivatives destined to give rise to leaf and
    stem progenitors at the active base of the meristem depicted
    in A. Nuclei are prominent in these meristematic cells al-
    though the mitochondria are now of a more typical appearance as
    small, rounded organelles 0.3 - 1.0 microns in diameter. The
    proplastids are much more numerous and are approximately the
    same size as the mitochondria.

C. Cells of the oldest portion of the most immature leaf of the
    shoot apex of A. These cells are perhaps one day "older" than
    the cells pictured in B. Nuclei are still prominent and cell
    vacuolation is minimal. The mitochondria are the same as in B.
    but the proplastids have enlarged and accumulated considerable
    membranous material (dark streak inside the organelles) as well
    as some paracrystalline material. These proplastids will not
    green for one to two weeks, dependent on the emergence of the
    leaf.

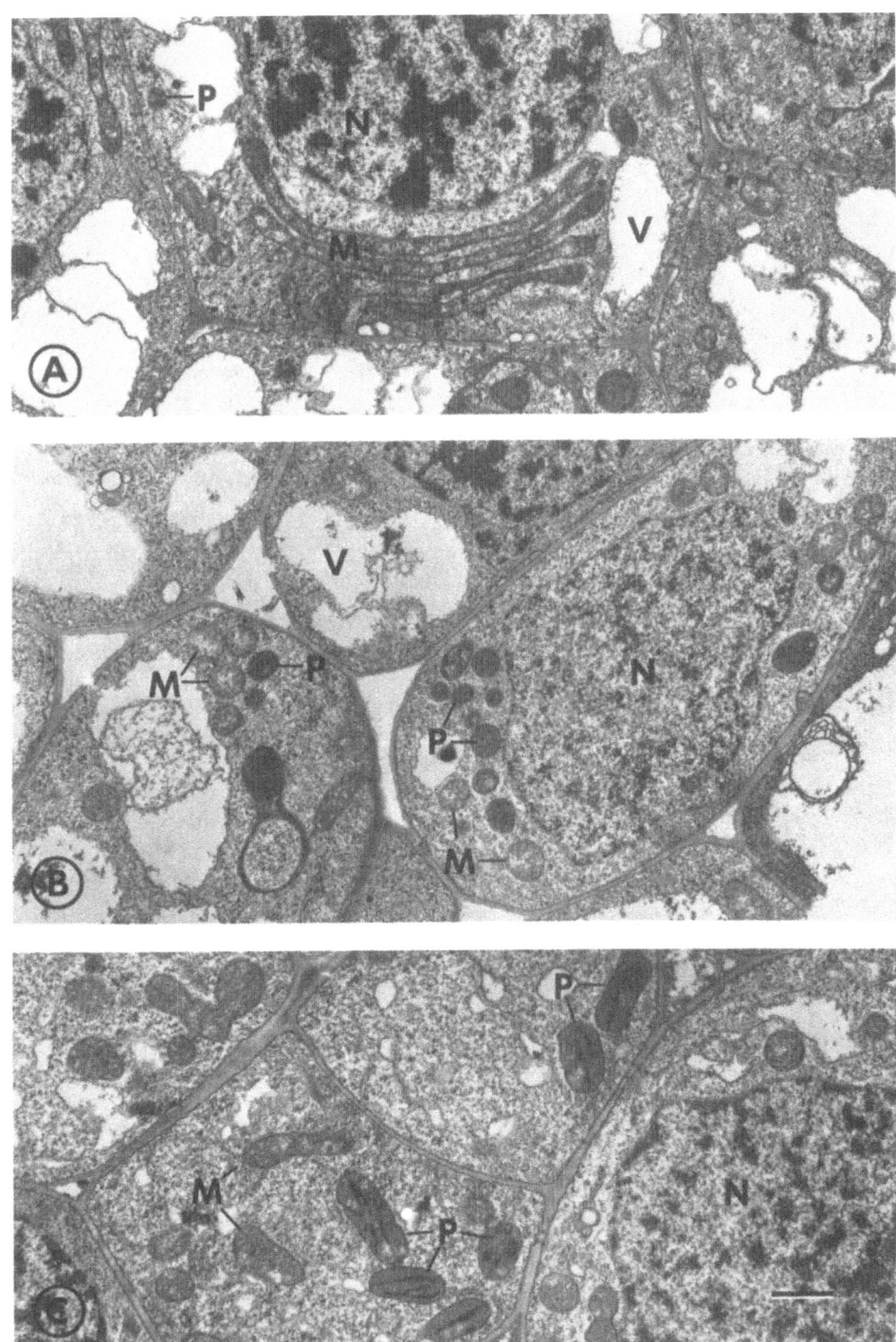

Table I.   Phenotypes of Plastids Affected by <u>Iojap</u>,
           <u>Chloroplast</u> <u>Mutator</u> and <u>Japonica 1$\!$2</u>

PLASTID TYPE

	<u>iojap</u>-affected	<u>cm</u>-affected	<u>japonica</u>-affected
PHENOTYPE			
Color	albino-cream	albino-yellow	albino-yellow
Stripe Width			
largest observed	clonal width	1/2 plant	1/8 plant
typical	clonal-subclonal	clonal	clonal
smallest	streaks, individual plastids	?	?
Time of Earliest Action	in leaf or stem progenitors; many late events	in meristem and many later events	in leaf and stem progenitors or in meristem; few late events
RuBPCase[+]	not detectable	present	?
Plastid DNA[*]	normal	present	?
Plastid ribosomes[*]	not detectable	present	?
Ability to Incorporate amino acids into TCA precipitable material[*]	not detectable	present	?

[+]from ref. 9; the presence of RuBPCase was tested utilizing antibody to tobacco holoenzyme in a double diffusion assay in which
5% of normal RuBPCase would have been detected and by assay
with anti-small subunit antibody

[*]from ref. 14 and unpublished results

of the plastid. Similarly, pigment-deficient mutants such as cm
may contain RuBPCase (**9**). These results suggest that pigment
production and photosynthetic enzyme accumulation are independently
regulated.

Some speculation is also possible concerning the coordinate
expression of nuclear and plastid gene products. Although the
state of differentiation of the cell, and hence nuclear gene
activity, appears to coordinate the differentiation of the plastid,
both new plastids as well as nuclear gene products are associated
with the transition from etioplast to green plastid (19). There
is some evidence to suggest that nuclear activities, such as the
synthesis of the small subunit of RuBPCase, are required to activate
plastid genes such as the large subunit of RuBPCase (20). However,
plastid regulation of nuclear gene activity cannot be a priori
excluded. Iojap-affected plastids accumulate only a limited
number of proteins: there is no evidence of RuBPCase large or
small subunit nor of some proteins found in etioplasts (V. Walbot,
unpublished data). The absence of protein synthetic activity in
the plastid may preclude the acquisition of some nuclear-coded
gene products by failing to provide a protein uptake or processing
machinery and/or by failing to produce products which regulate the
expression of nuclear genes. Loss of positive control of nuclear
transcription from the plastid might be invoked to explain the
extremely deficient nature of ribosome-deficient plastids. A less
extreme albinism is found in plants affected by cm or certain
other pigment-deficient mutations; the plastids in these plants
are found to contain RuBPCase and, by inference, plastid ribosomes,
and they generally appear to possess more nuclear gene products
(Walbot, unpublished data).

If nuclear gene expression is dependent on regulatory signals
from the plastids, nuclear mutations affecting response to such
signals would be expected. Such mutants would fail to differentiate
normally because of an inability of particular nuclear genes or
developmental programs to be expressed appropriately. Plastids
which fail to differentiate normally in maize often persist in a
pale or other abnormal state in adult leaves (e.g. zebra stripe
and other conditions depicted in ref. 7,8), suggesting that the
plant does not correct misdifferentiation post facto even when
environmental conditions become more favorable. Thus, there may
exist a restricted time period during which aspects of differentiation must occur. Mutations affecting the time of nuclear gene
expression could result in aberrant plastids as well, independent
of the ability of the plastid to produce appropriate regulatory
signals.

The nuclear genes we have studied which affect the ability of
plastids to differentiate act long before greening occurs. Chloroplast mutator mediated events can occur during the establishment

of the apical meristem, and both _japonica_ and _iojap_ events typically occur in the clonal progenitors of leaf and stem tissue. Cloroplast mutator is the least deficient class in terms of similarity to normal plastids (Table I and unpublished data) although the time of action is earliest; thus, the extent of tissue affected by plastid misdifferentiation is not an indicator of the severity of the deficiency phenotype.

For the future, it is hoped that examination of the time of action of genes affecting plastid differentiation combined with knowledge of the plastid and nuclear coded protein products present in affected plastids will allow us to order the major developmental steps in plastid differentiation and to define potential plastid and nuclear regulatory signals.

REFERENCES

1.   Kung, S.-D., 1977, Expression of chloroplast genomes in higher plants, Annu. Rev. Plant Physiol., 28: 401.
2.   Goodwin, T. W., 1971, Biosynthesis by chloroplasts, in: "Structure and Function of Chloroplast," M. Gibbs, ed., Springer-Verlag, New York.
3.   Ledbetter, M. C. and Porter, K. R., 1970, "Introduction to the Fine Structure of Plant Cells," Springer-Verlag, New York.
4.   Anstis, P. J. P. and Northcote, D. H., 1973, Development of chloroplasts from amyloplasts in potato tuber discs, New Phytol., 72: 449.
5.   Whatley, J., 1978, A suggested cycle of plastid developmental interrelationships, New Phytol., 80: 489.
6.   Miles, D., 1975, Genetic analysis of photosynthesis, Stadler Symp., 7: 135.
7.   Neuffer, M. G., Jones, L. and Zuber, M., 1968, "The Mutants of Maize," Crop Science Society of America, Madison.
8.   Coe, E.H. Jr. and Neuffer, M.G., 1977, The genetics of corn, in: "Corn and Corn Improvement," G.F. Sprague, ed., Amer. Soc. Agronomy, Inc., Madison, Wisconsin.
9.   Simpson, E., 1978, Ph.D. thesis, Washington University, St. Louis.
10. Rhoades, M. M., 1946, Plastid mutations, Cold Spring Harb. Symp. Quant. Biol., 11: 202.
11. Randolph, L. F., 1936, Developmental morphology of the caryopsis in maize, J. Agric. Res., 53: 881.
12. Coe, E. H. and Neuffer, M. G., 1978, Embryo cells and their destinies in the corn plant, in: "The Clonal Basis of Development," S. Subtelny and I. M. Sussex, eds., Academic Press, New York.
13. Sharman, B. C., 1942, Developmental anatomy of the shoot of Zea mays L., Ann. Bot., 6: 245.

14. Steffensen, D. M., 1968, A reconstruction of cell development in the shoot apex of maize, Amer. J. Botany, 55: 534.

15. Walbot, V. and Coe, E. H. Jr., 1979, Nuclear gene iojap conditions a programmed change to ribosome-less plastids, Proc. Natl. Acad. Sci. U.S.A., 76:2760.

16. Hagemann, R. and Börner, T., 1978, Plastid ribosome-deficient mutants of higher plants as a tool in studying chloroplast biogenesis, in: "Chloroplast Development," G. Akoyunoglou et al., eds., Elsevier/North Holland Biomedical Press, Amsterdam.

17. Goodenough, U. W. and Levine, R. P., 1970, Chloroplast structure and functions in ac-20, a mutant strain of Chlamydomonas reinhardii. III. Chloroplast ribosomes and membrane organization. J. Cell Biol. 44: 547.

18. Ellis, R. J., 1977, Protein synthesis by isolated chloroplasts, Biochim. Biophys. Acta, 463: 185.

19. Grebanier, A., Steinback, K. E. and Bogorad, L., 1979, Comparison of the molecular weights of proteins synthesized by isolated chloroplasts with those which appear during greening in Zea mays, Plant Physiol., 63: 436.

20. Feierabend, J. and Wildner, E., 1978, Formation of the small subunit in the absence of the large subunit of ribulose 1,5-bisphosphate carboxylase in 70S ribosome-deficient rye leaves, Arch. Biochem. Biophys., 186: 283.

This research was supported by a grant from the National Science Foundation to V.W. and by USDA-SEA support to E.H.C. This paper is journal series no. 8380 of the Missouri Agricultural Experiment Station.

PHYSICO-CHEMICAL AND RESTRICTION ENDONUCLEASE ANALYSIS OF

MITOCHONDRIAL DNA FROM HIGHER PLANTS

Francis QUETIER and Fernand VEDEL

Laboratoire de Biologie Moléculaire Végétale
associé au CNRS (LA 40) Université Paris-Sud
91405 - ORSAY (FRANCE)

The presence of DNA fibrils in plant mitochondria has been reported at the beginning of the 60s; the isolation of mt DNA molecules from higher plants has been achieved some years later and
most of the physico-chemical characteristics are now well established (Wells and Birnstiel 1969, Kolodner and Tewari 1972, Vedel and
Quétier 1974). The mt-DNA represent nearly 1% of the total cell DNA
and must be isolated from DNase-treated intact organelles. The mitochondria of all higher plants so far studied have been shown to
contain DNA quite distinguishable from both nuclear and chloroplastic DNAs. The G+C content of this mt-DNA, as determined by CsCl
buoyant density, is remarkably constant at 1.706 g.ml^{-1} whatever
the plant genus and the level of evolution. This constancy is rather puzzling when compared to the buoyant densities of animal mt DNAs
which range from 1.686 g.ml^{-1} to 1.711 g.ml^{-1}. The study of both
thermal denaturation profiles and analytical CsCl banding patterns
achieved at different molecular weights do not reveal any heterogeneity (Vedel and Quétier 1974) while interactions with synthetic
homopolyribonucleotides suggest that d-A and d-C rich clusters are
present. The only divergent parameters to be reported till 1974
were the length of DNA molecules examined under electron microscope
and the Cot value which respectively ranged from 10 to 30 μm and
from 74.10^6 d to 140.10^6 d. The dispersion within each of these parameters and the discrepancy between them remained unclear till recently, although they indicated that the mt DNA from higher plants
is about 6 times larger than the animal mt DNA.
    The use of restriction endonucleases brought   new informa-
tions on these mt-DNAs from 1976 on (Vedel, Quétier and Bayen 1976;
Levings and Pring 1976 ; Quétier and Vedel 1977). These informations
deal with i) the organisation of the mitochondrial genome of hi-

gher plants and ii) the possibility to use directly the mt DNA
restriction patterns as genetic markers of the cytoplasms.

ANALYSIS BY RESTRICTION ENDONUCLEASES

Whereas the hydrolysis of chloroplastic (cp) DNA by restriction endonucleases has led to strongly confirm the characteristics
obtained by other approaches, the use of these enzymes on mt DNA of
higher plants resulted in very surprizing results. The cp DNA gives
about 30 bands when digested by EcoR I, only a twenty by Bam I and
less than 10 by Sal I ; in each case, the sum of the molecular
weigths of the restriction fragments in within 98% of the native cp
DNA molecular weight ($\simeq$ 90.10^6 d) when band multiplicities are taken into account. The distribution of cp-DNA molecules in a given
higher plant is thus "uniclonal".

To the contrary, the mt DNA displays about 60 bands when hy-
drolysed by EcoR I, a forty by Bam I and more than 20 by Sal I ; in
these cases, the sum of the fragment sizes remains quite larger
than the native molecular weight, even when band multiplicities are
not taken into account. These results have been obtained for a variety of plants (potato, cucumber, virginia creeper, wheat.Vedel,
Quétier and Bayen 1976, Quétier and Vedel 1977) and have been extended to maize and soybean by the group of Levings and Pring (Levings and Pring 1976, Sisson, Pring and Levings 1978). This heterogeneity of the population of mt DNA molecules could be explained *a*
priori by the presence of truly different molecules or by the occurence of some methylated bases along some of the molecules. The
latter hypothesis is poorly documented : whereas it is generally
agreed that cp DNA does not contain unusual bases, no data is available on mt DNA. Attemps to label mt DNA of liquid cultured cells
of virginia creeper by ^{14}C methylmethionine were unsuccessfull but
permeation barriers could be involved (Vedel and Quétier, unpublished results). Efforts have thus been focused on a carefull reexamination of the length distribution of mt DNA molecules by electron
microscopy.

SIZE AND SHAPE OF mt DNA MOLECULES

Striking differences have been reported in the literature.
The first reports were by Wolstenholme and Gross (1968) and Mikulska
et al (1970) who observed only linear molecules of 10 to 20 µm.
Kolodner and Tewari (1972), using very mild lysis conditions, isolated 55% of pea, spinach and lettuce mt DNAs as 30 µm circular molecules, 1% as mini-circles of 3 to 5 µm and the remaining part as
linear molecules (2 to 27 µm). The mt DNA molecules extracted from
potato tubers (Vedel and Quétier 1974) were linear and up to 28 µm
long.

The reexaminations achieved by the group of Levings and Pring
with experimental procedures identical or very close to that of

Kolodner and Tewari revealed 7 discrete classes of circular molecules in soybean (5.9 - 10 - 12.9 - 16.6 - 20.4 - 24.5 - 29.9 µm
Synenki, Levings and Shah 1978) and at least 3 discrete classes in
maize (16 - 22 - 30 µm - Shah *et al* 1976, 1978). In these two samples the presence of minicircles (<5 µm) was noticed. Recent observations in our laboratory (Sévignac *et al*, unpublished results) on
liquid cultured cells of virginia creeper and on etiolated or green
wheat seedlings showed also an heterogeneous length distribution.
The mt DNA population ranges from 5 to 30 µm for both linear and
circular molecules and a small amount of minicircles is also present. However, the occurence of discrete classes is less obvious in
these samples.

All these experiments have been carried out in such conditions
that a contamination by allien DNA (other plant organelles, bacteria, fungi, mycoplasma) is either very unlikely or completely ruled
out. The main conclusion is that the mitochondrial genome of higher
plants may be composed of several molecules, a situation analogous
to the nuclear information which is encoded on multiple chromosomes.

CIRCULARITY AND REPLICATION OF mt DNA

It should be emphasized that the percentage of circles recovered after osmotic shocks or CsCl-dye purification is highly variable ; whereas pea, spinach and lettuce leaves were reported to contain up to 55% of circular molecules (Kolodner and Tewari 1972)
other materials show smaller contents in spite of the extremely careful procedures used ; (these same procedures give a high yield
of supercoiled cpDNA in control experiments). The general occurence
of only circular molecules *in vivo* thus cannot be ascertained yet.
Moreover, the possibility cannot be ruled out that the mt DNA of
higher plants occurs predominantly as linear molecules and that cir-
cularisation takes place only in specific conditions, i.e for replication, by means of covalent linkage or cohesive ends. A support
to this hypothesis is brought by the unique DNA molecule associated
with the cytoplasmic male sterility in maize ; this DNA species is
linear but exhibits terminal inverted repeats (Pring, Levings, Hu
and Timothy 1977).

As to the replication of mt DNA in higher plants, preliminary
observation in our laboratory showed both Cairns and rolling circles configuration, for wheat and virginia creeper (Sévignac *et al*,
unpublished results) ; the replication of mt DNA appears similar to
that of cp DNA (Kolodner and Tewari 1975).

THE POPULATION OF MITOCHONDRIA IN A GIVEN PLANT

Several physiological reports support the idea that the population of mitochondria is heterogeneous on functional basis. The
possibility that several classes of mitochondria could contain each
a distinct genome has been studied. The DNase-treated mt pellet ex-

tracted from etiolated wheat seedlings has been fractionated by sucrose density gradient (Quétier and Vedel, unpublished results) and
by ludox silica-sol gradients (Lejeune and Jubier unpublished results)
The mt population was fractionated into 5 bands by sucrose gradients
which were individually recovered. The mt DNA isolated from each
band have been analysed by Sal I : the restriction patterns are rigorously identical each other and identical to that of the unfractionated mitochondrial pellet. Silica-sol gradients led to more
than 5 bands but the results were exactly identical.

STABILITY OF THE mt GENOME

    We have checked that the mt DNA restriction pattern is identi-
cal for different organs of a given plant (hypocotyles and cotyledons of cucumber, Quétier and Vedel 1977). On the other hand, potato tubers, calluses initiated with these tubers and liquid cultured
cells developed from such calluses display identical restriction
patterns. Moreover, the comparison of mt DNA of virginia creeper
after 2 years of liquid culture did not show any difference. It can
be therefore concluded that the mt genome is rather stable and that
in vitro culture do not induce detectable drifts.

INFORMATIONAL CONTENT OF mt DNA

    This part is poorly documented, although extremely attractive
since the mt DNA of higher plant appears up to 6 times larger than
the animal mt DNA. The only data lies in the relation between cytoplasmic male sterility in maize and the occurence of a specific
band in restriction patterns (Pring *et al* 1977). Very recent
hybridization experiments in our lab have shown that mt DNA of
wheat codes the heavy and the light ribosomal mitochondrial RNAs
(Lejeune, Jubier and Quétier, unpublished results).

mt DNA RESTRICTION PATTERNS CAN BE DIRECTLY USED AS GENETIC MARKERS
OF THE CYTOPLASM

    The mt DNAs isolated from different plants can be specifically
identified on the basis of their restriction patterns ; this is
true for plants belonging to different genera ; wheat, potato, virginia creeper, cucumber (Quétier and Vedel 1977) and for different
species within a genus (Vedel *et al*, 78, Levings and Pring 77).
This specific identification has appeared as an efficient tool in
several lines of research :

    - the study of the wheat phylogeny has been reexamined accor-
      ding to cp and mt restriction patterns and some crosses have
      been cleared. It should be emphasized that mt DNA is quite
      more sensitive than the cp DNA, the latter appearing more
      conservative during evolution (Vedel, Quétier, Dosba and
      Doussinault 1978).

- the study of the cytoplasmic male sterility revealed that the mt DNA is involved in this phenomenon (see Levings and Pring 1978 for maize and Vedel and Quétier 1978 for wheat).

- the segregation of mt DNAs in plants regenerated from fused protoplasts (cybrides)

CONCLUSIONS ON THE STRUCTURE AND ORGANISATION OF THE MT GENOME OF HIGHER PLANTS

It is clear that the mt DNA is heterogeneous but the organisational features of the mt genome remain unknown. Several facts appears somewhat contradictory mainly the diversity of both length distribution and restriction patterns compared to the highly conserved buoyant density of mt DNA and therefore the G+C content. This last point is likely to be related to specific functions of mitochondria in higher plants.

REFERENCES

Kolodner, R. and Tewari, K.K., 1972, Proc. Natl. Acad. Sci. US, 69 : 1830.
Kolodner, R. and Tewari, K.K., 1975, Nature, 256 : 708.
Levings, C.S. and Pring, D.R., 1976, Science, 193 : 158.
Levings, C.S. and Pring, D.R., 1977, J. Heredity, 68 : 350.
Levings, C.S. and Pring, D.R., 1978, Stadler Symposium, vol. 10, University of Missouri, Columbia, USA
Mikulska, E., Odintsova, M.S. and Turisheva, M.S., 1970, J. Ultrastruct. Res. 32 : 258.
Pring, D.R., Levings, C.S., Hu W.W.L., and Timothy W.W.L., 1977, Proc. Natl. Acad. Sci. US, 74 : 2904.
Shah, D.M., Levings, C.S., Hu, W.W.L. and Timothy, D.H., 1976, Maize Genet. Coop. Newsletter, 50 : 94
Shah, D.M., Levings, C.S., Hu, W.W.L. and Timothy, D.H., 1978, Stadler Symposium, vol. 10, University of Missouri, Columbia USA.
Sisson, V.A., Brim, C.A. and Levings, C.S., 1978, Stadler Symposium, vol. 10, University of Missouri, Columbia USA.
Synenki, R.M., Levings C.S., and Shah, D.M., 1978, Plant Physiol., 61 : 460.
Quétier , F. and Vedel, F., 1973, Biophys. Biochem. Res. Comm., 54: 1326.
Quétier, F. and Vedel, F., 1977, Nature, 268 : 365.
Vedel, F. and Quétier, F., 1974, Biochim. Biophys. Acta, 340 : 374.
Vedel, F., Quétier, F. and Bayen, M., 1976, In Colloquium "Acides nucleiques et Synthèses des protéines chez les végétaux, Strasbourg 1976, CNRS ed.

Vedel, F., Quétier, F., Dosba, F., and Doussinault, G., 1978,
          Plant Science Lett., 13 : 97.
Vedel, F., and Quétier, F., 1978, Physiol. Veget., 16 : 411.
Wells, R. and Birnstiel, M., 1969, Biochem. J., 112 : 777.
Wolstenholme, D.R., and Gross, N.J., 1968, Proc. Nat. Acad. Sci. US.,
          61 : 245.

MITOCHONDRIAL GENOME EXPRESSION IN HIGHER PLANTS

C. J. Leaver and B. G. Forde

Department of Botany
University of Edinburgh
Edinburgh, EH9 3JH, Scotland

INTRODUCTION

Since the discovery that mitochondria contain DNA a considerable amount of effort has been directed towards the analysis of mitochondrial DNA (mt DNA) and its gene products. The majority of the work has been carried out with the yeast, <u>Saccharomyces cerevisiae</u>, although there have been important contributions from those working with <u>Neurospora</u> and animal cells (see Borst and Grivell, 1978, and articles in Bandlow et al, 1977). The successful application of mutant and recombinant analysis, coupled with physical techniques including the use of restriction enzymes, has allowed the identification and mapping of many of the genes located on yeast mtDNA (Borst and Grivell, 1978; Schweyen et al, 1978, Butow and Strausberg, 1979).

Since the initiation of the work on mitochondrial biogenesis it has become obvious that the formation and maintenance of the functional organelle during cell growth and differentiation is dependent upon both nuclear and mitochondrial genetic information. This genetic information is expressed through separate protein synthesising systems in the cytoplasm and mitochondrion, and although the role of the mitochondrial system is small, it is essential. Thus, the nucleo-cytoplasmic system codes for and makes, ca. 90% of the proteins and all the lipids and carbohydrates of the mitochondria. The relatively small mitochondrial genome codes for most, if not all, of the mitochondrial RNAs and between 5-10% of mitochondrial protein. The mitochondrial translation system, like that of the chloroplast, has several functional similarities with the procaryotic system, although certain significant differences exist (Schatz and Mason, 1974). Its role appears to be limited to the

synthesis of 10-12 rather hydrophobic polypeptides. Most of the
mitochondrially synthesised polypeptides which have been identified
in yeast and Neurospora are components of three enzyme complexes
which are vital for the assembly of the functional inner mitochondrial membrane. These include the three largest of the seven subunits of cytochrome c oxidase, the apoprotein of the seven polypeptides of cytochrome bc_1 complex and two to four (depending on the
organism) of the nine subunits of the oligomycin-sensitive ATPase
complex. One additional mitochondrial translation product has been
identified in yeast and Neurospora, as the polymorphic var-1 protein
associated with the small subunit of mitochondrial ribosomes. (See
Fig. 1).

The remainder of the mitochondrial proteins, namely those of
the outer membrane, the matrix space and a large proportion of the
inner membrane, are encoded in nuclear DNA, synthesised on 80S cytoplasmic ribosomes and subsequently transferred into the mitochondrial
structure. Thus, the respiratory enzyme complexes described above,
which are responsible for key steps in the process of oxidative
phosphorylation and the generation of ATP, are built up of a combination of mitochondrially and cytoplasmically synthesised polypeptides.
The structural and functional integration of each group of subunits

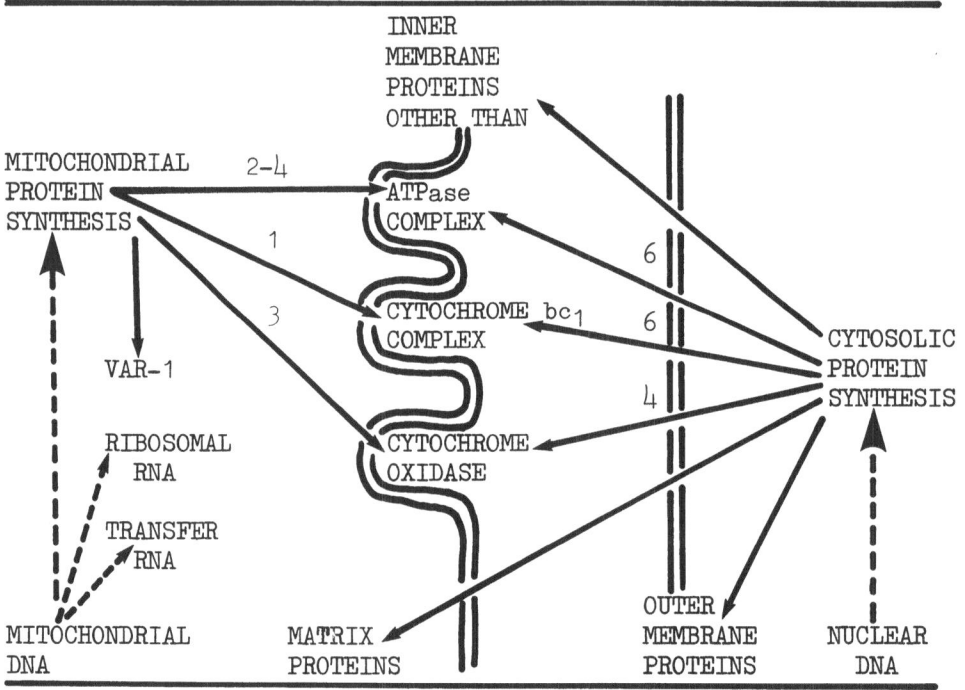

Fig. 1.   The contributions of nuclear and mitochondrial genetic
           systems to mitochondrial biogenesis in yeast and Neurospora.

into the membrane requires the coordinate synthesis of the other (Schatz and Mason, 1974).

The relatively small size of yeast mtDNA (a covalently closed circular duplex with contour length 25μm, molecular weight 50×10^6 and about 75,000 base pairs), coupled with its amenability to genetic and physical analysis, has made it an ideal system for studying genome organisation, expression, recombination and evolution. The detailed physical and genetic maps which have been obtained are in good agreement and more detailed information can be found in recent reviews (Borst and Grivell, 1978; Bandlow et al, 1977).

A few important findings concerning the organisation of the yeast mtDNA are perhaps worth a mention. In wild type yeast approximately 50% of the mtDNA contains stretches of nearly pure dAT ('spacer DNA') and also shorter clusters of dGC (Bernardi, 1979). Restriction enzyme analysis followed by RNA-DNA hybridisation has led to location and mapping of the genes for the mitochondrial rRNA species and about 20 mitochondria-specific tRNA's.

Of particular interest has been the finding that the gene for the large rRNA and some other protein-coding genes, seem to be 'split' with DNA sequences in the structural gene which are not expressed in the final gene product (Bernardi, 1978). Until now the presence of such 'intervening sequences' have only been shown in eucaryotic and viral genomes and their discovery in mtDNA (which was previously thought to be of a procaryotic nature) makes it a simple model system for the study of the function and evolution of intervening sequences and of the splicing mechanism used to remove these sequences from precursor RNAs.

A more fundamental question, which is being asked by an increasing number of scientists interested in both mitochondrial and chloroplast biogenesis, concerns the interaction and coordination between the two genetic systems in organelle biogenesis, (see Ellis et al and Schmidt and Chua in this volume for example). More specifically: 1) how are cytoplasmically synthesised mitochondrial proteins transported into the mitochondrion and functionally integrated to form a respiratorily competent organelle? 2) what is the mechanism which ensures the specific incorporation of a product of cytoplasmic protein synthesis into its destined organellar location (Chua and Schmidt, 1979)? 3) how is mitochondrial genome expression regulated by genetic, physiological and environmental factors during mitochondrial biogenesis and differentiation. Some progress has been made in answering the first of these questions (see Chua and Schmidt, 1979, Schatz, 1979 and chapters by Schmidt et al and Ellis et al in this volume), while the others still provide exciting problems for the future. The present consensus of opinion has been formulated as two very similar hypothesis by Schatz (1979) and Chua and Schmidt (1979). They propose that most cytoplasmically synthesised mitochondrial proteins are made on free 80S ribosomes as a precursor, with 20-60 'extra' amino acids (so called 'transit peptides'), presumably at the N-terminus. The precursor is then released into the cytosol and migrates to a specific receptor on a region of the mitochondrial surface in which the outer and inner membranes are in close contact or perhaps even fused. The transit peptide(s) are presumed to contain specific information to ensure not only specific transport into the mitochondrion, but also into a specific site within the organelle.

The interaction of the precursor with the receptor opens up a pore across both membranes through which the polypeptide can diffuse. Undirectional transport could then be driven by irreversible proteolytic processing of the precursor in the matrix ('vectorial processing') and/or by the additional driving force created by the assembly of the mature subunits into an oligomeric unit (such as the F_1-ATPase complex). It has been shown that this 'vectorial processing' is dependent upon energy, although whether ATP is necessary for transmembrane movement, for processing, or for both, is still under investigation (Schatz, 1979 and N. Nelson and G. Shatz personal communication).

While most of the cytoplasmically synthesised mitochondrial polypeptides studied so far are initially synthesised as larger precursors, there are exceptions e.g. the mitochondrial adenine nucleotide transporter. (N. Nelson and G. Shatz, personal communication). This could be explained by suggesting that this polypeptide is simply inserted into the membrane as opposed to the case in which a polypeptide destined for an inner membrane complex must, in fact, first of all be imported into the matrix space in order to be properly assembled with its 'partner proteins'.

PLANT MITOCHONDRIAL DNA

Plant mitochondrial DNA has been isolated as a circular molecule with a maximum contour length of 30μm and a kinetic complexity variously estimated as between $70-140 \times 10^6$ daltons (see accompanying chapter by Quetier and Vedel). It is thus at least seven times the size of animal mitochondrial genome (10×10^6 daltons) and perhaps twice as large as the yeast mitochondrial genome (50×10^6 daltons). There is no evidence that the yeast mitochondrial genome, which is about five times larger than the animal mitochondrial genome, contains any additional information content i.e. all known gene products encoded in yeast mtDNA are also encoded in animal mtDNA (Schatz and Mason, 1974; Bernardi, 1979). The difference in size of the mtDNAs raises the question as to why the plant mitochondrial genome is apparently so large and whether it contains additional genetic information not found in other mtDNAs.

With this question in mind our earlier work was directed towards the characterisation of plant mitochondrial ribosomes and their component rRNAs. In fact, the sedimentation value obtained for plant mitochondrial ribosomes differs from other mitochondrial ribosomes and is distinct from both cytoplasmic and chloroplast ribosomes (see Leaver and Pope, 1976 for review). The component mitochondrial rRNA species which sediment at 24-26S, 18-18.5S and 5S corresponding to molecular weights estimated as 1.12-1.26 x 10^6, 0.69-0.78 x 10^6 and 0.39 x 10^5 (see also Leaver and Pope, 1976) are thus larger than the homologous mitochondrial rRNAs from other species. Of particular interest was our identification of a unique mitochondrial-5S rRNA which has not been detected in the mitochondrial ribosomes of yeast or animal cells. (Leaver and Harmey, 1976; Cunningham et al, 1976).

Restriction endonuclease digestion of purified plant mtDNA yields a surprisingly large number of fragments of differing molar ratios (see chapter by Quetier and Vedel). The pattern obtained is much more complex than predicted from known physical size and is unlikely to be due to partial methylation (Bonen and Gray, pers. commun.).

Bonen and Gray (pers. commun.) have hybridised individual mitochondrial RNA species to restriction endonuclease treated mtDNA and shown that the 26S and 18S rRNAs hybridise to different DNA fragments. This finding was confirmed with each of eight different restriction endonucleases and indicates that the 26S and 18S rRNA genes are not closely linked. The 18S and 5S rRNAs both hybridise to the same mtDNA bands suggesting that their genes are physically close together. In contrast the tRNA genes appear to be scattered around the genome since bulk wheat mitochondrial tRNAs hybridise with many DNA fragments. The work of Bonen and Gray is the first direct evidence for the presence of specific genes on plant mtDNA.

PRODUCTS OF PLANT MITOCHONDRIAL PROTEIN SYNTHESIS

In recent years we have used an indirect approach to answering the question as to whether or not plant mitochondria can synthesise (and by extrapolation, contain the genetic information for) additional polypeptides to those synthesised by mitochondria from other organisms. Mitochondria have been isolated from a number of plant tissues and optimal conditions established for radioactive amino acid incorporation into protein. (Leaver and Pope, 1976; Forde et al, 1978). Incorporation is dependent upon the mitochondria being intact, possessing coupled oxidative phosphorylation (or an added energy-generating system), and displays characteristic sensitivity towards inhibitors of respiration and organellar protein synthesis (See Fig. 2 and Leaver and Pope, 1976).

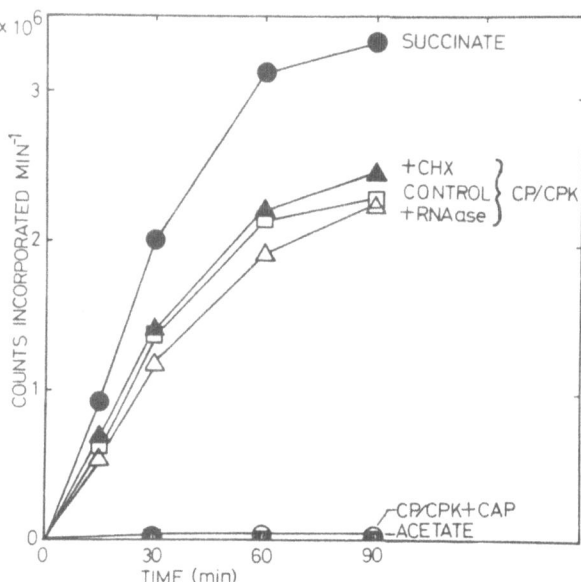

Fig. 2. Time course of incorporation of $[^{35}S]$ -methionine into
         protein by isolated maize mitochondria.

Mitochondria were isolated from etiolated shoots and incubated at
$25°C$ in a medium containing $[^{35}S]$ -methionine (Forde et al, 1978).
The following additions were made to the basic incorporation medium:
10mM Na succinate and 6mM ADP (-●-), 20mM Na acetate (-■-), or an
energy-generating system consisting of 8mM creatine phosphate, 25µg
of creatine phosphokinase and 6mM ATP (-□-). Included in some incubations with the energy generating system were: ribonuclease A at
40 µg/ml (-△-), cycloheximide at 20 ug/ml (-▲-) or chloramphorical at
200 µg/ml (-○-).

     We have analysed the radioactively labelled products of mito-
chondrial translation by SDS-polyacrylamide gel electrophoresis
followed by autoradiography, and compared them with those polypeptides synthesised in yeast mitochondria (Fig. 3). Plant mitochondria
appear to synthesise at least 18 polypeptides compared to the 8-10
synthesised by yeast mitochondria (Schatz and Mason, 1974; although
see Douglas and Butow, 1976). The translation products of plant
mitochondria have molecular weights ranging from $8-54 \times 10^3$ and with
only one major exception almost all are membrane-bound (B.G. Forde,
J. Forde and C.J. Leaver, unpublished observation). The one exception, a polypeptide of molecular weight 54×10^3, is easily solubilised
from broken mitochondria by mild treatment. By immunoprecipitation
with monospecific antibodies prepared against yeast cytochrome
oxidase subunits two of the products have been tentatively identified
as subunits I and II of cytochrome c oxidase (Fig. 3).

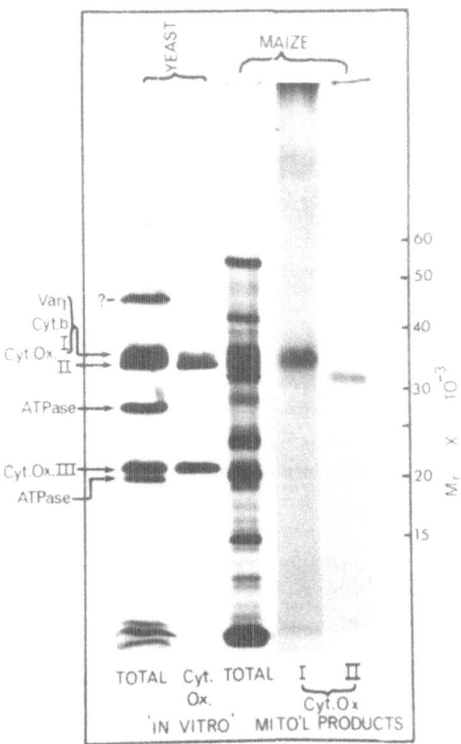

Fig. 3. Comparison of polypeptides synthesised by mitochondria from yeast and maize.

Maize mitochondrial translation products were labelled by incubating mitochondria from etiolated shoots for 90 min in a medium containing [35S]-methionine and an energy generating system (Forde et. al, 1978). For immunoprecipitation of individual polypeptides the labelled maize mitochondrial proteins were first dissociated at 100°C in 2% (w/v) sodium dodecyl sulphate (SDS) then mixed with antisera raised against purified subunits I and II of yeast cytochrome oxidase and formaldehyde-fixed Staphylococcus aureus cells. The translation products of yeast (S. cerevisiae) mitochondria were labelled with [35S]-methionine in a similar manner and cytochrome oxidase was purified using antiserum raised against the holoenzyme. The only subunits of yeast cytochrome oxidase to be labelled by this procedure are the three largest (I, II and III). All protein samples were solubilised in SDS and electrophoresed in 15% (w/v) polyacrylamide slab gels. Radioactively labelled polypeptides were detected by autoradiography of the dried gel.

An essentially similar spectrum of polypeptides is synthesised by mitochondria _in vivo_ when plant cells are pulse labelled with [^{35}S]-methionine in the presence of cycloheximide (Leaver and Pope, 1976). This gives us confidence in the validity of using the _in vitro_ approach to the study of mitochondrial translation products.

Our observations that 18 polypeptides are synthesised by plant mitochondria may in fact be an underestimate as it is likely that a number of labelled mitochondrial polypeptides may have the same electrophoretic mobility. In addition there are also a number of minor labelled polypeptides, although other explanations for these, such as aggregation on gels, degradation or precursor proteins, cannot be disregarded.

If it is assumed that plant mtDNA contains one cistron for each of the three rRNAs and 20 tRNAs, and taking into account the estimated number and molecular weight of polypeptides synthesised by plant mitochondria, it is possible to estimate the total required coding capacity (Table 1). The value obtained is larger than the corresponding value calculated for yeast mtDNA and allows us tentatively to suggest that plant mtDNA may contain a larger number of genes. However, even this estimate still leaves 80% of the potential coding capacity of plant mtDNA unaccounted for.

Table 1.   Information Content of Plant Mitochondrial DNA

Mitochondrial gene product	Molecular weight ($\times 10^{-6}$)	Molecular Weight of duplex DNA required for coding ($\times 10^{-6}$)
Ribosomal RNAs: Large	1.15	2.30
Small	0.78	1.56
5S	0.039	0.08
Transfer RNAs (20)	0.56	1.12
Translation products (About 20 polypeptides in the mol. wt. range 8-54 $\times 10^{3}$)	0.52	10.30
	Total	15.36

Total mt DNA available ca. 74 $\times 10^{6}$

$\therefore$ Percentage accounted for = 20.8%

MITOCHONDRIAL BIOGENESIS DURING PLANT DEVELOPMENT

In higher plants there are a number of developmental changes known to be linked to changes in mitochondrial structure and function. Two notable examples are the sequence of events occurring: a) in embryos and cotyledons during seed maturation and germination and, b) during mitochondrial biogenesis in excised plant storage tissue. Both these systems seem ideally suited to the study of the coordination of synthesis and assembly of mitochondrial proteins during mitochondrial differentiation. In addition they also allow an investigation of the physiological and other factors controlling mitochondrial biogenesis and turnover, and the way in which biogenesis is linked to the mitochondrion's main metabolic function of respiration and geared to the development of the plant cell.

Germination

We have used cotyledons from germinating seeds of the field bean (Vicia faba L.) for a preliminary investigation of the development of mitochondrial protein synthesis during the early stages of rehydration and germination (B.G. Forde, L.K. Dixon, J. Forde and C.J. Leaver, manuscript in preparation). On contact with water the respiration rate of the cotyledons, which is initially negligible, increases rapidly for about 18 hours. This phase, coinciding with the period of imbibition, is followed by a second phase of increasing respiration rate which begins after 40 hours. Mitochondria isolated from dry seeds possess very little respiratory activity and show no evidence of respiratory control. Within the first 24 hours of rehydration, however, there is a six-fold increase in respiratory activity (expressed per cotyledon/mg protein) and a rapid restoration of both respiratory control and efficiency of oxidative phosphorylation (measured by the ADP/O ratio).

The protein synthetic activity of mitochondria from dry and cold imbibed seeds is barely detectable and increases only slightly during the first 24 hours even when supported by an energy generating system, with the major increase (approximately 10 fold) occurring during the second day of development. Thus during the initial 24 hour period when much of the recovery of mitochondrial respiratory activity occurs the mitochondrion is apparently defective in its protein synthesising ability.

Analysis of the translation products of mitochondria isolated from cotyledons at various stages of germination shows that mitochondria from 96 hour cotyledons synthesise about 12 major polypeptides. Only half of these, however, are synthesised as major products by mitochondria from dry or 12 hour imbibed seed. The relative rates of synthesis of the remaining 6 polypeptides increase mainly between 12 and 48 hours.

Ultrastructural examination of cotyledon parenchymal tissue
shows distinct changes in the mitochondrion during germination.
Between 12 and 48 hours the surface area of the inner membrane
clearly increases, as reflected by an increase in the number of
cristae, and the matrix becomes more electron dense, containing many
more ribosome-like particles. By 48 hours the mitochondria appear
to be fully developed, although the number of mitochondrial profiles
is still increasing.

We can conclude from the above results that during the initial
24 hours of germination, the mitochondria of the cotyledon are
apparently defective in their protein synthesising ability. It is
during this period, however, that a number of important mitochondrial
changes occur. These include a recovery of mitochondrial integrity,
an increase in the number of cristae and a six-fold increase in the
respiratory activity of the isolated mitochondrial fraction. It is
therefore unlikely that the initial functional and structural development of the organelle is dependent upon protein synthesis by mitochondrial ribosomes. We are currently examining whether the lag in
development of protein synthesising ability in mitochondria from
germinating seeds is due to a requirement for transcription of new
mitochondrial messenger RNA, to deficiencies in the translational
apparatus of mitochondria from dry seeds or is dependent upon de novo
synthesis of cytoplasmic proteins.

Excised Plant Storage Tissue

The rapid increase in respiration rate which is induced by
slicing and aging of Jerusalem artichoke tuber tissue is followed by
active biogenesis of the mitochondria, as demonstrated by a significant increase in the protein content of the isolated mitochondrial
fraction and by an enhancement of specific respiratory activity.
Aging of the tissue resulted in a 3-fold increase in the rate of in
vitro incorporation of $[^{35}S]$-methionine into mitochondrial protein.
It was found, however, that this increase could be entirely accounted
for by a marked depletion of methionine in the free amino acid pool
of the mitochondrial fraction (Forde et al, 1979a).

We therefore conclude that aging of artichoke tissue produces
no significant increase in the ability of the isolated mitochondria
to synthesise protein. This contrasts with the situation in vivo,
where it has been suggested that wounding of plant storage tissue
activates mitochondrial protein synthesis (Sakano and Asahi, 1971).
Clearly then, the factors which regulate mitochondrial translation
during aging are not carried over into the purified mitochondrial
fraction; they may therefore be assumed to reside in the cytoplasm.
One possible mechanism by which cytoplasmic factors could control
mitochondrial protein synthesis is by the availability of the cytoplasmically synthesised 'partner' proteins which are required by
many of the polypeptides of mitochondrial origin to form functional

enzyme complexes. Rapid synthesis of protein in the cytoplasm,
including proteins destined for the mitochondrion, is known to be
initiated after wounding of storage tissue. It could therefore be
this response which leads indirectly to activation of mitochondrial
protein synthesis in vivo.

     Analysis of the products of protein synthesis by the isolated
mitochondria demonstrated that aging had no major qualitative effect
on the range of polypeptides labelled in vitro. As aging proceeded,
however, significant changes were noted in the relative rates of
synthesis of three out of the eight major polypeptides synthesised
in vitro (Fig. 4).   That these changes could be observed using
isolated mitochondria suggests that they reflect a response to aging
which is intrinsic to the organelle rather than one which is under
the direct control of cytoplasmic factors.

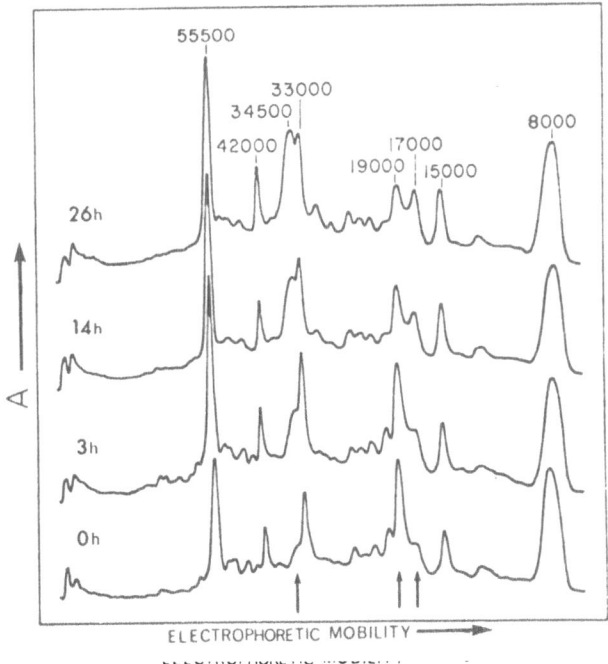

Fig. 4.   Polypeptides synthesised by mitochondria isolated during
          aging of Jerusalem artichoke tuber discs.

Aging of storage tissue discs, procedures for isolating and purifying
mitochondria and for labelling the mitochondrial translation products
with [^{35}S] -methionine have been described (Forde et al, 1979). The
labelled mitochondrial proteins were solubilised in SDS, electrophoresed in a 15% (w/v) polyacrylamide slab gel. An autoradiograph
obtained from the dried gel was scanned optically using a densitometer.

MITOCHONDRIAL TRANSLATION PRODUCTS AND CYTOPLASMIC MALE-STERILITY
IN MAIZE

Cytoplasmically-inherited male-sterility has until recently been
used extensively in the breeding of hybrid maize. In 1970, however,
an epidemic of southern corn leaf blight resulted in losses to the
U.S. maize crop estimated at over one billion dollars (Tatum, 1971).
The epidemic was caused by a new race of the fungal pathogen <u>Helmin-
thosporium maydis</u>, designated race T. At that time over 80% of the
U.S. maize crop carried the same T-type male-sterile cytoplasm and
it was this cytoplasm which conferred extreme susceptibility to the
fungus. As a result of the epidemic there was a greatly increased
awareness that important genetic variation can be inherited through
the cytoplasm.

Our interest in cytoplasmic male-sterility in maize arose
because several lines of circumstantial evidence pointed to the mitochondrion as the carrier of the genes controlling both male-sterility
and susceptibility to <u>H. maydis</u> race T. Ultrastructural studies of
the sterile anthers from plants with T cytoplasm (cms-T plants) have
found that mitochondrial degeneration is the first indication of
abnormality in pollen development (Warmke and Lee, 1977). Restriction endonuclease analysis of the cytoplasmic DNAs has revealed that
there are differences between the mtDNAs of normal (N) and cms-T
plants, although the respective chloroplast DNAs were indistinguishable (Levings and Pring, 1976; Pring and Levings, 1978). Normal and
cms-T mitochondria also differ in their response to the toxin produced by <u>H. maydis</u> race T. A number of studies have demonstrated that
race T toxin has a range of disruptive effects on isolated cms-T
mitochondria, although N mitochondria are resistant (Miller and
Koeppe, 1971; Gengenbach et al, 1973; Peterson et al, 1975;
Flavell, 1975). These disruptive effects indicate a site of action
on the inner mitochondrial membrane and they include uncoupling of
oxidative phosphorylation, irreversible swelling in KCl medium,
inhibition of malate-pyruvate oxidation and the activation of succinate cytochrome c reductase and cytochrome oxidase.

<u>Variant Mitochondrial Polypeptides</u>   We therefore attempted to estab-
lish a molecular basis for the properties of the T cytoplasm by looking for an alteration in the proteins synthesised on mitochondrial
ribosomes. We analysed the translation products of mitochondria
isolated from 4 day old etiolated shoots with N or T cytoplasm in the
same nuclear background (Fig. 5). The polypeptides synthesised by
the two types of mitochondria were found to be identical, except that
one 21,000 M_r polypeptide present in N mitochondria was missing from
cms-T mitochondria and was replaced by one of 13,000 M_r. This
variant 13,000-M_r polypeptide has been found to be characteristic of
cms-T mitochondria in all nuclear backgrounds tested (Forde et al,
1978; Forde and Leaver, 1979).

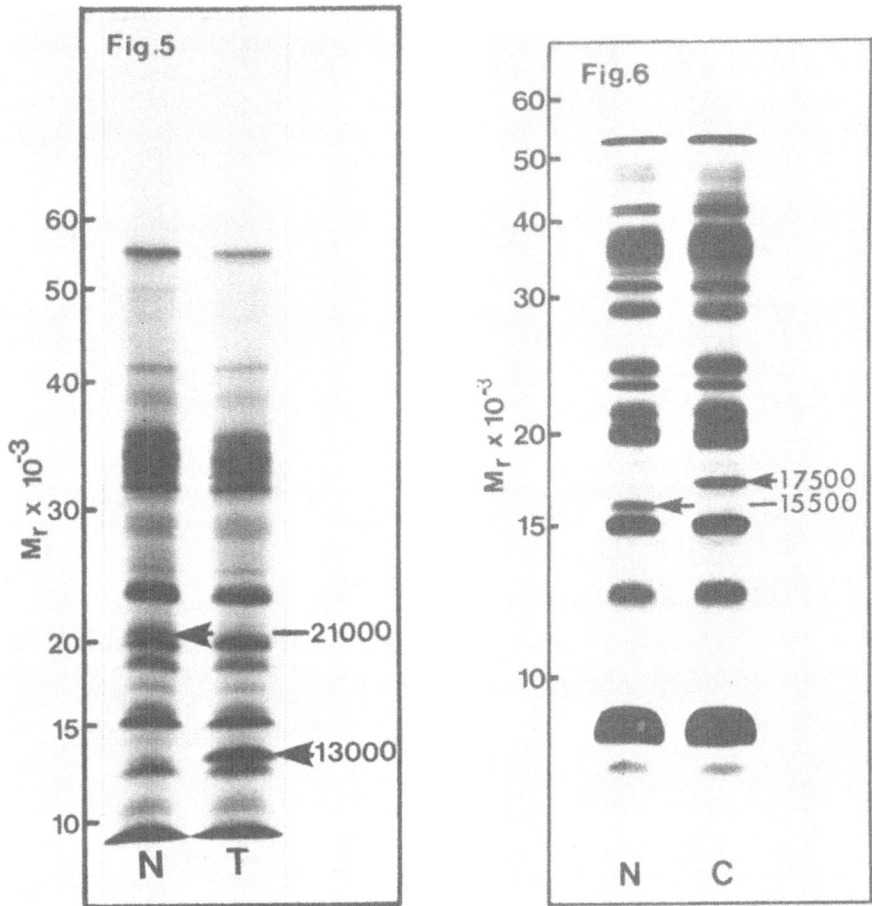

Fig. 5. Polypeptides synthesised by mitochondria from etiolated
maize shoots with N (normal) and T (male-sterile) cytoplasms.
Mitochondrial translation products were labelled with $[^{35}S]$ -methion-
ine as described in the legend to Fig. 3, solubilised in SDS and
electrophoresed on an 11-16% (w/v) polyacrylamide slab gel. Labelled
polypeptides were detected by autoradiography of the dried gel.

Fig. 6. Polypeptides synthesised by mitochondria from etiolated
maize shoots with N (normal) and C (male-sterile) cytoplasms.

The experimental procedure was as outlined in the legend to Fig. 5,
except that the polypeptides were fractionated on a 20% (w/v) polyacrylamide slab gel.

As well as the T cytoplasm there are two other forms of male-sterility in maize. The S and C cytoplasms were initially distinguished from the T cytoplasm because the nuclear genes (Rf genes) which could restore fertility to the male-sterile lines were different in each case (Duvick, 1965; Beckett, 1971; Gracen and Grogan, 1975). We have also studied the translation products of mitochondria from these cytoplasms and found that they too differ from those of N cytoplasm (Forde et al, 1978; Forde and Leaver, 1979). In cms-C mitochondria a 15,500 M_r polypeptide which is detected in N mitochondria has been replaced by one of slightly higher molecular weight (Fig. 6). Recent studies have shown that this variant polypeptide (molecular weight 17,500) is a soluble protein, although the polypeptide it replaces is membrane bound (B.G. Forde, J. Forde and C.J. Leaver, unpublished observations). It is not yet known whether these two polypeptides share a similar amino acid sequence.

Mitochondria from cms-S plants have been found to synthesise eight additional polypeptides, seven of which are of higher molecular weight than any of the translation products of N, cms-T and cms-C mitochondria (Fig. 7)

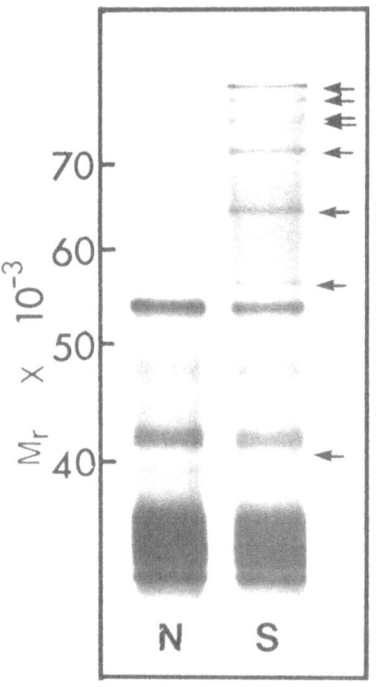

Fig. 7. Polypeptides synthesised by mitochondria from etiolated maize shoots with N (normal) and S (male-sterile) cytoplasms.

The experimental procedure was as outlined in the legend to Fig. 5, except that the mitochondrial polypeptides were fractionated on a 15% (w/v) polyacrylamide slab gel. Only the upper part of the autoradiograph is shown as the labelled polypeptides in the lower part of the autoradiograph were identical in both N and S.

Intriguingly, cms-S mitochondria also possess two plasmid-like DNA species not detectable in mitochondria from the other cytoplasms (Pring et al, 1977). One possibility is that the additional high molecular weight translation products which are characteristic of

cms-S mitochondria are encoded by these unique DNA species. Alternatively they may result from a fault in the processing of messenger
RNA or protein precursors.

Other sources of male-sterility In addition to the three male-
sterile cytoplasms already mentioned there have been about 80 separate discoveries of cytoplasmic male-sterility in maize (Duvick, 1965).
However, most of these have been classified as being either T-, C-
or S-type on the basis of which nuclear genes are able to restore
fertility (Beckett, 1971; Gracen and Grogan, 1974). In collaboration with the Plant Breeding Institute, Cambridge, we have analysed
the mitochondrial translation products from a total of 32 cytoplasmic
sources (Forde et al, 1979b). This study demonstrated that in each
case where a cytoplasm has previously been classified as T-, S- or
C-type by fertility restoration, the mitochondrial translation products are identical to those of cms-T, cms-S or cms-C plants respectively. Thus each of the four T-type cytoplasms synthesise the
variant 13,000 M_r polypeptide, each of the three C-type cytoplasms
synthesise the variant 17,500 M_r polypeptide and each of the 17 S-
type cytoplasms synthesise eight additional high molecular weight
polypeptides.

     Although we therefore find a direct correlation between a spec-
ific alteration in products of mitochondrial protein synthesis and a
particular form of cytoplasmic male-sterility, we have no direct
evidence that any of the variant polypeptides are involved in
triggering pollen abortion. Further circumstantial evidence may
however be obtained by examining the effect of Rf genes on the translation products of cms-T, -C and -S mitochondria. If the variant
polypeptides are somehow responsible for pollen abortion then the Rf
genes might restore fertility simply by suppressing their synthesis.

Restorer Genes   The effect of fertility restorer alleles on synthesis of the 13,000 M_r polypeptide by cms-T mitochondria is shown in
Fig. 8A. The nuclear backgrounds of the lines WF9 N, WF9 T and WF9
TRf were identical except that the TRf line was homozygous for the
dominant nuclear restorer alleles. The Rf alleles are seen to specifically suppress synthesis of the 13,000 M_r polypeptide in mitochondria isolated from 4 day old etiolated shoots. In addition, the
variant polypeptide is no longer detectable in gels stained for protein (Fig. 8B). Specific suppression of synthesis of the 13,000 M_r
polypeptide by restorer genes has been confirmed with five additional
lines of maize carrying Rf alleles derived from four different
sources (Forde and Leaver, 1979). However, in none of the restored
lines was synthesis of the 13,000 M_r polypeptide completely suppressed, nor was synthesis of the 21,000 M_r polypeptide even partially
recovered. Thus, even when fully restored to fertility, plants with
T cytoplasm can be distinguished from male-fertile plants which have
normal cytoplasm.

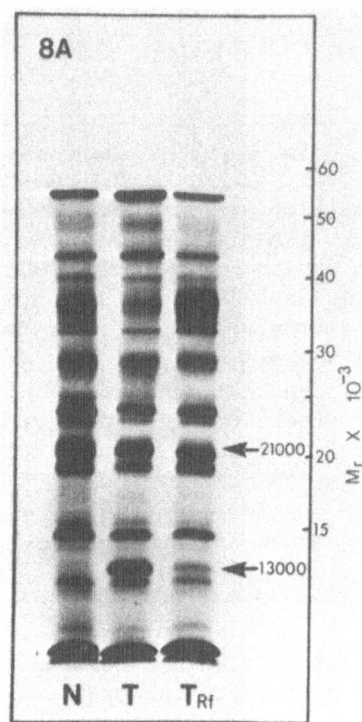
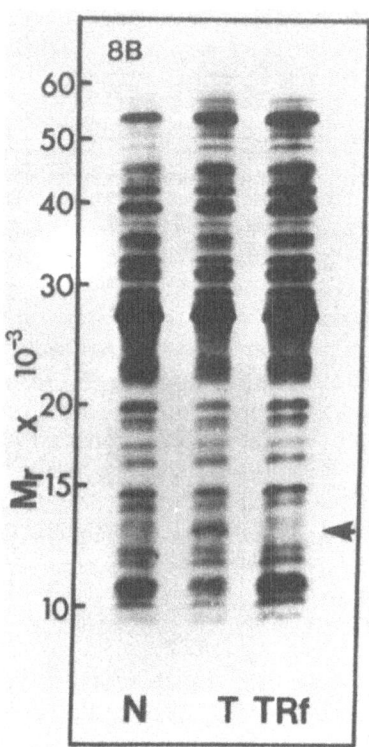

Fig. 8.   Effect of nuclear restorer genes (Rf) on proteins synthes-
          ised by T mitochondria.

A. Polypeptides synthesised by mitochondria from normal (N), male-
   sterile (T) and fertility-restored (TRf) forms of the maize line
   WF9. The translation products of mitochondria from etiolated
   shoots of the three closely related genotypes were labelled with
   [^{35}S]-methionine as described in the legend to Fig. 3. Mitochondrial proteins were then solubilised in SDS and electrophoresed
   in a 15% (w/v) polyacrylamide slab gel and the dried gel was
   autoradiographed.
B. Total mitochondrial proteins of the N, T and TRf forms of the
   maize line Pioneer inbred 4. Mitochondria were isolated and
   purified from etiolated shoots of the three genotypes, solubilised
   in SDS and electrophoresed in a 15% (w/v) polyacrylamide slab gel.
   The gel was then stained for protein with Coomassie brilliant blue.

        As well as being restored to fertility, it has also been found
that TRf lines contain mitochondria which are significantly less
sensitive to the H. maydis, race T toxin than mitochondria from non-
restored lines (Barratt and Flavell, 1975). In this case, as with
the experiments in Fig. 8A and B, we cannot rule out the possibility

that the genes involved are not the Rf genes themselves but are closely linked to them. If, however, it is possible to demonstrate with the help of well-characterised maize genotypes, that suppression of synthesis of the 13,000 M_r polypeptide and reduction in mitochondrial sensitivity to the toxin are under the control of the same nuclear restorer gene(s), this will provide strong circumstantial evidence that the variant polypeptide is involved in both male-sterility and suceptibility to H. maydis, race T toxin.

In contrast, we have been unable to detect any effect of restorer genes on synthesis of the variant polypeptides in cms-C and cms-S lines (Forde and Leaver, 1979). These experiments, like those with the T cytoplasm, were carried out using mitochondria from etiolated shoots. Therefore, it is still possible that the Rf genes which restore fertility to plants with S or C cytoplasms exert an effect on mitochondrial translation at a later stage, perhaps only in the developing anthers.

Conclusions   We have found three distinct cytoplasmically inherited variations in the products of mitochondrial protein synthesis, each of which is associated with a different form of cytoplasmic male-sterility in maize. The cytoplasmic mode of their inheritance and the site of translation on mitochondrial ribosomes strongly suggest that the variant polypeptides are coded for by mtDNA.

As demonstrated above (Figs. 8A and B), in addition to being controlled by cytoplasmic genes, the synthesis of the 13,000 M_r polypeptide is under the specific control of certain nuclear genes (Forde and Leaver, 1979). To our knowledge this is the first demonstration that a nuclear gene product can suppress synthesis of a protein which is made on organelle ribosomes.

An investigation of the alterations responsible for synthesis of the variant polypeptides, and the mechanism by which the synthesis of one such polypeptide is controlled by nuclear genes, should provide much-needed information on the biogenesis of the higher plant mitochondrion and on the nature of nuclear-mitochondrial interaction.

Acknowledgements: We wish to acknowledge the excellent technical assistance of Robin Oliver and Janice Forde. The [35S]-labelled yeast mitochondrial translation products, the purified yeast cytochrome oxidase and the antisera against subunits I and II of yeast cytochrome oxidase were kindly supplied by Drs. Schatz and Woodrow of the Biozentrum, Basel. The work described was supported by grants to C.J. Leaver from the Science Research Council (GRA 12373) and the Agricultural Research Council (AG 15/160).

REFERENCES

Bandlow, W., Schweyen, R.J., Wolf, K. and Kaudewitz, F., eds. (1977).
    Mitochondria 1977, De Gruyter, Berlin.
Barratt, D.H.P. and Flavell, R.B. (1975). Alterations in mitochon-
    dria associated with cytoplasmic and nuclear genes concerned with
    male sterility in maize. Theor. Appl. Genet. 45, 315-321.
Beckett, J.B. (1971). Classification of male-sterile cytoplasms in
    maize (Zea mays L.). Crop. Sci. 11, 724-727.
Bernardi, G. (1978). Intervening sequences in the mitochondrial
    genome. Nature (London) 276, 558-559.
Bernardi, G. (1979). The Petite Mutation in Yeast. TIBS 4, 197-201.
Borst, P. and Grivell, L.A. (1978). The Mitochondrial Genome of
    Yeast. Cell 15, 705-723.
Butow, R.A. and Strausberg, R.L. (1979). Biochemical genetics of
    mitochondrial biogenesis. TIBS 4, 110-113.
Chua, N.-H. and Schmidt, G.W. (1979). Transport of proteins into
    mitochondria and chloroplasts. J. Cell Biol. 81, 461-483.
Cunningham, R.S., Bonen, L., Doolittle, W.F. and Gray, M.W. (1976).
    Unique species of 5S, 18S and 26S ribosomal RNA in wheat mito-
    chondria. FEBS Letters 69, 116-122.
Douglas, M.G. and Butow, R.A. (1976). Variant forms of mitochondrial
    translation products in yeast: Evidence for location of deter-
    minants on mitochondrial DNA. Proc. Natl. Acad. Sci. U.S.A. 73,
    1083-1086.
Duvick, D.N. (1965). Cytoplasmic pollen sterility in corn. Advan.
    Genet. 13, 1-56.
Flavell, R.B. (1975). Inhibition of electron transport in maize
    mitochondria by Helminthosporium maydis race T pathotoxin.
    Physiol. Pl. Pathol. 6, 107-116.
Forde, B.G. and Leaver, C.J. (1979). Nuclear and cytoplasmic genes
    controlling synthesis of variant mitochondrial polypeptides in
    male-sterile maize. Submitted for publication.
Forde, B.G., Oliver, R.J.C. and Leaver, C.J. (1978). Variation in
    mitochondrial translation products associated with male-sterile
    cytoplasms in maize. Proc. Natl. Acad. Sci. U.S.A. 75, 3841-3845.
Forde, B.G., Oliver, R.J.C. and Leaver, C.J. (1979a). In vitro study
    of mitochondrial protein synthesis during mitochondrial biogenesis
    in excised plant storage tissue. Plant Physiol. 63, 67-73.
Forde, B.G., Oliver, R.J.C., Leaver, C.J., Gunn, R.E. and Kemble, R.J.
    (1979b). Classification of cytoplasms in maize. I. Electrophor-
    etic analysis of variation in mitochondrially synthesised pro-
    teins. Submitted to Genetics.
Gengenbach, B.G., Miller, R.J. Koeppe, D.E. and Arntzen, C.J. (1973).
    The effect of toxin from Helminthosporium maydis (race T) on
    isolated corn mitochondria: swelling. Can. J. Bot. 51, 2119-
    2125.
Gracen, V.E. and Grogan, C.O. (1974). Diversity and suitability for
    hybrid production of different sources of cytoplasmic male
    sterility in maize. Agron. J. 65, 654-657.

Leaver, C.J. (1975). The biogenesis of plant mitochondria. In The Chemistry and Biochemistry of Plant Proteins, J.B. Harborne and C.F. Van Sumere, eds. Phytochemical Society Symposium, Series 11 (London: Academic Press) pp. 137-165.

Leaver, C.J. and Harmey, M.A. (1976). Higher-plant mitochondrial ribosomes contain a 5S ribosomal RNA component. Biochem. J. 157, 275-277.

Leaver, C.J. and Pope, P.K. (1976). Biosynthesis of plant mitochondrial proteins, In: 'Nucleic Acids and Protein Synthesis in Plants', L. Bogorad and J.H. Weil, ed., Plenum Press, New York.

Levings, C.S., III and Pring, D.R. (1976). Restriction endonuclease analysis of mitochondrial DNA from normal and Texas cytoplasmic male-sterile maize. Science 193, 158-160.

Miller, R.J. and Koeppe, D.E. (1971). Southern corn leaf blight: susceptible and resistant mitochondria. Science 173, 67-69.

Peterson, P.A., Flavell, R.B. and Barratt, D.H.P. (1975). Altered mitochondrial membrane activities associated with cytoplasmically inherited disease susceptibility in maize. Theor. Appl. Genet. 45, 309-314.

Pring, D.R. and Levings, C.S., III (1978). Heterogeneity of maize cytoplasmic genomes among male-sterile cytoplasms. Genetics 89, 121-136.

Pring, D.R., Levings, C.S.,III, Hu, W.W.L. and Timothy, D.H. (1977). Unique DNA associated with mitochondria in the "S"-type cytoplasm of male-sterile maize. Proc. Natl. Acad. Sci. U.S.A. 74, 2904-2908.

Sakano, K. and Asahi, T. (1971). Biochemical studies on biogenesis of mitochondria in wounded sweet potato root tissue II. Active synthesis of membrane-bound protein of mitochondria. Plant Cell Physiol. 12, 427-436.

Schatz, G. (1979). How mitochondria import proteins from the cytoplasm. FEBS Letters 103, 203-211.

Schatz, G. and Mason, T.L. (1974). The biosynthesis of mitochondrial proteins. Ann. Rev. Biochem. 43, 51-87.

Schweyen, R.J., Weiss-Brummer, B., Backhaus, B. and Kaudewitz, F. (1978). The genetic map of the mitochondrial genome in yeast. Mol. Gen. Genet. 159, 151-160.

Tatum, L.A. (1971). The southern corn leaf blight epidemic. Science 171, 1113-1116.

Warmke, H.E. and Lee, S.L.J. (1977). Mitochondrial degeneration in T cytoplasmic male-sterile corn anthers. J. Hered. 68, 213-222.

GENETICS OF NITROGEN FIXATION IN THE BACTERIUM

KLEBSIELLA PNEUMONIAE

Ray Dixon

Agricultural Research Council
Unit of Nitrogen Fixation, University of Sussex
Brighton, Sussex, BN1 9RQ, U.K.

The reader may be somewhat surprised to find a contribution concerning a prokaryotic organism in a symposium devoted to the organisation and expression of plant cell genomes. It is well-established however that nitrogen-fixing micro-organisms contribute directly or indirectly to the nitrogen economy of higher plants and it is well known that some nitrogen-fixing bacteria are specifically able to form beneficial symbiotic associations with certain plants. In view of the importance of biological nitrogen fixation in the nitrogen cycle and the increasing demands on world agriculture, it is not surprising that there is considerable interest in the genetic manipulation of nitrogen fixation, particularly in regard to cereal crops. The aim of this paper is to provide the plant cell biologist with an overview of genetic research in a well-studied nitrogen fixing prokaryote; future potential for genetic engineering of nitrogen fixation (nif) can then perhaps be seen in its proper perspective.

KLEBSIELLA PNEUMONIAE

K.pneumoniae is an enteric bacterium, found in ubiquitous habitats including milk, water, soil, the human intestine and in association with higher plants. Not all representatives of this species are capable of fixing nitrogen and those which are able to reduce dinitrogen do so only at low oxygen tensions. Hence, expression of the nitrogen-fixation (Nif) phenotype as measured by growth in the absence of a source of fixed nitrogen, can only be assessed under anaerobic conditions.

K.pneumoniae has many genetic properties in common with Escherichia coli and Salmonella typhimurium and it is therefore not

surprising that the majority of genetic studies on nitrogen fixation
have been carried out with this organism. Genetic transfer can be
affected by generalised transduction with phage P1 (Streicher, et al.
1971) by R-factor mediated conjugation (Dixon and Postgate, 1971) or
by transformation with small amplifiable plasmids (Cannon et al.,
1977) and this organism is readily amenable to mutagenesis with both
chemical and biological mutagens.

THE NIF GENE CLUSTER

     All known genes specific to the regulation and synthesis of
nitrogenase (the enzyme complex which reduces dinitrogen) are
clustered near to the histidine operon in K.pneumoniae. The close
linkage of the nif and his genes has been exploited in the construction of self-transmissible plasmids which carry the nif region of
the K.pneumoniae genome (Cannon et al., 1976; Dixon et al., 1976).
pRD1 (formerly called RP41) contains the replication, transfer and
antibiotic resistance genes of the P incompatibility group plasmid
RP4 and the chromosomal genes gnd rfb his nif shiA. This plasmid
has been useful for complementation analysis and deletion mapping
of the nif region and the study of nif expression in other bacteria.

     A large number of nif point mutants have been isolated using a
wide variety of chemical mutagens including nitrosoguanidine,
alkylating agents such as diethyl sulfate and the frameshift
mutagens, ICR-170 and ICR-191 (Dixon et al., 1977; MacNeil et al.,
1978b; Dixon et al., 1979). In addition many insertion mutations
in nif genes have been generated using bacteriophage Mu (MacNeil
et al., 1978a; MacNeil et al., 1978b; Elmerich et al., 1978) and
by mutagenesis with the transposons Tn5, Tn7 and Tn10 (Merrick
et al., 1979). Complementation analysis between Nif⁻ derivatives
of the plasmid pRD1 and nif mutations on the K.pneumoniae chromosome
has allowed the identification of 13 complementation groups; nifQ,
nifB, nifA, nifF, nifM, nifV, nifS, nifN, nifE, nifK, nifD, nifH
and nifJ (Dixon et al., 1977; Elmerich et al., 1978; MacNeil et al.,
1978b; Merrick et al., 1978). A further cistron nifU has recently
been identified (Merrick et al., 1979).

     Translocatable genetic elements can excise imprecisely from
their point of insertion generating various chromosomal rearrangements including deletions (for review see Kleckner, 1977). It has
been possible to utilise this property of insertion elements to
obtain a series of overlapping deletions with end-points in each nif
cistron (Bachhuber et al., 1976; MacNeil et al., 1978a,b; Elmerich
et al., 1978; Merrick et al., 1978). A fine structure map of the
nif region has been constructed by scoring recombination between
point mutations and deletions; around 1000 nif mutations have been
mapped to date into approximately 100 deletion intervals (MacNeil
et al., 1978b; Merrick et al., 1979). The gene order in the nif
region is shown in Fig. 1.

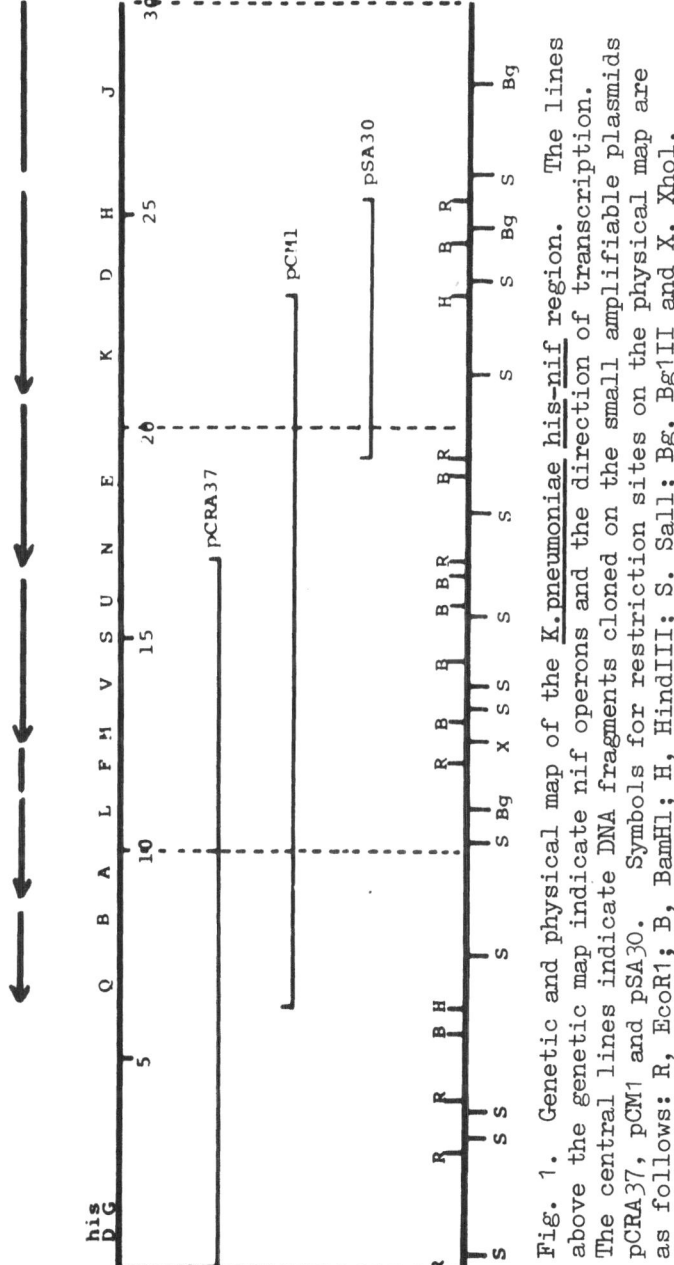

Fig. 1. Genetic and physical map of the K.pneumoniae his-nif region. The lines above the genetic map indicate nif operons and the direction of transcription. The central lines indicate DNA fragments cloned on the small amplifiable plasmids pCRA37, pCM1 and pSA30. Symbols for restriction sites on the physical map are as follows: R, EcoR1; B, BamH1; H, HindIII; S. Sal1; Bg, Bg1III and X, XhoI.

Although most mutations in the nifA region can be allocated
to a single complementation group, there is evidence to suggest that
this region contains two genes which have been designated nifA and
nifL (Kennedy, 1977). The mutations in nifL isolated by MacNeil
et al., (1978b) were shown to be polar; two were amber suppressible
point mutations, the remainder were Mu insertions. Most of the
strains carrying Mu insertions were able to revert to a Nif$^+$ phenotype whereas Mu inserts in nifA or in any other nif complementation
group were unable to do so. Since Mu-induced mutations do not
normally revert, Nif$^+$ revertants can only be obtained by extragenic
suppression of the original insertion or by a polarity relief
mutation. This suppression of nifL insertions led MacNeil et al.,
(1978b) to suggest that nifL may not be essential for growth on
dinitrogen.

Fragments of nif DNA derived by restriction digestion of
plasmid pRD1, have been cloned on small amplifiable plasmids. Three
of these cloned fragments are shown in Fig. 1: pCRA37 carries an
EcoR1 fragment extending from hisD to nifN; pCM1 contains a HindIII
fragment extending from nifQ to nifK and pSA30 carries an EcoR1
fragment with one end-point within nifE and the other between nifH
and nifJ (Cannon et al., 1977; Cannon et al., 1979). In addition,
the entire nif region has been cloned by insertion of two HindIII
restriction fragments from pRD1 into a single plasmid vector
(Pühler et al., 1978). The nif gene cluster has been estimated to
be 24 Kb in length, which is in agreement with the physical size of
nif DNA estimated from co-transduction data (Kennedy, 1977). The
restriction map of this region (Cannon et al., 1979; Riedel et al.,
1979) is also shown in Fig. 1.

Bacteriophage Mu and transposon-induced insertion mutations
in the nif gene cluster have been physically mapped by comparing
size changes in restriction fragments due to the presence of the
translocatable element. Total DNA was extracted from the insertion
mutants, digested with a given restriction endonuclease, electrophoresed on agarose gels and hybridised with nick-translated plasmid
DNA using the Southern blotting technique; 86 nif insertion
mutations were assigned physical locations using this procedure
(Reidel et al., 1979). The physical and genetic maps of the nif
gene cluster are virtually in complete agreement and this
correlation is of great value when specific restriction fragments
with known genetic function are sought e.g. the cloning of promoter
regions.

TRANSCRIPTIONAL ORGANISATION OF THE NIF CISTRONS

Translocatable genetic elements frequently generate polar
mutations and a large number of polar mutants in the nif gene
cluster have been used to determine the operon structure of the 15
nif genes. The transcriptional polarity of each insertion mutation

has been determined by complementation analysis and where possible,
by analysis of nif-specific polypeptides on SDS polyacrylamide gels.
These data suggest that the cluster is organised into 7 transcriptional units comprising 5 polycistronic and two monocistronic operons
(Elmerich et al., 1978; MacNeil et al., 1978; Merrick et al., 1979).
All polycistronic operons are transcribed in the same direction as
the histidine operon as shown by the arrows in Fig. 1. DNA-RNA
hybridisation studies also indicate that the majority of nif genes
are transcribed from right to left as shown in Fig. 1. Separated
single strands of cloned plasmid DNA representing his genes and six
out of seven nif operons were hybridised with his-specific and nif-
specific transcripts synthesised in vivo; all transcripts hybridised
to the same strand of plasmid DNA indicating that the nif m-RNA's
detected were transcribed in the same direction as the histidine
operon (Jannsen et al., 1979).

NIF GENE PRODUCTS

     Nitrogenase the nitrogen-fixing enzyme complex, consists of
two component proteins:
     (a) Kp1, is a tetrameric protein (M.W. 218,000 daltons)
consisting of two non-identical pairs of sub-units with molecular
weights of 56,000 and 60,000 daltons. This protein contains 32 Fe
atoms and 2 Mo atoms per molecule. A low molecular weight co-factor
containing Fe, Mo and S (FeMo-co) can be isolated from this protein;
the co-factor is essential for nitrogenase activity and may contain
the active site of the enzyme.
     (b) Kp2 is a dimeric protein (M.W. 68,000 daltons) with
identical sub-units and contains a single Fe_4S_4 centre.
     Both proteins are extremely oxygen sensitive. ATP, Mg^{2+} and
a reductant such as sodium dithionite are required for nitrogenase
activity in vitro. Besides reducing dinitrogen the enzyme will
reduce other substrates which are stereochemically similar to the
dinitrogen molecule e.g. acetylene (for reviews see Eady and Smith,
1979; Mortenson and Thorneley, 1979).

     The biochemical phenotype of nif mutants has been determined
on the basis of enzyme activity in vivo and in vitro using acetylene-
reduction as a measure of nitrogenase activity. Cell-free extracts
prepared from mutant strains have been tested for Kp1 and Kp2
activity and for re-activation of Kp1 with purified FeMo-co. In
addition nif-specific polypeptides have been detected on one- and
two-dimensional polyacrylamide gels and by examining mutants for
the presence of immunological cross-reacting material to Kp1 and
Kp2 (St John et al., 1975; Dixon et al., 1977; Roberts et al., 1978).
A summary of the molecular weights of nif-encoded proteins and their
probable function is given in Table 1.

Table 1. Products of the nif gene cluster

Gene	Polypeptide molecular weight (daltons)	Function
nifJ	120,000	electron input to nitrogenase
nifH	35,000	structural gene for Kp2
nifD	56,000	α sub-unit of Kp1
nifK	60,000	β sub-unit of Kp1
nifE	46,000[a]	synthesis or processing of FeMo-co
nifN	50,000[a]	"
nifU	N.D.	Unknown
nifS	18K?	Unknown
nifV	N.D.	Unknown
nifM	N.D.	Kp2 processing
nifF	17,000[a]	Electron transport
nifL	N.D.	Regulation?
nifA	N.D.	Activator for nif transcription
nifB	N.D.	synthesis or processing of FeMo-co
nifQ	N.D.	Unknown

a = tentative assignments   N.D. not determined

as electron donor. Since sodium dithionite by-passes the normal
electron transport pathway, it is possible that nifF and nifJ
mutants are defective in electron transport (Roberts et al., 1978).
Pyruvate can provide electrons for acetylene reduction by nitrogenase
in wild-type extracts but not in crude-extracts of nifF and nifJ

NifJ and nifF mutants have low levels of acetylene reducing activity in vivo but significant activity in vitro with dithionite mutants which suggests that these mutants are blocked in the electron transport pathway to nitrogenase (Hill and Kavanagh, 1979). Pyruvate supported activity can be restored to nifF but not nifJ mutants by addition of partially purified Azotobacter chroococcum flavodoxin, indicating that nifF may code for an electron transport protein. Hill and Kavanagh (1979) also observed that nifJ mutants have lower than the wild-type level of active Kp1, suggesting that nifJ may have a second role in maintaining the activity of this protein. NifJ determines a 120,000 dalton polypeptide; strains with insertion mutations in nifJ lack this product (Elmerich et al., 1978; Merrick et al., 1978) and some nifJ point mutations alter the electrophoretic mobility of this polypeptide (Roberts et al., 1979). The nifF product has been tentatively assigned to a 17,000 dalton polypeptide (Roberts et al., 1979).

NifH, nifD and nifK are the structural genes for nitrogenase. The gene products have been identified on the basis of charge changes in mutant polypeptides; nitrogenase activity in vitro can be restored to these mutants by addition of the appropriate purified nitrogenase component protein. NifH determines Kp2; nifD and nifK code for the 56,000 and 60,000 dalton sub-units of Kp1 respectively.

Strains carrying nifE, nifN and nifB mutations lack Kp1 activity, which can be restored by addition of purified K.pneumoniae FeMo-co. These genes are thought to be involved in synthesis or processing of FeMo-co. The nifE and nifN products have been assigned to polypeptides with molecular weights of 46,000 and 56,000 respectively (Roberts et al., 1978).

NifM mutants lack Kp2 activity and it is thought that this gene product may be involved in processing of Kp2. Relatively little is known about the products or function of the other genes in the nifUSVM operon. Mutations in nifV and nifQ result in a 'leaky' Nif⁻ phenotype and it is particularly difficult to assess the function of these gene products.

Strains carrying nifA mutations lack all identifiable nif-specific proteins; this pleiotropic phenotype is common to all nifA mutants suggesting that the nifA gene product has a regulatory role and is required for derepression of the nif gene cluster. Complementation tests indicate that nifA can activate nif derepression in trans suggesting that the nifA product is a positive activator for nif transcription. Since the nifL mutations characterised so far show polarity on nifA it is difficult to assign a function to the nifL gene product. Further analysis of the role of nifL requires the isolation of strains with non-polar mutations in this gene.

REGULATION OF THE NIF GENE CLUSTER

   Ammonia efficiently represses nitrogenase synthesis (Tubb and
Postgate, 1973). Nif-specific m-RNA's are not detected in ammonia
grown cultures (Jannsen et al., 1979) and all identifiable nif-
specific polypeptides are absent in cultures grown in the presence
of ammonia (Eady et al., 1978; Roberts et al., 1978). Ammonia
repression of nif can be explained in terms of repression of the
level of glutamine synthetase, a protein which presumed to act as a
positive control element for nif and for other genes involved in
nitrogen metabolism (Tubb, 1974; Streicher et al., 1974). Ausubel
et al., (1976; 1977) have isolated nif mutants (designated nifT)
which are independent of glutamine synthetase regulation. The
precise location of these mutations is unknown, but one nifT mutant
has been shown to be cis dominant, suggesting that it carries a
mutation in a regulatory site. Since mutations which allow nif
transcription to proceed independently of GS activation presumably
occur at a single nif locus, Ausubel et al., (1977) have argued that
glutamine synthetase may act at only a single site within nif (e.g.
the nifA or nifL promoter(s)) and that the other nif operons are
positively controlled by a specific nif activator protein (e.g. nifA
product). Strains carrying nifT mutations are still susceptible to
ammonia repression indicating that repression by ammonia is not
solely mediated by glutamine synthetase. These results suggest that
nif is also regulated by a specific repressor; none of the nif genes
have so far been implicated in this process. Shanmugam and Morandi
(1976) observed that some mutants defective in ammonia assimilation,
were capable of producing nitrogenase in the presence of ammonia
but not in the presence of amino acids. Products of ammonia
assimilation i.e. amino acid(s) may therefore act as nif co-
repressor(s).

   Oxygen not only irreversibly inactivates nitrogenase but also
represses synthesis of nitrogenase polypeptides (Eady et al., 1978).
Although the structural genes, nifH, nifD and nifK are clearly
subject to regulation by oxygen, it is not yet known whether all
nif transcripts are subject to this regulation.

   Other nif genes besides nifA and nifL have been implicated in
genetic control of nitrogen fixation, although the evidence at
present is rather tentative. Roberts et al., (1978) found that
several strains carrying point mutations in nifH contained
apparently normal levels of Kp2, but lacked detectable levels of
other nitrogenase polypeptides. These strains were not polar in
complementation tests and it was suggested that nifH could have a
regulatory function. NifS mutations have a similar direct or
indirect effect on derepression of nif-specific polypeptides.
Finally, Roberts et al., (1978) observed that all nifJ mutants
lacked detectable levels of the 17,000 dalton protein (nifF product)
in steady state conditions; however this effect may be an artefact

since it is not observed in pulse-labelled cultures of nifJ mutants.

More detailed studies of nif regulation will require the isolation of nif-specific regulatory mutants and techniques for examining transcription from each nif promoter, both in vivo and in vitro. Current research in this area is therefore directed towards the construction of nif-lac fusions using the techniques of Casadaban (1976), detection of specific nif transcripts using cloned DNA probes and the cloning and sequencing of nif promoters.

ACKNOWLEDGEMENTS

I would like to thank my colleagues at the A.R.C. Unit of Nitrogen Fixation for useful discussions and for allowing me to quote their work prior to publication. I also thank Mike Merrick for constructive criticism of the manuscript.

REFERENCES

Ausubel, F.M., Margolskee, R.F. and Maizels, N. (1977). Mutants of Klebseilla pneumoniae in which expression of nitrogenase is independent of glutamine synthetase control, in: "Recent Developments in Nitrogen Fixation", W. Newton, J.R. Postgate, C. Rodriguez-Barrueco, eds. Academic Press, London.

Ausubel, F.M., Reidel, G., Cannon, F., Peskin, A. and Margolskee, R. (1977). Cloning nitrogen fixing genes from Klebsiella pneumoniae in vitro and the isolation of nif promotor mutants affecting glutamine synthetase regulation, in: "Genetic Engineering for Nitrogen Fixation", A. Hollaender, ed. Plenum Press, New York.

Bachhuber, B., Brill, W.J. and Howe, M.M. (1976). The use of bacteriophage Mu to isolate deletions in the his-nif region of Klebsiella pneumoniae. J.Bacteriol., 128:749

Cannon, F.C., Dixon, R.A. and Postgate, J.R. (1976). Derivation and properties of F-prime factors in Escherichia coli carrying nitrogen fixation genes from Klebsiella pneumoniae. J.gen. Microbiol., 93:111

Cannon, F.C., Reidel, G. and Ausubel, F.M. (1977). Recombinant plasmid that carries part of the nitrogen fixation (nif) gene cluster of Klebsiella pneumoniae. Proc. Natl. Acad. Sci. U.S.A., 74:2963

Cannon, F.C., Reidel, G.E., and Ausubel, F.M. (1979). Overlapping sequences of Klebsiella pneumoniae nif DNA cloned and characerised. Mol. Gen. Genet. (in press)

Casadaban, M.J. (1976). Transposition and fusion of the lac genes to selected promotors in Escherichia coli using bacteriophage Lambda and Mu. J. Mol. Biol., 104:541

Dixon, R.A. and Postgate, J.R. (1971). Transfer of nitrogen fixation genes by conjugation in Klebsiella pneumoniae, Nature, 234:47

Dixon, R.A., Cannon, F.C. and Kondorosi, A. (1976). Construction
    of a P plasmid carrying nitrogen fixation genes from Klebsiella
    pneumoniae, Nature, 260:268
Dixon, R., Kennedy, C., Kondorosi, A., Krishnapillai, V. and Merrick,
    M. (1977). Complementation analysis of Klebsiella pneumoniae
    mutants defective in nitrogen fixation. Mol. Gen. Genet.,
    157:189
Dixon, R., Merrick, M., Filser, M., Kennedy, C. and Postgate, J.R.
    (1979). Transcriptional organisation of the Klebsiella
    pneumoniae nif gene cluster, in: "Proceedings of the 3rd
    International Symposium on Nitrogen Fixation", W.H. Orme-Johnson
    and W. Newton, eds., University Park Press, Baltimore
Eady, R.R. and Smith, B.E. (1979) Physico-chemical properties of
    nitrogenase and its components in: "A Treatise on Dinitrogen
    Fixation, Sections I and II: Inorganic and physical chemistry
    and biochemistry", R.W.F. Hardy, F. Bottemely and R.C. Burns.
    John Wiley and Sons.
Eady, R.R., Issack, R., Kennedy, C., Postgate, J.R. and Ratcliffe,
    H.D. (1978). Nitrogenase synthesis in Klebsiella pneumoniae:
    comparison of ammonium and oxygen regulation. J.gen.Microbiol.,
    104:277
Elmerich, C., Houmard, J., Sibold, L., Manheimer, I., Charpin, N.
    (1978). Genetic and biochemical analysis of mutants induced
    by bacteriophage Mu DNA integration into Klebsiella pneumoniae
    nitrogen fixation genes. Mol. Gen. Genet., 165:181
Hill, S. and Kavanagh, E. (1979). Some biochemical properties of
    nifF and nifJ mutants relevant to electron transport to
    nitrogenase in Klebsiella pneumoniae (in preparation)
Jannsen, K.A., Reidel, G.E., Ausubel, F.M. and Cannon, F.C. (1979)
    Transcriptional studies with cloned nitrogen fixing genes,
    in: "Proceedings of the 3rd International Symposium on
    Nitrogen Fixation", W.H. Orme-Johnson and W. Newton, eds.,
    University Park Press, Baltimore
Kennedy, C. (1977). Linkage map of nitrogen fixation (nif) genes
    in Klebsiella pneumoniae, Mol. gen. Genet., 157:199
MacNeil, T., Brill, W.J. and Howe, M.M. (1978a). Bacteriophage
    Mu-induced deletions in a plasmid containing the nif (N$_2$-
    fixation) genes of Klebsiella pneumoniae. J.Bacteriol., 134:247
MacNeil, T., MacNeil, D., Roberts, G.P., Supiano, M.A. and Brill,
    W.J. (1978b). Fine-structure mapping and complementation
    analysis of nif (nitrogen fixation) genes in Klebsiella
    pneumoniae. J.Bacteriol., 136:1
Merrick, M., Filser, M., Kennedy, C., and Dixon, R. (1978). Polarity of
    mutations induced by insertion of transposons Tn5, Tn7 and Tn10
    into the nif gene cluster of Klebsiella pneumoniae. Molec.
    gen. Genet., 165:103
Merrick, M., Filser, M., Dixon, R., Elmerich, C., Sibold, L. and
    Houmard, J. (1979). The use of translocatable genetic elements
    to construct a fine-structure map of the Klebsiella pneumoniae
    nitrogen fixation (nif) gene cluster (in preparation)

Mortenson, L.E. and Thorneley, R.N.F. (1979). Structure and function of nitrogenase. Ann.Rev.Biochem. (in Press)

Pühler, A., Burkardt, H.J., Klipp, W., Wohlleben, W. (1978). Identification of the gene region for nitrogen fixation from Klebsiella pneumoniae on the P-type plasmid RP41, in: "Plasmids and other extrachromosomal genetic elements". K. Timmis and W. Goebel, eds., Max Planck Institute for Molecular Genetics, Berlin

Reidel, G.E., Ausubel, F.M. and Cannon, F.C. (1979). Physical map of chromosomal nitrogen fixation (nif) genes of Klebsiella pneumoniae. Proc. Natl. Acad. Sci. U.S.A. (in Press)

Roberts, G.P., MacNeil, T., MacNeil, D. and Brill, W.J. (1978) Regulation and characterisation of protein products coded by the nif (nitrogen fixation) genes of Klebsiella pneumoniae J.Bacteriol., 136:267

Shanmugam, K.T. and Morandi, C. (1976). Amino acids as repressors of nitrogenase biosynthesis in Klebsiella pneumoniae, Biochim. Biophys.Acta., 437:322

Streicher, S.L., Gurney, E. and Valentine R.C. (1971). Transduction of nitrogen fixation genes in Klebsiella pneumoniae. Proc. Natl. Acad. Sci. U.S.A., 50:1174

Streicher, S.L., Shanmugam, K.T., Ausubel, F., Morandi, C. and Goldberg, R.B. (1974). Regulation of nitrogen fixation in Klebsiella pneumoniae: evidence for a role of glutamine synthetase as a regulator of nitrogenase synthesis. J.Bacteriol., 120:815

St. John, R.T., Johnston, H.M., Seidman, C., Garfinkel, D., Gordon, J.K., Shah, V.K. and Brill, W.J. (1975). Biochemistry and genetics of Klebsiella pneumoniae mutant strains unable to fix N_2. J.Bacteriol., 121:759

Tubb, R.S. (1974). Glutamine synthetase and ammonium regulation of nitrogenase synthesis in Klebsiella. Nature, 251:481

Tubb, R.S. and Postgate, J.R. (1971). Control of nitrogenase synthesis in Klebsiella pneumoniae, J.gen.Microbiol., 79:103

EXPRESSION OF HOST GENES DURING SYMBIOTIC NITROGEN FIXATION

Desh Pal S. Verma

Department of Biology
McGill University
Montreal, Quebec, Canada

INTRODUCTION

The Rhizobium-legume association which results in the development of root nodule structures is restricted by a narrow host and
bacterial range for each species (Pueppke et al. 1978). This suggests
a highly specific interaction between the two organisms. Besides
infection, the development of the root nodule is also influenced by
both bacterial (Maier and Brill 1976) and plant genes (Holl and
LaRue 1976). Mutation in any of these organisms may abort the
development of the nodule structure and thus affect the process of
symbiotic nitrogen fixation. Although it is apparent that the
specific expression of both plant and bacterial genes is required
during this process, it is not known how many structural genes of the
host or bacteria are involved in the development of the symbiosis
necessary for nitrogen fixation in legumes. While some progress has
been made on Rhizobium genetics, and the control of nitrogenase
enzyme in bacteria, our understanding of the genetic contribution of
the host to this process is very limited.

The role of the host in root nodule-symbiosis has been recognized for some time (Nutman 1956). Through classical genetic
experiments several plant genes are linked to nodulation in soybean
(Caldwell 1966, Caldwell and Vest 1968) as well as in other legumes
(Holl and LaRue 1976), however, to date, besides leghaemomoglobin
(Baulcombe and Verma 1978) and Nodulin-35 (Legocki and Verma 1979a)
no other plant gene product "specific"[1] to nodules has been reported.

---

[1]The term "specific" is used for those proteins which are present in
nodules but are undetectable in control root tissue.

Both of these proteins are major components representing about 30%and 4% respectively of the total soluble protein in soybean root nodules. There is an apparent correlation between leghaemoglobin and effectiveness of the nodules in fixing nitrogen while Nodulin-35 appears to be produced in both effective and ineffective nodules of soybean (Legocki and Verma 1979a).

Estimation of the total number of structural genes being expressed in the soybean root nodules suggests that besides leg-haemoglobin, which is a major component in nodule RNA, there is a middle abundant class of mRNA comprised of about 500 sequences whichare significantly more abundant in nodule than in control root tissue (Auger et al. 1979). Heterologous hybridizations of DNA complementary to root RNA with nodule RNA and vice versa suggest that the RNA populaions of the two tissues are substantially homo-logous, however, a marked shift in abundances of certain sequences is observed. Analysis of the soluble proteins and the major in vitro translation products of root and nodule mRNA suggests that there may be several "nodule-specific" proteins whose synthesis appear to be influenced by Rhizobium.
A coordinated expression of the host and bacterial genes is required for the development of symbiosis leading to effective nitrogen fixation. An understanding of the induction of specific plant genes which are involved in root nodule symbiosis and their correlation with those in Rhizobium is essential for modulating the capacity for nitrogen fixation. It is, at present, restricted by the availability of the molecular probes to measure the expression of host genes.

LEGHAEMOGLOBIN

Origin and the number of leghaemoglobin genes: Leghaemoglobin is a myoglobin-like protein. It is synthesized exclusively in the nitrogen fixing root nodules developed due to the symbiotic association of Rhizobium with legume roots and represents about 25-30% of the total soluble protein in nodules. This protein is synthesized on free polysomes in the host cell cytoplasm (Verma and Bal 1976). Its intracellular location is restricted to the cyto-plasm of the infected cells of the nodules (Verma et al. 1978a). In soybean, there are three major leghaemoglobins (Verma et al. 1979). Although all three are very similar in their structure (Ellfolk and Sievers 1974,Nicola 1975, Hurrell and Leech 1977), the differences in their primary sequence show that they are different gene products.
Indirect evidence suggested that leghaemoglobin may be incoded by the host genome (Dilworth 1969, Cutting and Schulman 1971, Verma et al. 1974). A direct demonstration of this has been obtained by purification of a leghaemoglobin complementary DNA (Lb-cDNA) and itshybridization to soybean DNA from uninfected tissue (Baulcombe and Verma 1978). Lb-cDNA was prepared by hybridization of a

complementary DNA synthesized to 9S poly(A)+ nodule mRNA with its
template and isolation of the rapidly hybridizing component of this
reaction (up to a Rot of 3 x 10^{-3}). This purified Lb-cDNA, when
hybridized to soybean embryo DNA and the DNA from free living
Rhizobium (Fig.1a) showed positive hybridization only with plant
DNA demonstrating the presence of leghaemoglobin genes solely in
the plant genome.

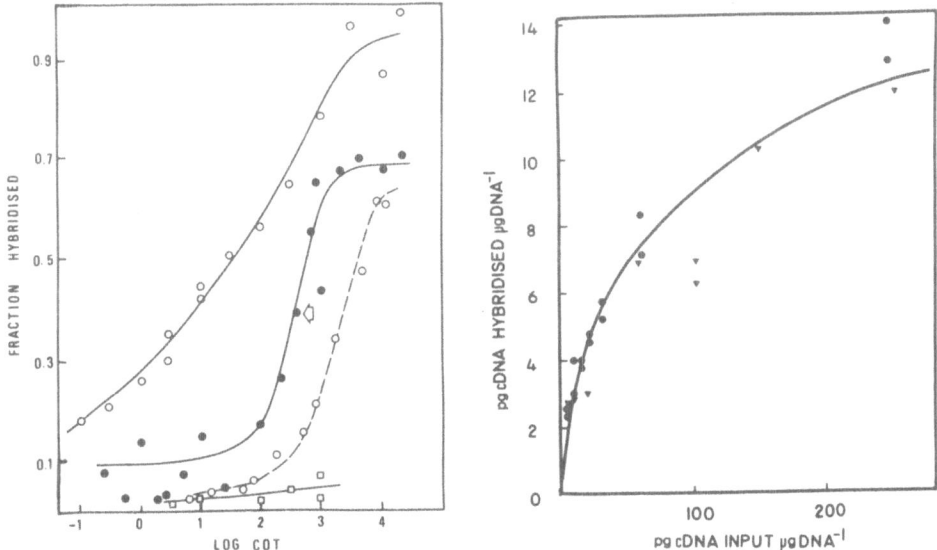

Fig. 1. a, Kinetics of hybridization of leghaemoglobin-cDNA
with soybean embryo DNA (●) and Rhizobium DNA (□). The hybridization of control root-cDNA with soybean DNA (O-O) and renaturation
kinetics of soybean DNA (O) is also shown for comparison.
Rhizobium DNA: Lb-cDNA was 10^6:1 and soybean DNA: Lb-cDNA or total
root-cDNA was 5 x 10^6:1.
      b, Saturation hybridization of soybean DNA from embryo (●)
and from nodule (▼) with leghaemoglobin-cDNA. (Data is redrawn from
Baulcombe and Verma, 1978).

      Analysis of the hybridization kinetics (Fig. 1a) showed that
the $Cot\frac{1}{2}$ value (330) for Lb-cDNA hybridization is somewhat faster
than the unique sequence DNA (see also hybridization of control
root cDNA) indicating that leghaemoglobin genes may be repeated in
soybean genome. A more accurate estimate of the repetition of
leghaemoglobin genes was obtained by titrating increasing amounts
of Lb-cDNA with fixed amount of soybean DNA from both uninfected
and infected tissues. The results in Fig. 1b show that about 12 pg
of Lb-cDNA were hybridized with 1 ug soybean DNA. This represents
about 40 copies of leghaemoglobin genes per haploid genome (1.4 pg)
and its value does not change during nodulation. It is not known

whether all 3 major leghaemoglobin genes (Verma et al. 1979) are
repeated equally. Since the synthesis of LbS (electrophoretically
slow moving component of Lb) and LbF (fast moving, includes LbC_1 +
LbC_2, Appleby et al. 1975) changes during root nodule development
(Verma et al. 1979), leghaemoglobin genes may be regulated differentially. A 40-fold reiteration in leghaemoglobin genes coupled with
a high level of ploidy in the infected cells may be responsible, in
part, for the abundant synthesis of this protein in nodules. Since
the uninfected plants do not synthesize detectable leghaemoglobin,
induction of its genes appear to be under the control of Rhizobium.

Induction of leghaemoglobin is independent of nitrogenase: Although
a direct correlation between the presence of leghaemoglobin and
nitrogenase activity has been established in legume root nodules,
little is known about the coordination of induction of these
molecules in the host and Rhizobium respectively. This correlation
has been primarily based upon the assay of leghaemoglobin as
hemochromogen which parallels nitrogenase activity in nodules. By
using antibodies against leghaemoglobin and the analysis of the in
vitro translation product of the nodule polysomes (Verma et al.
1979), we can detect synthesis of Lb protein three to four days
before nitrogenase which appears on the 10th day after infection
(Fig. 2a). Similar results have been obtained in pea root nodules
(Bisseling et al. 1979). Bergersen and Goodchild have detected

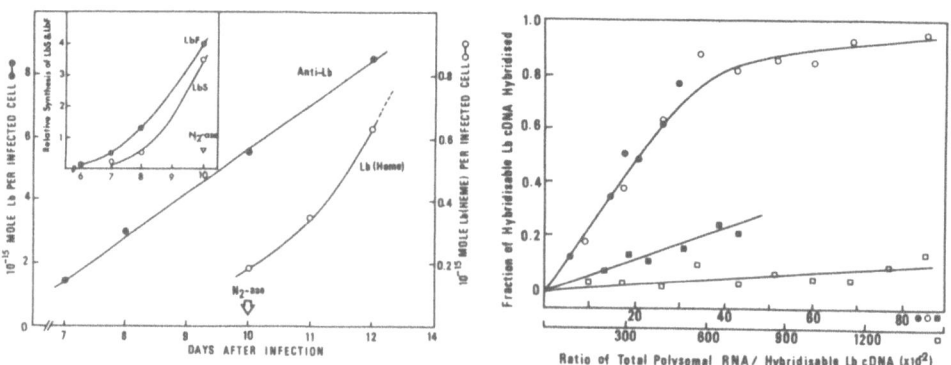

Fig. 2. a, Appearance of leghaemoglobin in relation to nitrogenase activity during root nodule development. Leghaemoglobin was
determined by hemochromogen (Lb-heme), by immunoelectrophoresis
(anti-Lb) and by in vitro translation of polysomes (data from Verma
et al. 1979).
       b, Determination of leghaemoglobin sequences in developing root
nodules as measured by hybridization of Lb-cDNA with polysomal RNA
from (O) 21 day; (●), 13 day; (■), 10 day and (□) 8 day nodules
(data from Auger et al. 1979).

leghaemoglobin in soybean nodules two days prior to the appearance
of acetylene reduction (nitrogenase) activity. These observations
suggest that the two molecules may be induced through independent
mechanisms.

To determine further the dependence of leghaemoglobin induction
on nitrogenase, we used several mutant strains of Rhizobium which
form nodules that are unable to fix nitrogen (ineffective). One of
these strains, SM5 (Maier and Brill 1976), has a mutation in the
nitrogenase proteins and is thus unable to synthesize functional
nitrogenase. Nodules developed with this mutant as well as with
other ineffective strains, such as SM3 and 61A24, were assayed both
for the presence of leghaemoglobin and the appearance of the

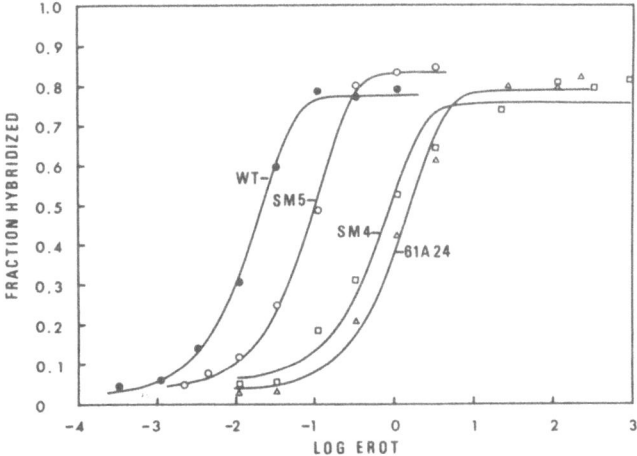

Fig. 3. Hybridization of Lb-cDNA with polysomal poly A(+) RNA
from root nodules developed by WT effective (strain 61A76) and
ineffective (strains SM4, SM5 and 61A24) Rhizobium.

leghaemoglobin mRNA. The data in Fig. 3 suggest that all ineffective nodules tested contain leghaemoglobin mRNA sequences in
their poly A(+) polysomal RNA, although at a much reduced level as
compared to the wild type effective nodules. Among the ineffective
nodules (produced by different Rhizobium mutants) the concentration
of Lb sequences varies over several orders of magnitude. This,
along with the observation that Lb appears before nitrogenase during
nodule development, suggests that the induction of Lb-sequences is
independent of nitrogenase and the magnitude of Lb induction is
controlled by some factor(s) of Rhizobium origin. In some ineffective nodules (e.g. developed by strain 61A24) only one of the
major leghaemoglobins (LbF) is synthesized (Verma and Haugland,
unpublished data). Accumulation of leghaemoglobin is apparently

dependent on the effectiveness of the nodules in fixing nitrogen.
In soybean LbF accumulates over LbS, even though the synthesis of
the latter predominates in mature nodules (Verma et al. 1979). This
suggests different half lives for the Lb components. Since there is
a temporal difference in the induction of the two leghaemoglobins in
soybean and their synthesis changes during root nodule development,
they may perform different functions in root nodule symbiosis.

NODULIN-35

Presence of Nodulin-35 in effective and ineffective nodules:
Analysis of soluble cytoplasmic proteins from uninfected roots and
nodules developed by various mutants of Rhizobium japonicum by
sodium dodecyl sulfate - polyacrylamide gel electrophoresis shows
the presence of several polypeptides specific to nodules. Among
these is a polypeptide of about 35,000 MW which is present in all
nodules developed by effective as well as ineffective strains of
Rhizobium. This peptide, referred to as "Nodulin-35" (Legocki and
Verma 1979a) represents about 4% of the total cytoplasmic protein.
Purification of this peptide was accomplished by preparative SDS-gel
electrophoresis and antiserum was raised to this protein. Reaction
of the anti-Nodulin-35 serum with soluble proteins from various
ineffective nodules as well as from control root tissue showed
(Fig. 4) that this protein is "nodule-specific" and is present in
all nodules irrespective of their effectiveness in fixing nitrogen.
Lack of this protein in free-living Rhizobium and bacteroids

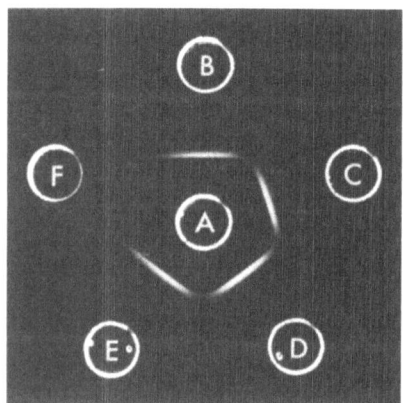

Fig. 4.   Immunodiffusion assay of the anti-Nodulin-35 serum (a)
with the cytoplasmic proteins from root nodules developed by (b)
effective (61A76); (c, d, and e) ineffective (SM4, SM5 and 61A24)
strains of Rhizobium and (f) control root. (Data from Legocki and
Verma 1979).

suggests that it may be of plant origin. This conclusion is
supported by the specific immunoprecipitation   of the in vitro
translation product of nodule polysomes with antibodies against
Nodulin-35 as well as inhibition of its synthesis by cycolheximide.
Further characterization of Nodulin-35 will demonstrate the role of
this peptide in the development of the root nodule symbiosis.
Synthesis of a molecular probe (cDNA) to Nodulin-35 mRNA will allow
the measurements on the coordination of induction of this protein
with leghaemoglobin.This will contribute to our understanding of the
role of Nodulin-35 in the establishment and/or maintenance of
symbiosis.

OTHER.HOST GENES INVOLVED IN SYMBIOTIC NITROGEN FIXATION

Estimation of the total number of structural genes active in
nodules: Classical genetic experiments have suggested the
involvement of several host genes (Holl and LaRue 1976) in effective

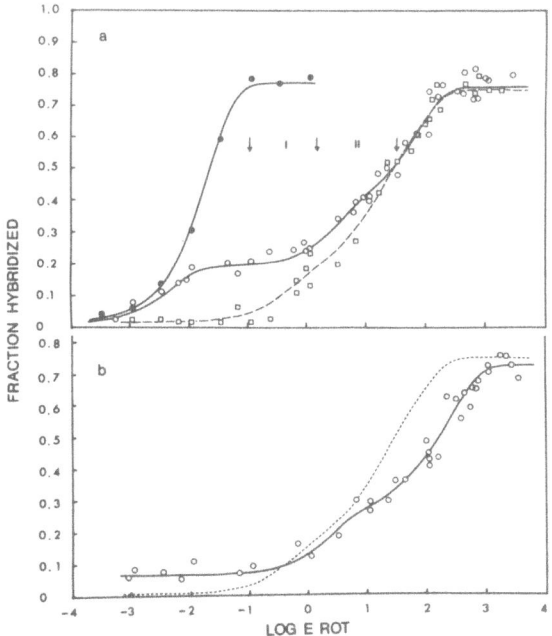

Fig. 5.   a, Homologous hybridization of nodule-cDNA (o) and
Lb-cDNA (●) with nodule polysomal poly(A)+ RNA and root-cDNA (□)
with root-polysomal poly(A)+ RNA. The two components of the middle
abundant sequences are marked and were used for purification of
these sequences (see fig. 6).

    b, Cross hybridization of nodule-cDNA with root RNA (o) and
root-cDNA with nodule-RNA shown by dotted line (redrawn from Auger
et al. 1979).

nodulation. Some of these genes appear to be involved in the recognition of the bacteria while others may be responsible for the development and function of the nodules. In order to estimate the the number of structural genes active in nodules we determined the total sequence complexity of the polysomal poly(A)+ RNA populations from uninfected root and the mature nodule tissues by RNA-cDNA hybridizations (Auger et al. 1979). If the molecular mechanism of root nodule development involved repression of several sequences or induction of some new sequences, this could be detected by using homologous and heterologous hybridizations with these two tissues.

Such experiments are shown in Fig. 5a and b. The poly(A)+ RNA sequences can be resolved into three abundance classes in both uninfected root and in nodule tissue. The most abundant component in nodule has been characterised (Baulcombe and Verma 1978, Auger et al. 1979) to contain primarily Lb sequences. Similarities between the mRNA's of the three major leghaemoglobins which are structurally very similar would permit extensive cross hybridizations and thus it may appear as a single kinetic component. This component is not detectable in control root tissue and increases during root nodule development (Fig. 2b). The middle abundant component, which comprises of about 500 sequences in nodules, is much more abundant than that in roots while the final transitions of both tissues contain almost the same number of sequences and constitute similar fractions of the cDNA. Due to the difficulties in accurately determining the complexity of the final transition (Bishop et al.1974, Rayffel 1976, Young et al. 1974, Hynes et al. 1977), it appears that, within the experimental limitations of this technique, both uninfected root and nodule tissues contain about the same number of structural genes (18,000-22,000) in their polysomal poly(A)+ RNA populations.

The homologies of structural genes between the two tissues was determined by heterologous hybridizations. Both tissues showed a substantial sequence homology though a shift in abundance in many sequences is observed. The apparent sequence homology of nodule RNA with root cDNA may be due in part to the fact that only about 50% of the nodule cells are infected with Rhizobium. The uninfected cells would contain mRNA sequences of root RNA. The kinetics of the root cDNA - nodule RNA hybridization suggests that most sequences which are present in the root are also present in nodules.

In the other heterologous reaction of that between nodule-cDNA and root RNA, the results indicate that most nodule sequences are also present in root, but at lower concentrations. This data does not, however, rule out the possibility that a small number of "nodule specific" sequences exist in nodule RNA but due to the lack of specific probes for these sequences, they are not detectable in this analysis.

Expression of the abundant mRNA sequences during root nodule
development: From the homologous hybridization reaction of
nodule-cDNA with nodule RNA it is apparent that besides leghaemoglobin there is a class of middle abundant sequences which are more
abundant in nodule than in root tissue. In order to examine the
mode of expression of these sequences, the total nodule-cDNA was
fractionated by repeated hybridization, removing the most and least
abundant cDNAs (Auger, Legocki and Verma, manuscript in preparation). Hybridization of this middle-abundant nodule-cDNA with
nodule and root RNA resolved the middle abundant sequences into two
components. These two subclasses, mid-I and mid-II, are composed of
about 30 and 1200 sequences respectively, representing 2400 and 75
copies per cell in nodules (Fig. 6a). Evidently purified middle
abundant cDNA contained some of the rare sequences since the total
complexity of middle abundent components is about 500 in the unfractionated cDNA. Similar kinetics are obtained with root tissue
though these sequences are 20-30 fold less abundant in the root.

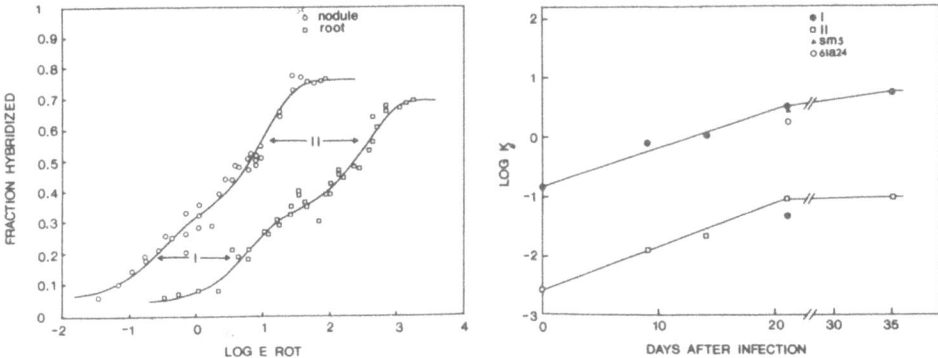

Fig. 6. a, Hybridization of mid-cDNA probe to nodule and root
RNA. The mid-cDNA was prepared by hybridizing total nodule cDNA up
to a Rot of 8×10^{-2} and removing the hybridized fraction (which
contains Lb-sequences). The hybridization was further continued up
to a Rot of 6.8 and the unhybridized fraction was removed by
hydroxylapatite chromatography. This cycle of hybridization was
repeated 3 times to obtain a mid-cDNA probe.
b, Increase in abundance of mid-RNA sequences during root
nodule development. The RNA from different age nodules was
hybridized with mid-cDNA probe and the rate constant of the reaction
was determined. Similar analysis was performed on RNA from 21 day
ineffective nodules developed by Rhizobium mutants SM5 (▲) and
61A24 (O). (Data from Auger, Legocki and Verma, manuscript in
preparation.)

To determine the temporal sequence of increase in abundance of these mRNAs, the mid-cDNA probe was hybridized to poly A(+) polysomal RNA from developing nodules. Both components showed a parallel increase in concentration up to 21, days by which time they reached apparent saturation (Fig. 5b). Analysis of these sequences in ineffective nodules developed by <u>Rhizobium</u> mutants showed the increase in abundance of most of these sequences although at a slightly reduced level (Fig. 6b).

IDENTIFICATION OF "NODULE-SPECIFIC" PROTEINS (NODULINS):

Antiserum raised to total soluble proteins from nodules (antinodule serum), when reacted with soluble proteins from control root and nodules in tandem-crossed immunoelectrophoresis (Fig. 7), showed that there are several proteins which are present in nodules but are not detectable in root tissue. About 9 major precipitation lines can be detected (Fig. 7C). Most of these proteins, including Nodulin-35 (Legocki and Verma 1979a), are synthesized on 80S type host polysomes. In order to obtain an antiserum specific to these proteins, the anti-nodule serum was absorbed by root proteins and the remaining fraction of antibodies was considered "nodule-specific" (Legocki and Verma 1979b).

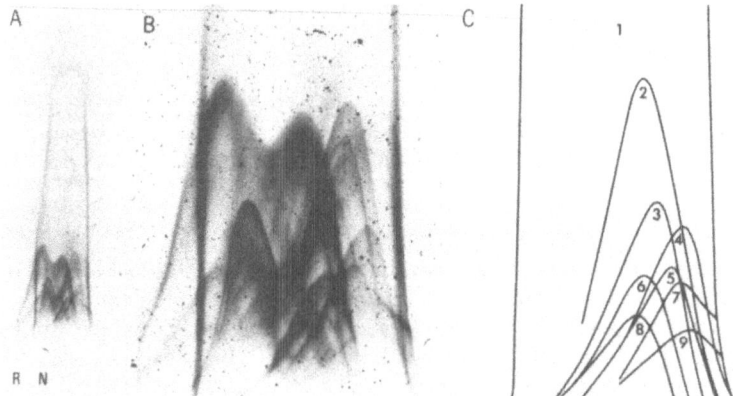

Fig. 7. Tandem-crossed immunoelectrophoresis of soluble cytoplasmic proteins from uninfected root (R) and root nodules (N) with the antibodies against root nodule proteins. Panel B is an enlargement of A; panel C outlines the precipitation arcs formed by the "nodule-specific" proteins (data from Legocki and Verma 1979b).

Immunoprecipitation of the <u>in vitro</u> translation product of the 80S-type nodule polysomes with this antiserum and analysis of the

resulting immunoprecipitate on 2-D gels (O'Farrell 1975) showed that
there are several "nodule-specific" peptides referred to as Nodulins
(Fig. 8). These proteins are not detectable in the translation product of the control root polysomes (Legocki and Verma 1979b). A
minimum number of 20 can be assigned to these proteins. It is
interesting to note that most of these proteins are between
12-20,000 MW and thus their mRNA would be of 8-10S size. Due to the
overlap with leghaemoglobin (16,000 MW),these sequences would contribute to the heterogenity of the 9S RNA fraction observed earlier
(see Baulcombe and Verma 1978).

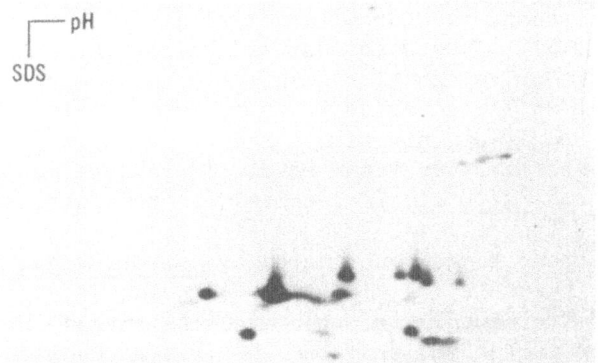

Fig. 8. Two dimensional gel electrophoresis of the immunoprecipitated in vitro translation product of nodule polysomes with
"nodule-specifc" antibodies (data from Legocki ad Verma 1979b). In
order to visualize polypeptides other than leghaemoglobin, translation was carried out in the presence of ^{35}S-methionine (leghaemoglobin does not contain methionine residues, Ellfolk and Sievers
1971, Nicola 1975).

There appears to be a direct correlation between the appearance
of "nodule-specific" proteins and the increase in abundance in
certain sequences in the nodule. Attempts were made to prepare
"nodule-specific"-cDNA by immunoprecipitation of polysomes using
"nodule-specific" antiserum. RNA isolated from these polysomes was,
reverse transcribed and used as a probe for hybridization to nodule
RNA. The preliminary data suggest that "nodule-specific" sequences
may be coded for by the abundant population of mRNA. This data does
not rule out the possibility;however, that some "nodule-specific"
sequences may still be present in the least abundant class since all
proteins are not immunoprecipitated with equal efficiency
and their mRNA sequences are not represented in this "nodule-specific"
probe. In any event, it is apparent that during the development of
root nodule symbiosis, in addition to leghaemoglobin, a family of
peptides (Nodulins) are produced by the host as a result of
infection with Rhizobium.

OVERVIEW

        The contribution of the host in root nodule symbiosis is well
recognized, although the specificity of this interaction is not
fully understood. Among the host genes which are involved in this
process are some genes which are responsible for primary interaction
at the cell surface of the root hair with Rhizobium. These genes
are "constitutively" expressed in legume roots providing a specificity for recognition by bacteria. Certain mutants of soybean such
as Rj_1 (Caldwell 1968) are unable to recognize Rhizobium and thus
may lack the genes responsible for this specificity. Similarly some
mutants of Rhizobium (e.g. strains SM-1, SM-2) which can not form
nodules on soybean, are capable of fixing nitrogen in culture (Maier
and Brill, 1976). Thus both host and Rhizobium genes are involved
in the process of recognition. Infection of the host cell takes
place through an invagination process (Nutman 1956) where bacteria
become enclosed in an infection thread lined by the host cell wall
and the plasmalemma (see Verma et al. 1978a). A cooperative action
of the host and Rhizobium is required for the infection of the host
cell in which two cell wall hydrolysing enzymes, pectinase and
cellulase, appear to be contributed by Rhizobium and the plant
respectively (Verma et al. 1978b). Following infection, the
bacteria are "released" from the infection thread into a membrane
vesicle which is derived from the host plasma membrane and retains
several characteristics of this membrane during subsequent
development of the nodule (Verma et al. 1978a).

        Development of the nodule does not assure a functional
symbiosis since mutation in the host as well as in Rhizobium may
produce an ineffective nodule which is unable to fix nitrogen. The
ineffectiveness may be due to the abortive development of the nodule
and is influenced by several host genes which are apparently under
bacterial influence. There may be two sets of such genes, those
involved in the structural development of the nodule and those related to its function. While there is a tight correlation between
the presence of leghaemoglobin and the effectiveness of the nodules
in fixing nitrogen, Nodulin-35 is produced in all effective as well
as ineffective nodules. Studies on the induction of leghaemoglobin
genes suggest that the expression of Lb genes is not tightly coupled
with that of nitrogenase. A different level of expression of Lb
genes was found in various ineffective nodules. Since Lb does not
accumulate significantly in ineffective nodules, stability of Lb
mRNA or its product may play an important role in controlling the
level of expression of leghaemoglobin genes.

        The complex interdependence of the host and Rhizobium suggests
that there must be other host genes which are involved in symbiotic
nitrogen fixation, however no other plant gene product has yet been
identified. The fact that there is a dramatic shift in the middle
abundant class of mRNA in root nodules suggests that some of the

"nodule-specific" proteins may be coded for by this class of mRNA. An antibody probe prepared to "nodule-specific" proteins (Legocki and Verma 1979b) immunoprecipitates several peptides from in vitro translation product of nodule mRNA. Further characterization of these products will help to determine the total number of the host genes which are obligatory for symbiotic nitrogen fixation. It is not known if all the "nodule-specific" sequences are under a common control by Rhizobium or a cascade of events is involved in the development of the nodule leading to the establishment of a "functional symbiosis". The fact that both effective and ineffective nodules contain most of the abundant sequences suggests that induction of these sequences follows by infection with Rhizobium and is independent of nitrogenase activity, however, the magnitude of induction as well as their maintenance appears to depend on the concomitant induction of nitrogenase in bacteria. Failure to induce nitrogenase may result in a parasitic/saprophytic state in the nodules and senescence may ensue.

Since Rhizobium can fix nitrogen nonsymbiotically, it is for the benefit of the host that a symbiosis be established with bacteria for effective utilization of the fixed nitrogen (Verma et al 1978b) . Considering the stringent requirements of the nitrogenase (high energy and low pO_2) it may not be possible to transfer nitrogen-fixation genes from bacteria directly to other plant cells. However, it is likely that transfer of those plant genes which are responsible for development of symbiosis with Rhizobium to other plants that can not associate with Rhizobium, may confer the capacity of nitrogen fixation to these plants.

Acknowledgements

This study was supported by grants from the Rockefeller Foundation and the Natural Sciences and Engineering Research Council of Canada. I would like to acknowledge the contribution of R. Legocki, S. Auger, L. Lacroix and D. Baulcombe to this study. I wish to thank W. Brill for providing Rhizobium japonicum mutants (SM3 to SM5), K. Newrock and D. Sullivan for reading and C. Neubrandt for typing this manuscript.

REFERENCES

Appleby, C.A., Nicola, N.A., Hurrell, J.G.R., and Leach, S.J. 1975. Biochemistry 14: 4444-4450.

Auger, S., Baulcombe, D., and Verma, D.P.S. 1979. Biochem. Biophys. Acta (in press).

Baulcombe, D. and Verma, D.P.S. 1978. Nucleic Acids Research 5: 4141-4153.

Bergersen, F.J. and Goodchild, D.J. (1973). Aust. Jour. Biol. Sci. 26: 741-756.

Bishop, J.O., Morton, J.G., Rosbash, M., and Richardson, M. 1974. Nature 250: 199-204.

Bisseling, T., Van Den Bos, R.C.; Weststrate, M.W., Hakkaart, M.J. and VanKammen, A. (1979). Biochem. Biophys. Acta 562: 515-526.

Caldwell, B.E. 1966. Crop Science 6: 427-428.

Caldwell, B.E. and Vest, G. 1968. Crop Science 8: 680-682.

Cutting, J. and Schulman, H.M. 1971. Biochem. Biophys. Acta. 229: 58-62.

Dilworth, M.J. 1969. Biochem. Biophys. Acta 184: 432-441.

Ellfolk, N. and Sievers, G. 1971. Acta. Chem. Scand. 25: 3535-3534.

Goldberg, R.B., Hoschek, G., and Kamalay, J.C. 1978. Cell 14: 123-131.

Holl, F.B. and Larue, A. 1976. Proc. Ist Internat. Symp. Nitrogen Fixation. Vol. 2: 391-399.

Hurrell, G.R. and Leach, S.J. 1977. FEBS Lett. 80: 23-26.

Hynes, N.E., Groner, B., Sippel, A.E., Nguyen-Hun, M.C., and Schütz, G. 1977. Cell 11: 923-932.

Legocki, R. and Verma, D.P.S. 1979. Science (in press).

Legocki, R. and Verma, D.P.S. 1979. Proc. Nat. Acad. Sci. (submitted)

Maier, R.J. and Brill, W.J. 1976. J. Bact. 127: 763-769.

Nicola, N.A. 1975. Ph.D. Thesis. Univ. of Melbourne.

Nutman, P.S. 1956. Biol. Rev. Cambridge Philos. Soc. 31: 109-151.

O'Farrell, P.H. 1975. J. Biol. Chem. 10: 4007.

Puepplke, S.G., Bauer, W.D., Ecegstra, K., and Ferguson, A.L. 1978. Plant Physiol. 61: 779.

Ryffel, G.V. 1976. Eur. J. Biochemistry 62: 417-423.

Verma, D.P.S., Nash, D.T., and Schulman, H.M. 1974. Nature 251: 74-77.

Verma, D.P.S. and Bal, A.K. 1976. Proc. Natl. Acad. Sci. USA, 73: 3843-3847.

Verma, D.P.S., V. Kazazian, V. Zogbi and A.K. Bal (1978a). J. Cell Biol. 78: 919-936.

Verma, D.P.S., V. Zogbi and A.K. Bal (1978b) Plant Sci. Letters 13: 137-142.

Verma, D.P.S., Ball, S. Guérin, C.W., and Wanamaker, L. 1979. Biochemistry 18: 476-483.

Young, B.D., Harrison, P.R., Gilmour, R.S., Birnie, G.D., Hell, A, Humphries, S., and Paul, J. 1974. J. Mol. Biol. 84: 555-568.

THE Ti-PLASMIDS OF AGROBACTERIUM TUMEFACIENS AND THEIR ROLE IN CROWN GALL FORMATION

J. Schell[#¶] and M. Van Montagu[#]

[#] Laboratory for Genetics, State University-Gent (Belgium)
[¶] Max-Planck-Institut für Züchtungsforschung-Köln (BRD)

INTRODUCTION. Crown Gall : Genetic Engineering in Function of a Parasitic Mode of life.

A.tumefaciens strains induce tumors (crown galls) on a wide range of dicotyledonous plants[1]. These tumors contain stably transformed - genetically altered - cells that are selfproliferating and graftable[2].

The capacity to induce tumors is genetically determined by large plasmids : the Ti-plasmids[3]. Removal of the Ti-plasmid from oncogenic strains results in the loss of oncogenicity* whereas introduction of Ti-plasmids in non-oncogenic Agrobacterium strains produces oncogenic derivatives[4,5].

Several observations have allowed a fairly precise explanation of this neoplastic transformation.It was demonstrated[6] that Ti-plasmids carry genes that somehow determine the specificity of synthesis of so-called "opines"[7,8] by Crown-gall cells. Other Ti-plasmid genes allow Agrobacteria to use specific opines as sole Carbon, Nitrogen and energy source[4,5,6,7,9]. The correlation between the specificity of opine synthesis in transformed plant cells and the specificity of opine catabolism by the transforming bacteria[8], could thus be explained genetically, provided that one assumed that the Ti-plasmid genes specifying opine synthesis were transferred to the plant cells as a result of -or

---

* Oncogenicity is used here in a broad sense as the capacity[4] of an Agrobacterium strain to induce crown-gall tumors on normally susceptible host plants.

concomitant with- the transformation event. These observations therefore provided genetic evidence in favour of a model involving the Ti-plasmid in a DNA-transfer from bacterium to plant.

The correctness of this model was demonstrated by hybridisation experiments indicating that a small segment of the Ti-plasmid showed specific homology with DNA extracted from transformed plant cells[10,11,12]. The demonstration that the T-DNA (i.e. the Ti-plasmid DNA segment present in transformed plant cells) is actually responsible for the synthesis of the specific opine by the transformed plant cells and probably also for the maintenance of their tumorous condition, was based on a combination of hybridization data and genetic mapping of various functions on the Ti-plasmid[11,12,13]. The genetic mapping has also borne out the conclusion that opine synthesis and opine catabolism are determined by distinct genetic loci on the Ti-plasmid[9]. All these and other observations can be fitted into a biological model of "genetic colonisation"[11]. It is obvious that opine synthesis and catabolism has played a central role in the evolution of Ti-plasmids. Most Ti-plasmids fall into one of three groups depending on whether they code for octopine or nopaline metabolizing enzymes[6,14,15] and depending on their host-range[16]. These different types of Ti-plasmids are clearly of distinct evolutionary origin since they differ markedly in DNA sequences. Only DNA segments involved in the determination of oncogenicity and conjugational transfer seem to be relatively well conserved in most Ti-plasmids[13,15].

Furthermore opines have been found to specifically induce the conjugational transfer properties of Ti-plasmids[17,18]. It is therefore reasonable to assume that Ti-plasmids have evolved as a natural means for Agrobacterium to genetically "engineer" or "colonize" plant cells, thus illustrating a new type of a parasitic mode of life. By the transfer of a specific genetic information carried as a defined segment of the Ti-plasmid - the so-called T-DNA - Agrobacteria manage to force proliferating plant cells to synthesize products (opines) that only homologous Ti-plasmid harbouring bacteria can utilize as sole Carbon, Nitrogen and energy source. In the last couple of years much of the research on the Crown-gall system has centered around three major problems. This review will briefly report the progress that was achieved with regard to these questions.

THE GENETIC AND FUNCTIONAL ORGANISATION OF Ti-PLASMIDS

The general aim of these studies was to provide a definitive proof for the involvement of specific plasmid-borne genes in the determination of the properties of Crown-gall plant tissues and a first step towards the elucidation of the precise mechanism of

tumorous transformation. It was hoped that it would be possible
to identify the genes and later the gene-products that are directly or indirectly involved in this phenomenon. The approach was to
isolate and map mutations inactivating the functions that were
supposedly controlled by Ti-plasmid genes. A very efficient way
to obtain mutant Ti-plasmids was by insertion of antibiotic
resistance transposons[19,20]. No selection for mutant phenotypes
has to be devised and a straightforward selection for Ti-plasmids
carrying the transposon was sufficient to produce a collection of
mutant Ti-plasmids from which most of the desired phenotypes were
obtained by screening a few hundred mutants[15,20].

    The location of the site of insertion on a physical map of
the Ti-plasmid was determined by isolating the mutant plasmid on
a preparative scale and by analysing the fragmentation pattern of
the digest with different restriction endonucleases. The mapping
of a given insertion or deletion can be deduced from the restriction fragment(s) which disappear(s) or change(s) mobility. In
several cases the mapping was confirmed by analysing heteroduplex
molecules formed with a reference plasmid marked by two well
mapped insertions under the electron-microscope. Recently[21], a
more efficient way was developped to analyse large numbers of
mutant colonies using the Southern blot hybridisation technique.
A filter containing the fragments obtained from a restriction
endonuclease digest of the total DNA (chromosomal and plasmid) of
the strain to be analysed, was hybridised with the radioactive
probes, which consisted of the pure Ti-plasmid DNA and the drug
transposon DNA.

    A further advantage of transposon insertion mutagenesis is
the fact that the site at which the transposon is inserted becomes the starting point for the formation of deletions. Starting
with a given pTi::Tn mutant plasmid it was possible, by selection
or screening for the loss of one or more plasmid encoded phenotypes, to obtain deletions of varying sizes. The extent of the
deletion could than be determined with the same methods as described above for the mapping of insertions.

    The physical map on which both the insertion and deletion
mutants were localized, had been previously constructed by molecular cloning in pBR322 of large DNA fragments derived from
partial Hind III digests of these plasmids. Restriction endonuclease digestion of the different clones and Southern blot hybridisations, established the map order.

    A map for Hind III, Hpa I, Sma I, Kpn I, Eco RI, Xba I and
Bam I of both an octopine[22a] and a nopaline[23] type of Ti-plasmid
was thus constructed.

TABLE 1. Localization of some Ti-plasmid encoded functions by
          transposon insertion mutagenesis.

Phenotypes		Approximate map position in Md*	
		pTiC58	pTiAch5
Ape	Exclusion of phage AP1[22b,c25]	70	9-11
Agr^S	Sensitivity to the pAT-K84 encoded agrocin[25,26]. Only <u>Agrobacterium</u> strains harbouring a nopaline plasmid show this$_R$ phenotype,the other strains are Agr^R.	85-87	---
Arc	Arginine catabolism[27]. Is part of a complex, inducible pathway. Deletion mutants constitutive for arginine catabolism (Arc^C) have been isolated[22c,e].	6-7	12
Agc	Agropine catabolism, an octopine tumor specific sugar derivative[28,22c].	---	50 51
Noc	Nopaline catabolism. An inducible function, possibly under both negative and positive control. Mutants can be grouped in several classes (according to the aspect of colony growth on nopaline as sole nitrogen or sole carbon source). The necessary enzymes are encoded by nopaline Ti-plasmids. Nopaline is an inducer of its own catabolic pathway. Noc constitutive mutants are able to catabolize octopine[7].	6-7 and 11-12	---

* The length of the octopine Ti-plasmid pTiAch5 and of the nopaline Ti-plasmid as measured under the electron microscope[22f] and as obtained from the sum of the fragment lengths from a single restriction endonuclease digest[22a,d], correspond sufficiently well to propose a genome size of 120 Md for pTiAch5 and of 132 Md for pTiC58. A common Sma I site, located in the T-DNA region was chosen as the zero coordinates for both plasmids. All mapping data given refer to distances in Md from this common restriction site.

Table 1 (cont'd)

Phenotypes		Approximate map position in Md*	
		pTiC58	pTiAch5
Nos	Nopaline synthesis in Crown-gall tumors. This function is encoded by nopaline Ti-plasmids only[6].	0-1	---
Occ	Octopine catabolism. An inducible pathway involving many functions. Octopine is an inducer. Nopaline is not catabolized by the octopine Ti-plasmid encoded enzymes[7,29].	---	24-26
Tra	Functions responsible for the conjugative phenotype of the plasmid. The transfer genes are repressed in the wild type plasmid. Conditions for induction of transfer have been determined and Tra constitutive mutants have been isolated[30].	70-80 and 14-19	15 20
Onc	Oncogenicity or the capacity for induction of Crown-galls on at least tobacco, peas and sunflower.	129-132 115 106-108 99-101 74	115-117 97-102 85-90 63-64
Onch	Host range effect on oncogenicity. Tumors are formed on Kalanchoë and/or potatoes, but not on tobacco, peas and sunflower.	128	9 and 105
Orc	Catabolisme of ornithine. It can be distinguished from Arc since $Arc^+ Orc^-$ mutants were found[22e].		28

In tables 1 and 2 we have summarized the data for the different Ti-plasmid encoded functions for which mutations were obtained and mapped. It appears from these studies that the Ti-plasmids have been assembled functionally starting from a number of different building blocks. Although little information is as yet available on the number of different genes required for the determina-

TABLE 2. Localization of some Ti-plasmid encoded functions by
          deletion mapping.

Approximate coordinates of the deletions in Md		Mutant phenotype
pTiC58	Δ 99 - 13	Onc^- Noc_R^- Orc^- Apr^-
	Δ 64 - 131	Onc^- Agr^R Ape^- tra^-
	Δ 35 - 110	Onc^- Agr^R Ape^- tra^-
	Δ 4 - 16	Noc^- Orc^C
	Δ 3 - 16	Noc^- Orc^-
	Δ 3 - 12	Noc^- Arc^C
pTiAch5	Δ 105 - 17	Onc^- Ape^-
	Δ 85 - 118	Onc^-

The "viability" of these deletion plasmids suggests that the
origin of replication must be in the region 16 to 35.

tion of a particular phenotype, it is nevertheless possible to
make some generalizations.

Oncogenicity functions

     As indicated in fig. 1 the Onc mutations were found to be
distributed over the entire Ti plasmid. These findings underline
that there is probably a diversity of functions involved in
determining consecutively the contact between the bacteria and
the plant[31,32,33], the transfer of the Ti plasmid via a plant
receptor structure, the transfer of Ti plasmid DNA to the plant
nucleus, the restructuring of the T-DNA into a replicating structure and finally the expression of functions that interfere with
the plant cell metabolism so as to create the tumor cell phenotype.

     By definition the Onc mutants which map within the T-DNA
identify those genes supposedly involved in the establishment
and/or maintenance of the tumorous growth of the plant cells ; in
nopaline plasmids these genes span roughly 5 Md. It is stricking
to note that this region also corresponds to the major area of
homology between nopaline and octopine Ti plasmids. To date there

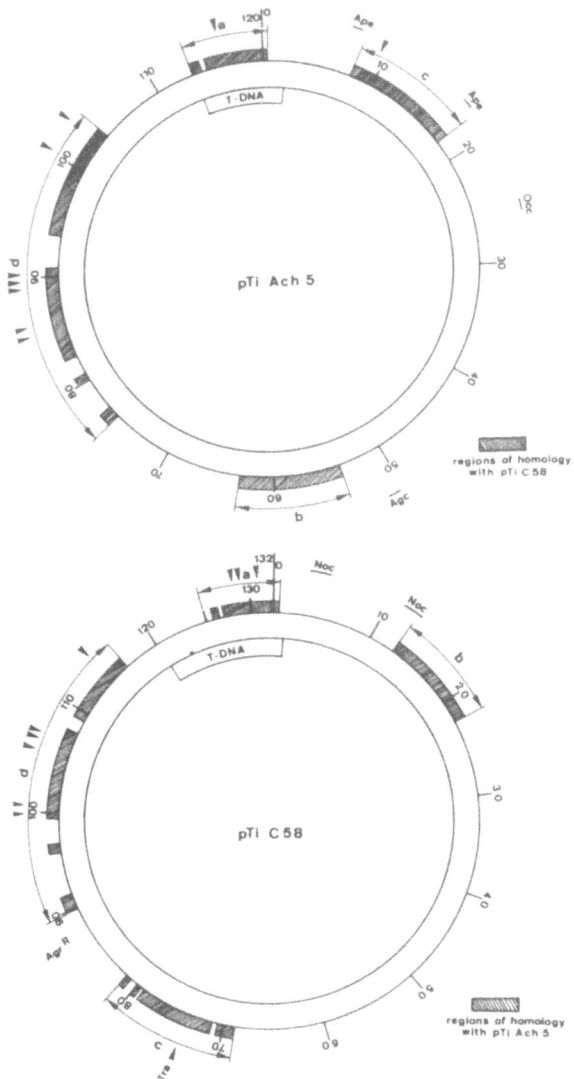

Fig. 1. Functional organization of an octopine (pTiAch5) and a nopaline (pTiC58) plasmid. The length of the plasmids (in Md) is indicated along the outer circle and is used as map coordinates. A Sma I restriction site in a segment common to both plasmids is chosen as zero point. The sequences common to both types of Ti-plasmid are indicated by a heavy line and are labeled a to d. The sites of transposon insertions resulting in Onc⁻ phenotypes are indicated by an arrow.

is no biochemical information about the products of this region. It is however well established that the Onc region of the T-DNA is actively transcribed in the plant cell[34,228]. Interestingly some Onc mutants (both within and outside of the T-DNA region) suggest a host specific interaction. For example, some mutants affect functions which are dispensible for tumor formation on Kalanchoë but not on other plants. There is no information about the products encoded by Onc regions outside of the T-DNA except that several of these DNA fragments cloned in E. coli synthesize proteins in minicells.

Opine biosynthesis

Ti-plasmid insertion mutants producing an Onc^+Nos^- phenotype were isolated[11,12,15,20]. Agrobacteria carrying such mutant Ti-plasmids induce tumors in which no nopaline synthesis can be detected. It is important to stress that these mutations were found to map within a particular fragment of the T-DNA region of the Ti-plasmids (see further) thus providing evidence for the direct and active involvement of this DNA segment in the control of opine synthesis in transformed plant cells. Furthermore these observations demonstrated that opine synthesis and determination of tumorous growth must be functionally independent but are controlled by DNA segments adjacent to one another on the T-DNA.

Opine catabolism

The functions involved in providing Agrobacterium with the capacity to use octopine or nopaline as sole nitrogen and/or sole carbon source can be analyzed due to the availability of a large set of insertion and deletion mutants in these functions[20,22]. Furthermore, these regions of the Ti-plasmid have been isolated and propagated by molecular cloning techniques. Cloned DNA fragments have been used to study the expression of these regions in Agrobacterium[221,j]. Preliminary results point to the existence of a rather complex operon structure probably containing both negative and positive controlling elements. For the octopine plasmid pTiB6 there clearly is a common element in the regulation of octopine catabolism and conjugal tranfer[7,29]. These mapping data have also borne out the conclusion that opine synthesis and opine catabolism are determined by distinct genetic loci on the Ti-plasmid.

When taken together with hybridisations studies[15,22d], comparing various segments of octopine and nopaline Ti-plasmids (see fig. 1), these genetic studies have clearly indicated that the genes allowing Agrobacteria to catabolize opines appear to have determined the evolution of these plasmids and hence of the Crown-gall phenomenon. It is interesting to note that these genes do not appear to have a single origin but must instead be of diverse origins since little or no homology was found in these genes between octopine and nopaline type plasmids.

T-DNA, ITS SIZE, LOCALISATION AND POSSIBLE ASSOCIATION WITH PLANT DNA

When it became obvious that the opine synthesizing capacity, expressed by the transformed plant cells, was actually a trait determined by the Agrobacterium Ti-plasmid, it was evident to consider the possibility that a segment of the Ti-plasmid became part of the transformed plant genome. A first indication was obtained by solution DNA hybridisation studies. Some restriction fragments of the octopine plasmid pTiB6806 did indeed show an accelerated rate of reassociation after addition of DNA from a particular tobacco Crown-gall tumor[10]. A more complete picture was obtained when Southern gel blotting hybridisations were performed with the Crown-gall DNA. Total tumor DNA was digested with a variety of restriction enzymes and used to drive the hybridisation of nick translated cloned fragments of the Ti-plasmid[11,12,22h]. Some generalizations may be advanced from this approach. In the case of all tumors induced by the nopaline plasmids pTiT37 and pTiC58 the T-DNA corresponds to one large 15.6 Md continuous segment of the Ti-plasmid. Cloned fragments representing the whole of the Ti-plasmid were used as radioactive probes in this Southern gel blotting hybridisations. Only fragments overlapping with the thus defined T-DNA region showed positive and reproducible hybridisations. The T-DNA is not integrated in plastid DNA (M.-D. Chilton, personal communication) but appears to be located in the nucleus[22k]. Several copies of the T-DNA can be present as seen in fig. 2 where border fragments of the T-DNA were used as probes in hybridisations to Southern blots of tumor DNA. In this tumor line one can observe that there are at least three different right borders and three different left borders. The molecular weights of these fragments suggest that they are composed of T-DNA linked to some other DNA, possibly plant DNA.

As can be seen in fig. 3 an Eco RI digest of a cloned tobacco teratoma T37 Crown-gall DNA indicated the presence of two distinct "right end border" composite fragments of respectively ±9.0 and 6.0 Md. Both these composite fragments were recently cloned[22m], using bacteriophage lambda as the initial cloning vector. A carefull analysis by Southern blot hybridisations of restriction endonuclease fragments derived from these cloned composite fragments is summarized in fig 4. As can be seen the 6.0 Md fragment indeed consisted of the "right end" border of the T-DNA linked to a very small sequence of DNA not present as such

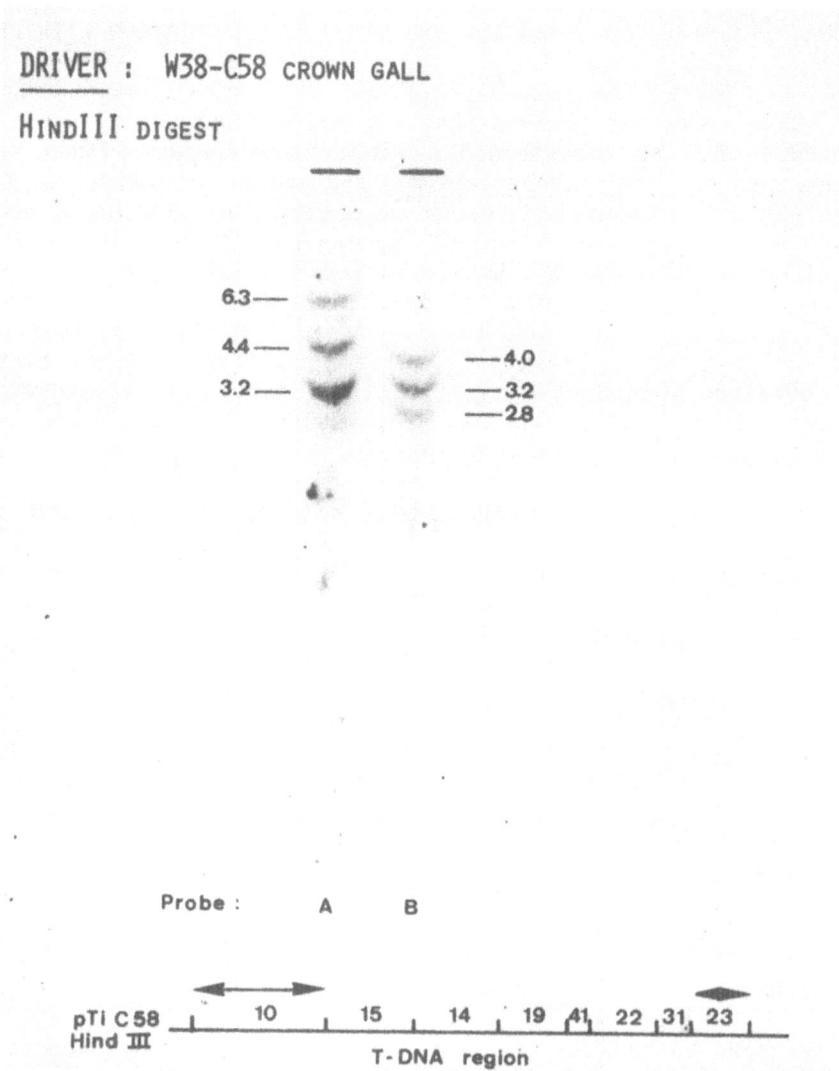

Fig 2. A "genomic blot" of Wisconsin 38 tobacco crown gall DNA, digested with Hind III restriction enonuclease, was hybridised to the nick-translated T-DNA border fragments H-10 (lane A) and H-23 (lane B). The autoradiogram demonstrates the existence of at least three border fragments.

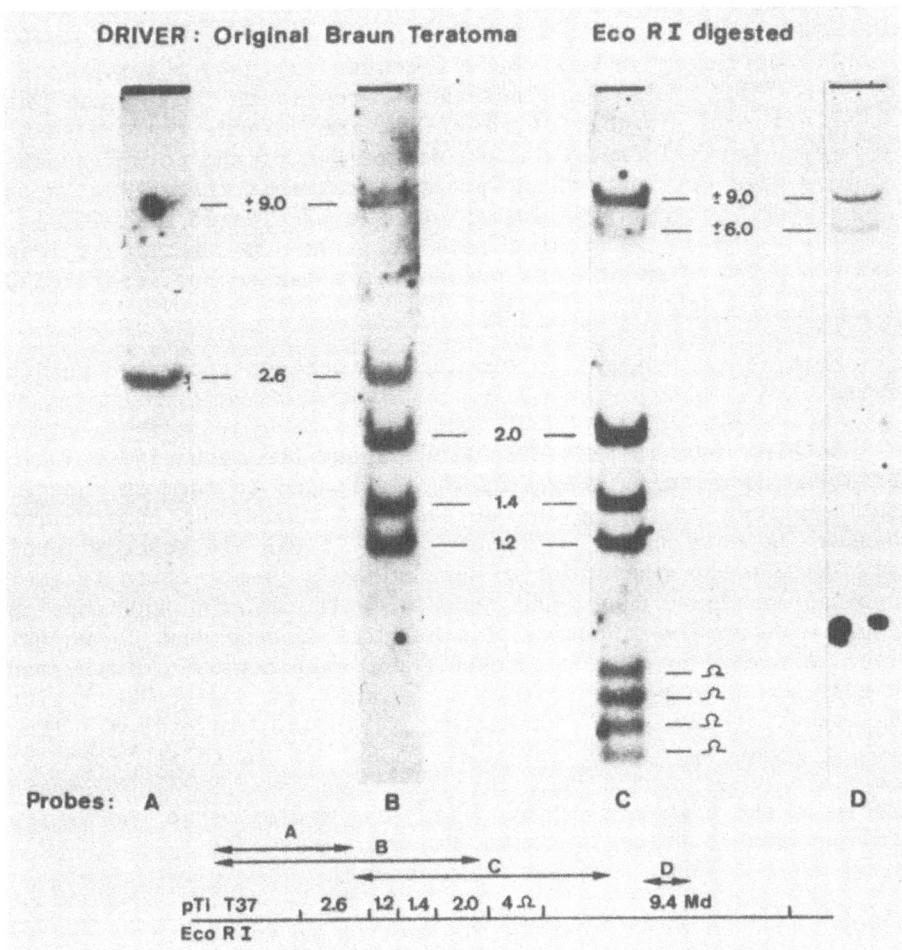

Fig 3. A "genomic blot" of cloned T37 tobacco teratoma from Dr. A. Braun's collection, digested with Eco RI was hybridised to nick-translated T-DNA fragments A, B, C, D.

in Ti-plasmid *. The 9.0 Md fragment gave unexpected results. Indeed it not only contained the expected "right end" border of the T-DNA (the Hind III fragments 23,31,22 and 41) but also part of the "left end" border (part of Hind III fragment 10). Furthermore a sequence of about 0.3 Md was found that could not be detected as such either in the Ti-plasmid Hind III fragment 23 or in the Hind III fragment 10.

Two different models could account for these observations :
1°   The T-DNA forms an independant replicon, "right end" and "left end" borders of the T-DNA are joined, rearrangements (e.g. partial duplications) of the DNA of the border sequences have occured and explain the fragments that do not occur as such in the T-DNA segment of the Ti-plasmid DNA.
2°   The T-DNA is integrated in some plant DNA so that at least two T-DNA segments are present as a tandem but separated by a 0.3 Md sequence of plant DNA.

Work is in progress to distinguish between these two possibilities.

Nothing much is presently known about the mechanism by which the T-DNA is either integrated or rearranged to form an independent replicon in the plant nucleus. A number of observations however indicate that the "ends" of the T-DNA are somehow involved. Within the resolution of our analysis (restriction fragment mapping) we found the T-DNA ends of both octopine and nopaline Ti-plasmids to be identical in several independent Crown-gall lines initiated on tobacco, petunia and arabidopsis. Furthermore inserts of transposons within a segment of the T-DNA region, non-essential for tumor formation, yield transformed cell lines with a T-DNA which has the same "ends" as found in Crown-gall containing wild-type T-DNA. The insert in this DNA is located at precisely the same site within the T-DNA region as in the mutated Ti-plasmid with which the tumor line was initiated.

---

* Note added in press. During the meeting in Edinburgh we reported that a subclone of the 6.0 Md fragment hybridized with repeated DNA sequences in non transformed tobacco callus DNA. Further analysis revealed that this hybridisation might have been caused by a small (0.15 Md) Eco RI fragment that was co-ligated with the 6.0 Md fragment into the λ-vector. The Eco RI site separating the 6.0 Md fragment from the 0.15 Md fragment appears to be partially insensitive to Eco RI digestion explaining why it was unexpectedly still present in the subclone. For the time being we therefore have to withdraw the conclusion that T-DNA would be integrated in repeated plant DNA sequences.

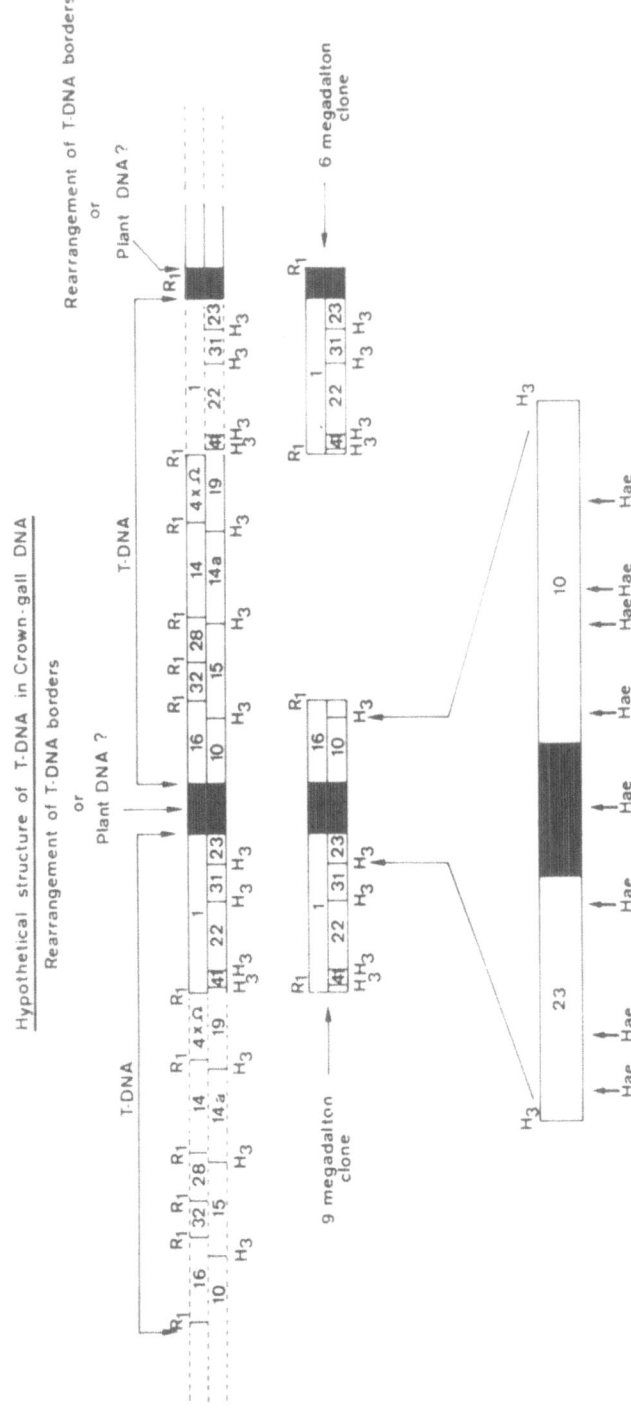

Fig. 4. Model of T-DNA in plants.
In solid lines the arrangement of the internal restriction endonuclease fragment of the T-DNA is illustrated. It is important to note that this arrangement of T-DNA fragments is identical both in the Ti-plasmid DNA and in the DNA of tobacco Crown-gall. Underneath the structure of the T-DNA Eco RI "right end" border composite fragments of ± 9.0 and 6.0 Md is illustrated. These composite fragments were isolated from T37 tobacco Crown-gall DNA by molecular cloning.

By genetic means (insertion and deletion mutants) is was established that the T-DNA consists of at least three different functional units. A central segment of about 5.0 Md is conserved in all types of Ti-plasmids. Mutations in this segment invariably result in the loss of capacity to produce plant tumors[20].

To either side of this conserved DNA segment, we found DNA sequences that are different in different types of Ti-plasmids[22d]. Mutations in these non-conserved segments of the T-DNA do not result in the loss of capacity to induce tumors. Mutations in what has been called the right side of the T-DNA (± 1.5 Md), do however result in Ti-plasmids that induce tumors in which no opine synthesis occurs, thus identifying the DNA sequence involved in opine synthesis.

Possible aberrant phenotypes of plant tumors induced with Ti-plasmids with a mutation in the "left" end of the T-DNA, have not yet been carefully studied.

CAN THE Ti-PLASMID BE USED AS A HOST-VECTOR SYSTEM FOR PLANTS?

With the realization that Ti-plasmids are in fact a natural host-vector system able to promote transfer, integration and expression of "foreign" DNA in plants, we asked whether of not the properties of this system could be used to introduce, at will, foreign DNA sequences in various plants. Recently[11,12,15] we were able to report that this was indeed the case. The general strategy used was to incorporate, by in vivo recombination, a bacterial transposon, Tn7 into the 1.5 Md segment of the T-DNA that codes for opine synthesis. Subsequently we were able to demonstrate that tobacco tumor-lines induced with such Tn7 harbouring Ti-plasmids, did indeed contain a T-DNA consisting of the whole of the T-DNA (15 Md) of the wild-type plasmid together with the whole of the Tn7 DNA (9.5 Md) still inserted in its original site in the T-DNA. Whether or not the T-DNA is transcribed in these plant cells, has not yet been established.

An important question is whether or not plant cells containing a T-DNA segment can be used to breed new plants. Recent observations[35] have demonstrated that whole plants can be regenerated from transformed teratoma cultures. In these studies tobacco teratoma derived tumor shoots were isolated and grafted to cut stem tips of normal tobacco plants of a morphologically distinct cultivar. This way, shoots were obtained that developed quite normally and ultimately flowered and set viable seed. It was found that the leaves of these grafts were normally organized but were still transformed in the sense that, when such specialized cells were isolated and planted on a basic culture medium, they grew as Crown-gall cells and synthesized nopaline. These tissues were studied by Southern blotting hybridisations and it was found that they still contained T-DNA from the TiT37 plasmid [22,25] (and M.P. Gordon, personal communication). In fact we were able to demonstrate that the T-DNA segment found in tissues derived from organized leaves of the teratoma grafts, was the same as the T-DNA segment found in the original teratoma tissues with which the grafts were initiated. No gross rearrangement of the T-DNA had therefore occurred during the differentiation of tumor cells to organized plant cells. It was thus demonstrated that it is possible to create new plants containing a specific DNA segment using the Ti-plamsid as a host-vector.

These promising results will now be followed by experiments where non transposon DNA and possibly plant genes are inserted into the T-DNA by in vivo recombination. It is a real challenge to put this natural genetic engineer, the Ti-plasmid, at work to introduce new traits into plants.

ACKNOWLEDGEMENTS

The authors wish to thank their collaborators[22] for providing the information presented in this article.
This work was supported by grants from the "Kankerfonds van de A.S.L.K." and from th " Fonds voor Wetenschappelijk Geneeskundig Onderzoek" (no. 3.0052.78).

REFERENCES

1. M. De Cleene and J. De Ley, The host range of Crown gall, Bot. Rev., 42:389 (1976)
2. A.C. Braun and P.R. White, Bacterial sterility of tissues derived from secondary crown gall tumors, Phytopathology, 33: 85 (1943)
3. I. Zaenen, N. Van Larebeke, H. Teuchy, M. Van Montagu, and J. Schell, Supercoiled circular DNA in crown-gall inducing Agrobacterium strains J. Mol. Biol., 86:109 (1974).
4. N. Van Larebeke, G. Engler, M. Holsters, S. Van den Elsacker, I. Zaenen, R.A. Schilperoort, and J. Schell, Large plasmid in Agrobacterium tumefaciens essential for crown-gall inducing ability, Nature, 252:169 (1974)
5. B. Watson, T.C. Currier, M.P. Gordon, M.-D. Chilton, and E.W. Nester, Plasmid required for virulence of Agrobacterium tumefaciens, J. Bacteriol., 123:255 (1975).
6. G. Bomhoff, P.M. Klapwijk, C.H.M. Kester, R.A. Schilperoort, J.P. Hernalsteens, and J. Schell, Octopine and nopaline synthesis and breakdown genetically controlled by a plasmid of A. tumefaciens, Mol. gen Genet., 145:177 (1976).

7. A. Petit and J. Tempé, Isolation of Agrobacterium Ti-plasmid
    regulatory mutants, Molec. gen. Genet., 167:147 (1978).
8. A. Petit, S. Delhaye, J. Tempé, and G. Morel, Sur les guani-
    dines des tissus de crown gall. Mise en évidence d'une
    relation biochimique spécifique entre les souches
    d'Agrobacterium et les tumeurs qu'elles induisent, Physio.
    vég., 8:205 (1970).
9. P. Klapwijk, P. Hooykaas, H. Kester, R.A. Schilperoort, and
    A. Rörsch, Isolation and characterization of Agrobacterium
    tumefaciens mutants affected in the utilization of octopine,
    octopinic acid, and lysopine, J. Gen. Microbiol., 96:155
    (1976)
10. M.-D. Chilton, H.J. Drummond, D.J. Merlo, D. Sciaky, A.L.
    Montoya, M.P. Gordon, and E.W. Nester, Stable incorporation
    of plasmid DNA into higher plant cells: the molecular basis
    of crown gall tumorigenesis, Cell, 11:263 (1977).
11. J. Schell, M. Van Montagu, M. De Beuckeleer, M. De Block, A.
    Depicker, M. De Wilde, G. Engler, C. Genetello, J.P.
    Hernalsteens, M. Holsters, J. Seurinck, B. Silva, F. Van
    Vliet, and R. Villarroel, Interactions and DNA transfer
    between A. tumefaciens, the Ti-plasmid and the plant host,
    Proc. R. Soc. Lond. B 204:251 (1979).
12. J. Schell, The use of the Ti-plasmid as a vector for the
    introduction of foreign DNA into plants, Proc. IVth Int.
    Conf. Plant Path. Bact., I.N.R.A.-Angers (1978).
13. A. Depicker, M. Van Montagu, and J. Schell, Homologous DNA se-
    quences in different Ti-plasmids are essential for oncoge-
    nicity, Nature, 275:150 (1978).
14. A. Montoya, M.-D. Chilton, M.P. Gordon, D. Sciaky, and
    E.W. Nester, Octopine and nopaline metabolism in
    Agrobacterium tumefaciens and crown-gall tumors cells:
    role of plasmid genes, J. Bacteriol., 129:101 (1977).
15. M. Van Montagu and J. Schell, The Ti-plasmids of A. tumefa-
    ciens, in: "Plasmids of medical environmental and commer-
    cial importance", K. Timmis, and A. Pühler, eds.,
    Elsevier/North Holland Medical Press, Amsterdam, (1979).
16. M. Tomashow, C. Panagopoulos, M.P. Gordon, and E.W. Nester,
    Host range of A. tumefaciens is determined by the Ti-plas-
    mid, in press (1979).
17. A. Kerr, P. Manigault, and J. Tempé, Transfer of virulence in
    vivo and in vitro in Agrobacterium, Nature, 265:560 (1977).
18. C. Genetello, N. Van Larebeke, M. Holsters, A. Depicker, M.
    Van Montagu, and J. Schell, Ti-plasmids of Agrobacterium
    as conjugative plasmids, Nature, 265:561 (1977).
19. J. Brevet, D.J. Kopecko, P. Nisen, and S.N. Cohen, Promotion
    of insertions and deletions by translocating segments
    of DNA carrying antibiotic resistance genes, in: "DNA, in-
    sertion elements, plasmids, and episomes", A.I. Bukhari,
    J.A. Shapiro, and Adhya, S.L., Eds, Cold Spring Harbor
    Laboratory, New York (1977).

20. M. Holsters, F. Van Vliet, B. Silva, M. De Block, A. Depicker, P. Dhaese, M. Van Montagu, and J. Schell, Genetic and functional organization of nopaline Ti-plasmids, submitted to Plasmid.

21. P. Dhaese, H. De Grève, H. Decraemer, J. Schell and M. Van Montagu, Rapid mapping of transposon insertion and deletion mutations in the large Ti-plasmids of A. tumefaciens, submitted to Nucleic Acids Research

22. Unpublished work of the authors' laboratory:
  a G. De Vos, et al., in preparation.
  b J.P. Hernalsteens, et al., in preparation.
  c H. De Grève et al., in preparation
  d A. Depicker et al., in preparation
  e M. De Block et al., in preparation
  f G. Engler et al., in preparation.
  g J. Seurinck et al., in preparation.
  h M. De Beuckeleer et al., in preparation.
  i E. Van Haute et al., in preparation.
  j J. Leemans et al., in preparation.
  k L. Willmitzer et al., in preparation.
  m P. O'Farrell et al., in preparation.

23. A. Depicker, M. De Wilde, G. De Vos, R. De Vos, M. Van Montagu, and J. Schell, Molecular cloning of the overlapping segments of the nopaline Ti-plamsid pTiC58 DNA as a means to restriction endonuclease mapping, submitted to Plasmid (1979).

24. J.G. Ellis, A. Kerr, M. Van Montagu, and J. Schell, Agrobacterium: genetic studies on agrocin 84 production and the biological control of crown gall, Physiol. Plant Path., in press (1979).

25. N. Van Larebeke, C. Genetello, J.P. Hernalsteens, A. Depicker, I. Zaenen, E. Messens, M. Van Montagu, and J. Schell, Transfer of Ti-plasmids between Agrobacterium strains by mobilization with conjugative plasmid RP4, Molec. gen. Genet., 152:119 (1977).

26. G. Engler, M. Holsters, M. Van Montagu, J. Schell, J.P. Hernalsteens, and R.A. Schilperoort, Agrocin 84 sensitivity: a plasmid determined property in Agrobacterium tumefaciens, Molec. gen. Genet., 138:345 (1975).

27. J. Ellis, A. Kerr, J. Tempé, and A. Petit, Arginine catabolism: a new function of both octopine and nopaline Ti-plasmids of Agrobacterium, Molec. gen. Genet., 173:263 (1979).

28. J.L. Firmin and R.G. Fenwick, Agropine - a major new plasmid-determined metabolite in crown gall tumours, Nature, 276:842 (1978).

29. P.M. Klapwijk, T. Scheuldermon, and R.A. Schilperoort, Co-ordinated regulation of octopine degradation and conjugative transfer of Ti-plasmids in Agrobacterium tumefaciens: Evidence for a common regulatory gene and separate operons, J. Bacteriol., 136:775 (1978).

30. A. Petit, J. Tempé, A. Kerr, M. Holsters, M. Van Montagu, and
    J. Schell, Substrate induction of conjugative activity of
    Agrobacterium tumefaciens Ti-plasmids, Nature, 271:570
    (1978).
31. B.B. Lippincott, M.H. Whatley, and J. Lippincott, Tumor induc-
    tion by Agrobacterium involves attachment of the bacterium
    to a site on the host plant cell wall, Plant Physiol.,
    59:388 (1977).
32. A. Matthysse, P. Wyman, and K. Holmes, Plasmid-dependent at-
    tachment of Agrobacterium tumefaciens to plant tissue cul-
    ture cells, Infect. Immun., 22:516 (1978).
33. M.H..Whatley, J.B. Margot, J. Schell, B.B. Lippincott, and
    J.A. Lippincott, Plasmid and chromosomal determination of
    Agrobacterium adherence specificity, J. Gen.Microbiol.,
    107:395 (1978).
34. W.B. Gurley, J.D. Kemp, M.J. Albert, D.W. Sutton, and J.
    Callis, Transcription of Ti-plasmid derived sequences in
    three octopine-type crown gall tumor lines, Proc. Natl.
    Acad. Sci. US, 76:2828 (1979).
35. R. Turgeon, M.N. Wood, and A.C. Braun, Studies on the reco-
    very of Crown gall tumor cells, Proc. Natl. Acad. Sci. USA,
    73:3562 (1976).

LOCATION AND FATE OF pTi T37 DNA IN REVERSION OF

CROWN GALL TERATOMA

M.-D. Chilton, R.K. Saiki, F.-M. Yang,
K. Postle, A.L. Montoya, E.W. Nester,
F. Quetier*, and M.P. Gordon**

Department of Microbiology and Immunology, and
Department of Biochemistry**, University of
Washington, Seattle, WA 98195, USA, and
Laboratoire de Biologie Moleculaire Vegetale*,
Universite Paris-Sud, 91405 Orsay, France

INTRODUCTION

Crown gall is a plant cancer that is incited by
inoculation of wound sites with Agrobacterium tumefaciens.
The resulting tumor cells are neoplastic and maintain
altered growth characteristics in vitro in defined medium
without exogenously supplied hormones (cytokinins
and auxins) that are required by normal tissues. In
addition, tumor cells posses the unique ability to synthesize high levels of basic amino acid derivatives
called opines, e.g. octopine (1) and nopaline (2,3),
compounds not found in normal plant tissue (4-6).

Large tumor-inducing (Ti) plasmids in Agrobacterium
tumefaciens (7) carry genetic information required for
oncogenicity (8-10). Ti plasmid genes dictate which opine
if any, will be manufactured by the tumor (5,11-13) and
separate Ti plasmid genes confer on the bacterium
the ability to catabolize these compounds (5,9-12,14,15).
Octopine-catabolizing strains incite octopine-synthesizing
tumors, while nopaline-catabolizing strains incite
nopaline-producing tumors (16). Because the bacterium,
not the plant, determines which opine the tumor will
produce, these compounds were suggested to result from the
action of foreign genes in the tumor cell (16).

It is now clear that axenic (bacteria-free) crown
gall tumor tissue does indeed contain foreign DNA: a part
of the Ti plasmid of the oncogenic bacterium, now called
T-DNA, is maintained in the tumor cell at 1-20 copies
(17-22). Two types of experimental evidence support this
conclusion: 1) the rate of renaturation of specific labeled
fragments of Ti plasmid DNA ("probe") is accelerated
by high concentrations of tumor ("driver") DNA (17-20).
Other fragments of Ti plasmid are unaffected . Driver DNA
from salmon or normal plant tissue does not accelerate
the renaturation of the Ti plasmid fragments. The
acceleration caused by tumor DNA is ascribed to copies of
part of the Ti plasmid present in the transformed plant
cells. 2) A second technique has recently confirmed and
extended this conclusion. Tumor DNA, when cleaved with
a restriction endonuclease and fractionated by horizontal
agarose gel electrophoresis, is separated into ca. 10^6
fragments, among which a few contain the foreign (Ti plasmid) DNA of interest. By preparing Southern blots and
hybridizing with appropriate labeled Ti plasmid DNA
fragments, one can detect the T-DNA-containing tumor
fragments (21-24). From the molecular weights of the
"boundary fragments" containing the edges of the T-DNA
region, it is concluded that T-DNA must be attached to
some other DNA, presumably that of plant (22,24)

Ti plasmids are diverse, as evidenced by their
restriction endonuclease cleavage patterns ("fingerprints") (25) and their DNA homology characteristics (26,
27). In contrast, a 5.5×10^6 dalton segment of DNA is
common to all wide host range Ti plasmids (28,29) and
is now known to form a part of the T-DNA of every tumor
line examined thus far (20-22,28). This DNA (at least in
part) codes for functions essential to oncogenicity since
the insertion (cointegration) of RP4 plasmid into this
portion of pTi C58 renders the Ti plasmid nononcogenic
(23). If this highly conserved element of T-DNA codes for
products that maintain the transformed phenotype, it
should be transcribed in tumor cells (30). The highly
conserved portion has been found to be transcribed at a
low level in all tumor lines examined thus far (29,31).

Although T-DNA always includes the highly conserved
region of the Ti plasmid, it also includes a variable
extent of the adjacent (32) Ti plasmid DNA, at least
in the case of octopine-type tumors (19). Thus far, we
have never found as T-DNA in the plant cell any separate
part of the Ti plasmid distant from the highly conserved
DNA region. These observations suggest that Ti plasmid

incorporation into the plant genome may occur by some
type of orderly process, rather than by random incorporation of diverse Ti plasmids. A second possibility is
that DNA insertion is random, but only those cells that
incoporate T-DNA (including the conserved element)
are phenotypically transformed.

Normal-appearing plants have been regenerated from
cloned crown gall tumor tissue (33-35). Braun and his
collaborators brought about regeneration of a cloned
Havana tobacco T37-incited tumor line by grafting
tumorous shoot primordia to apices of decapitated Turkish
tobacco plants. Rarely, shoots of increasingly normal
Havana-type morphology arose (33,34). One such shoot
that was fertile and set seed has been studied in
considerable detail. Several apparently normal tissues
of this shoot (leaf, flower petal, and filament) posses
two tumorous traits: they synthesize nopaline, and
they return to tumorous growth habits when planted on agar
(33,34). Tissue culture lines from these tissues
retain copies of T-DNA in an amount similar to that of
the parental line (20). In contrast, tissue from a
haploid plant derived from the anther of the grafted
shoot was completely normal and lacked T-DNA (20). All
fl progeny of the grafted shoot were also normal and a
callus line from one was found to be free of T-DNA (20).
Braun and collaborators found that very rarely the entire
grafted shoot was completely normal (34); T-DNA studies
have not been performed on such tissue.

Post-meiotic tissues of the teratoma grafted shoot
thus appear to have lost T-DNA. Such loss may also have
occured in vegetative cells. Because the original
tumor line was cloned, one must conclude that tumor cells
can, infrequently, produce non-tumorous daughter cells.
Meiosis may play an active role in the loss or elimina-
tion of T-DNA. Alternatively, spontaneous segregation of
normal daughter cells from tumorous parents could
produce a chimeric graft shoot, with only normal cells
contributing meiotic products.

Segregation of T-DNA in this T37 tumor line would be
easily explained if T-DNA were incorporated into an
extrachromosomal element in the plant cell. We report
here experiments that address this possibility. Chloroplast and mitochondrial DNA were isolated, characterized,
and tested for T-DNA content. In addition, we have tested
the possibility that T-DNA may be attached to other DNA.
Studies of the arrangement of T-DNA in cloned T37 tumors

indicate that they have a distinctive pattern of organization.

MATERIALS AND METHODS

Tumor DNA isolation

    Tumor and normal tobacco leaf DNAs were isolated as described previously (17). Before restriction enzyme cleavage, the DNAs were further purified by banding in CsCl-ethidium bromide gradients. Dye was removed (25) and the DNAs dialyzed exhaustively.

Chloroplast and mitochondrial DNA isolation

    The method of Quetier and collaborators (38,39) was adapted as follows: tumor callus tissue was disrupted by manually chopping with razor blades. Chloroplast and mitochondrial fractions were prepared by differential sedimentation, treatment with pancreatic DNase, washing, sedimentation through sucrose step gradients, and further washing. After lysis and pronase digestion, plastid DNA was isolated by CsCl-ethidium bromide density gradients. Dye and CsCl were removed (25) and the DNA was phenol extracted and dialyzed.

Preparation of Southern blots

    DNA digestion and agarose gel electrophoresis were performed as described previously (25). Blots were prepared essentially by the method of Southern (36). Probe DNAs were labeled with α-^{32}P-deoxyribonucleotide triphosphates (New England Nuclear, 300 Ci/mMol) by nick translation (37) to a specific activity of 90-110 x 10^6 cpm/ μg DNA. Hybridization conditions were: 3x SSC, 0.2% bovine serum albumin, 0.2% polyvinylpyrrolidone 360, 0.2% Ficoll 500, 0.1% sodium dodecylsulfate, 20mM Tris pH 7.4, 20 μg/ml denatured salmon sperm DNA, and 0.025 μg/ml denatured labeled probe DNA; 67°; 48 hours.

RESULTS

    Normal tobacco leaf DNA (5 μg), total T37-incited

tumor DNA (BT37, 5 µg) and isolated BT37 chloroplast
and mitochondrial DNAs (0.5 µg of each) were cleaved
with Bst I and fractionated by horizontal agarose gel
electrophoresis. The fragment pattern of the plastid
DNAs (Fig. 1) indicates that both are essentially free
from cross-contamination. Furthermore, there is little
evidence of contamination by nuclear DNA in either
preparation. A Southern blot was prepared from the gel
and hybridized with labeled pTi T37 (see Methods). Sites
of hybridization were determined by autoradiography
(Fig. 2).

Fig. 1. Bst I fragments of plastid DNA

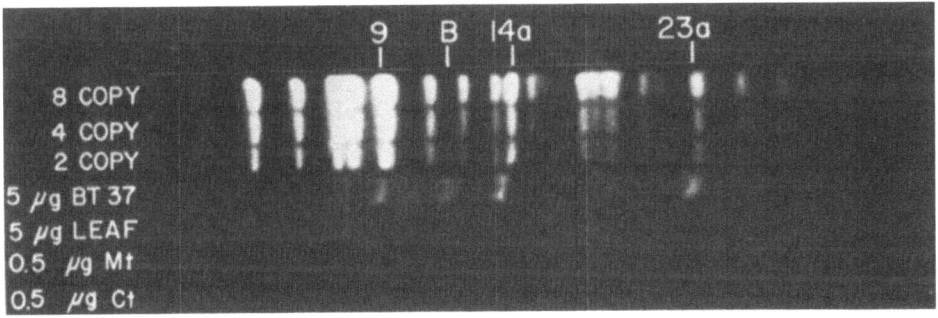

Fig. 2. Autoradiogram of Southern blot of plastid DNAs.

While T-DNA fragments are detectable in total BT37 tumor
DNA under these conditions, none is visible in the chloroplast or mitochondrial DNA tracks. Overexposure of
this autoradiogram (not shown) revealed no trace of
hybridization to plastid DNA fragments. T-DNA fragments
are detectable in nuclear DNA isolated from BT37 tumor
tissue at an intensity comparable to that for total
DNA from the same tissue (data not shown).

     The T-DNA fragments detected by Southern analysis
of BT37 tumor DNA (Figure 2) yield information about
the organization of T-DNA in this tumor line. We have
previously reported (20) that T-DNA of the BT37 cloned
tumor line extends from Sma I fragment 2 to Sma I
fragment 7 on the restriction map of pTi T37 (shown in

part in Fig. 3). The map position of Bst I digest
fragments of pTi T37 (Fig. 3) indicates that T-DNA
extends over Bst I fragments 23a, 9 and 14a, with ends
in 6 and 3a. The autoradiogram of Fig. 2 indicates
that indeed fragments 23a, 9, and 14a are excised intact
from tumor DNA. The identification of these fragments
has been confirmed by other experiments (not shown)
using cloned specific Bst I fragments or pTi T37

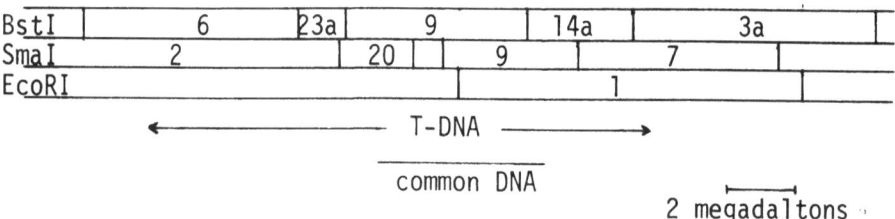

Fig. 3.   Fragment map of T-DNA region of pTi T37.

deletion mutants as a source of labeled probe DNA. One
novel band not corresponding to any Bst I fragment of
pTi T37 is visible in the BT37 tumor DNA track of
Fig. 2 (labeled "B"). This presumptive boundary fragment
did not hybridize when cloned Bst I fragment 3a was
used as a probe. We predict that it should be homologous to Bst I fragment 6; however this has not yet been
tested. Fragment 3a probe failed to hybridize detectably
with any fragment of tumor DNA. We conclude that T-DNA
extends only a short distance to the right of Bst I
fragment 14a.

     Digests of tumor DNA with other restriction enzymes
(Eco RI, Sma I, Hpa I) are more useful for visualizing
the right-hand boundary fragments of T-DNA because these
enzymes do not cut near the edge of T-DNA. Figure 4
shows that BT37 tumor DNA when digested with Eco RI
yields two presumptive right-hand boundary fragments,
both of lower molecular weight than Eco RI fragment 1
of the Ti plasmid. DNA from several tumor tissue lines
derived from a plant regenerated from BT37 tumor tissue,
when analyzed by these methods, exhibits two right-hand
boundary fragments, at least one of which does not match
those of BT37 tumor. We have also noted differences
between BT37 and these regenerate lines when we analyzed
their T-DNA by the driver DNA technique (20).

Fig. 4.   Analysis of Eco RI boundary fragments of T-DNA
          in BT37 tumor and BL2, a tumor line derived
          from the bottom leaf of a plant regenerated
          from BT37 tumor.

DISCUSSION

     The sensitivity of our analysis of chloroplast and
mitochondrial DNA can be calculated by the following
reasoning. Tumor cell total DNA does not contain more
than 10% chloroplast or mitochondrial DNA, based on the
restriction enzyme digest patterns in which bands ascrib-
able to plastid DNA fragments are not a dominant feature
(21,22). If all T-DNA were in chloroplast or mitochondrial DNA, the T-DNA must then be at least ten-fold
enriched in the pure plastid DNA sample. Therefore,
0.5 μg of pure tumor cell plastid DNA should yield at
least as intense a hybridization reaction as seen with
5 μg of total tumor cell DNA. The fact that no hybridization is detectable to plastid DNA fragments in the
experiment of Fig. 2 indicates that T-DNA in this tumor
line is not in the plastids. T-DNA is recovered in
good yield with nuclear DNA. These results are consistent with the view that T-DNA is attached to nuclear
DNA.

     Analysis of the fragmentation patterns of T-DNA by
Bst I (Fig. 2) indicates the continuity of T-DNA: all
three of the Bst I fragments that map on the original
Ti plasmid between the left and right T-DNA boundaries
are maintained intact in the tumor cell. In addition,
new T-DNA-containing fragments are detected in both the
Bst I and Eco RI digests (Figs. 2 and 4). Isolation of
these boundary fragments by molecular cloning techniques
will allow analysis of the DNA to which T-DNA is attached.
They may be junction fragments containing both T-DNA
and plant nuclear DNA. Alternatively, they could consist
of junctions between tandemly arranged copies of T-DNA,
or fusion fragments generated by T-DNA rearrangement.
If T-DNA during tumor induction excised from the Ti
plasmid in a manner analogous to lambda phage excision
from a lysogen, a fusion of "left" and "right" edges of

T-DNA would be created which could be preserved in the
T-DNA copies detected in the tumor. Experiments are
now underway that will test these models.

BT37 tumor DNA exhibits one T-DNA boundary fragment
that differs from that of the tumors derived from the
BT37 regenerated plant. The explanation of this difference is not clear. The possibility of misidentification
of the T37 tumor line must be considered. Detailed
analysis of the boundaries of T-DNA in these lines is
in progress.

The finding that T-DNA of BT37 tumor tissue is not
in chloroplast or mitochondrial DNA rules out the
simplest explanation of T-DNA loss during regeneration:
loss cannot be ascribed to segregation of organelles
carrying T-DNA. If T-DNA is indeed attached to plant
nuclear DNA, its loss must involve either recombination
or excision. Somatic crossing over could give rise to
segregation of cells with no T-DNA; however the several
copies of T-DNA (20) would have to be genetically linked
in order to explain their complete loss by this
mechanism. Excision of T-DNA copies could occur by a
reversal of the process of their insertion. The insertion and excision of lambda bacteriophage and the transposition and excision of transposable elements provide
precedent for such phenomena. Alternatively the
insertion and loss of T-DNA may occur by a novel
mechanism, whose rules will be discernable from analysis
of DNA sequence at the sites of insertion and excision.

ACKNOWLEDGMENTS

This research was supported by research grant
number NP194 from the American Cancer Society and
CA13015 from the National Institutes of Health. We
are indebted to Dr. Ross Bowman for excellent technical
assistance.

REFERENCES

1. A. Menage and G. Morel, C.R. Acad. Sci., 259:4795
    (1964)
2. A. Goldman, J. Tempe, and G. Morel, C.R. Seances Soc.
    Biol. Paris, 162:630 (1968)
3. A. Goldman, D.W. Thomas, and G. Morel, C.R. Acad. Sci.,
    268:852 (1969)
4. E. Holderbach and R. Beiderbeck, Phytochemistry,
    15:955 (1976)
5. M.P. Gordon, S.K. Farrand, D. Sciaky, A.L. Montoya,
    M.-D. Chilton, D.J. Merlo, and E.W. Nester, in:
    "A Symposium on the Molecular Biology of Plants,"
    I. Rubenstein, ed., Academic Press, New York (1976)
7. I. Zaenen, N. Van Larebeke, H. Teuchy, M. Van Montagu,
    and J. Schell, J. Mol. Biol., 86:109 (1974)
6. J.D. Kemp, Biochem. Biophys. Res. Comm., 69:816 (1976)
8. M.-D. Chilton, S.K. Farrand, F. Eden, T.C. Currier,
    A.J. Bendich, M.P. Gordon, and E.W. Nester, in:
    "Modification of the Information Content of Plant
    Cells," R. Markham et al., eds., North Holland
    Publishing Co., Amsterdam (1974)
9. N. Van Larebeke, G. Engler, M. Holsters, S. Van den
    Elsaker, I. Zaenen, R.A. Schilperoort, and J. Schell,
    Nature, 252:169 (1974)
10. B. Watson, T.C. Currier, M.P. Gordon, M.-D. Chilton,
    and E.W. Nester, J. Bacteriol., 123:255 (1975)
11. J. Schell, M. Van Montagu, A. DePicker, D. De Waele,
    G. Engler, C. Genetello, J.P. Hernalsteens, M. Holsters
    E. Messens, B. Silva, S. Van den Elsacker, N. Van
    Larebeke, and I. Zaenen, in: "Proceedings of a
    Symposium on the Molecular Biology of Plants," I.
    Rubenstein, ed., Academic Press, New York (1976)
12. A.L. Montoya, M.-D. Chilton, M.P. Gordon, D. Sciaky,
    and E.W. Nester, J. Bacteriol., 129:101 (1977)
13. G. Bomhoff, P.M. Klapwijk, H.C. Kester, R.A.
    Schilperoort, and J. Schell, Molec. Gen. Genet.,
    145:171 (1976)
14. M.-D. Chilton, S.K. Farrand, R. Levin, and E.W. Nester,
    Genetics, 83:609
15. P.M. Klapwijk, P.J.J. Hooykaas, H.C.M. Kester,
    R.A. Schilperoort, and A. Rorsch, J. Gen. Microbiol.,
    96:155 (1976)
16. A. Petit, S. Delhaye, J. Tempe, and G. Morel, Physiol.
    Veg., 8:205 (1970)
17. M.-D. Chilton, M.H. Drummond, D.J. Merlo, D. Sciaky,
    A.L. Montoya, M.P. Gordon, and E.W. Nester, Cell,
    11:263 (1977)
18. M.-D. Chilton, M.H. Drummond, M.P. Gordon, D.J. Merlo

A.L. Montoya, D. Sciaky, and E.W. Nester, Microbiology, 1978:135 (1978)

19. D.J. Merlo, R.C. Nutter, A.L. Montoya, D.J. Garfinkel, M.H. Drummond, M.-D. Chilton, M.P. Gordon, and E.W. Nester, submitted to Molec. Gen. Genet.

20. F.-M. Yang, A.L. Montoya, D.J. Merlo, M.H. Drummond, M.-D. Chilton, E. W. Nester, and M.P. Gordon, submitted to Molec. Gen. Genet.

21. R.K. Saiki and M.-D. Chilton, in preparation

22. M.T. Thomashow, unpublished data

23. A. DePicker, M. Van Montagu, and J. Schell, Nature, 275:150 (1978)

24. J. Schell (and 22 collaborators), in:" Nucleic Acids in Plants," T. Hall and J. Davies, eds., in press

25. D. Scaiky, A.L. Montoya, and M.-D. Chilton, Plasmid, 1:238 (1978)

26. T.C. Currier and E.W. Nester, J. Bacteriol., 126:157 (1976)

27. M.H. Drummond and M.-D. Chilton, J. Bacteriol., 136:1178 (1978)

28. M.-D. Chilton, M.H. Drummond, D.J. Merlo, and D. Sciaky, Nature, 275:147 (1978)

29. W. Gurley and J.D. Kemp, Proc. Nat. Acad. Sci. USA, in press

30. M.H. Drummond, M.P. Gordon, E.W. Nester, and M.-D. Chilton, Nature, 269:535 (1977)

31. A. Lederboer, Ph.D. Thesis, University pf Leiden (1978)

32. M.-D. Chilton, A.L. Montoya, D.J. Merlo, M.H. Drummond, R.C. Nutter, and E.W.Nester, Plasmid, 1:254 (1978)

33. A.C. Braun, and H.N. Wood, Proc. Nat. Acad. Sci. USA, 73:496 (1976)

34. R. Turgeon, H.N. Wood, A.C. Braun, Proc. Nat. Acad. Sci. USA, 73:3562 (1976)

35. M.D. Sacristan and G Melchers, Molec. Gen. Genet., 152:111 (1977)

36. E.M. Southern, J. Mol. Biol., 98:503 (1975)

37. T. Maniatis, A. Jeffery, and D.G. Kleid, Proc. Nat. Acad. Sci. USA, 72:1184 (1975)

38. F. Vedel, F. Quetier, and M. Bayen, Nature, 263:440 (1976)

39. F. Quetier and F. Vedel, Nature, 268:365 (1977)

CROWN GALL TRANSCRIPTION OF TI PLASMID — DERIVED SEQUENCES

W.B. Gurley, J. Callis, and J.D. Kemp

SEA, USDA, Department of Plant Pathology

University of Wisconsin, Madison, WI   53706, USA

The Ti plasmid of *Agrobacterium tumefaciens* (E.F. Sm. &
Town.) Conn. is a unique natural example of the expression of
prokaryotic DNA sequences by eukaryotes. The mechanism whereby
A. tumefaciens transfers and stably associates plasmid DNA with
plant cells may serve as a model for the artificial introduction
of foreign genes into plant genomes. For octopine tumors, the
region of the plasmid transferred to the plant cell (T-DNA) includes *Sma* I fragments 3b, 10c[1], and 16a[2]. Furthermore,
Drummond et al.[3] and Ledeboer[4] have shown that some plasmid-
derived sequences are transcribed by these tumors. Both studies
involved B6-derived *Agrobacterium tumefaciens* strains and their
respective tumors incited in tobacco. In spite of the close similarities in the plasmids and host plants used, differences seem
to exist in the pattern of T-DNA transcription by tumor cells.
We have conducted an analysis of T-DNA transcription using
several Ti plasmid/host plant combinations in order to better
evaluate the extent of variability of tumor transcription[2].

In order to analyze in detail the transcription of T-DNA, a
physical map of restriction endonuclease sites of *Eco*Rl and
*Hin*d III was constructed for the region of pTi-15955 adjacent to
and including *Sma* I fragments 3b and 10c. The identification and
ordering of the major restriction fragments was determined by
molecular weight analysis of fragments showing overlapping
regions of sequence homology. The mapping of several of the
minor fragments was achieved by analysis of partial restriction
endonuclease digestion products of total pTi-15955 DNA. Both
approaches involved the hybridization of ^{32}P-labeled cRNA probes
synthesized *in vitro* by *E. coli* RNA polymerase holoenzyme.

TUMOR TRANSCRIPTION

     The transcription of Ti plasmid-derived sequences by crown
gall tumor cells was determined by the hybridization of total
^{32}P-labeled (*in vivo*) tumor RNA with plasmid DNA sequences[2]. The
crown gall tumor lines used in this study are identified in
Table 1. Ti plasmid (pTi-15955) DNA used in these hybridization
experiments was digested with either *Eco*R1, *Sma* I, or *Hin*d III,
fractionated by agarose gel electrophoresis, and bound to nitrocellulose filters by the Southern procedure[5]. Southern filters
of the recombinant lambda bacteriophages CH3A-B and CH3A-C were
used for a detailed analysis of tumor transcription of sequences
mapping within *Eco*R1-generated fragments B and C of pTi-15955.
The results of the hybridization of tumor RNA with Southern filters prepared from total pTi-15955 DNA are presented in Figure 1.
The results of incubating ^{32}P-RNA isolated from the habituated
sunflower line HSSS are also presented in Figure 1 (Lane a). The
absence of autoradiographic bands indicates a lack of RNA
homology with Ti plasmid DNA in this type of neoplastic tissue.

     A qualitative summary of the tumor transcript analysis is
presented in Figure 2. There are four major sites of T-DNA
transcription in octopine-type tumor cells. These sites of tumor
transcription include: 1) the right terminus of *Eco*R1 fragment
A (7.6x10^{6}D), 2) the right terminus *Eco*R1 fragment B (4.5x10^{6}D),
and 3) both the left and right terminal portions of *Eco*R1 fragment C (3.7x10^{6}D). All of these sites of tumor transcription are
included in a sector of the Ti plasmid of approximately 13x10^6
daltons. Regions within this sector showing no sequence homology
with tumor RNA are either not present in the plant, nontran-
scribed components of T-DNA, or transcribed at very reduced
levels relative to the other sites.

     The same general regions of T-DNA are transcribed by the two
sunflower tumor lines PSCG-15955 and PSCG-B6[4]. However, differences between the two tumor lines are evident in the intensities of the autoradiographic bands corresponding to certain of
the plasmid restriction fragments. For example, the order of
autoradiographic intensities for the *Eco*R1-generated fragments
is C>A>B>D,E for PSCG-15955 ^{32}P-RNA and B>C>A>D,E for PSCG-B6[4]
^{32}P-RNA.

TABLE 1. Crown Gall Tumor Lines.

Tumor tissue	Plant	Inciting bacteria
PSCG-15955	sunflower	15955
PSCG-B6[4]	sunflower	B6
E1	tobacco	B6-806

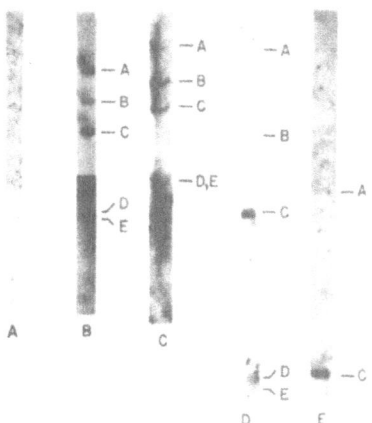

FIGURE 1. Hybridization of ^{32}P-RNA isolated from habituated and
    tumor tissue with *Eco*R1-generated restriction fragments of
    total pTi-15955 DNA. RNA was isolated from the following
    tissue lines: (a) habituated sunflower, HSSS; (b) PSCG-
    15955; (c) PSCG-B6[4]; (d-e) E1.

The pattern of T-DNA transcription in the tobacco tumor line
E1 is quite distinctive from that seen in the sunflower tumors.
RNA homologous to *Eco*R1 fragments A and B is clearly present in
the two sunflower lines, but is barely detectable in E1 ^{32}P-RNA
(Figure 1, d-e). In addition, hybridization with multiply restricted phage CH3A-C DNA reveals a striking difference in the
transcription of the right terminal portion of *Eco*R1 fragment C
(map section 11) in that the autoradiographic band associated
with this region is much more intense using E1 ^{32}P-RNA relative
to ^{32}P-RNA from the sunflower tumor lines.

If autoradiographic intensities are actually proportional to
quantities of homologous transcript present, then the transcription of plasmid-derived sequences by tumor cells appears to be
both nonuniform across the T-DNA and variable among tumor lines.
The cause of this apparent variability is unknown but may
possibly be influenced by differences in T-DNA organization and
copy number in the plant. This effect may explain the apparent
abundance of transcript homologous to map section 11 in E1 tissue since Merlo et al. (personal communication) has shown that
at least a portion of *Hin*d III fragment b is present in excess
over *Hin*d III fragment a in this tissue.

Although there is no direct evidence that T-DNA codes for
functional peptides, it is tempting to speculate that map section
9 contains the structual gene for octopine synthase. This prediction is based on the observation that all three tumor lines
contain nearly equal amounts of both octopine[6] and octopine

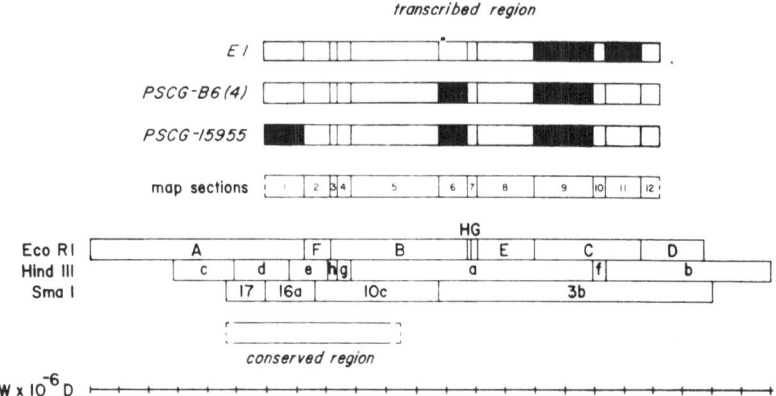

FIGURE 2. Restriction endonuclease map of T-DNA of pTi-15955[2]. The conserved region is according to Chilton et al.[7]. Regions of tumor transcription are designated by shaded blocks. The intensity of shading is directly proportional to the relative autoradiographic intensities.

synthase (unpublished). Assuming comparable levels of octopine synthase mRNA, then it seems reasonable that the transcription site showing the most uniform hybridization among the tumor lines should code for this enzyme. Map section 9 which is located on *Eco*R1 fragment C seems to meet this requirement.

One of the major sites of T-DNA transcription maps within sequences which are common to most, if not all, Ti plasmids[7,8,9]. The common DNA sequences which overlap with T-DNA on the nopaline-type plasmid pTi-C58 appear to be essential for onco-genicity[9]. The role these conserved sequences play in tumor induction and maintenance is unclear. The results of our tran-scription analysis[2] suggest that conserved sequences (*Sma* I frag-ment 16a, *Hin*d III fragment d) are transcribed in variable amounts by octopine-type tumors. Evidence suggesting transcrip-tion of conserved DNA is seen in Figure 1 as autoradiographic bands corresponding to *Eco*R1 fragment A. The finding that RNA homologous to conserved sequence DNA may be present in all three tumor lines suggests the possibility that transcription of these sequences may be required in maintaining the neoplastic state of these plant cells.

Many models for the possible function of T-DNA in tumor pro-duction and opine synthesis can be constructed. Several of the more general models are presented in Figure 3. These models are based in varying degrees on three assumptions: 1) plasmid sequences are stably associated with plant cells, 2) transcrip-tion of these sequences occurs in the plant, and 3) RNA

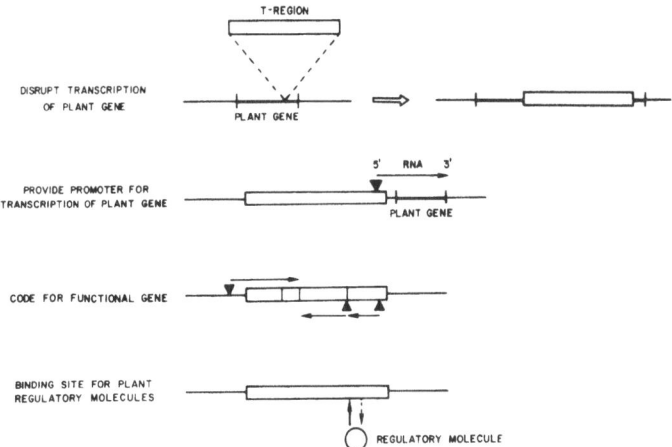

FIGURE 3. Possible functions of T-DNA.

transcribed from T-DNA can be translated to form functional peptides. Although none of these models can be excluded at this
time, those involving transcription of T-DNA by plant cells seem
quite plausible.

LOCATION OF TI PLASMID PROMOTER SEQUENCES

    A fundamental question concerning the expression of T-DNA by
plant cells relates to the mechanism whereby the prokaryotic sequences are correctly transcribed in an eukaryotic subcellular
environment. Information regarding the nature and location of
promoter sequences involved in T-DNA transcription is required
for a complete understanding of this process.

    The filter binding assay first developed by Jones and Berg[10],
has been refined and modified to serve as a powerful technique
for the localization of bacterial promoter sequences. It has
been used in the identification of restriction fragments containing promoter sequences in the bacteriophages T3, T5, fd, fl,
and M13[11,12,13]. It has also identified promoter bearing
fragments in several lambda bacteriophages containing *E. coli*
chromosomal genes such as lac5[14] and rif[d]18[15]. The assay procedure basically involves the trapping of RNA polymerase/DNA
restriction fragment complexes on nitrocellulose filters.
Restriction fragments retained by the filters are eluted, precipitated, and fractionated by gel electrophoresis. The assay
can be used to locate both open promoter complexes and actual
initiation, or ternary, complexes. For the open promoter assay,
RNA polymerase holoenzyme and restricted DNA are incubated
together at 37°C in a suitable buffer and salt solution before
passage through a nitrocellulose filter. The initiation assay
differs in that nucleotide triphosphates are added after the open

promoter complexes have formed. The mixture is passed through a
nitrocellulose filter after 30 sec, and the filter rinsed with a
high salt (0.6 M KCl) wash to disassociate any open promoter
complexes that may be present. Only ternary complexes are stable
enough to be retained under these conditions.

The filter binding assay can be utilized in determining the
location of *in vitro* promoter sequences on pTi-15955 DNA (unpublished). The filter-retained fragments resulting from the incubation of *E. coli* RNA polymerase holoenzyme with *Hin*d III
restricted pTi-15955 DNA are shown in Figure 4. In this experiment the concentration of KCl was variable from 10mM to 200mM in
order to analyze relative promoter strengths. Restriction fragments retained at the higher KCl concentrations contain either
more than one promoter or a single strong promoter. In general,
most of the strongest *in vitro* RNA polymerase binding sites are
localized on those restriction fragments mapping within T-DNA.
These same fragments are also retained in the assay for ternary
complexes indicating that most of the polymerase binding sites
are capable of *in vitro* initiation. The filter binding assay
has been used with *Eco*Rl-restricted plasmid DNA with the same
general result indicating a strong bias for the localization of
strong *in vitro* promoters within T-DNA.

A detailed analysis of *in vitro* promoter localization within
T-DNA was conducted using multiply restricted lambda phage DNA
from CH3A-B and CH3A-C. A summary of these results together with
results obtained using total Ti plasmid DNA is presented in
Table 2. Not only are strong *in vitro* promoters preferentially
located within T-DNA; but within this region of the plasmid,

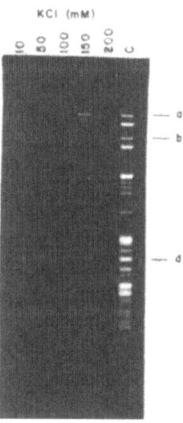

FIGURE 4. Filter binding assay for *Hin*d III restriction frag-
     ments of pTi-15955 containing *E. coli* RNA polymerase bind-
     ing sites. T-DNA fragments are designated. Lane C (con-
     trol) contains total pTi-15955 DNA.

TABLE 2. Tumor Transcription and *E. coli* RNA Polymerase
in vitro Promoters within T-DNA.

Map section	Tumor transcription*	*In vitro* promoters
1	+	+
2	−	?
3	−	−
4	−	−
5	+	+
6	+	+
7	−	?
8	+	+
9	+	+
10	+	−
11	+	+
12	+	+

*(−)Autoradiographic band on Southern filter not detectable.

promoters seem to be present only on those restriction fragments
actively transcribed by plant tumor cells.

Our results regarding the location of promoter sequences on
the Ti plasmid confirm and expand the initial observations of
Ledeboer[4] concerning the preferential transcription of T-DNA
in vitro by *E. coli* RNA polymerase. The significance of the
localization within T-DNA of sequences capable of functioning as
in vitro promoters is unclear, although it is tempting to speculate that these sequences serve as promoters in the plant. This
possibility seems worth consideration in view of the sequence
similarities between prokaryotic promoters[16] and the potential
promoter regions of several eukaryotic structual genes[17]. An
understanding of the transcriptional requirements of T-DNA in
plant cells may provide insight regarding eukaryotic gene regulation at the molecular level.

LITERATURE CITED

1. Chilton, M-D, M.H. Drummond, D.J. Merlo, D. Sciaky,
   A.L. Montoya, M.P. Gordon, E.W. Nester. *Cell* 11,263 (1977).
2. Gurley, W.B., J.D. Kemp, M.J. Albert, D.W. Sutton, J. Callis.
   PNAS 76,2828 (1979).
3. Drummond, M.H., M.P. Gordon, E.W. Nester, M-D. Chilton.
   Nature 269,535 (1977).
4. Ledeboer, A.M. Dissertation, Leiden, Holland (1978).
5. Southern, E.M. *J. Mol. Biol.* 98,503 (1975).
6. Kemp, J.D. *Biochem. Biophys. Res. Commun.* 69,816 (1976).
7. Chilton, M-D., M.H. Drummond, D.J. Merlo, D. Sciaky.
   Nature 275,147 (1978).

8. Drummond, M.H., M-D. Chilton. *J. Bacteriol.* 136,1178 (1978).
9. Depicker, A., M. van Montagu, J. Schell. *Nature* 275,150 (1978).
10. Jones, O., P. Berg. *J. Mol. Biol.* 22,199 (1966).
11. Takeya, T., M. Takanami. *Biochem.* 13,5388 (1974).
12. Gabain, A., H. Bujard. *Mol. Gen. Genet.* 157,301 (1977).
13. Seeburg, P., H. Schaller. *J. Mol. Biol.* 92,261 (1975).
14. Jones, B.B., H. Chan, S. Rothstein, R.D. Wells, W. Rez nikoff. *PNAS* 74,4914 (1977).
15. Taylor, W.E., R.R. Burgess. *Gene* 6 (1979) (in press).
16. Scherrer, G.E.F., M.D. Walkinshaw, S. Arnott. *Nucleic Acid Res.* 5,3759 (1978).
17. Gannon, F., K. O'Hare, F. Perrin, J.P. LePennec, C. Benoist, M. Cochet, R. Breathnach, A. Royal, A. Garapin, B. Cami, P. Chambon. *Nature* 278,428 (1979).

CROWN GALL SPECIFIC GENE PRODUCTS:

OCTOPINE AND NOPALINE SYNTHASE

J. D. Kemp, E. Hack and D. W. Sutton

SEA, USDA and Department of Plant Pathology
University of Wisconsin-Madison
Madison, WI   53706   U.S.A.

INTRODUCTION

The crown gall disease appears to be a unique interaction
between a prokaryote {Agrobacterium tumefaciens (E. F. Sm. & Town.)
Conn.} and a eukaryote (dicotyledons plants). This interaction results in the transer of bacterial DNA to the plant cell[1] where it
is stably integrated[2] and possibly expressed[3,4]. This being the
case, the crown gall system certainly demonstrates that genetic
engineering at the molecular level is possible. However, the crown
gall system per se may never be a useful genetic engineering system
since it has evolved for the specific purpose of causing plant
tumors. Nevertheless, an understanding of how this system operates
at the molecular level may allow us to design a vehicle with just
those characteristics we desire. One desirable characteristic of
a useful system is that the transferred DNA is expressed, i.e., a
functional, structural protein is produced. It is still not certain
whether or not the crown gall system fulfills this characteristic.

The small piece (5-10 $\times$ 10^{6}D) of DNA (T-DNA) transferred to the
plant cell during transformation resides on the large tumor-inducing
(Ti) plasmid of A. tumefaciens. The majority of the Ti plasmids
can be assigned to one of two groups (octopine-type or nopaline-
type) depending on whether the tumors they incite synthesize the
crown gall specific compounds octopine or nopaline[5,6]. Since the
Ti plasmid, rather than the plant species, determines whether
octopine or nopaline are produced in the tumor cells, the obvious
candidates for structural gene products of T-DNA are the enzymes
that cayalyze the synthesis of these unusual compounds.

Since the original discovery that tumor tissue synthesizes RNA that hybridizes to T-DNA[3] a detailed transcript analysis has been performed on three octopine-type crown gall tissue cultures[4]. The study shows there are four major sites of T-DNA transcription. The intervening silent areas and the differences in hybridization intensity between the four sites suggests that there may be promoters within the T-DNA that are active in the plant cell. We have begun an investigation of the location of promoter sequences within T-DNA by analyzing the in vitro interaction of E. coli RNA polymerase with pTi DNA[7]. The strongest bacterial promoters, with one exception, map within the T-DNA and there appears to be a promoter on the 2×10^6 D fragment we believe is involved in the control of octopine synthesis in the tumor tissue.

It should now be a relatively straightforward project to demonstrate that the enzymes from tumor tissue that catalyze the synthesis of octopine and nopaline are coded for by a gene on the T-DNA. First, the enzymes are purified and characterized. Then, the 2×10^6 D T-DNA fragment is transcribed and translated, and the in vitro translation product compared to the enzyme. This paper will limit itself to the first part of the project; the purification and characterization of two crown gall specific enzymes. The first enzyme is octopine synthase purified from primary sunflower crown gall tissue cultures (PSCG) incited by octopine utilizing strains of the bacteria (B_6 or 15955), and the second is nopaline synthase purified from PSCG-C58 or PSCG-T37, tissues incited by the nopaline utilizing strains C58 or T37.

PURIFICATION OF OCTOPINE AND NOPALINE SYNTHASES

Two enzymes, one from PSCG-B_6 (octopine synthase) and one from PSCG-T37 (nopaline synthase) were purified to homogeneity by similar procedures [8,9] (Table I); the yields ranged from 25-35%. Standard enzyme assay[8] consists of: a phosphate buffer, NADPH, arginine, pyruvate (octopine synthase) or α-ketoglutarate (nopaline synthase), and enzyme. The reaction is followed by the decrease in A_{340}. The greatest purification for either enzyme is achieved by affinity chromatography on Cibacron Blue F3GA-agarose. This gel has recently been used in the purification of several pyridine nucleotide-dependent dehydrogenases; it appears that the blue dye ligand binds to the nucleotide binding site of these enzymes.

Both enzymes are unstable and rapidly lose their activity if they are subjected to mechanical stress or aeration or if 2-mercaptoethanol is not included in all buffers. In addition, glycerol is generally included in all buffers used in the purification of octopine synthase and NADPH helps stabilize nopaline synthase.

Table I.   Purification Scheme

Octopine Synthase	Nopaline Synthase
1. Both tissues homogenized in an equal volume of buffer[a] followed by centrifugation. Supernatant is crude extract.	
2. 45-70% $(NH_4)_2SO_4$ fraction	40-55% $(NH_4)_2SO_4$ fraction
3. Eluted from DEAE with a linear gradient of chloride (80-200 mM)	Eluted from hydroxylapatite with 0.1 M PO_4 (pH 6.8)
4. Eluted from Cibacron Blue with 0.1 mM NADPH	Eluted from Cibacron Blue with 0.1 mM NADPH
5. Eluted from hydroxylapatite with 10 mM PO_4 (pH 6.3)	Gel-filtration on Sephacryl S-200

[a] Buffer: 0.1 potassium phosphate (pH 7.4), 0.5 M sucrose, 1 mM EDTA, 10 mM 2-mercaptoethanol.

The purification scheme in Table I will routinely give enzyme preparations that are greater than 98% homogeneous as judged by scans of SDS-polyacrylamide gels (Figure 1). The purity of purified octopine synthase and the proteins identity with octopine synthesizing activity were confirmed when a single protein coeluted with activity from Sephacryl G-200 (separation by molecular weight) and when a single protein band comigrated with activity on cellulose acetate membranes (separation by charge). Thus it is apparent that the protein band observed on SDS gel electrophoresis is the enzyme octopine synthase: The enzyme is not merely a minor contaminant of the purified preparation.

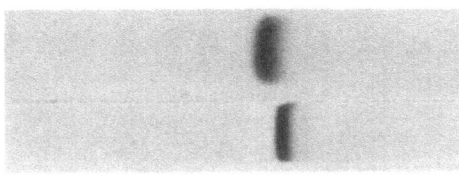

NOPALINE   SYNTHASE

OCTOPINE   SYNTHASE

Figure 1.   SDS-polyacrylamide disc gels of purified nopaline and octopine synthases stained with Coomassie Blue.

NUMBER OF SYNTHASES

Of the seven documented "opines," four are N^2-(1-carboxyethyl)-amino acids (octopine, octopinic acid, lysopine, and histopine) and two are N^2-(1,3-dicarboxypropyl)-amino acids (nopaline and ornaline)[9]. The early research on the synthesis of the N^2-(1-carboxyethyl)-family using crude enzyme preparations from crown gall tissues suggested to some that multiple enzymes were present. Goldmann-Menege[10] proposed a different enzyme for each amino acid and one set specific for NADPH and one for NADH. Birnberg et al.[11] concluded that there were two octopine synthesizing activities, one specific for NADPH and one for NADH. We reported[12], as did Otten et al.[13] that crown gall tissues contain a single enzyme activity that can utilize either NADPH or NADH in the synthesis of the N^2-(1-carboxyethyl)-family.

The strongest evidence that a single enzyme could account for all the N^2-(1-carboxyethyl)-amino acid synthesizing activities found in crude extracts of crown gall tissues came with the complete purification of octopine synthase. First, the purified enzyme is active with all amino acids that can be converted to N^2-(1-carboxyethyl)-derivatives by crude extracts. Second, the purified enzyme can utilize NADH as well as NADPH, even though it was eluted from Cibacron Blue with NADPH. Third, the apparent Michaelis constants and the relative maximum velocities for each reaction were the same for both crude enzyme preparations and purified enzyme (Table II). Finally, arginine and ornithine activities with NADPH or NADH coincide on gel filtration (separation by molecular weight) and on DEAE cellulose chromatography (separation by charge). Activities for the other amino acids were not checked but the purified enzyme has the same ratio of activities as the crude (see V_{max}, Table II). Thus there is no evidence for the existence of more than one N^2-(1-carboxyethyl) amino acid synthesizing enzyme.

Table II. Kinetic Constants for N^2(1-carboxyethyl) Amino Acid Synthesis by Purified and Crude (numbers in parenthesis) Octopine Synthase.

Amino Acid	Km (mM) amino acid	pyruvate	V (in % Varg)
L-arginine	8.8 (8.5)	2.0 (2.1)	100
L-glutamine	4.7 (3.6)	14.0 (14.4)	42 (37)
L-histidine	6.8 (6.0)	3.6 (2.3)	75 (70)
L-lysine	0.8 (0.9)	0.17 (0.18)	10 (9)
L-methionine	11.2 (10.0)	10.4 (10.2)	127 (95)
L-ornithine	1.8 (1.8)	1.0 (1.1)	25 (27)

Earlier we argued from kinetic evidence[14], as have others[15], that both members of the N^2-(1,3 dicarboxypropyl)-family are synthesized by a single enzyme. Again the strongest evidence for a single enzyme comes from experiments with the purified enzyme[8]. Those experiments include the following: 1) purified nopaline synthase uses either NADPH or NADH to synthesize either nopaline or ornaline as does crude preparations; 2) all of the activities copurify with no change in their kinetic parameters; and 3) the specific enzymic activities do not change upon further attempts at purification.

The seventh recognized "opine," agropine[16], has not been completely characterized. We do know, however, that it is not synthesized by octopine or nopaline synthases. Another possible "opine" may be the uncharacterized lysine derivative we described in nopaline crown gall tissues[17]. It also is not synthesized by octopine or nopaline synthases.

TISSUE SPECIFICITY

Model[18] concluded that octopine and nopaline were products of bacterial gene expression since these compounds are unique to crown gall tissues and since their presence in tumors is determined by the inciting bacterial rather than the plant. Over the years there have been various claims that octopine is present in normal tissue but none of these claims have been substantiated. We surveyed various plant tissues for "opine" synthesizing activities (Table III) and have concluded that N^2-(1-carboxyethyl)-synthesizing activity is present only in octopine type crown gall tissues and N^2-(1,3 dicarboxypropyl)-synthesizing activity is present specifically in nopaline type crown galls. One laboratory found octopine activity in embryonic Pinto beans[19]. We could not repeat that result (Table III).

Since both octopine and nopaline synthases now exist as discrete proteins, we can ask the question whether or not the proteins themselves are crown gall tissue specific. To answer this question proteins were purified from HSSS tissue by both purification procedures (Table I). In neither case was there a protein band on SDS-polyacrylamide gels observed at the position where the synthases migrate. In order to ensure that the synthases could have been isolated from HSSS if they had been present, proteins were again purified but this time authentic synthase was included. In both cases the added synthase was recovered. A similar purification experiment showed that PSCG-B$_6$ tissue does not contain nopaline synthase. From these experiments we conclude that nopaline synthase is specific for nopaline type crown gall tissues and octopine synthase is specific for octopine type crown gall tissues.

Table III.   "OPINE" Synthesizing Activities in Various Tissues

Tissue[a]	Octo-pine	Histo-pine	Lyso-pine	Octopinic Acid	Nopaline	Ornaline
PSCG-B$_6$	100[b]	82	16	42	<1	<1
PSCG-15955	100	89	18	42	—	—
E-1	100	90	20	44	—	—
PSCG-T37	—[c]	—	—	—	100[b]	27
PSCG-C58	<1	<1	<1	<1	100	20
PSCG-AT1	<1	<1	<1	<1	<1	<1
HSSS	<1	<1	<1	<1	<1	<1
Embryonic Bean	<1	—	—	—	<1	—

[a] PSCG: Primary sunflower crown gall tissue culture, bacterial strain listed after hyphen; HSSS: Habituated sunflower from normal sunflower tissue culture; E-1: Tobacco crown gall from M. D. Chilton, Univ. of Washington, Seattle, WA; Pinto beans were germinated for three days under aseptic conditions.

[b] Octopine activity or Nopaline activity taken as 100.

[c] Not assayed.

PROPERTIES OF THE SYNTHASES

a. Kinetic Properties. The nopaline synthase protein is present in PSCG-C58 tissue at a concentration of 12 µg of synthase/gm fresh weight of tissue, whereas, octopine synthase protein is present in PSCG-B$_6$ tissue at a concentration of only 2 µg of protein/gm fresh weight. Both purified enzymes have very similar specific activities in vitro so one might expect to find the levels of nopaline in PSCG-C58 to be higher than the levels of octopine in PSCG-B$_6$. We actually find about 200 times as much nopaline in PSCG-C58 than octopine in PSCG-B$_6$. A 200-fold difference is reasonable when the enzyme Km's are considered. The octopine synthase Km for arginine (8.8 mM, Table II) is 12 times higher than the nopaline synthase Km for arginine (0.74) mM) and the octopine synthase Km for pyruvate (2.1 mM, Table II) is 2.3 times the nopaline synthase Km for α-ketoglutarate (0.9 mM).

Information about the mechanism of the synthase catalysed reactions can be obtained from initial-rate data. If a three-substrate enzyme-catalysed reaction shows Michaelis-Menten behavior with respect to each substrate, i.e., if plots of the reciprocal initial rate against the reciprocal substrate concentration, are linear then the initial-rate behavior can be described by the empirical equation of Dalziel[20]. His equation was obtained by equilibrium treatment of a random-order quaternary-complex mechanism. Other mechanisms can be considered as special cases of this equation in which one or more of the parameters may be zero. If any of the parameters are zero, the mechanism will be either ordered or enzyme-substitution (ping pong). This will be reflected in the characteristics of the primary and secondary reciprocal plots. These characteristics include a family of parallel plots or a family of plots that all intersect on the ordinate.

A "ping pong" mechanism can be ruled out for both enzymes since all of the primary plots intersect in the left quadrant. It is much harder to discriminate between the random and the ordered mechanism because the former requires that all secondary plots intersect at a point other than the ordinate. This appears to be the case for nopaline synthase. Thus the kinetics are consistent with a ter-bi rapid-equilibrium random-order mechanism.

The kinetics are not as clear for octopine synthase. The simplest interpretation of our kinetic data is that octopine synthase has an ordered mechanism: NADPH binds first, followed by the amino acid, and then pyruvate. However, a partial random order for only arginine cannot be excluded at present. In an ordered mechanism the amino acid must bind before pyruvate for two reasons. First, the lysine analog, ε-aminocaproate, is a competitive inhibitor with respect to arginine but a non-competitive inhibitor with respect to pyruvate. If pyruvate were to bind before arginine, inhibition with respect to pyruvate would have been uncompetitive. Second, the Km for pyruvate is dependent on the amino acid (Table II).

b. Physical Properties. Octopine synthase is a monomeric enzyme with a native molecular weight of 39,000 as determined by gel filtration on Sephadex G-100 and a denatured molecular weight of 38,000 as determined by SDS-polyacrylamide gel electrophoresis. Nopaline synthase is a tetrameric enzyme with a native molecular weight of 158,000 as determined on a column of Ultrogel AcA34, and a subunit molecular weight of 40,000 as determined by SDS-polyacrylamide gels.

The amino acid composition of the two synthases are remarkably similar. Each monomer is composed of 360-370 amino acid residues with the variation between the two proteins limited to a total of 60 residues (30 in each protein) comprising eight different amino acids. If one protein evolved from the other and if the eight amino acids are grouped in pairs that involve only a single base replacement, then the differences can be minimized.

c. Peptide Map Comparisons. Preliminary data suggests that the octopine synthase protein is very different from nopaline synthase even though their amino acid compositions are similar. Finally, octopine synthase isolated from tobacco is identical to the synthase from sunflower; this is consistent with octopine synthase being coded for by a plasmid gene.

REFERENCES

1. M.-D. Chilton, M. H. Drummond, D. J. Merlo, D. Sciaky, A. L. Montoya, M. P. Gordon, and E. W. Nester, Cell 11:263 (1977).
2. J. Schell, "Proceedings of the IV International Conference on Plant Pathogenic Bacteria," Station de Pathologie Vegetale et Phytobacteriolgie, ed., Gilbert-Clarey, Angers (1978).
3. M. H. Drummond, M. P. Gordon, E. W. Nester, and M.-D. Chilton, Nature 269:535 (1977).
4. W. B. Gurley, J. D. Kemp, M. J. Albert, D. W. Sutton, and J. Callis, Proc. Natl. Acad. Sci. USA 76 (6):2828 (1979).
5. G. Bomhoff, P. M. Klapwijk, H. C. M. Kester, R. A. Schilperoort, J. P. Hernalsteens, and J. Schell, Mol. Gen. Genet. 145:177 (1976).
6. A. L. Montoya, M.-D. Chilton, M. P. Gordon, D. Sciaky, and E. W. Nester, J. Bacteriol. 129:101 (1977).
7. W. B. Gurley, J. Callis, and J. D. Kemp, in these proceedings.
8. J. D. Kemp, D. W. Sutton, and E. Hack, Biochem, in press.
9. J. D. Kemp, E. Hack, D. W. Sutton, and M. El-Wakil, in "Proceedings of the IV International Conference on Plant Pathogenic Bacteria," Station de Pathologie Vegetale et Phytobacteriologie, ed., Gilbert-Clarey, Angers (1978).
10. A. Goldmann-Menege, Plant Sci. Lett. 10:49 (1977).
11. P. R. Birnberg, B. B. Lippincott, and J. A. Lippincott, Phytochem. 16:647 (1977).
12. E. Hack and J. D. Kemp, Biochem. Biophys. Res. Commun. 78:785 (1977).
13. L. Otten, D. Vregdenhil, and R. A. Schilperoort, Biochem. Biophys. Acta 485:268 (1977).
14. D. W. Sutton, J. D. Kemp, and E. Hack, Plant Physiol. 62:363 (1978).
15. I. M. Scott and J. L. Firmin, Phytochem. 17:1103 (1978).
16. J. L. Firmin and R. G. Fenwick, Nature, 276:842 (1978).
17. J. D. Kemp, Plant Physiol. 62:26 (1978).
18. A. Petit, S. Delhaye, J. Tempé, and M. G. Morel, Physiol. Vég. 8:205 (1970).
19. J. A. Lippincott, Chai-Cheng Chang, V. R. Creaser-Pence. P. R. Birnberg, S. S. Rao, J. B. Margot, M. H. Whatley, B. B. Lippincott, in: "Proceedings of the IV International Conference on Plant Pathogenic Bacteria," Station de Pathologie Vegetale et Phytobacteriologie, ed., Gilbert-Clarey, Angers (1978).
20. K. Dalziel, Biochem. J. 114:547 (1969).

STRUCTURE OF PLANT VIRAL GENOMES

L. HIRTH

Laboratoire de Virologie, Institut de Biologie

Moléculaire et Cellulaire, 15 rue Descartes Strasbourg

One of the main subjects in virus research in the past few years has been viral gene identification, localisation, and expression, especially in the case of virus infecting eucaryotic cells. Indeed viruses are a good tool for providing information concerning the structure and the regulation of the expression of genes, in general. RNA plant viruses, in view of the small number of genes they contain, are a material of choice to contribute to the solution of these problems. In the first part of this paper, the distribution of the genes of RNA plant viruses is discussed in relation to the mechanism of their expression.

In this regard plant viruses can be divided into three main groups.

The first contains those viruses, all of whose genetic information is contained in a single piece of RNA. The inoculation of this RNA triggers the infectious process, but not all the genes are expressed directly ; rather, subgenomic pieces of messenger RNA arise from the genomic RNA. Two examples will be briefly examined : TMV and TYMV. In vitro translation of TMV and TYMV genomic RNA gives rise to three proteins in the case of TMV (1,2) and to two proteins in the case of TYMV (3). No coat protein, which is obtained in great amounts in vivo, is observed. The 35 Kd protein in the case of TMV and of the coat protein in both viruses is translated only from subgenomic RNA.

In the case of TMV two subgenomic RNAs were isolated by several authors and studied in detail by the group of Zaitlin (2). It is worth noting that, in contrast with the 35 Kd protein gene, the coat-protein gene of the common strain of TMV is not coated but is found as a naked RNA in the infected plants.

In the case of TYMV, only one subgenomic RNA coding for coat protein was isolated from virus preparations, suggesting that this subgenomic RNA is coated (3). A similar situation, but with some variation, exists for all members of the tymovirus group (4). It is worth noting that the subgenomic RNA of TYMV is capped. This rules out the possibility that this RNA arises from an accidental cleavage of the genomic RNA (5). However, so far nothing is known concerning the mechanism of formation of this subgenomic RNA.

The second group of viruses consists of those having a multipartite genome. In this case several RNAs carry the various genes necessary to the expression of the pathogenic power. The simplest case is that of the tobraviruses, an example of which is tobacco rattle virus, in which two RNA components are individually translated and carry information, one for RNA replication, and the other for the coat protein. In the case of more complicated multi-component viruses, such as AMV, BMV, CMV, and related viruses, all the information is contained in several pieces of RNA, one of them carrying two genes. One of these two genes codes for the coat protein but, as in the case of TMV or TYMV, this gene is translated only from a subgenomic RNA arising by an unknown mechanism from the bicistronic RNA (6).

In the third category of plant viruses are found those viruses whose genome consists of two or three pieces but with each RNA translated as a single protein chain which is then apparently processed. This type of virus, such as cowpea mosaic virus and tomato black ring virus, show several other characteristics shared by the picornaviruses : (1) there is poly A at the 3'end of the RNA and (2) the 5'end is not capped but seems to have a covalently bound protein (7,8). It was shown recently that the RNA of the satellite of tobacco necrosis virus has a free 5'end (9).

Whatever the type of virus considered, the various RNAs generally seem to behave in a monocistronic fashion, and when the RNA contains several genes, a subgenomic RNA corresponding to each is produced, from which a unique protein is translated. This strategy, initially discovered with TMV and AMV in the case of plant viruses, seems to be a general one and is also used for the translation of the genes of animal viruses. Why this complicated translation strategy is used is unknown.

In addition to the genetic information necessary for the multiplication of the virus some viruses have a "satellite RNA" which is unable to replicate by itself and needs the presence of a "helper". This satellite RNA can be coated by a protein whose synthesis is controlled either by the satellite (10) or by the helper (11). Some satellite RNAs can be translated in vitro ; this is the case for the satellite of tobacco necrosis virus, in which the only product of translation is the coat protein itself, and also for tomato black ring virus. However, some satellites -such as that of CMV- are not translated or are only very poorly translated. The origin and the role of these satellite RNAs are not known but can be related in some cases to a specific aspect of the expression of symptoms.

Whatever the type of RNA virus considered, one of the main problems is to understand how the genes are expressed and how this expression is regulated. It is well known that coat protein is produced in quantities several hundred times greater than other viral proteins. The coat protein messengers are clearly much more efficient in vitro and in vivo than are the other genomic or sub-genomic messengers. Structural studies of the part of the viral RNAs playing a strategic role in gene translation might be expected to contribute to a solution of this problem and could be extended to the general problem of the mechanism of the translation of eucaryotic mRNAs.

Furthermore, the knowledge of the genetic map of these RNA viruses is essential to an understanding of why some initiation sites are "closed" and how the subgenomic RNAs are synthesized. Some recent results obtained in the author's laboratory and in others in this field are presented or discussed here.

In our laboratory regions of several plant viral RNA suspected to play a strategic role in translation and in replication have been sequenced. We were first interested in the 5' noncoding region of several viral RNAs ; this region contains the signal for recognition of ribosomes and perhaps also for other functions such as maturation of RNAs.

TMV genomic RNA possesses a 68-nucleotide-long sequence at the 5'end before the first functional AUG (12). This sequence is capped, completely devoid of G, and has some unusual characteristics such as AAC tracts (seven of them) and a sequence UUACCUUAC which is repeated twice (Figure 1).

"AAC" TRACT

m⁷GpppGUAUUUUUACAACAAUUACCAACAACAACAAACAACAAACAACAUUACAA

UUACUAUUUACAAUUACAAUG

Figure 1. Complete nucleotide sequence of the 5'end of TMV RNA.

If we compare this sequence with the corresponding sequence of genomic RNA of TYMV (Figure 2), it is clear that this latter sequence is longer but contains several G units (13). However, the first twenty nucleotides are devoid of G. No AAC tract or sequence repetition is observed. The functional AUG is probably the second one, in position 93 (13). In the case of genomic RNA 3 of AMV, which contains two genes (the gene close to the 3'end is the coat-protein gene), the nontranslated part is very long and the first AUG is 98 nucleotides from the cap (Figure 3). It is not

firmly established that this AUG is the first functional one, but
this sequence definitely contains several G and no repetitions (14).
The cap is a general feature of these RNAs.

```
                10        20        30        40        50        60
m⁷GpppGUAAUCAACUACCAAUUCCAGCUCUCUUUUGACAACUGGUCUUAUACCCACUUCCGUACA
              C
       70        80        90       100       110
CUUGCAACCCUCGUAAGACAAUUGCAAAUGAGUAAUGGCCUUCCAAUUAG
                                 Met Ala Phe Gln Leu
```

Figure 2. Nucleotide sequence of the 5'-terminal fragment of
          TYMV genome RNA.

```
          5          10         15         20         25         30
m⁷GpppGUUUUCAUCUUACACACGCUUGUGCAAGAUAG

     35     40     45       50       55       60       65       70
UUAAUCAUUCCAAUUCAACUCAAUUAACGUUUUUUACA

   75       80       85       90       95  98
GUGUAAUUCGUACUUUUUGUAAGUAUG
```

Figure 3. Nucleotide sequence of the 5'terminal fragment of
          alfalfa mosaic virus RNA-3.

Concerning the subgenomic RNAs, the completion of the sequence of the subgenomic RNA of TYMV (5) has shown that the 5' nontranslated part of the RNA is short (19 N) and contains 2 G (Figure 4), in contrast with the nontranslated part of BMV RNA 4, which does not contain G but is rich in U, and with that of AMV RNA 4, which is relatively long but does not contain any G. Recently, the complete sequence of subgenomic TMV RNA was established, and the nontranslated part of the 5'end was found to contain only 9 nucleotides and no G (15) (Figure 5).

m^{7}GpppAAUAGCAAUCAGCCCCAACAUGGAAAUCGACAAA

Figure 4. Nucleotide sequence of the 5'end of TYMV coat protein mRNA.

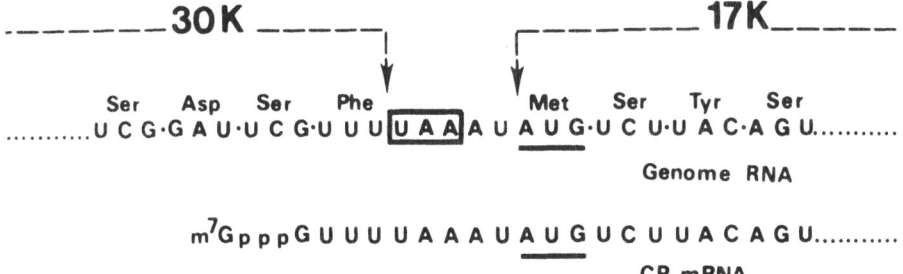

Figure 5. Sequence around the initiation codon of the TMV coat protein cistron.

The common characteristic of all these 5'OH nontranslated
sequences is the absence of a stable secondary structure. Concerning
the sequence itself, it is difficult to find any specific differences or any common sequences long enough to justify speculation
concerning their ability to be more or less efficiently translated.

However, some remarks can be made. Firstly, these sequences
have a cap and an AUG in common. But the cap is not the same for
all the RNA examined. For example, TYMV subgenomic RNA has a cap
of the type m^{7}GpppAp, instead of the sequence m^{7}GpppGp found in
the other subgenomic and genomic RNAs.

Although this difference between subgenomic and genomic RNAs
in the case of TYMV could be evoked to justify the differences in
the level of translation of both RNAs, that is not the case for
AMV, whose four RNAs are capped in the same manner (16). A similar
observation can be made concerning the AUGG sequence that is found
in several eucaryotic messenger RNAs and which may play a role in
the affinity of the initiator tRNA for the initiation site. As
indicated in Fig. 6, several subgenomic RNAs which are efficiently
translated do not possess this sequence.

TMV	ACA AUGG
AMV RNA 4	AUC AUGA
BMV RNA 4	AUA AUGU
TYMV coat protein	AUC AUGG
TYMV genome	GUA AUGG
αglobin	ACC AUGG
βglobin mouse	AUC AUGG
βglobin rabbit	AGA AUGG
βglobin human	ACC AUGG
ovalbumin	ACC AUGG
VSV N	CAA AUGU
VSV Ns	AUC AUGG
VSV G	ACU AUGA

Figure 6. Sequence near initiation codons of several
          eucaryotic mRNAs.

An interesting question is whether some complementarity
exists between the 18 S RNA of eucaryotic ribosomes and certain
regions of the untranslated part of various plant viral RNAs, like
that suggested by Shine and Delgarno for RNA phages and procaryotic

16 S ribosomal RNA. Apparently this is not the case unless, contrary to Shine and Delgarno, the initiator AUG is included within the sequence to be considered. If we include the initiator codon in the case of TYMV, the AAUGG (68-71) sequence is complementary to the $_{OH}$AUU-ACC of the 3'end of 18 S ribosomal RNA from eucaryotic cells.

The same type of situation exists in the case of genomic TYMV RNA. But this is not the case for other plant viral RNAs, and hence it is difficult to relate specific sequences of plant viral mRNAs to their ability to bind to 18 S ribosomal RNA. Perhaps such interactions do not play a role in the mRNA-ribosome recognition process in eucaryotes.

But an unexpected sequence homology was found between the 5'end of rabbit β-globin and the corresponding part of TYMV RNA - about 80 % homology exists (13) (Figure 7). This surprising feature is still unexplained but recently some sequence homologies were found between the introns of adenovirus 2 and ovalbumin (Chambon, personal communication).

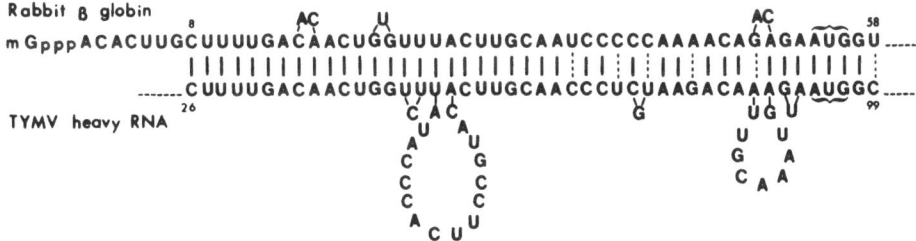

Figure 7. The regions of the TYMV genome RNA sequence that display homology with the 5' terminus of rabbit β-globin mRNA are indicated by the dotted line.

The great variety in the length and in the sequence of the part of viral RNA which apparently contains the recognition site for ribosome does not prove that there is no homology in the tertiary structure. Indeed, if we refer to the noncoding part of the 3'end of certain plant viral RNAs, it is known that this part of the molecule binds an amino acid : TYMV binds valine very effi-ciently in vitro and also in vivo, although the primary and secon-dary structure of the 3'end of this RNA is relatively unlike that of a true valyl-tRNA, as is seen in Fig. 8. But the possible folding of this part of the molecule that can bind valine mimics the main characteristics of tRNA.

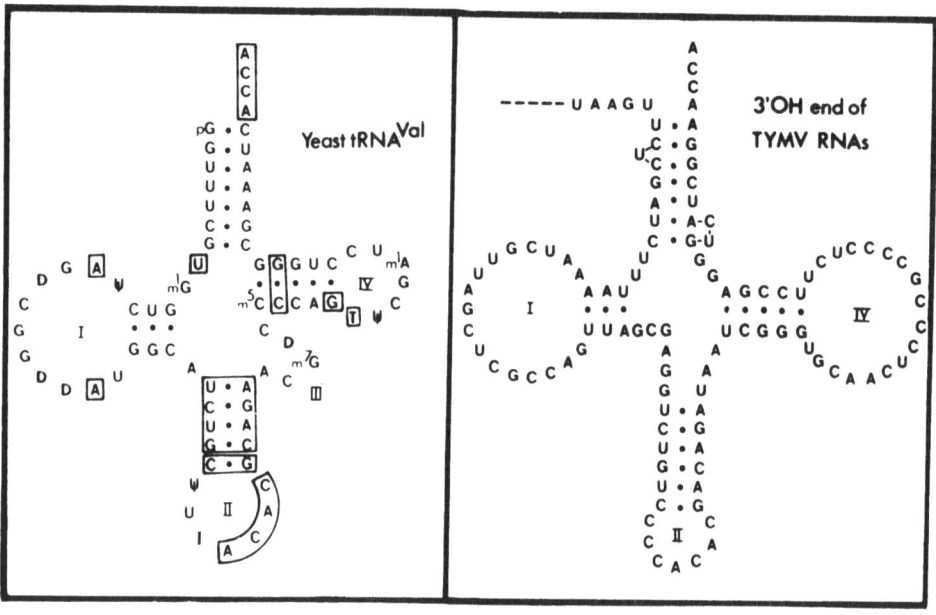

Figure 8. Sequence comparison of the 3'OH end of TYMV RNAs and of yeast tRNA^{Val}. The nucleosides of tRNA^{Val} common to the putative cloverleaf structure of the 3'OH end of TYMV RNAs are in boxes.

Recently the sequence of the 3'end of TMV has been determined (15) and Figure 9 shows that no real analogy exists between this structure and a true tRNA except the terminal CCA and the UUCG corresponding to TψCG of the Tψ loop of all tRNAs. However, this part of the TMV molecule can bind histidine very efficiently (R. Giegé, personal communication).

Although several plant viruses can bind an amino acid at their 3'end, others have neither a tRNA-like structure nor a poly A sequence. This is the case for AMV ; from Fig. 10 it is clear that the 3'ends of RNA 1, 2, and 3 are identical over a relatively short length (26 nucleotides)(17) and even when there are relatively large differences in primary structure, the same type of stable secondary structure is present at least through 100 nucleotides. However, it is clear that the homology of structure between the three genomic RNAs of AMV is less than in the case of the three genomic RNAs of

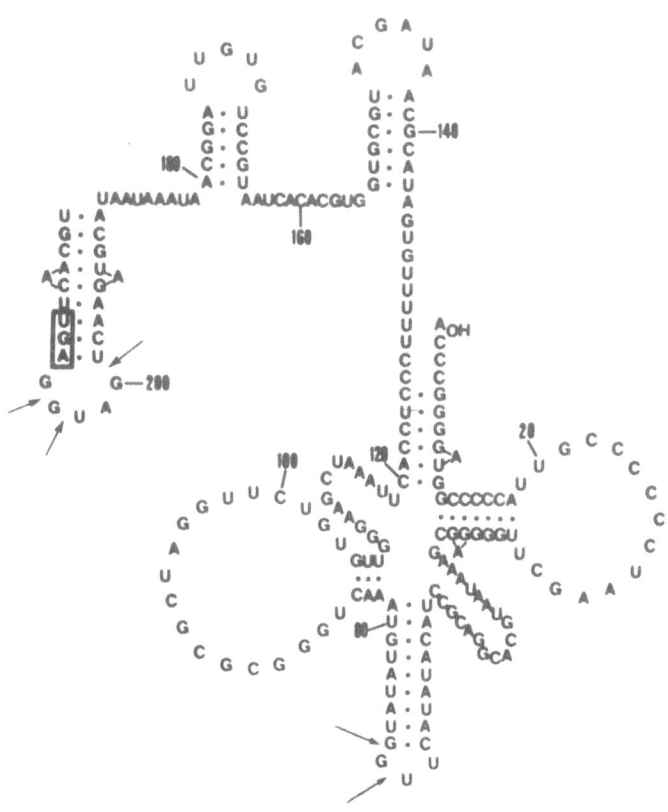

<u>Figure 9.</u> Nucleotide sequence to the right of the coat protein
           cistron of TMV. Sequence has been folded into a possible
           secondary structure.

BMV that efficiently bind tyrosine to the same extent. It could be
of interest to investigate whether this particular structure is
related to the affinity of AMV RNAs for the coat protein of the
virus. It can be concluded that in the case of multicomponent
viruses, the 3'ends of the genomic RNAs show a certain degree of
sequence homology.
    Another example of the difficulty of relating structural
features to the efficiency of translation is the case of satellite
RNAs. For example the CMV satellite RNA, whose complete sequence
is known (Fig. 11) has every characteristic necessary for translation to take place -cap, AUG, termination codon, etc.- suggesting
the potential existence of two peptides, but it is in fact poorly

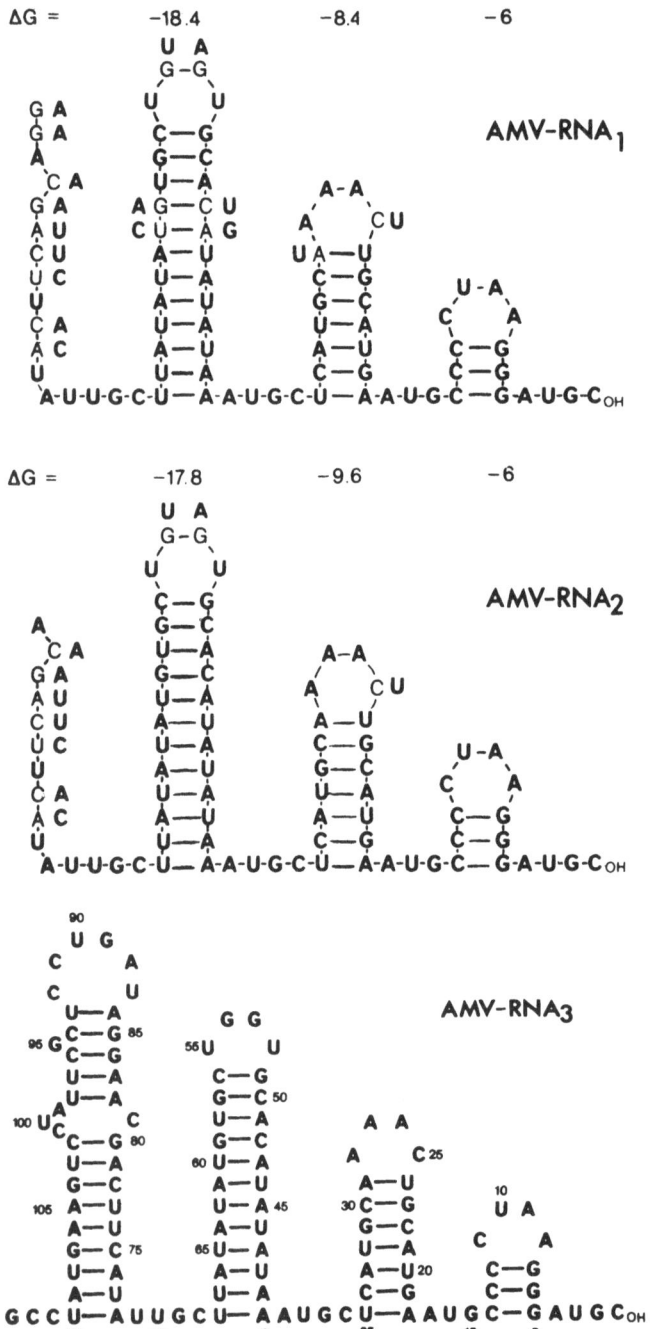

Figure 10. Comparison of the 3'OH terminal sequences of RNA 1, 2 and 3 of AMV. The continuous chain corresponds to RNA 3. Substitutions which occur in RNA 1 and 2 are indicated.

                          met   glu   asn   cys   ala   glu   gly   leu   tyr
$_m^7G_{ppp}$GUUUUGUUUG   AUG   GAG   AAU   UGC   GCA   GAG   GGG   UUA   UAU

 leu   arg   glu   asp   leu   ser   leu   gly   gly   val   gly   tyr
 CUG   CGU   GAG   GAU   CUG   UCA   CUC   GGC   GGU   GUG   GGA   UAC

 leu   pro   ala   lys   ala   gly
 CUC   CCU   GCU   AAG   GCG   GGU   UGAGUGAUGUUCCCUCGGACUGG

                                  met   ser   ala   thr   leu   ser   thr
 GGACCGCUGGCUUGCGAGCU   AUG   UCC   GCU   ACU   CUC   AGU   ACU

 thr   leu   ser   phe   glu   pro   pro   leu   ser   leu   leu   ala
 ACA   CUC   UCA   UUU   GAG   CCC   CCG   CUC   AGU   UUG   CUA   GCA

 glu   pro   gly   thr   trp   phe   ala   asp   thr   met   asp   phe
 GAA   CCC   GGC   ACA   UGG   UUC   GCC   GAU   ACU   AUG   GAU   UUU

 leu   lys   lys   his   ser   val   arg   trp   tyr   glu   ser
 CUA   AAG   AAA   CAC   UCU   GUU   AGG   UGG   UAU   GAG   UCA   UGA

 CGCACGCAGGGAGAGGCUAAGGCUUAUGCUAUGCUGAUCUCCGUGAA

 UGUCUAUCAUUCCUCUGCAGGACCC$_{OH}$

Figure 11. Complete sequence of the satellite RNA of CMV :
         a possible coding scheme.

translated in vitro, if at all (18). It is a very short RNA (385 N)
which should be efficiently translated, but perhaps a minimal
critical length is necessary for efficient translation.
       From the above consideration, it is difficult to relate the
recognition of the initiation site for translation of a eucaryotic
mRNA to a particular nucleotide sequence or structure, and it is
also difficult to understand why the ribosome recognize initiation
sites of subgenomic RNAs but do not recognize them when they are
included in a genomic RNA. Kosak (19) has suggested that ribosomes
recognize the free 5'end of any mRNA, regardless of the nature and
the length of the sequence. The cap is not absolutely necessary
and the 40 S subunits of the ribosome runs along the nontranslated
sequence as far as the first AUG, which is obligatorily the initiation codon. In this "scanning hypothesis", cistrons near the 3'end
of the viral mRNA are not recognized. L. Pinck and M. Pinck in my
laboratory have tested this hypothesis. They found that in suitable
conditions, AMV RNA 3 (which contains the coat-protein cistron in

addition to the cistron for a 35 Kd protein) binds 2 ribosomes but that coat protein is not translated. The sequence of the part of the RNA protected by the ribosomes in the case of RNA 4 was analyzed and was compared with the protected sequences in the case of RNA 3. In the latter case, two types of protected sequence were detected. One corresponds to the nontranslated part of the 5'end of RNA 3 preceding the 35 Kd protein gene, and the other corresponds to the nontranslated part of the coat-protein cistron, but it is clear that the protected sequence is different from that obtained in the case of the subgenomic RNA 4. This strongly suggests that the positioning of the ribosome is not the same in both cases. This incorrect positioning of the ribosomes could explain why elongation does not occur. It could be due to the sequence preceding the RNA 4 sequence and/or to the absence of the cap, which seems play an important role (but not an obligatory role) in the recognition process.

Controls have shown that the isolated fragments are really ribosome-protected fragments.

These experiments seem to indicate that ribosomes are able to bind to specific sites but in such a position that no elongation can occur. This observation suggests that some specific structure is recognized by the ribosomes. This is contrary to the hypothesis of Kosak, who claimed that a specific sequence does not play a role in the process of recognition of an mRNA by eucaryotic ribosomes. It is of course necessary to check whether this observation is an isolated one or is really some general feature. If this virus is an exception, it is one more which does not fit well with the "scanning model", in addition to avian sarcoma virus (20), turnip yellow mosaic virus (13), and simian virus 40 (19).

It is of course surprising to suggest some recognition specificity of eucaryotic mRNA by eucaryotic ribosomes in view of the preceding considerations on the absence of common characteristics in leader sequences of these mRNAs, but I wish to emphazise the fact that in several cases, and especially in that of the tRNA-like structure of the 3'end of plant viral RNA, very different sequences seem to be recognized specifically by the same enzyme. Similar spatial configuration could probably arise from different sequences.

Concerning the regulation of the expression of the gene, TYMV offers a subject of speculation. It is clear from sequence analysis of the 5'end and the 3'end of genomic and subgenomic RNA (Fig. 12) that 9 bases are complementary (13). Base pairing between the two terminal parts of the same molecule causes a large loop to be formed ; such base-pairing could account for the loop structure for this RNA detected 12 years ago in my lab by physico-chemical means. But if the base-pairing occurs between the 5'end of a genomic RNA molecule and the 3'end of a subgenomic RNA, it is clear that the translation of the large RNA could be stopped while that of the subgenomic RNA could continue very actively. This is of course only a hypothesis, but it could be of interest to examine whether this type of base-pairing occurs in other plant viruses showing a subgenomic system.

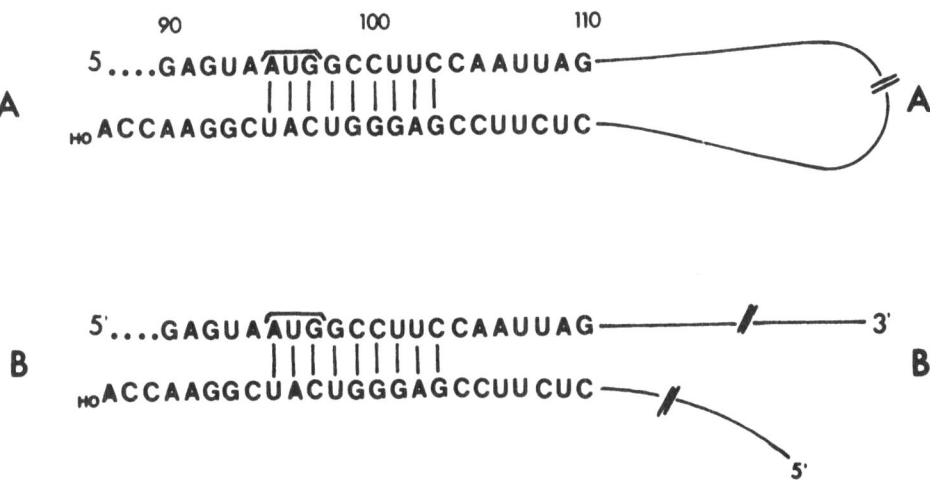

Figure 12. Possible interaction between the 5' and 3'extremities of TYMV genome RNA (A) or between the 5'extremity of TYMV genome RNA and the 3'extremity of the coat protein mRNA (B).

    In conclusion, the knowledge of the sequences of strategic regions of plant viral RNA is very useful and has allowed discussion of the general problem of translation of eucaryotic mRNAs. It is clear that the problem is more complicated than expected and that sequence studies are not enough to show how ribosomes recognize the mRNA at the right place and why some genes are more efficiently translated than others. Studies of the spatial conformation of mRNAs will probably help to elucidate the matter.

REFERENCES

1. Hunter, T.R., Hunt, T., Knowland, J. and Zimmern, D. (1976)
    Nature 260, 759-764.
2. Beachy, R.N. and Zaitlin, M. (1977) Virology 81, 161-169.
3. Klein, C., Fritsch, C., Briand, J.P., Richards, K.E., Jonard,
    G. and Hirth, L. (1976) Nucleic Acids Res. 3, 3043-3061.
4. Szybiak, U., Bouley, J.P. and Fritsch, C. (1978) Nucleic Acids
    Res. 5, 1821-1831.
5. Guilley, H. and Briand, J.P. (1978) Cell 15, 113-122.
6. Mohier, E., Hirth, L., Le Meur, M.A. and Gerlinger, P. (1976)
    Virology 71, 615-618.
7. El Manna, M.M. and Bruening, G. (1973) Virology 56, 198-206.
8. Harrison, B.D. and Barker, H. (1978) J. Gen. Virol. 40, 711-
    715.
9. Leung, D.W., Browing, K.S., Heckmann, J.E., RajBhandary, U.L.
    and Clark, J.M. (1979) Biochemistry 18, 1361-1366.
10. Kassanis, B. (1967) Nature 214, 178-180.
11. Murant, A.F., Mayo, M.A., Harrison, B.D. and Goold, R.A. (1973)
    J. Gen. Virol. 19, 275-278.
12. Richards, K., Guilley, H., Jonard, G. and Hirth, L. (1978)
    Eur. J. Biochem. 84, 513-519.
13. Briand, J.P., Keith, G. and Guilley, H. (1978) Proc. Natl.
    Acad. Sci. USA 75, 3168-3172.
14. Pinck, L. and Pinck, M. to be published.
15. Guilley, H., Jonard, G., Kukla, B. and Richards, K.E. (1979)
    Nucleic Acids Res. 6, 1287-1308.
16. Pinck, L. (1975) FEBS Letters 59, 24-28.
17. Pinck, L. and Pinck, M. In preparation.
18. Richards, K.E., Jonard, G., Jacquemond, M. and Lot, H. (1978)
    Virology 89, 395-408.
19. Kosak, M. (1978) Cell 15, 1109-1123.
20. Shine, J., Czernilofsky, A.P., Friedrich, R., Bishop, J.M. and
    Goodman, H.M. (1977) Proc. Natl. Acad. Sci. USA 74, 1473-1477.

ABBREVIATIONS

        TMV   : tobacco mosaic virus
        TYMV   : turnip yellow mosaic virus
        AMV   : alfalfa mosaic virus
        BMV   : brome mosaic virus
        CMV   : cucumber mosaic virus
        VSN   : vesicular stomatitis virus

TRANSLATION OF PLANT VIRUS RNAs

Lous Van Vloten-Doting and Lyda Neeleman

Department of Biochemistry

State University of Leiden, Leiden, The Netherlands

INTRODUCTION

In vitro translation of plant virus RNAs not only yields valuable information about the proteins encoded by these RNAs, but it also gives information about the protein synthesizing machinery of the host. In the present paper we only discuss the results obtained with plant viruses with a genome consisting of single stranded RNA(s) with the plus polarity (genomic RNA has the same polarity as the messenger RNA). The viruses are discussed according to the type of structure present at the 5' terminus, the 3' terminus and the strategy employed to express the genetic information. In the last paragraph the possible functions of the virus coded proteins are discussed.

I. STRUCTURE AT THE 5' TERMINUS

Three different types of structures have been found at the 5' termini of plant virus RNAs: "cap", genome linked protein or (p)ppX. For the plant viruses studied so far, all RNAs (genomic as well as subgenomic) of one particular virus have always the same type of structure at their 5' termini.

A. $m^7G^{5'}ppp^{5'}X^{(m)}pY^{(m)}p\ldots$ "cap"

Like most eukaryotic mRNAs a number of plant virus RNAs (Table I) have been shown to be capped at their 5' terminus. It is remarkable that all caps found to date in plant viruses are devoid of the methyl groups at the ribose moiety of nucleotides X and Y. The role of cap-structures in the biosynthesis of proteins has been studied extensively and was recently reviewed[24]. A brief summary at the present state of knowledge is given here.

The presence of the cap facilitates the formation of the initiation complex, probably by enhancing the affinity of the RNA for one of the initiation factors. In addition the cap protects the RNA (at least in *in vitro* systems) from digestion by exonucleases. The presence of the cap is not obligatory for the initiation of protein synthesis. The role played by the cap during initiation of protein synthesis is more important for some RNAs than for others. Furthermore the presence of the cap is more important at low mRNA concentration than at high mRNA concentration, at low salt concentration than at high salt concentration, in the wheat germ cell free system than in the reticulocyte cell free system. It is possible that this latter difference is due to the different ionic conditions used in these two systems. It is unknown if the differences found in *in vitro* systems do play a role *in vivo*.

B. Genome linked protein

The 5' terminus of the RNAs from CPMV and SBMV have been shown to be linked to a protein, while there is evidence that the same is true for two viruses of the nepogroup (Table I). Since protease K digestion of the protein linked to CPMV-RNAs did not change the infectivity nor the *in vitro* messenger function of these RNAs, the intact protein is not required for these functions [86]. This is in contrast to the finding that the infectivity of the nepoviruses is lost upon protease K treatment[38], no effect of this treatment of the messenger function of these RNAs has been reported. The effect of protease treatment of SBMV-RNA has not yet been studied.

The genome linked protein of poliovirus is not required for infectivity[61]. For this virus it has been found that the virion RNA and the nascent chains are protein linked, but the RNA present in the polysomes is not[61].

At present it cannot be excluded that in the cell free system (and maybe also *in vivo*) the protein is removed from the RNA prior to translation. It therefore seems premature to conclude that plant ribosomes are capable of translating protein linked RNA.

C. (p)ppX

To date only two plant virus RNAs have been found which are neither "capped" nor "protein-coupled" at their 5' terminus (Table I). Although it cannot be ruled out that *in vivo* these RNAs become "capped", it has been shown that STNV-RNA is not capped prior to translation *in vitro*[52,73].

Table I. Terminal Structures of Plant Virus RNAs

Type of genome	Virus		Terminal structure 5'	Terminal structure 3'	Virus group[a]	Number of members
monopartite	TMV	tobacco mosaic virus	cap 44,96	tRNA his[62]	tobamo	10
	CcTMV	cowpea strain of TMV	-	tRNA val[5]		
	TYMV	turnip yellow mosaic virus	cap 45	tRNA val[169]	tymo	16
	EMV	egg plant mosaic virus	-	tRNA val[169]		
	PMV	papayo mosaic virus	cap 2	-	potex	8
	TEV	tobacco etch virus	-	poly A [37]	poty	34
	SBMV	southern bean mosaic virus	protein[28]	pX_{OH} 75		5
	CarMV	carnation mottle virus[b]	-	-		4
	TNV	tobacco necrosis virus	ppX 51	-		1
bipartite	TRV	tobacco rattle virus[c]	cap 1	-	tobra	2
	BSMV	barley stripe mosaic virus	cap 4	poly A [3]	hordei	3
	BPMV	bean pod mottle virus	-	poly A 82	como	10
	CPMV	cowpea mosaic virus	protein[15,86]	poly A 22		
	RCMV	red clover mosaic virus	-	poly A 64		
	TBRV	tomato black ring virus	protein[38]	-	nepo	10
	TRSV	tobacco ringspot virus	protein[38]	-		
tripartite	BMV	brome mosaic virus	cap 14	tRNA tyr[35]	bromo	3
	CCMV	cowpea chlorotic mottle virus	-	tRNA tyr[47]		
	CMV	cucumber mosaic virus	cap 87	tRNA tyr[47]	cucumo	3
	AMV	alfalfa mosaic virus[d]	cap 67	pX_{OH} 34	ilar	14
	TSV	tobacco streak virus	-	pX_{OH} 34		
satellites	STNV	satellite of TNV[e]	ppX 51			
	CARNA	satellite of CMV[f]	cap 54	pX_{OH} 72		
	TRSV 3	satellite of TRSV·	-	-		

[a] Grouping of viruses is according to[23] except SBMV and CarMV which are grouped according to[41].

[b] The 3' terminus is pX_{OH}[79], there are no reports about aminoacylation.

[c] It has been published that only the largest TRV-RNA is capped. However, later evidence suggest that both RNAs are capped. The 3' terminus is pX_{OH}[50], there are no reports about aminoacylation.

[d] AMV is still classified separately[23], however, there is increasing evidence that AMV belongs to the ilar group[33,90].

[e] The 3' terminus is pX_{OH}[40a], there are no reports about aminoacylation.

[f] CARNA 5 is CMV associated RNA 5[54] also called SAT RNA[57].

II. STRUCTURE AT THE 3' TERMINUS

Three different types of structures have been found at the 3' termini of plant virus RNAs: poly A tail, tRNA-like or pX_{OH}. It has been consistently found that all RNAs (genomic as well as subgenomic) of one particular virus have the same type of structure at their 3' termini.

A. Poly A tail

In contrast to most eukaryotic mRNAs only a few plant virus RNAs have been shown to carry a poly A tail (Table I). It is not yet known if the poly A is virus coded (the minus strand carries a poly U track) or that it is added to the viral RNA after replication by host or viral-coded enzymes. For BSMV[3] as well as for TEV [37] it has been found that part of the RNA molecules is poly A plus and part is poly A minus. For both viruses the infectivity of the poly A plus RNA and the poly A minus RNA was about equal[3,37]. For picorna viruses (which have a virus coded poly A tail[94], it was found that the poly A tail was required for infectivity[29,85].

The poly A tail of globin mRNA has been shown to be involved in mRNA stability, possibly by protecting the coding sequences from exonuclease digestion[55]. Theoretically it is possible that the encapsidated virus RNA is not, while the virus RNA functioning *in vivo* as messenger is polyadenylated. However, it has been shown for AMV that the virus RNAs present in polysomes of infected tobacco leaves do not bind to oligo dT columns[8]. Thus if the viral mRNA fraction is polyadylated the tracks are extremely short.

Virus RNAs, in contrast to cellular mRNAs, should not only function as messenger, but also serve as template for RNA replication. It is conceivable that the poly A tail interferes with the latter function.

B. "tRNA-like" structure

A feature found upto now only for virus RNAs (both animal as well as plant) is the presence of a "tRNA-like" structure, which can be charged with an amino acid, at the 3' terminus. It has been suggested[36] that the tRNA-like structure plays a role in the replication of virus RNA.

However, it is also possible that the tRNA-like structure plays a real tRNA role in the translation of virus RNAs during the infection cycle. It has been shown[42] that in oocytes the tRNA-like structure of TYMV-RNA can be liberated and charged with valine. During *in vitro* protein synthesis the valine bound to the viral RNA can be donated to the growing peptide chain[32]. This is in contrast to the finding that neither the intact tyrosylated BMV-RNA nor the tyrosylated 3' fragment of 160 nucleotides was capable of transferring tyrosine to peptide material during protein synthe-

sis[34]. It is possible that the tRNA-like structure of BMV has to
be further matured to become active. Yeast tRNAs have been shown
to contain intervening sequences which are removed during maturation[63]. One could speculate that the other viral RNAs (poly A and
pX_{OH}) do contain internal tRNA-like structures as has been found
for EMC[53].
Recently it has been shown that along virus RNAs the (*in vitro*)
translation at some points slows down or is even completely arrested. These points could correspond to amber codons (as is suggested for TMV[66]) or to codons for which the corresponding tRNA is
only present in limiting amounts (as is suggested for AMV[88]). It
is possible that the liberated virus coded tRNA is required to
translate these difficult codons.

C. pX_{OH}

It has been shown that the RNAs from several viruses, which
do not contain a poly A tail, cannot be charged with an amino acid
(Table I). For AMV and TSV this was rather surprising since these
viruses have a lot in common with BMV and CMV[90], the latter two
can both be charged with tyrosine.

CONCLUSIONS FROM SECTION I AND II

From Table I it is evident that there is no correlation between the type of structure present at the 5' terminus with the
type of structure present at the 3' terminus. Nor is there a correlation between one of these terminal structure and the type of
genome.

III. FIDELITY OF TRANSLATION

In vitro translation of plant virus RNAs only gives useful
information about the proteins directed by these RNAs *in vivo* when
the *in vitro* translation occurs with fidelity. It has been shown
for several viruses that the coat protein directed by the subgenomic coat protein messenger is identical to the coat protein formed
in vivo[16], and there in. Thus the translation of these small messengers seems to be reliable. A more difficult question to answer
is the fidelity of the translation of the larger RNAs, especially
the question whether the proteins are terminated at the correct
position. Only for TMV has the *in vitro* and the *in vivo* produced
non-coat protein been analysed[81]. The peptide patterns obtained
after cyanogen bromide cleavage of the two proteins (M.W. 130 000)
were very similar, suggesting that these proteins are similar or
even identical. These results prove that *in vivo* TMV-RNA is also
translated into a 130 000 protein. However, they do not prove that
this large protein is the functional active protein.
For several viruses the proteins induced after infection have
been studied. The size of these proteins has been compared to the

size of the proteins directed by the viral RNAs *in vitro*. However,
these results should be considered with care, since it has not been
proven that the proteins seen are indeed virus coded and not virus
induced host proteins. Based on size comparison, it has been suggested that the proteins directed *in vivo* by TMV-RNA[76], satellite
RNA of TBRV[27], all four RNAs of CCMV[77] and BMV[78] and the two RNAs
of CPMV[18] are identical to the largest products directed by these
RNAs *in vitro*.

Another approach to the question of fidelity is the comparison
of the products directed by the same RNA in different cell free
systems[16], and therein. The main difference between products synthesized in the different eukaryotic systems seems to be the point
of termination. This difference might be due to a difference in the
capacity of the systems used to suppress termination[66]. Premature
termination could take place at "difficult" points along the RNA.
From the results obtained with paromomycine it is evident that mis-
matching enhances the possibility to proceed along these points
(Fig. 1). Paromomycine was able to suppress in yeast *in vivo* as
well as *in vitro* all three stop codons[65,84]. From the fact that
under the direction of several plant virus RNAs the largest product made in the presence of paromomycine is of the same size as
the largest product made in the absence of paromomycine follows
that the termination at this point is not suppressed. This could
mean that either not all three stop codons are suppressed in reticulocytes or at the end of the cistron there are several succee-
ding termination codons. It is rather remarkable that even in the

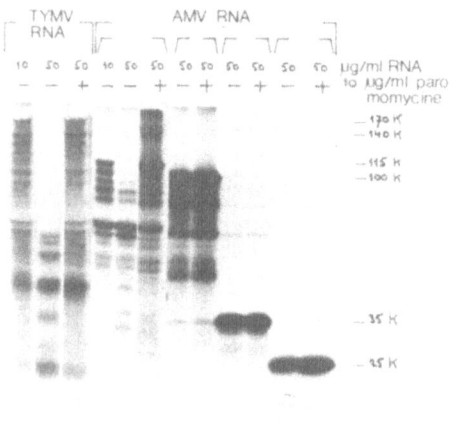

Fig. 1. Comparison of the products directed by TYMV and AMV-RNA
        1, 2, 3 and 4 (concentrations indicated in µg/ml) in the
        absence (-) and in the presence (+) of 10 µg/ml paromomy-
        cine in the mRNA dependent rabbit reticulocyte cell-free
        system. (Figure courtesy R.G.L. Van Tol).

presence of paromomycine AMV-RNA 3[x] in reticulocytes does not direct a significant amount of a "read-through" product of 65 000.
It has been shown previously[74],[89] that in wheat germ "read-through"
can take place on this RNA. A difference in the point of initiation
of protein synthesis has been found upon translation of AMV RNA
4 in a cell free system of *E. coli*. Besides initiation on the normal AUG codon initiation took also place on a UUU codon located
close to the cap[10],[48]. Initiation on this latter point resulted in
the formation of a protein which was identical to the coat protein
but did contain 12 additional amino acids at the N terminus[10],[91].

IV. STRATEGY OF THE EXPRESSION OF THE INFORMATION

In vitro translation of one purified plant virus RNA often
yields more than one protein. In several cases it has been demonstrated that the products represent overlapping chains[16], and therein. Most of the numerous bands found in the wheat germ system are
probably artefacts, since variation in the incubation conditions
often leads to products with a higher molecular weight. However,
also in the reticulocyte lysate (which yields a "clearer" picture)
one purified RNA often directs the synthesis of two or three main
products. The sum of these proteins exceeds the coding capacity of
the messenger. For a number of virus RNAs[6],[66],[88] it has been shown
that these proteins represent overlapping chains with the same N
terminus. Since it is rather difficult to find *in vivo* virus-coded
proteins other than the coat protein it is hard to determine which
of the products found *in vitro* represent the "real" product. At
present we cannot decide whether these partial translates are due
to dual use of the genetic information or are merely artefacts of
the *in vitro* system. For the bacteriophage Qβ it has been shown
that the "read-through" protein A1, which arises by suppression
of a stop codon, is an essential protein[40].

Theoretically the three letter coding system offers the possibility of coding three chemically different proteins on one nucleotide sequence. For some bacteriophages double, and even triple use
of the same nucleotides has been reported[83], and therein. To date
there is no evidence that plant virus RNAs are read in more than
one reading frame.
The different RNAs from the same virus apparently do not all have
to employ the same strategy of expression. For instance the two
largest RNAs of the viruses with a tripartite genome are, at least
in vitro, translated into full-lengths products, which might be
processed later on, while only the 5' half of the smallest genomic
RNA is translated directly, the 3' half is translated from a subgenomic messenger[16], and therein.

---

[x] The RNAs of viruses with a multipartite genome are numbered in order of decreasing molecular weight.

A. Functionally monocistronic: expression of the internal cistron is mediated by a subgenomic mRNA

Like most eukaryotic mRNAs the majority of the plant virus RNAs studied so far are functionally monocistronic (only one initiation site per RNA molecule). If the information for a second (or a third) cistron is present, that part of the RNA is amplified *in vivo*[16]. It is not yet known if these subgenomic mRNAs arise by partial transcription or by processing of the genomic RNA. The sequence at the 5' terminus of the subgenomic mRNA of plant viruses is different from that of the corresponding genomic RNA[9,30,31,48,68,71]. This is in contrast to the situation found in several DNA animal viruses where the same leader sequence is spliced onto several mRNAs[11].

Although there is no doubt that the main translation product is derived from the cistron located near the 5' terminus, small amounts of the translation products from the internal cistron are often found. In most cases this has been explained by the assumption that the genomic RNA is slightly contaminated with subgenomic RNA(s). However, the question whether the internal cistron is absolutely not accessible for translation or is only expressed at a much lower efficiency remains open.

B. Monocistronic, the primary product is a "polyprotein"

In vitro, the two RNAs from CPMV both direct the synthesis of a large product, corresponding approximately to the total of genetic information present in the RNAs[18]. The two products do not react with antiserum against CPMV capsids[17]. No products corresponding in size to one or both of the viral capsid proteins were formed[18]. Up to now no subgenomic messenger for the coat protein have been found in the virus preparation nor in the host[17]. These results suggest (but do not prove) that the primary translation products have to be processed into functional units. The results obtained with TBRV[27] suggest that the translation strategy of this virus has much in common with that of CPMV.

The post translational cleavage of the polio virus polyprotein is well documented[13]. It is remarkable that the plant viruses employing this mechanism, resemble polio virus in having a protein linked to the 5' terminus of the genome and (in the case of CPMV) a poly A tail to the 3' terminus. However, there is not necessarily a correlation between these structural features and the mode of translation. The largest protein directed *in vitro* by SBMV-RNA, which is also linked to a protein (Table I), corresponds to only about half of the information present, and the virus preparations do contain a small messenger for the coat protein[75].

C. Di-or polycistronic

At present it is often suggested that initiation by eukaryotic

ribosomes is completely restricted to the AUG codon located close
to the 5' terminus[49]. However, it might be an oversimplification
to confine the translation of all eukaryotic mRNAs to one model.
We have already seen above that at the 3' or the 5' terminus of a
mRNA at least three different structures can be found. The polycistronic messenger has proven to be very succesful in prokaryotes.
Furthermore it has been shown that eukaryotic cell free systems are
capable of translating polycistronic prokaryotic messengers[16,50].
So it is quite conceivable that although the majority of eukaryotic
messengers are functionally monocistronic, polycistronic messengers
will be found. Indeed there are already several reports suggesting
that such polycistronic RNAs (from plant as well as from animal viruses) do exist.

Ribosome binding studies have shown that more than one ribosome can be bound to some plant virus RNAs under initiation conditions. However, in it self this is not sufficient to conclude that
such an RNA has two open initiation sites. It has been shown recently[25] that the two ribosomes on TMV-RNA are both present in the
leader sequence. The same has been suggested, but not proven for
TYMV-RNA[25]. However, the two ribosomes on AMV-RNA 3[59,74,91] protect
two different pieces of RNA against digestion with RNases[68]. One of
these is identical to the 5' terminus of the subgenomic mRNA for
the coat protein. Furthermore *in vitro* translation of this RNA
suggest that the coat protein cistron, which is located internally,
is translated at a very low efficiency[58]. It is possible that the
low rate of protein synthesis initiation at this internal cistron
is a reminiscence of the evolutionary older situation of a dicis-
tronic messenger.

In vitro translation of the intact CarMV-RNA (which had been
purified under denaturing conditions) yield three unrelated proteins[79]. Together these three proteins could represent the total
genetic information present in this RNA. It is remarkable that limited phosphorylysis of the 3' terminus with polynucleotide phos-
phorylase caused the selective disappearance of the largest protein.
From these results it was concluded[79] that the 3' terminus might
have a regulatory function.

In vitro translation of TRV-RNA 2 yields two proteins which
are partly unique. The relative ratio of these two products is influenced by the concentration of Mg-ions[26]. At present the homogeneity of the TRV-RNA 2 has not been rigorously proven. However,
the effect of Mg-ions on the ratio of the two products is rather
similar to the effect of Mg-ions on the ratio of the two initiating
peptides directed by polio RNA[46].

In vitro translation of TNV-RNA (purified under denaturing
conditions) yields mainly coat protein and two larger proteins[80].
However, it has not yet been shown that these larger products are
unrelated to the coat protein, so the possibility that they represent "read-through" products cannot be ruled out.

V. COMPETITION BETWEEN RNAs

One of the interesting questions is how a virus manipulates its host cell so that this cell starts to manufacture new virus particles at its own expense. Although plant viruses do not show the phenomenon of host protein synthesis "shut-off", they are produced in massive amounts in the infected cell.

One way to ensure the preferential synthesis of virus proteins could be a higher affinity of the viral RNA compared to the host mRNA   for one or more of the components of the protein synthesizing machinery. Up to now only few studies in this field have been reported. It has been found that the poly A plus RNA isolated from wheat germ is in a wheat germ cell free system a less efficient messenger than several plant virus RNAs[39]. It has been postulated [39,43] that there is an (initiation) factor which recognizes specific sequences in the mRNA other than the cap or the AUG.

Most studies on messenger RNA competition have been performed with different RNAs from the same virus. In these studies the subgenomic coat protein messenger is mostly preferentially translated [7,16 and therein,74,88,95]. It has been shown that the subgenomic RNA from BMV binds more efficiently to ribosomes than the three genomic RNAs[70]. From competition experiments between the AMV-RNAs it was concluded that, at least in the reticulocyte cell free system, RNA 1 and RNA 3 require an (initiation) factor which is not required for the translation of RNA 2[88].

In competition experiments between STNV-RNA and several other RNAs it was shown that the capped mRNAs were out-competed by the non-capped STNV-RNA[39].

When evaluating this type of experiments, one must be aware of the possibility that the preferential translation of one RNA is due to the *in vitro* lability of a factor required for the translation of the competing RNA species.

VI. FUNCTION OF VIRUS CODED PROTEINS

At present the only virus coded protein with a known function is the coat protein. It has been assumed that RNA viruses have to code for their own RNA replicase. However, it is now well documented that healthy plants do contain such an enzyme[12]. To date no convincing evidence has been presented for the involvement of a virus coded protein in the RNA replicase. That virus coded proteins do play a role during the process of RNA replication is evident from the fact that two *ts* mutants of TMV have been described in which the RNA replication is thermosensitive[19,20]. Besides the RNA replication, virus coded proteins other than coat protein are involved in the transport of virus through the host [21,60].

Studies with *ts* mutants from alfalfa mosaic virus have shown that each of the two largest genomic RNAs consists of at least two complementation groups[92]. Apparently the large primary product coded for by these RNAs has two active sites and/or is processed *in vivo*.

From this study we can conclude that proteins encoded by the AMV genome perform at least 5 different functions.

ACKNOWLEDGEMENTS

We thank our collegues Gied Jaspars, John Bol, Ruud Van Tol and Bob Tank for critical reading of the manuscript. We are grateful to Tineke Rutgers and Jeffrey Davies for providing us with research results prior to their publication.

This research was supported by the Netherlands Foundation for Chemical Research (SON), which is subsidized by the Netherlands Organization for the Advancement of Pure Research (Z.W.O.).

REFERENCES

1. M. Abou Haidar, and L. Hirth, 5'-Terminal structure of tobacco rattle virus RNA: Evidence for polarity of reconstitution, Virology 76:173 (1977).

2. M. Abou Haidar, and J.B. Bancroft, The structure of the 5'-terminus of papaya mosaic virus RNA, J. Gen. Virol. 39: 559 (1978).

3. A.A. Agranovsky, V.V. Dolja, V.M. Kavsan, and J.G. Atabekov, Detection of polyadenylate sequences in RNA of barley stripe mosaic virus, Virology 91:95 (1978).

4. A.A. Agranovsky, V.V. Dolja, V.K. Kagramanova, and J.G. Atabekov, The presence of a cap structure at the 5'-end of barley stripe mosaic virus RNA, Virology 95:208 (1979).

5. R.N. Beachy, M. Zaitlin, G. Bruening, and H.W. Israel, A genetic map for the cowpea strain of TMV, Virology 73:498 (1976).

6. C. Benicourt, J. Pere, and A.-L. Haenni, Translation of TYMV RNA into high molecular weight proteins, FEBS Lett. 86:268 (1978).

7. C. Benicourt, and A.-L. Haenni, Differential translation of turnip yellow mosaic virus mRNA *in vitro*, Biochem. Biophys. Res. Commun. 84:831 (1978).

8. J.F. Bol, C.E.G.C. Bakhuizen, and T. Rutgers, Composition and biosynthetic activity of polyribosomes associated with alfalfa mosaic virus infections, Virology 75:1 (1976).

9. J.P. Briand, G. Keith, and H. Guilley, Nucleotide sequence at the 5' extremity of turnip yellow mosaic virus genome RNA, Proc. Natl. Acad. Sci. U.S.A. 75:3168 (1978).

10. A. Castel, B. Kraal, P.R.M. Kerklaan, J. Klok, and L. Bosch, Initiation of polypeptide synthesis with various NH_2-blocked aminoacyl-tRNAs under the direction of alfalfa mosaic virus RNA 4, Proc. Natl. Acad. Sci. U.S.A. 74:

        5509 (1977).

11. L. Chow, and T.R. Broker, The spliced structures of Adeno-
     virus 2. Fiber Message and the other late mRNAs, Cell
     15:497 (1978).

12. C.M. Clerx, and J.F. Bol, Properties of solublized RNA-
     dependent RNA polymerase from alfalfa mosaic virus-
     infected and healthy tobacco plants, Virology 91:453 (1978)

13. P.D. Cooper, Genetics of picornaviruses, in: "Comprehensive
     Virology", H. Fraenkel-Conrat and R.R. Wagner, eds.,
     Plenum Press, New York, vol. 9 (1977).

14. R. Dasgupta, F. Harada, and P. Kaesberg, Blocked 5'-termini
     in brome mosaic virus RNA, J. Virol. 18:260 (1976).

15. S. Daubert, G. Bruening, and R. Najarian, Protein bound to
     the genome RNAs of cowpea mosaic virus, Eur. J. Biochem.
     92:45 (1978).

16. J.W. Davies, Translation of plant virus ribonucleic acids
     in extracts from eukaryotic cell in: "Nucleic Acids in
     Plants", T.C. Hall and J.W. Davies, eds., CRC Press,
     West Palm Beach, vol. 2 (1979).

17. J.W. Davies, Personal Communication.

18. J.W. Davies, A.M.J. Aalbers, E.J. Stuik, and A. Van Kammen,
     Translation of cowpea mosaic virus RNA in a cell-free
     extract from wheat germ, FEBS Lett. 77:265 (1977).

19. W.O. Dawson, and J.L. White, Characterisation of a temp-
     erature-sensitive mutant of tobacco mosaic virus
     deficient in synthesis of all RNA species, Virology 90:
     209 (1978).

20. W.O. Dawson, and J.L. White, A temperature-sensitive mutant
     of tobacco mosaic virus deficient in synthesis of single-
     stranded RNA, Virology 93:104 (1979).

21. A. Dingjan-Versteegh, L. Van Vloten Doting, and E.M.J. Jas-
     pars, Alfalfa mosaic virus hybrids constructed by
     exchanging nucleoprotein components, Virology 49:716
     (1972).

22. M.M. El Manna, and G. Bruening, Polyadenylate sequences in
     the ribonucleic acids of cowpea mosaic virus, Virology
     56:198 (1973).

23. F. Fenner, Second report of the international committee on
     Taxonomy of viruses, Intervirology 7:1 (1976).

24. W. Filipowicz, Functions of the 5'-terminal m⁷G cap in
     eukaryotic mRNA, FEBS Lett. 96:1 (1978).

25. W. Filipowicz, and A.-L. Haenni, Binding of ribosomes to 5'-
     terminal leader sequences of eukaryotic messenger RNAs,
     Proc. Natl. Acad. Sci. U.S.A. In press (1979).

26. C. Fritsch, M.A. Mayo, and L. Hirth, Further studies on the
     translation products of tobacco rattle virus RNA in vitro
     Virology 77:722 (1977).

27. C. Fritsch, M.A. Mayo, and A.F. Murant, Translation of the
     satellite of tomato black ring virus in vitro and in
     tobacco protoplasts, J. Gen. Virol. 40:587 (1978).

28. A. Ghosh, and R. Dasgupta, Personal Communication.
29. N.O. Goldstein, I.U. Pardoe, and A.T.H. Burness, Requirement of an adenylic acid-rich segment for the infectivity of encephalomyocarditis virus, J. Gen. Virol. 31:271 (1976).
30. H. Guilley, and J.P. Briand, Nucleotide sequence of turnip yellow mosaic virus coat protein mRNA, Cell 15:113 (1978).
31. H. Guilley, G. Jonard, B. Kukla, and K.E. Richards, Sequence of 1000 nucleotides at the 3'-end of tobacco mosaic virus RNA, Nucleic Acids Res. 6:1287 (1979).
32. A.-L. Haenni, A. Prochiantz, O. Bernard, and F. Chapeville, TYMV valyl-RNA as an amino-acid donor in protein bio-synthesis, Nature New Biol. 241:166 (1973).
33. E.L. Halk, and R.W. Fulton, Stabilization and particle morphology of prune dwarf virus, Virology 91: 434 (1978).
34. T.C. Hall, Personal communication cited in 90.
35. T.C. Hall, D.S. Shih, and P. Kaesberg, Enzyme-mediated binding of tyrosine to brome mosaic virus ribonucleic acid, Biochem. J. 129:969 (1972).
36. T.C. Hall, and R.K. Wepprich, Functional possibilities for amino-acylation of viral RNA in transcription and trans-lation, Ann. Microbiol. 127A:143 (1976).
37. V. Hari, A. Siegel, C. Rozek, and W.E. Timberlake, The RNA of tobacco etch virus contains poly(A), Virology 92:568 (1979).
38. B.D. Harrison, and H. Barker, Protease-sensitive structure needed for infectivity of nepo-virus RNA, J. Gen. Virol. 40:711 (1978).
39. D. Herson, A. Schmidt, S. Seal, A. Marcus, and L. Van Vloten-Doting, Competitive mRNA translation in an in vitro system from wheat germ, J. Biol. Chem. In press (1979).
40. H. Hofstetter, H.J. Monstein, and C. Weissmann, The read-through protein A_1 is essential for the formation of viable Q particles, Biochem. Biophys. Acta 374-238 (1974).
40a. J. Horst, H. Fraenkel-Conrat, and S. Mandeles, Terminal heterogeneity at both ends of the satellite tobacco necrosis virus ribonucleic acid, Biochemistry 10:4748 (1971).
41. R. Hull, The grouping of small spherical plant viruses with single RNA components, J. Gen. Virol. 36: 289 (1977).
42. S. Joshi, A.-L. Haenni, E. Hubert, G. Huez, and G. Marbaix, In vivo aminoacylation and "processing" of turnip yellow mosaic virus RNA in Xenopus laevis oocytes, Nature 275: 339 (1978).
43. R. Kaempfer, H. Rosen, and R. Israeli, Translation control: recognition of the methylated 5'-end and an internal sequence in eukaryotic mRNA by the initiation factor that binds methionyl-tRNA$_f^{Met}$, Proc. Natl. Acad. Sci. U.S.A. 75:650 (1978).
44. J. Keith, and H. Fraenkel-Conrat, Tobacco mosaic virus RNA carries 5'-terminal triphosphorylated guanosine blocked

by 5'-linked 7-methylguanosine, <u>FEBS Lett</u>. 57:31 (1975).

45. C. Klein, C. Fritsch, J.-P. Briand, K.E. Richards, G. Jonard
     and L. Hirth, Physical and functional heterogeneity in
     TYMV RNA: evidence for the existence of an independent
     messenger coding for the coat protein, <u>Nucleic Acids Res</u>.
     3:3043 (1976).

46. F. Knauert, and E. Ehrenfeld, Translation of poliovirus RNA
     <u>in vitro</u>: studies on n-formyl-methionine-labelled poly-
     peptides initiated in cell-free extracts prepared from
     poliovirus infected HeLa cells, <u>Virology</u> 93:537 (1979).

47. R.J. Kohl, and T.C. Hall, Aminoacylation of RNA from several
     viruses: amino acid specificity and differential activity
     of plant, yeast and bacterial synthetases, <u>J. Gen. Virol</u>.
     25:257 (1974).

48. E.C. Koper-Zwarthoff, R.E. Lockard, B. Alzner-De Weerd,
     U.L. RajBandary, and J.F. Bol, Nucleotide sequence of 5'
     terminus of alfalfa mosaic virus RNA 4 leading into coat
     protein cistron, Proc. Natl. Acad. Sci. U.S.A. 74:5504
     (1977).

49.   M. Kozak, How do eukaryotic ribosomes select initiation
     regions in messenger RNA, <u>Cell</u> 15: 1109 (1978).

50. S. Legon, P. Model, and H.D. Robertson, Interaction of rabbit
     reticulocyte-ribosomes with bacteriophage f1 mRNA and of
     <u>Escherichia coli</u> ribosomes with rabbit globin mRNA.
     <u>Proc. Natl. Acad. Sci. U.S.A</u>. 74:2692 (1977).

51. J.A. Lesnaw, and M.E. Reichmann, Identity of the 5'-terminal
     RNA nucleotide sequence of the satellite necrosis virus
     and its helper virus: possible role of the 5'-terminus in
     the recognition by virus-specific RNA replicase, <u>Proc</u>.
     <u>Natl. Acad. Sci. U.S.A</u>. 66:140 (1970).

52. D.W. Leung, C.W. Gilbert, R.E. Smith, N.L. Sasavage, and
     J.M. Clark, Jr., Translation of satellite tobacco
     necrosis virus ribonucleic acid by an <u>in vitro</u> system
     from wheat germ, <u>Biochemistry</u> 15:4943 (1976).

53. I.J.D. Lindley, and N. Stebbing, Aminoacylation of enceph-
     alomyocarditis virus RNA. <u>J. Gen. Virol</u>. 34:177 (1977).

54. H. Lot, G. Jonard, and K. Richards, Partial characterization
     and evidence for no large sequence homologies with genomic
     RNAs, <u>FEBS Lett</u>. 80:395 (1977).

55. G. Marbaix, G. Huez, A. Burny, Y. Cleuter, E. Hubert,
     M. Leclerq, H. Chantrenne, H. Soreq, U. Nudel, and U.Z.
     Littauer, Absence of polyadenylate segment in globin
     messenger RNA accelerates its degradation in <u>Xenopus</u>
     oocytes, <u>Proc. Natl. Acad. Sci. U.S.A</u>. 72:3065.
     (1975).

56. T. Minson, and G. Darby, 3'-Terminal oligonucleotide frag-
     ments of tobacco rattle virus ribonucleic acids, <u>J. Mol</u>.
     <u>Biol</u>. 77:337 (1973).

57. D.W. Mossop, and R.I.B. Francki, The stability of satellite
     virus RNAs _in vivo_ and _in vitro_, _Virology_ 94:243 (1979).
58. L. Neeleman, T. Rutgers, and L. Van Vloten-Doting, Internal
     initiation of protein synthesis on RNA of eukaryotic
     virus?, in: "Translation of Natural and Synthetic Poly-
     nucleotides", A.B. Legocki, ed., University of Agriculture,
     Poznan.
59. L. Neeleman, and L. Van Vloten-Doting, Determination of the
     number of ribosomal binding sites on the RNAs of eukrayo-
     tic viruses, _Methods in Enzymology_ LX:410 (1979).
60. M. Nishigushi, F. Motoyoshi, and N. Oshima, Behaviour of
     a temperature sensitive strain of tobacco mosaic virus in
     tomato leaves and protoplasts, _J. Gen. Virol._ 39:53 (1978).
61. A. Nomoto, N. Kitamura, F. Golini, and E. Wimmer, The 5'-
     terminal structures of poliovirion RNA and polio virus
     mRNA differ only in the genome-linked protein VPg, _Proc.
     Natl. Acad. Sci. U.S.A._ 74:5345 (1977).
62. B. Oberg, and L. Philipson, Binding of histidine to tobacco
     mosaic virus RNA, _Biochem. Biophys. Res. Commun._ 48:927
     (1972).
63. P. O'Farrell, B. Cordell, P. Valenzuela, W.J. Rutter, and
     H.M. Goodman, Structure and processing of yeast precursor
     tRNAs containing intervening sequences, _Nature_ 274:438
     (1978).
64. P. Oxelfelt, Biological and Physicochemical characteristics
     of three strains of red clover mottle virus, _Virology_ 74:
     73 (1976).
65. E. Palmer, J.M. Wilhelm, and F. Sherman, Phenotypic
     suppression of nonsense mutants in yeast by aminoglycoside
     antibiotics. _Nature_ 277:148 (1979).
66. H.P. Pelham, Leaky UAG termination codon in tobacco mosaic
     virus RNA, _Nature_ 272:469 (1978).
67. L. Pinck, The 5'-end groups of alfalfa mosaic virus RNAs are
     m7G5' ppp5' Gp, _FEBS Lett._ 59: 24 (1975).
68. L. Pinck, and C. Fritsch, Characteristic nucleotide
     sequences in alfalfa mosaic virus, Abstract 3rd inter-
     national congress of plant pathology (1978).
69. M. Pinck, S.-K. Chan, M. Genevaux, L. Hirth, and H.
     Duranton, Valine specific tRNA-like structure in two
     viruses of turnip yellow mosaic virus group, _Biochemistry_
     54:1093 (1972).
70. J.W. Pyne, and T.C. Hall, Efficient ribosome binding of
     brome mosaic virus (BMV) RNA 4 contributes to its ability
     to outcompete the other BMV RNAs for translation, _Intervi-
     rology_ 11:23 (1979).
71. K.E. Richards, H. Guilley, G. Jonard, and L. Hirth, Nucleo-
     tide sequence at the 5' extremity of tobacco mosaic virus
     RNA 1. The noncoding region (nucleotides 1-68). _Eur. J.
     Biochem._ 84:513 (1978).

72. K.E. Richards, G. Jonard, M. Jacquemond, and H. Lot,
    Nucleotide sequence of cucumber mosaic virus-associated
    RNA 5, Virology 89:395 (1978).
73. R. Roman, J.D. Brooker, S.N. Seal, and A. Marcus,
    Inhibition of the translation of a 40S ribosome-Met-t-
    RNA$^{Met}_i$ complex to an 80S ribosome -Met-tRNA$^{Met}_i$ complex by
    7-methylguanosine 5'-phosphate, Nature 260-359 (1976).
74. A.S. Rutgers, In vitro and in vivo translation of the RNAs
    of alfalfa mosaic virus, Thesis, University of Leiden
    (1977).
75. A.S. Rutgers, and T. Salerno-Rife, Personal Communication.
76. F. Sakai, and J. Takebe, A non-coat protein synthesized in
    tobacco mesophyll protoplasts infected by tobacco mosaic
    virus, Molec. Gen. Genet. 188:93 (1972).
77. F. Sakai, J.W. Watts, J.R.O. Dawson, and J.B. Bancroft,
    Synthesis of proteins in tobacco protoplasts infected
    with cowpea chlorotic mottle virus, J. Gen. Virol. 34:285
    (1977).
78. F. Sakai, J.R.O. Dawson, and J.W. Watts, Synthesis of proteins
    in tobacco protoplasts infected with brome mosaic virus,
    J. Gen. Virol. 42:323 (1979).
79. R. Salomon, M. Bar-Joseph, H. Soreq, I. Gozes, and U.Z.
    Littauer, Translation in vitro of carnation mottle virus
    RNA: regulatory function of the 3' region, Virology 90:
    288 (1978).
80. M.S. Salvato, and H. Fraenkel-Conrat, Translation of tobacco
    necrosis virus and its satellite in a cell-free wheat germ
    system, Proc. Natl. Acad. Sci. U.S.A. 74:2288 (1977).
81. R. Scalla, P. Romaine, A. Asselin, J. Rigaud, and M. Zaitlin,
    An in vivo study of a non-structural polypeptide synthe-
    sized upon TMV infection and its identification with a
    polypeptide synthesized in vitro from TMV RNA, Virology
    91:182 (1978).
82. J.S. Semancik, Detection of polyadenylic acid sequences in
    plant pathogenic RNAs, Virology 62:288 (1974).
83. D.C. Shaw, J.E. Walker, F.D. Northrop, B.G. Barrell,
    G.N. Godson and J.C. Fiddes, Gene K, a new overlapping
    gene in bacteriophage G4, Nature 272-510 (1978).
84. A. Singh, D. Ursic, and J. Davies, Phenotypic suppression
    and misreading in Saccharomyces cerevisiae, Nature 277:
    146 (1979).
85. D.H. Spector, and D. Baltimore, Requirement of 3'-terminal
    poly(adenylic)acid for infectivity of poliovirus RNA,
    Proc. Natl. Acad. Sci. U.S.A. 71:2983 (1974).
86. J. Stanley, P. Rottier, J.W. Davies, P. Zabel, and A. Van
    Kammen, A protein linked to the 5' termini of both RNA
    components of the cowpea mosaic virus genome, Nucleic
    Acids Res. 5:4505 (1978).

87. R.H. Symons, Cucumber mosaic virus RNA contains 7-methylguan-
    osine at the 5' terminus of all four RNA species, Mol. Biol.
    Rep. 2:277 (1975).
88. R.G.L. Van Tol, and L. Van Vloten-Doting, Translation of
    alfalfa mosaic virus RNA 1 in the mRNA-dependent trans-
    lation system from rabbit reticulocyte lysates, Eur. J.
    Biochem. 93:461 (1979).
89. L. Van Vloten-Doting, Similarities and differences between
    viruses with a tripartite genome, Ann. Microbiol. (Inst.
    Pasteur) 127A:119 (1976).
90. L. Van Vloten-Doting, and E.M.J. Jaspars, Plant covirus
    systems: three-component systems, in: "Comprehensive
    Virology", H. Fr nkel-Conrat, and R.R. Wagner, eds.,
    Plenum Press, New York, vol. 11 (1977).
91. L. Van Vloten-Doting, J.F. Bol, L. Neeleman, T. Rutgers,
    D. Van Dalen, A. Castel, L. Bosch, G. Marbaix, G. Huez,
    E. Hubert, and Y. Cleuter, In vivo and in vitro trans-
    lation of the RNAs of alfalfa mosaic virus, in: "Nucleic
    Acids and Protein Synthesis in Plants", L. Bogorad, and
    J. H. Weil, eds., Plenum Publishing Corporation, New
    York (1977).
92. L. Van Vloten-Doting, J.A. Hasrat, E. Oosterwijk, P. Van't
    Sant, M. Schoen, and J. Roosien, Description and
    complementation analysis of 13 temperature sensitive
    mutants of alfalfa mosaic virus, Virology, Submitted for
    Publication.
93. E.G. Westaway, Strategy of flavivirus genome: evidence for
    multiple internal initiation of translation of proteins
    specified by Kunjin virus in mammalian cells, Virology
    80: 320 (1977).
94. Y. Yogo, and E. Wimmer, Sequence studies of poliovirus
    RNA. III. Polyuridylic acid and polyadenylic acid as
    components of the purified poliovirus replicative inter-
    mediate, J. Mol. Biol. 92:467 (1975).
95. W. Zagorski, Translational regulation of expression of the
    brome mosaic virus RNA genome in vitro, Eur. J. Biochem.
    86:465 (1978).
96. D. Zimmern, 5'-End group of tobacco mosaic virus RNA is
    m^{7}G5'ppp5'Gp, Nucleic Acids Res. 2:1189 (1975).

EXPRESSION OF THE CAULIFLOWER MOSAIC VIRUS GENOME IN TURNIPS

(BRASSICA RAPA)

S. H. Howell, J. T. Odell and K. R. Dudley

University of California, San Diego
Biology Department C016
La Jolla, California 92093, U.S.A.

Cauliflower mosaic virus (CaMV) DNA is a possible vehicle for introducing foreign DNA into plant cells. CaMV is one of the few plant viruses that has a double-stranded DNA genome and the DNA, isolated from the virus, is infective.[1] CaMV would probably have to be considered as a "nonintegrating" vehicle, since there is no evidence as yet that the viral DNA integrates into the host plant genome.[1] However, the possibility for using nonintegrating viral DNAs as molecular vehicles in animal cells has been made dramatically apparent this past year through the experiments of Mulligan et al. who used SV40 DNA to introduce and express rabbit β-globin cDNA in monkey cells.[2] These authors replaced a viral structural gene in SV40 DNA with foreign DNA thereby placing the inserted DNA element under viral genome control.

We are interested in using CaMV DNA as a vehicle in plant cells in much the same way as SV40 has been used in animal cells. To do so, we are in search of an appropriate site in the CaMV genome where foreign DNA could be added as an insert or a replacement of deleted CaMV DNA. Here we describe a CaMV gene which codes for a gene product which is produced in abundance in infected plants.

P66

When turnips (Brassica rapa) are infected with CaMV (Cabb B-J.I. isolate), there appears with the onset of visible symptoms, a new polypeptide in leaf extracts which we call P66 (66,000 MW). Although this protein is the only new protein that can be easily detected using one dimensional SDS-PAGE, it is not a protein of the isolated virus. Virus isolated by our procedures has a major

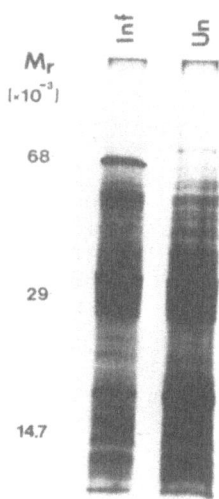

Fig. 1 Translation of poly[A] RNA (0.5 µg) from uninfected and
CaMV infected leaves in a wheat germ protein synthesizing
system. ^{35}S-methionine labeled translation products were subjected to SDS-PAGE and detected by fluorography.

coat protein (37,000 MW), a coat protein degradation product
(33,000 MW) and a minor species (82,000 MW). Tryptic fingerprints
of ^{125}I labeled P66 and the virus coat protein show that P66 is not
related to the viral coat protein in any obvious way. Furthermore,
we find that P66 is not precipitated by an antiserum raised against
the purified virus. What function P66 serves is not clear at this
point; however, preliminary cell fractionation experiments suggest
that it may be a viral inclusion body protein.

P66 can be easily detected by translating RNA from infected
leaves (Fig. 1). We have used the translation assay to determine
several properties about the RNA which codes for P66 -- whether
it is polyadenylated, its sedimentation properties and if it is
encoded by the CaMV genome. First, we have found that RNA, which
binds to oligo[dT] cellulose codes for P66 (Fig. 1) suggesting
that at least some P66 messages are polyadenylated. Second, when
poly[A] RNA from infected leaves is sedimented on a neutral
sucrose gradient, RNA coding for P66 sediments at about 18 S.[3]

Third, to show that P66 is a viral coded protein and not
simply a protein produced by the plant in response to virus infection, we performed hybridization arrested translation, HART (4).
In the HART procedure, we hybridized viral DNA to infected plant
leaf RNA prior to translation. If the RNA directing the synthesis

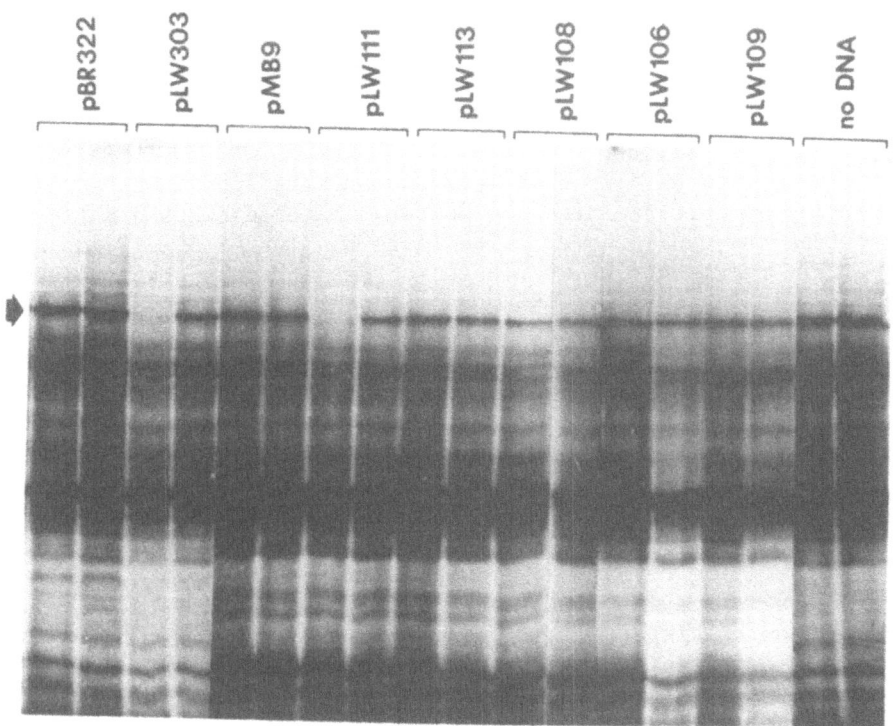

Fig. 2 Hybridization arrested translation of P66 synthesis by
   cloned CaMV DNA fragments. Total RNA (5µg) from infected
   leaves was hybridized to various recombinant plasmid DNAs (1µg)
   amd translated in a wheat germ protein synthesizing system.
   Translation products from each hybridization reaction involving
   a different plasmid are shown in paired channels where the
   right hand channel is a control in which the hybrid was melted
   prior to translation.

of P66 is encoded by viral DNA, then prior hybridization will specifically block the translation of P66. As a source of viral DNA
we used a recombinant plasmid, pLW303, which contains the entire
CaMV genome inserted into the Sal I site of pBR322. As shown in
Fig. 2, when infected leaf RNA is hybridized to pLW303 DNA the synthesis of P66 is suppressed. The suppression can be overcome by
melting the hybrid prior to translation. These results show that
P66 is, indeed, encoded by the CaMV genome.

To locate the site coding for P66 on the CaMV genome, we performed the HART procedure using plasmids carrying different seg-

ments of the CaMV genome. We found that PLW111 with continguous EcoR$_1$ b, c and d fragments blocked P66 synthesis, but that pLW113 and pLW108 with EcoR$_1$ c fragment, pLW106 with EcoR$_1$ a and d fragments and pLW109 with EcoR$_1$ a fragment did not. This demonstrates that P66 is largely encoded by the EcoR$_1$ b fragment (Fig. 6).

Large vs. Smaller Viral RNA Transcripts

We reported earlier that leaf protoplasts from infected plants synthesize in culture large and somewhat heterodispersed RNA derived from the CaMV genome.[5] The large RNA is 2-2.5 x 10^6 MW and presumably represents transcription of one strand of the entire genome. Finding the large RNA was an unexpected observation because we anticipated finding smaller discrete RNA species such as the one coding for P66 as discussed above. Furthermore, we have never been able to show that a large viral RNA species accumulates in leaves of infected plants.

We suspected that our protoplast preparations might be defective in producing small viral RNA in culture. Therefore, we attempted to label viral RNA transcripts in leaf discs and not in protoplasts. Figure 3 shows a comparison of viral RNAs labeled in protoplasts versus in leaf discs. The protoplast-labeled RNA is large and heterodispersed while the leaf disc RNA shows two discrete species which comigrate with 18S and 25S cytoplasmic ribosomal RNA. If RNA labeled in leaf discs is fractionated on oligo[dT] cellulose into poly[A] and nonpoly[A] RNA, it can be seen that an 18S species is the principal component of the viral poly[A] RNA fraction (Fig. 3). The nonpoly[A] RNA includes the 25S species and a slightly smaller 18S form, presumably a non polyadenylated form of the poly[A] 18S RNA.

Our infected leaf protoplast preparations are as efficient in synthesizing viral RNA as are leaf discs. Therefore, the protoplasts we prepare are probably defective in a step of RNA processing and not in viral RNA synthesis. The large RNA transcripts that we see may be initial transcripts or intermediates in the processing pathway. This suggestion is made more plausible from the results of RNA transcript mapping experiments described below in which we find that the 18S and 25S RNA are most likely composite transcripts formed by joining RNA segments from discontinuous regions of the CaMV genome.

Mapping RNA Transcripts on the CaMV genome

To map the two major RNA transcripts on the viral genome we hybridized radioactive infected leaf disc RNA to CaMV DNA restriction fragments. We reported previously[5] that the large viral RNA labeled in protoplasts hybridizes to all EcoR$_1$ restriction frag-

ments. When 18S and 25S RNAs labeled in leaf discs are hybridized
separately, a surprisingly similar pattern emerges as shown in
Fig. 4. First, 25S viral RNA hybridizes strongly to R_1-Sal fragment a_2 and R_1 fragments b and c, but not to R_1-Sal fragment a_1.
(DNA fragments are from pLW303 which is the entire CaMV Cabb
B-J.I. genome cut with Sal 1 and inserted into pBR322 at the Sal 1
site. The orientation of the CaMV insert into the plasmid is such
that $EcoR_1$ fragment a migrates on gels as an upper band containing
R_1-Sal fragment a_1 and a lower band containing R_1-Sal fragment a_2.
See Fig 6). 18S RNA on the other hand hybridizes strongly to R_1
fragment b and R_1-Sal fragment a_2, and more weakly to R_1 fragment c
and R_1-Sal fragment a_1.

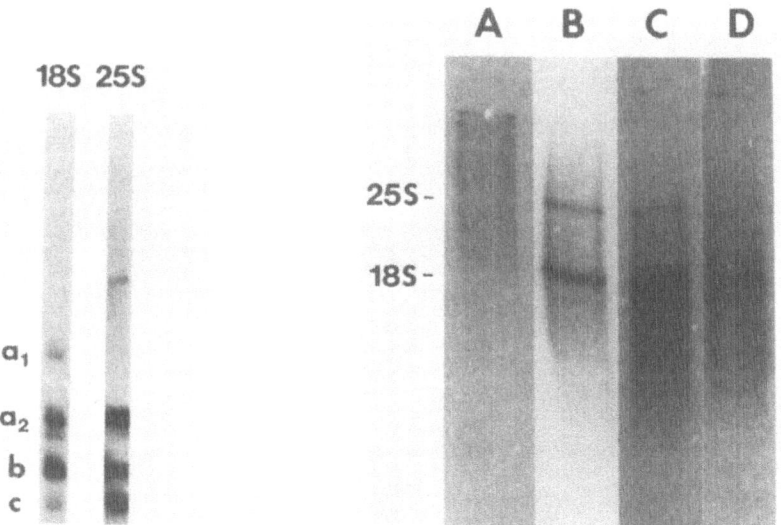

Fig. 3 Sizing analysis of ^{32}P-labeled viral RNAs. Labeled RNA
was hybridized to pLW303 DNA bound to nitrocellulose filters
and eluted with 70% formamide. Eluted RNA was subjected to
electrophoresis on methylmercury hydroxide containing agarose
gels. (A) RNA labeled in infected protoplasts, (B-D) RNA
labeled in leaf discs, (B) total RNA (C) poly[A] RNA, (D) non
poly[A] RNA.

Fig. 4 Southern filter hybridization of 18S and 25S RNA to $EcoR_1$
fragments of pLW303 DNA (containing a whole CaMV genome insert).
RNAs were labeled with $^{32}PO_4{}^{-3}$ in leaf discs and separated on
methylmercury hydroxide agarose gels.

What was truly unexpected was that these RNAs seemed to hybridize to widely separated regions of the CaMV genome. This
suggested that the 18S and 28S RNAs are derived from discontinuous
genome segments. We attempted finer mapping of the 18S RNA site(s)
by hybridizing to Bgl-R_1 fragments of CaMV DNA and various
plasmids. We found that Bgl-R_1 fragments b, c, e, f, i and R_1
fragment d hybridize but not Bgl-R_1 fragments a and g (Fig 6).

We also attempted to find the 3' terminus of the polyadenylated 18S RNA by synthesizing a cDNA probe from infected cell poly
[A] RNA. The ^{32}P-labeled cDNA was hybridized to $EcoR_1$ fragments
of CaMV DNA (Fig. 5). Larger cDNA fragments (0.3-0.6kb) isolated
on a sucrose gradient hybridized equally to R_1 fragments b and c,
but smaller cDNA fragments (<0.3kb) hybridized preferentially to
R_1 fragment b. This suggests that the polyadenylated 3' terminus
is near the $EcoR_1$ b/c junction and that the RNA extends from c to
its 3' terminus in b.

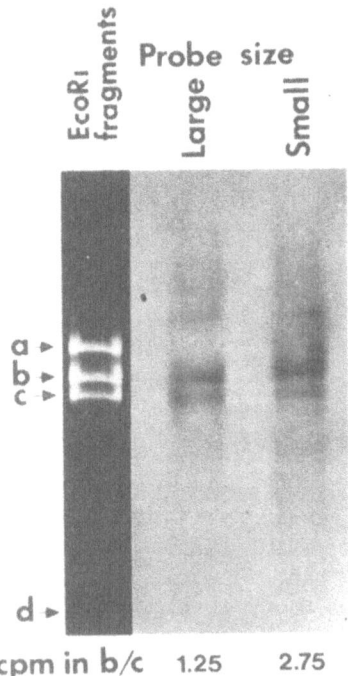

Fig. 5 Determination of 3' terminus and polarity of polyadenylated viral RNA. ^{32}P-labeled cDNA was synthesized from infected
cell poly[A] RNA and size fractionated on an alkaline sucrose
gradient. Copy DNAs <0.3kb (small) and >0.3kb (large) were
hybridized to CaMV (Cabb B-J.I.) DNA.

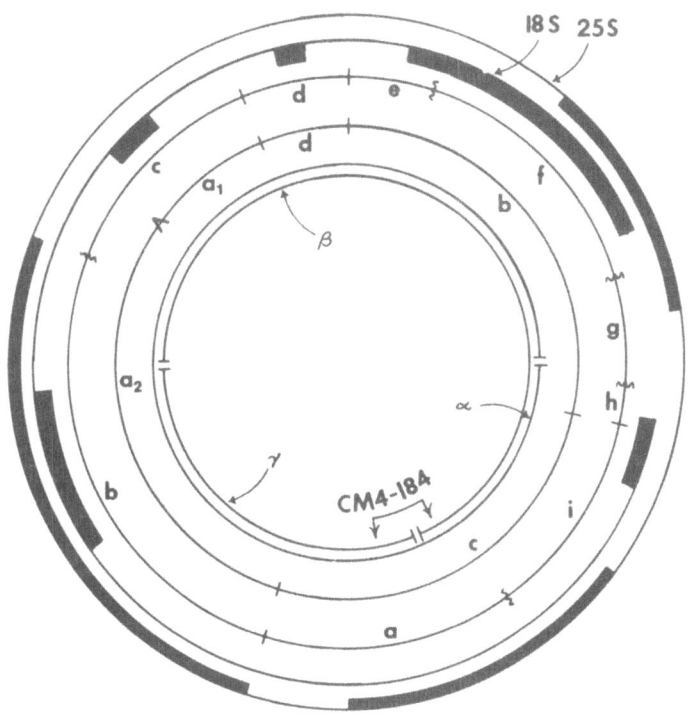

Fig. 6 Restriction site and transcription map of CaMV (Cabb B-
J.I.) genome. Inner double-line circle shows double-strand DNA
genome with site specific single-strand breaks. Single-strand
DNA fragments are designated α, β and γ. Region deleted in
CM4-184 isolate is indicated by arrows. Second circle from
inside shows EcoR$_1$ sites (|), Sal 1(Λ) site and R1-Sal fragments
a - d. Third circle from inside shows EcoR$_1$ (|), Bgl II (⅔)
sites and R$_1$Bgl fragments a-i. Outer circles show regions
(dark bars) to which 18S and 25S virus RNAs hybridize. Length
of bars was determined by the overall size of RNA transcript
and proportion of transcript hybridizing to various DNA restriction fragments. Ends of bars are not necessarily intended to
be exact sites of RNA discontinuity.

Perspective

If CaMV DNA is to be used as a vehicle for the introduction and expression of foreign DNA in plant cells, then the site for inserting foreign DNA in the viral genome must be carefully chosen. Following the example for SV40, an ideal site would be within an exon of a dispensible gene, conserving the viral RNA promoter, terminator and processing signals. An insertion within the coding region for P66 may be an ideal site.

This work was supported by the Science and Education Administration of the United States Department of Agriculture and by the National Science Foundation.

References

1) R. J. Shepherd, DNA Plant Viruses, Ann. Rev. Plant Physiol. 30: 405 (1979).
2) R. C. Mulligan, B. H. Howard, and P. Berg. Synthesis of rabbit β-globin, Nature 277, 108 (1979).
3) J. T. Odell and S. H. Howell, in preparation.
4) B. M. Paterson, B. E. Roberts and E. L. Kuff. Structural gene identification and mapping, Proc. Nat. Acad. Sci. U.S. 74: 4370 (1977).
5) S. H. Howell and R. Hull. Replication of cauliflower mosaic virus and transcription of its genome, Virology 86: 468 (1978).

A REEXAMINATION OF McCLINTOCK'S "CONTROLLING ELEMENTS" IN
MAIZE IN VIEW OF RECENT ADVANCES IN MOLECULAR BIOLOGY

P. Starlinger
Institut für Genetik der Universität Köln
Weyertal 121, 5ooo Köln 41

Introduction

Genetics has been studied in great detail long before the
molecules carrying genetic information were identified as
such, and before the biochemistry of their action began
to be elucidated. In the early days of this science,
geneticists understandably focused their attention on the
study of phenomena sufficiently stable to be investigated
repeatedly and to yield reproducible results. It is partly
for that reason that mutation was considered to be a very
rare event. Only rare mutation events ensured the stability
of the altered genotype which was necessary for its study
in successive generations.

It was, however, realized soon that not all genetic alterations were of this stability. Some changes occuring in
multicellular organisms frequently to the appearance of
isolated clones of cells with altered phenotypes. The
results of such frequent changes is a variegation. In
some instances, these variegation patterns turned up with
regularity.

A detailed study of these phenomena was difficult, to say
the least. About 35 years ago, Barbara McClintock under-
took the study of variegation phenomena, which occurred
in maize. Some of the variegations known best are colour
patterns on maize kernels. This study led to the formulation on the concept of the controlling element. The controlling element was thought to be different from a gene,
and it was, as its name says, thought to control the action

of a gene. Changes in this control, which are inherited
clonally are in many instances thought to be due to the
fact that the whole element can be transposed from one
location in the genome to another one. These changes in
position lead to visible alterations in the genes adjacent
to the controlling elements.

The study of the genetic behaviour of these elements led
to the discovery of several independent elements of this
type. All of them have certain similarities, but are
also clearly distinguished by other features. Their
different actions with regard to transposition and with
regard to the action at the point of insertion are very
complex. It is important to note that a certain controlling
system is often found to consist of two independently
transposable units, one of which is inserted near a gene
and influences the action of this gene. Changes in this
control are determined by the other member of the pair.
This second element is integrated far away or even on
another chromosome and therefore must act in position
trans, probably via a diffusible gene product.

All this complexity was revealed by the meticulous study
of the phenotypes exhibited on the kernels and sporophytes
of plants harbouring the controlling elements. No biochemical studies at the DNA level could be undertaken at
the time of the discovery of these elements (McClintock,
1951; 1956;, 1965; 1968; Peterson, 1953; 1965; 1976;
1978; review: Finacham and Sastry, 1974).

Much later, and completely unrelated to the maize work,
transposable DNA elements were identified in the bacterium
Escherichia coli and its plasmids and phages. Owing to
the easier manipulation of bacteria and to the progress
that had been made in the meantime in molecular biology,
the transposable DNA elements in bacteria were quickly
identified as DNA sequences, and characterized extensively
at this level (reviews: Cohen, 1976; Starlinger and
Saedler, 1976; Kleckner, 1977; Bukhari et al. 1977).

It was soon realized that these elements bear some similarity to the controlling elements in maize. These similarities are not confined to the capability of transposition, but are also observed in some detail of their
function (Starlinger and Saedler, 1972; Nevers and Saedler,
1977).

I will first try to summarize briefly the most obvious
features common to transposable elements both in maize
and in bacteria. These similarities are impressive and

show that transposition of genetic elements is a general genetic mechanism and not confined to just one of the biological kingdoms. I will then adress, however, some of the open questions that may now become amenable to experimental study. The extensive body of genetic data on controlling elements in maize shows that these elements must be more complex than transposable elements in bacteria. A reexamination of these known facts will undoubtedly help to formulate the biochemical experiments which are now becoming possible due to the rapid progress in the study of DNA structure and function in eucaryotes.

Comparsion of "controlling elements" in maize and transposable DNA elements in bacteria

With regard to some basic properties, the similarity between the two classes of elements is striking:

1) Both types of elements are transposable. This can be shown by the acquisition of new linkage relations to known genetic markers. In bacteria, the appearance of certain DNA sequences in new locations can also be demonstrated by physical methods.

2) Both in maize and in E. coli, there is a limited number of elements, which are transposable. Different elements, when integrated in the same location, exert similar effects. Otherwise, the elements are unrelated to each other. In bacteria, this can be shown by hybridization techniques or DNA sequencing. In maize, the unrelatedness has been demonstrated by physiological criteria.

3) All transposable element's exert effects at their site of integration (cis-effects). The simplest of these effects is the insertion of the element into a gene, abolishing its function and thus scoring as a null mutation. More complicated cis-effects will be described below.
   In addition, some of the elements are also capable to show effects in position trans.
   They are known best by their effects on the physiology of the bacteria. Many of them confer resistance to an antibiotic or a heavy metal. In addition, however, transposons encode proteins involved in their own transposition or in the regulation of this process (Gill et al., 1978).
   In maize, a transposable element inserted into a gene also scores as a null mutation of that gene. In

addition, some of the elements are capable to influence their own transposition to new locations.

4) The insertion of a transposable genetic element increases the probability of chromosomal aberrations in its vicinity. Several types of chromosomal aberrations have been described in bacteria. One class of aberrations is known to occur in the vicinity of transposable elements only. It is called "adjacent deletion". These deletions extend from the location of the element in either one direction to varying extents, but leave the element itself in its place (Reif and Saedler, 1975). By DNA sequence analysis it could be shown that the deletions start very exactly at the terminus of the element (Ohtsubo and Ohtsubo, 1978). In maize, secondary mutations occuring near the element Ds ("dissociator", because frequently chromosome breaks occur at the site of Ds) integrated between genes I (inhibitor of aleuron colour formation) and Sh ("shrunken" endosperm) on the short arm of chromosome 9 may be adjacent deletions. They have not been analysed at the DNA level yet (McClintock, 1953).

5) At their site of insertion, transposable elements influence the action of genes in their vicinity. In bacteria, three types of effects are known. Insertion into the continuity of a structural gene leads to inactivation of that gene. Insertion in a transcription unit most usually leads to a severe polar effect on the expression of distal genes. However, in certain instances, promoter sequences on an element may allow the expression of a gene, which is otherwise not possible. In maize, due to our incomplete knowledge about the structure of genes, the effects of transposable elements cannot be correlated with certainty to the site of integration and the types of signals carried on the elements. However, phenotypically, the three types of effects (null mutation, decrease and increase in expression) are also found.

In the properties listed, bacterial and maize transposable genetic elements are very similar. Maize elements, however, are more complex, and some of their properties have not been detected in bacterial elements.

1) In maize, controlling elements often (but not always) occur in pairs, which are functionally related to each other. An example are the two elements Ds ("dissociator") and Ac ("activator", because effects at Ds occur only, if Ac is present). Ds can be

transposed to the site of a gene, causing a mutation of that gene. This mutation can be reversed by the excision of Ds. Both insertion and excision of Ds are dependent on the presence of a second factor, Ac. Ac can be unlinked or linked to Ds, and it is transposable itself. If Ac is inserted into or near a gene, it exerts effects very similar to those caused by the integration of Ds. It is, however, independent of any other known factor for its own excision (McClintock, 1955). It is possible (but not at all proven) that Ds is part of Ac, capable of some, but not of all actions of Ac.

2) In many instances, the effect of a transposable element looks like a somatic mutation. The insertion of the element near a gene abolishes the function and the occasional reactivation of the gene (by excision of the element or by some other means) leads to the reappearance of the gene function, which is clonally inherited and shows up in both sporophytes or in kernels as a variegation pattern. These patterns vary both in the size of the spots forming the pattern and in the distribution of the spots on the kernels. This indicates that the time of action of the elements is controlled in a precise way.
It is important to understand that both precise control of the time, at which mutations are possible, and the occurance of these mutations at the programmed time in a stochastic manner are simultaneously observed with these elements. This complex behaviour has not been found in bacteria.

3) In some instances, controlling elements seem to be subject to outside influences, which have to be exerted in sequence, and can be separated by a long time interval. In some instances, a "presetting" event occurs before meiosis, and the two elements influencing each other can be separated at the subsequent meiosis. Still, the final event occurs many cell generations after meiosis and leads to the appearance of a certain variegation pattern. This separation in time resembles the two events in differentiation which are called "determination", and "terminal differentiation". It is not known, whether McClintock's "presetting" events and regular differentiation have anything in common. As the former events are accompanied by insertion and excision of genetic elements, probably DNA sequences, a common mechanism with differentiation would mean that the latter also can be linked to DNA rearrangements. It will clearly be important to study

"presetting events" with this question in mind.

4) A cis acting controlling element subject to the action
of an unlinked trans acting element has a certain
autonomy in its response to the latter. This is observed in the so called "changes in state" in which
the responsiveness of the cis acting element suddenly
changes in a certain cell, which is subsequently
reacting differently. As an example, the cis acting
element Ds responds to the action of the trans acting
element Ac either by producing chromosome breaks at
its site of insertion or by being excised from this
site. The latter effect can only be seen, if the
integration of Ds had led to the loss of expression
of a nearby gene, which is reactivated when the
element is excised. The frequency of the two responses
(chromosome breaks or excisions) is governed by the
properties of Ds itself. A so called "change in state"
can convert a Ds with a predominance of production of
chromosome breaks into a Ds, which is most frequently
excised. It is tempting to speculate that one of the
events is an intermediate step in the production of
the other event, and that the change in state influences the completion of a chain of reactions. It
is not yet known, however, whether the chromosome break
or the excision is the final step in such a chain of
events.
The changes of state described alter the response of Ds
towards Ac. In other instances, an alteration of the
influence of a transposable element on the gene, where the
element is inserted, has been observed. Ds can abolish the
function of genes in its vicinity, and reversion to
the wildtype phenotype of that gene can occur without
excision of that gene. The same is true for Ac. Is
the reappearance of gene function also due to a change
in state of the elements? In this case, the change in
state closely resembles the changes accompanying
unstable mutations occuring in IS elements in bacteria,
which lead to unstable expression of a gene formerly
mutated by the insertion of an IS element (Saedler
et al., 1972). Ghosal and Saedler (1978) could show
that these mutations consist of the addition of short
DNA sequences to the element IS2. These sequences
create a promoter site and thus allow the transcription
of an adjacent gene. Most probably, the short
sequences arise during faulty replication of the
elements themselves. It will be interesting to see,
whether changes in state of controlling elements in
maize are also accompanied by DNA sequence additions.

The mechanism of transposition

Homologous genetic recombination is a frequent event and
has been studied in detail at the molecular level in
procaryotes and lower eucaryotes. The situation is
different with regard to transposition. Biochemical studies in eucaryotes are completely absent. Even in bacteria,
no mutants have been isolated, and consequently, no
enzymes active in transposition have been described yet,
apart from a few promising experiments on the transposition
of the element Tn3 (Gill et al., 1978). Most recent advance
has come from DNA sequence studies. This advance can be
summarized by saying that transposition seems to be somehow related to DNA synthesis. What is not yet clear is the
extent to which DNA synthesis is necessary. Models have
been formulated, which go so far as to assume that
transposition is really a replication process, in which
the transposable element is copied once, and in which it
appears, at the end of the process, in two positions: in
its old one and at a new site (Grindley and Sherratt,
1978; Shapiro, 1979). The experimental evidence, while
not being a rigorous proof for these hypotheses, points
in this direction. No experiment supports the assumption
that the elements are excised from one position and re-
inserted at another one. If the number of transposons
integrated in a recipient plasmid is compared with the
number of donor plasmids that have lost the transposon,
the latter class does not account for the former one
(Bennett et al., 1977). If the transposition of IS4 in
E. coli to a recipient site in the galactose operon is
detected by Southern's blotting method, it is clearly
seen that the element is still retained in its old position (Klaer et al., 1979). These observations are
either compatible with the assumption that transposition
is a copying process, or else the transposition is lethal
for the donor chromosome (or chromatid) and that only
those donor structures survive, which have not partici-
pated in the event. An observation by Gill et al. (1978)
goes further than this by showing that a co-integrate
structure can be formed between a donor and a recipient
plasmid, which contains one copy each of the two plasmids,
but two copies of the transposon. This observation, while
being clear cut with regard to the copying of the transposon, is obtained in a situation, where the transposon
carries an internal deletion, and may not necessarily be
representative for the physiological reaction. Taking all
experiments together, the models assuming transposition
by copying are plausible for bacteria, though rigorous
proof has not yet been obtained.

Because of these recent findings on bacterial IS elements, it is interesting to see whether controlling elements in maize behave similarly. Are there instances, where an element is retained in its old position, while appearing in a new one? This question is difficult to answer. Segregation from a heterozygous condition can often be used as an explanation, when an element occurs in a new position and is not at the same time seen it its former position. There are, however, a few cases, where the analysis is more stringent. In these cases, two kinds of transpositions can be seen which are different from those known to occur in bacteria.

1) The transposition of either Ds or Ac from an old position to a new one is accompanied by the loss from the old position. This is seen most clearly upon transposition of Ds from its standard position proximal to gene Wx on the short arm of chromosome 9 to a new position on the same chromosome, located more distally. In this case the presence of Ds can be either inferred from its genetic effects, or it can be observed by cytogenetical methods, because Ds produces breaks at its site of insertion. In the transposition mentioned above it is clearly stated, that the breaks now occur at the new position only, and not longer occur at the former one (McClintock, 1949).

2) Another, more complicated type of transposition occurs with the larger, trans-acting elements. These have been detected originally by the insertion at the vicinity of the gene, the inactivation of which led to a visible phenotype. Very often these genes affect aleuron colour. Insertion of the transposable element leads to loss of colour and its subsequent excision leads to the reappearance of colour and thus to a variegation of the kernel. The excision event has been investigated in subsequent plant generations. At first, it appeared to be dependent on a factor located at the mutated gene. In later generations, the factor responsible for the excision seemed to have been transposed to a new location. Here it was indistinguishable in its action from its behaviour at the former site. It may have been transposed in its entirety. Still, something had been left at the original location, which caused the gene still to be inactive, and which had to be removed, before wildtype activity of the gene was restored. If this were proven by molecular studies, it might indicate the replica-

tion of a part of the element which remains at the old site, while the complete element is transposed to the new site (McClintock, 1956b; Peterson, 1961). The study of this question will undoubtedly be important in order to find out, whether transposition mechanisms in plants and in bacteria are basically similar.

What is the stochastic event leading to transposition or excision at Ds?

The events associated with the integration or excision of the transposable elements are called mutations, because they occur rarely and in a stochastic manner. The insertion events are most probably caused by the action of enzymes. Why, then, is the event so rare? Are the enzymes present at all times, but designed in such a way that they act only rarely? This is a little hard to assume, because the interaction between the enzyme and its DNA site is, at least in bacteria, specific enough (as can be judged from the presence of preferential integration sites) to postulate a relatively strong binding interaction. If however, binding does occur frequently during the life time of a bacterium, why then is transposition not occuring more often? Is it, because binding is not sufficient and the catalytic action is much rarer than binding, because it needs a specific conformation, which is rarely attained? Do several factors have to come together at one site of the DNA to bring about the necessary effect?

We can discuss an alternative: It is conceivable that the enzymes, when present, act quite efficiently. However, due to some repression, they are synthesised only rarely and are present only in a tiny fraction of the cells. These questions can at present be posed but not be answered in bacteria. There is, however, convincing evidence in maize that the stochastic event necessary to cause a mutation at Ds has not occured in the Ds element itself. McClintock has described a situation, where one copy each of Ds was carried on the short arm of the two chromosomes 9, but the elements were carried in none-homologous positions. On one chromosome 9, Ds was carried in the vicinity of gene C, giving rise to a mutation of this gene. Excision of Ds from this position causes the reappearance of aleuron colour. The other chromosome 9 carried the recessive allele of C, and Ds was ocated in its standard position proximal to gene Wx. The two copies of Ds were in different states.

The state of Ds inserted near gene C were such, that its
favorable action was the excision an thus the reappearance
of the phenotype of the dominant allele C (coloured
aleuron). The state of Ds in its standard position predominantly caused chromosome breaks. These led to the
appearance of an acentric chromosome fragment and
consequently to the loss of Wx. Due to the presence of
the recessive allele wx in the other chromosome the waxy
phenotype is found in the progeny cells. In the absence
of Ac, both Ds elements are stable. Consequently, the
kernels are homogeneous clourless and have normal starch.
Introduction of Ac leads to the occasional mutation at
Ds. Coloured patches appear, and can now be investigated
for the presence of normal starch in the endosperm
underlying the aleuron in the coloured area. It was found
that in most instances, the region of colour exactly
overlaps the region of altered starch (McClintock, 1949).

It is hard to avoid the conclusion, that in the same cell
both Ds elements were affected. This alteration must
occur with high efficiency. Otherwise, sectors would be
expected, in which only one or the other copy of Ds was
affected. Such sectors occured but only as an exception.

It was stated above, that Ac is capable of exerting the
same mutation effects on itself, which it exerts on Ds.
If, in the same cell, action of Ac on two indepent Ds is
observed, action on Ac itself also is expected. This
expectation has been born out by observation (McClintock,
1949).

A simple explanation for this phenomenon would be the
stochastic activation of Ac. If, for instance, Ac produces a repressor for transposition genes, stochastic
inactivation of the repressor gene may allow for the
synthesis of a transposition enzyme in large quantities.
This enzyme might now be able to act on every Ds present
in the cell with so high an efficiency that in most
instances observed, both of them are altered in a heritable fashion. Of course, inactivation of a repressor is
only one possible explanation. Formally activation of Ac
is equally possible.

Ac shows a peculiar dosis response. Two doses of Ac, far
from enhancing the effect, rather delay it in time. Three
doses of Ac, or even four doses (which are obtainable in
a homozygotic condition in the triploid endosperm) lead
to such a delay of Ac action, that a kernel carrying four

Ac may not be distinguishable from a kernel with no Ac,
because its development has ended, before Ac has had a
chance to work (McClintock, 1951). Could this be explained by the assumption that Ac is producing a repressor, and that its transposition activity can only
be seen, after this repressor has been removed? If
removal of the repressor occurs by a mutation, and if
the two copies of Ac elements remain inactive, as long,
as one of them produces its normal level of repressor,
a delayed action of Ac on transposition of Ds might be
expected. Closer inspection, however, reveals difficulties with this interpretation. If the inactivation of
the repressor is a rare event, the two events inactivating
the two repressor genes might be separated by several cell
divisions. The first inactivation would produce a clone
of cells, in which inactivation of the one remaining Ac
repressor gene would produce an event at Ds. Outside of
this sector, inactivation of one Ac repressor gene would
leave the other one intact and should therefore not
show Ac action as a Ds-linked alteration. Sectors of
this kind have not been observed as a consequence of the
dose effect of Ac. The rare colour patches occurring
late are distributed evenly over the whole kernel. This
behaviour is not easily incorporated into the above model.

Another hypothesis to explain the delayed action of
higher doses of Ac is possible. If a repressor is made
early in development only, and subsequently decays slowly,
the transposition effect of Ac might be expressed only,
after the repressor level has fallen below a certain
threshold. In this case, it would have to be explained,
however, why an increase in the dose of Ac by a factor
of two delays Ac action by more than two generations.
Clearly, a better knowledge of the biochemistry of these
processes is necessary.

An understanding of the events described above may become
important, if it would support the concept that sequence
rearrangements at the DNA level, which are clonally
inheritable and thus have the characteristics of mutations,
can nevertheless be caused by (enzymatic?) reactions with
high precision and efficiency. Such events may become
important for our understanding of differentiation.

What is the origin of transposable DNA elements?

In bacteria, this question has a simple answer. Is-elements in E. coli are present in one or several copies in
the E. coli chromosome. From here, they can be transposed
to new locations. If it is true, as was discussed above,

that they are retained in their original positions, it
has to be explained, why there is not a steady accumulation of these elements. An independent mechanism must
be present, which leads to a decrease in the number of
IS-elements. Precise excision without reinsertion at a
new position has been observed. We do not know the
exact conditions governing the equlibrium between transposition and excision and we also do not know, whether
these two processes are completely independent of each
other. At least, however, we know, where ISelements
appearing in new positions originate.

Transposons are not found regularly in E. coli. They are
introduced on a plasmid, and can be transposed from here
to the bacterial chromosome.

In both cases, clear observations are possible, because
DNA hybridization techniques allow the detection of IS
elements in their normal locations, even if no physiological effects can be ascribed to them. The situation is
more difficult in maize. Here, the presence of the
elements can be detected only, when they function. If a
cis-acting element, like Ds, is transposed into a gene,
this gene is inactivated. Ds may have been present in
some other position before the transposition event, but
it could not be detected there. Detection should be easier,
however, in the case of the trans-acting elements, like
Ac or another element, Spm (suppressor-mutator). Both of
these were observed first in the progeny of cells, in
which the breakagefusion bridge cycle had been introduced
deliberately (McClintock, 1946, 1951b). The element En
(enhancer), which is identical to Spm was first observed
in material that had been exposed to the radioactive
fallout after an atomic explosion (Peterson, 1960).

In these cases, the functional elements were first
observed in strains which had undergone an unphysiological
treatment. Before these treatments, the elements had not
been detected. Most probably, however, they did not
originate de novo. Had they been silent completely previously, or were their normal rôles confined to very
specific times during development and possibly to an
influence on the action of a selected set of genes only?
If this were so, it must be assumed that the activation
of Ac which causes events at Ac itself and at two
differently located copies of Ds must leave the precursors
of the known Ac and Ds unaffected. Otherwise, additional
responses to Ac would be expected. Did Ac and Ds undergo
a special "change in state" upon transposition to the
sites, where they were detected first? Or are Ac and Ds

and the other elements dispensable for the plants, and, apart from their known locations, present only in sites, where they exert no profound influence on the physiology of the plants? It will be interesting to answer these questions.

Concluding remarks

Some of the questions posed in the preceding paragraphs could be answered, if Ac and Ds could be studied at the DNA level. Due to the developement of techniques suited to the isolation of genes from complex eucaryotic genomes, such possibilities should now exist. If Ds or another controlling element is known to integrate into a gene, this gene could be isolated, and adjacent DNA sequences could be compared in the wild type and in the mutants caused by the integration of the element. Sequences unique to the mutants would be good candidates for DNA contained in the element. They might be isolated by modern cloning techniques.

Experiments in this direction are presently under way in our laboratory, using the gene Sh, encoding endosperm sucrose synthetase (Chourey and Nelson, 1976), and mutants caused by the integration of Ds in the vicinity of this gene (McClintock, 1952; 1953).

References:

Bennet, P.M., Grinsted, J., and Richmond, M.H. 1977.
Molec. Gen. Genet. 154:2o5

Bukhari, A.I. , Shapiro, J.A., Adhya, S.L. (eds.) 1977.
DNA Insertion Elements, Plasmids and Episomes (Cold
Spring Harbor Laboratory, Cold Spring Harbor, N.Y.)

Chourey, P.S., and Nelson, O.E. 1976. Biochem. Genet.
14:1o41

Cohen, S.N. 1976. Nature (London) 263:731

Fincham, J.R.S., and Sastry, G.R.K. 1974. Ann. Rev. Genet.
8:15

Ghosal, D., and Saedler, H. 1978. Nature (London) 275:611

Gill, R., Heffron, F., Dougan, G., and Falkow, S. 1978.
J. Bacteriol. 136:742

Grindley, N.D.F., and Sherratt, D. 1978. Cold Spring Harbor Symp. Quant. Biol. 43 in press

Klaer, R., Pfeifer, D., and Starlinger, P. in preparation

Kleckner, N. 1977. Cell 11:11

McClintock, B. 1946. Carnegie Inst. Yearb. 45:176
-            1949. Carnegie Inst. Yearb. 48:142
-            195o. Carnegie Inst. Yearb. 49:157
-            1951. Cold Spring Harbor Symp. Quant.Biol.
                                            16:13
-            1951b.Carnegie Inst. Yearb. 5o:174
-            1952. Carnegie Inst. Yearb. 51:212
-            1953. Carnegie Inst. Yearb. 52:227
-            1955. Carnegie Inst. Yearb. 54:245
-            1956. Cold Spring Harbor Symp. Quant.Biol.
                                            21:197
-            1956b.Carnegie Inst. Yearb. 55:323
-            1965. Brookhaven Symp. Biol. 18:162
-            1968. Develop. Biol. Suppl. 1:84

Nevers, P., and Saedler, H. 1977. Nature (London) 268:1o9

Ohtsubo, H., and Ohtsubo, E. 1978. Proc. Natl. Acad. Sci.
USA 75:615

Peterson, P.A. 1953. Genetics 38:682
-             196o. Genetics 45:115
-             1965. Am. Natur. 99:391
-             1976. Molec. Gen. Genet. 149:5
-             1978. in: Maize Breeding and Genetics,
                    Walden, D.B., ed. John Wiley and
                    Sons Inc. p. 6o1

Reif, H.-J., and Saedler, H. 1975. Molec. Gen. Genet. 137:17

Saedler, H., Besemer. J., Kemper, B., Rosenwirth, B., and Starlinger, P., 1977. Molec. Gen. Genet. 115:258

Shapiro, J.A. 1979. Proc. Natl. Acad. Sci. USA 76:1933

Starlinger, P., and Saedler, H. 1972. Biochemie 54:177

Starlinger, P., and Saedler, H. 1976. Curr. Top. Microbiol. Immunol. 75:111

VIROIDS: BIOLOGY, STRUCTURE AND POSSIBLE FUNCTIONS

Heinz L. Sänger

Arbeitsgruppe Pflanzenvirologie
Justus-Liebig-Universität
D-6300 Giessen, West Germany

INTRODUCTION

The discovery of viroids in 1971/72 as free infectious RNA molecules of low molecular weight and hence the smallest disease agents presently known is an interesting example of serendipity in modern biology, because viroids were found although not actually sought for. In fact, they finally turned out to be completely different from what they were expected to be when the efforts started to characterise the elusive virus-like agents to the corresponding plant diseases. After the early scepticism and even rejection viroid research has now become a rather interesting and well accepted task. Considerable progress has been made in elucidating the structure of viroids and the wealth of information which has recently accumulated renders viroids the most thoroughly studied RNA molecules of today. It is the aim of this report to review the present knowledge on the biology and the structure of viroids and to discuss some of its implications with respect to the functions of viroids which are still largely enigmatic.

BIOLOGY OF VIROIDS

Viroid Diseases

The eight viroid-incited plant diseases presently known are listed in Table 1. The names of these diseases reflect the characteristic disease symptoms caused by the corresponding viroids. In susceptible host plants they may range from malformations and discolorations of leaves and fruits to retardation of plant growth in general and even death of the whole plant may occur. The viroid nature of these diseases has been established by isolating and

Table 1: The viroid diseases presently known and the
               abbreviations of the corresponding viroids.

Viroid disease	Viroid	References[+]
Potato spindle tuber	PSTV	1, 2
Citrus exocortis	CEV	3, 4
Chrysanthemum stunt	CSV	5, 6
Chrysanthemum chlorotic mottle	ChCMV	7, 8
Cucumber pale fruit	CPFV	9, 10
Coconut "cadang-cadang"	CCCV	11, 12
Hop stunt	HSV	13, 14
Avocado sunblotch	ASBV	15

[+]References:

1. Diener, 1971 b; 2. Singh and Clark, 1971; 3. Sänger,
1972; 4. Semancik and Weathers, 1972 b; 5. Hollings and
Stone, 1973; 6. Diener and Lawson, 1973; 7. Romaine and
Horst, 1975; 8. Horst and Romaine, 1975; 9. Van Dorst
and Peters, 1974; 10. Sänger et al., 1976; 11. Randles
1975; 12. Randles et al., 1976; 13. Sasaki and Shikata,
1977 a; 14. Sasaki and Shikata, 1977 b; 15. Thomas and
Mohamed, 1979.

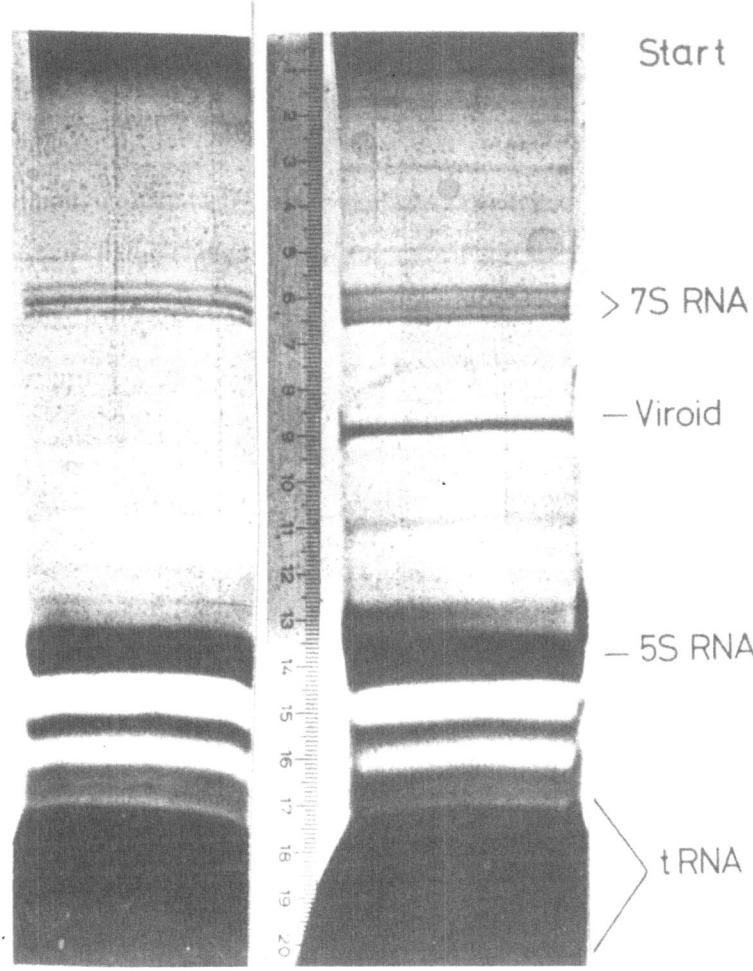

Fig. 1.   Separation of 25 mg 2M LiCl-soluble RNA from
           healthy (left lane) and 25 mg viroid (CPFV)-
           infected (right lane) tomato leaf tissue on a
           non-denaturing 5 % polyacrylamide slab gel (6 mm
           thick). The RNA bands are made visible by stain-
           ing with methylene blue. Note the extra band of
           viroid RNA in the right lane which usually com-
           prises less than 0.001 % of the total cellular
           nucleic acids from which purification is started.

Fig. 2.   Coconut plantation in the Philippines with palm trees
           killed by the viroid of the cadang-cadang disease (CCCV).

Fig. 3.   Symptoms of growth retardation and epinasty as caused by
           the viroid of the exocortis disease of citrus (CEV);
           left: healthy citrus plant.

partially characterizing the infectious viroid RNA from diseased plants. In particular the presence and the appropriate size and the infectivity of this RNA must be demonstrated by polyacrylamide gel electrophoresis of nucleic acid preparations directly extracted from diseased plant tissue with the aid of phenol (Fig. 1) followed by bioassay. However, the absence of a visible viroid band cannot be considered as a reliable proof for the absence of the agent. Since viroid RNA may represent only a minute fraction (usually less than 0.001 %) of the total nucleic acids directly extracted from diseased plants, it may be beyond the level of detection after staining the gel. Therefore, in critical tests gel analysis has to be completed by appropriate bioassays. All viroid diseases are of economic importance and the effect of a viroid infection may be quite dramatic as in the case of cadang-cadang, which causes disastrous losses in coconut palms in the Philippines. It may be estimated that probably more than 20 million trees were killed from 1926 to 1978, which makes the disease is the main threat to coconut production in the Philippines (Randles, 1975) (Fig. 2).

The Question of Viroid "Species"

In view of the different viroid diseases and the host specificity of the corresponding agents the question of similarities and differences between these different viroids arises. Fingerprint analysis carried out with purified viroids after in vitro labelling with ^{125}I (Dickson et al., 1975) and after in vitro 5'-phosphorylation with γ-^{32}P-ATP and T4 phage induced polynucleotide kinase (Gross, Domdey and Sänger, 1977) showed that PSTV from tomato, CSV from cineraria and CEV from Gynura differ significantly from each other in their oligonucleotide patterns and hence their primary sequence. These findings support the concept of individual viroid species and invalidate the previous assumption (Semancik and Weathers, 1972c; Singh and Clark, 1973b; Semancik et al., 1973a) that for example PSTV and CEV are independent isolates of the same pathogen, because they produce similar symptoms in certain hosts. Moreover, fingerprint analysis shows that each of the three viroid preparations investigated consists of a single molecular species of RNA and not of a mixture of several RNAs of about the same length, but of different nucleotide sequence(Fig.4).

Experimental Transmission

All viroids described so far are experimentally transmissible with mechanical means using crude tissue homogenates. The success of transmission is largely increased if measures are taken to eliminate the action of nucleases, which inactivate viroid infectivity quite rapidly by degrading the nuclease-sensitive viroid RNA molecule. Correspondingly, tissue homogenates are usually freshly prepared with cold and slightly alkaline buffers in the presence of the nuclease-absorbing clay mineral bentonite and immediately

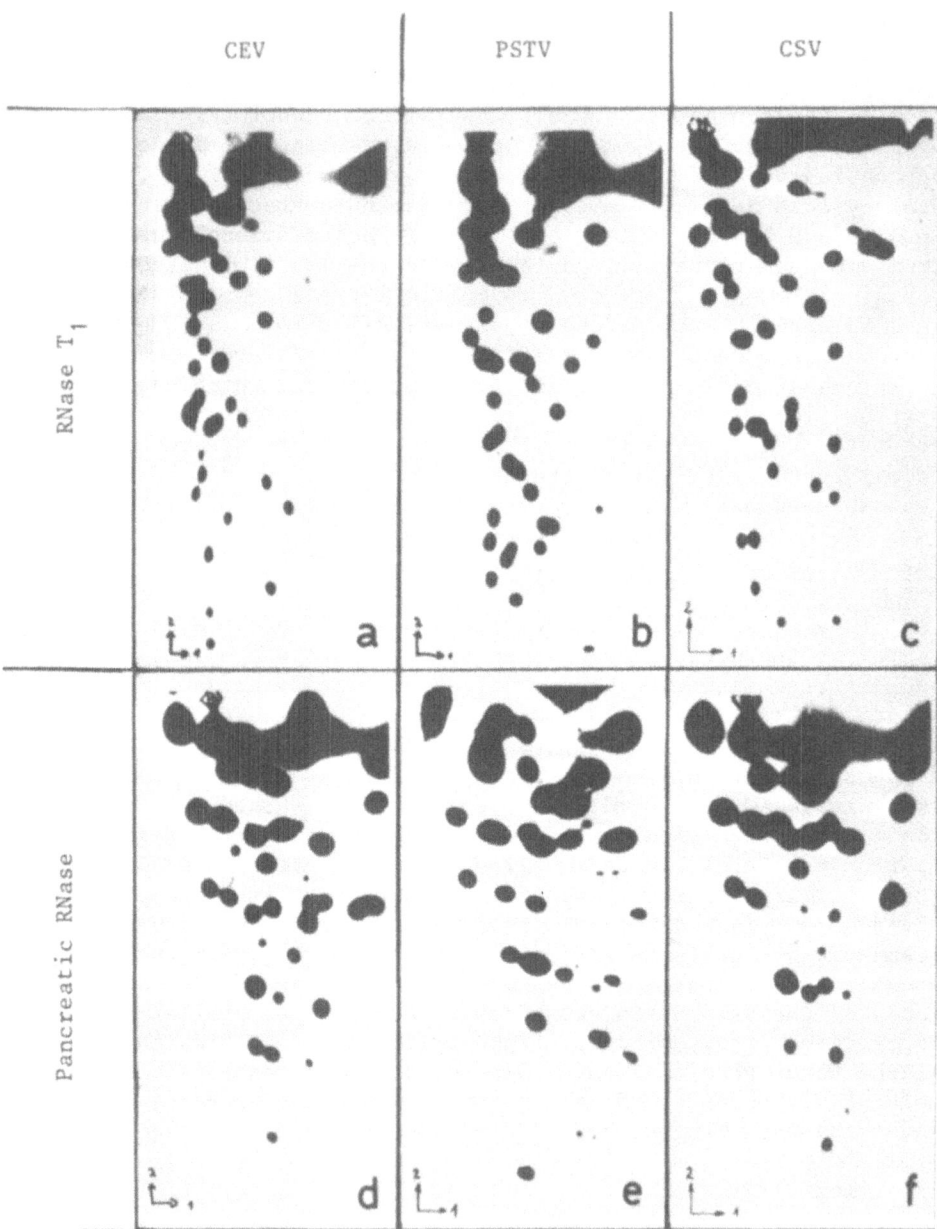

Fig. 4.   Oligonucleotide fingerprints after complete digestion
           of three viroids with RNase T1 and pancreatic RNase,
           CEV = Citrus exocortis viroid from Gynura aurantiaca,
           PSTV = Potato spindle tuber viroid from tomato, ChSV =
           Chrysanthemum stunt viroid from Cineraria. The lower
           half of each pattern represents large oligonucleotides
           which are specific for each viroid "species".

used for inoculation. Under these conditions standard inoculation
techniques as used with conventional plant viruses or their RNAs
may be applied. Thus, rubbing of the inoculum onto carborundum-
dusted leaves is used experimentally in many viroid-host combinations. In certain host plants stem-puncturing or stem-slashing
with razor blades or scalpels is more efficient in viroid transmission. Grafting has played an important role in the early experimental transmission of the citrus exocortis disease (Fig. 3).

Experimental Host Range

The experimental host range of different viroids differs considerably. PSTV is able to systemically infect numerous plant
species of several families (Singh, 1973), whereas ChCMV seem to
infect chrysanthemum only. CCCV has, so far, only been transmitted
successfully to a few other palm species and ASBV seems to infect
only avocados. The host range of the other viroids is limited to
certain species of a few specific plant families, respectively.
Thus, CEV can be transmitted from citrus to the herbaceous composite Gynura aurantiaca and to different solanaceous hosts including tomato, potato, petunia and scopolia. CPFV infects many
species of the cucurbit family, but it may also be transmitted to
certain solanaceous hosts.

An interesting problem in this context is the influence of
the host on the specificity of viroids as reflected in their primary sequence. To clarify this point the fingerprints of PSTV and
CEV after propagation in tomato and Gynura aurantiaca, respectively,
were compared (Dickson et al., 1978). It was found that both
viroids displayed the essential features of their characteristic
fingerprints irrespective of the host plant, in which they were
propagated. Some minor variations were observed but their nature
and origin has not yet been elucidated. These results clearly show
that the infecting viroid must serve as a template from which the
progeny viroid is copied and that the host plant has no pronounced
influence on the primary sequence of the viroid replicating in it.

Expression of Symptoms

In herbaceous host plants such as potato, tomato, Gynura,
chrysanthemum and cucumber the first symptoms of a generalised
viroid infection usually occur in newly developing leaves 2 - 8
weeks after inoculation with PSTV, CEV, CSV, ChCMV and CPFV,
respectively. On the other hand, symptoms may take several months
and even years to develop as in the case of CEV in certain citrus
varieties, CCCV in coconut palms and ASBV in avocado. However,
it should be noted here that especially in many wild host plant
species, viroid infection and replication does not produce any
recognisable disease symptoms. Thus, viroids may often be latent
in certain hosts and cause only disease, when transmitted to

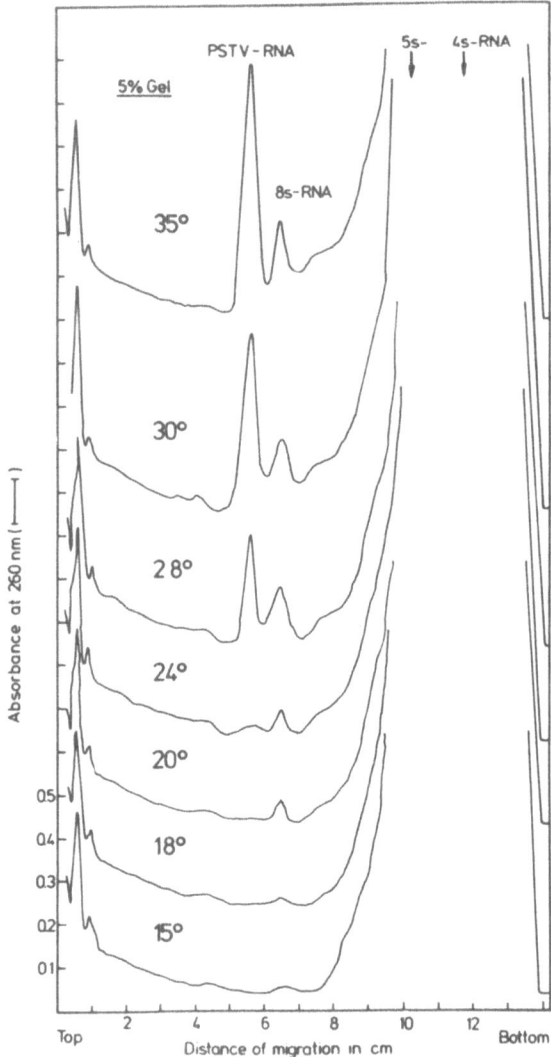

Fig. 5.   Electrophoretic analysis on 5 % polyacrylamide gels of
           "2M LiCl-soluble RNA" from PSTV-infected tomatoes
           grown for four weeks at different temperatures. On each
           gel the RNA-equivalent from 50 g leaf tissue was run.
           The increase of the concentration of PSTV-RNA with
           increasing growth temperature is evident.

cultivated plant species and varieties. The expression of disease
symptoms is, of course, specified by both the host plant and the
viroid. For example, three different tomato cultivars may respond
to an infection with the common strain of PSTV with severe, mild
or no symptoms (Mühlbach and Sänger, 1977; Mühlbach et al., 1977).
On the other hand, viroid strains producing severe, intermediate
and mild symptoms in a given host have been reported for PSTV
(Fernow, 1967) CEV (Weathers and Calavan, 1961; Calavan et al.,
1964) and ChCMV (Horst, 1975). Fingerprint analysis has shown that
such severe and mild strains of PSTV are uniquely characterised by
the presence and absence of oligonucleotide particular spots, from
which it was estimated that they differ in 2 - 10 of their nucleotides (Dickson et al., 1979). Therefore, the marked contrast between severe and mild symptoms caused by PSTV in a given host plant
can be related to specific differences in the nucleotide sequence
of the pathogen. If symptoms are produced, their expression is
critically dependent on a threshold concentration of viroid and on
environmental conditions. High temperatures, high light intensities
and inoculation with highly infectious inocula generally increase
their severity of symptoms and the rapidity of their appearance.
This is evidently based on the inordinate temperature dependence
of viroid replication, which differs strikingly from the replication of conventional plant viruses. At 18 - 20 $^{\circ}$C, where most
plant viruses are replicating optimally, the replication of PSTV,
CEV, CSV and CCMV in tomato, Gynura and chrysanthemum, respectively, is usually so low that no stainable viroid band can be detected
in gel analysis. Rising the growth temperature of infected plants
to 32 - 35 $^{\circ}$C may cause a more than five hundredfold increase in
viroid yields (Sänger and Ramm, 1975)(Fig.5).Under these conditions
severe symptoms usually appear 10 - 14 days after inoculation,
whereas it may take more than three months at 18 - 20 $^{\circ}$C. The
influence of a threshold viroid concentration for symptom expression can also be observed after inoculation with highly dilute
inocula as in end point dilution experiments. In this case clear
cut symptoms may only be recognised, when apparently healthy
looking test plants like tomato, Gynura aurantiaca and chrysanthemum are decapitated. The axillary buds of infected plants which
now start to grow exhibit pronounced symptoms of viroid infection
which can be directly related to the high concentration of viroid
which accumulates in this tissue.

Interference Between Viroids

The availability of common hosts for several viroids allowed
cross-protection tests among them. In such tests a viroid is used
for challenge inoculation on a previously viroid-infected plant
to probe for possible interference in symptom expression. It was
found that CEV, CSV and a severe and a mild strain of PSTV exhibit
cross protection in a variety of combinations. Interestingly,
ChCMV did not protect against these viroids in chrysanthemum.

Alternatively, PSTV did not protect against ChCMV in this host (Niblett et al., 1978). Although the actual mechanism of this phenomenon is not known one must assume, that the interfering viroids affect a common biological process in symptom expression although they differ significantly from one another in their fingerprints and hence their primary structure. The observation that a threshold concentration of viroid is required for symptom expression complicates the interpretation of these results. The protecting viroid may partially inhibit the replication of the challenging viroid so that this threshold concentration might not be reached. Therefore, further experiments are needed to determine the level of replication by the individual viroids in doubly infected cross-protected plants.

Viroid Transmission Under Natural Conditions

PSTV and CSV are known to spread rapidly to neighbouring plants by the combination of wounding and contact of foliage as occurring during culturing measures and handling in the field and in greenhouses, respectively. Vertical transmission through seed and pollen has been reported for PSTV in certain solanaceous host species, which causes serious problems in obtaining healthy plant material by breeding. Early work indicated that PSTV could also be accidently transmitted at low frequencies in the field by biting insects and by aphids. Most viroid diseases, however, are inadvertently spread by man via the vegetative propagation of viroid-infected but symptomless plants through tubers (potato), cuttings (chrysanthemum) or grafting (citrus). Thus, CEV has been disseminated inadvertently into all citrus growing areas of the world by distributing and grafting citrus varieties, which were symptomless carriers of the agent. In addition, CEV may also be transmitted by contaminated tools during pruning and cutting to neighbouring trees in a citrus plantation. The mode of natural transmission of the cadang-cadang disease of the coconut palm, however, which is propagated only by seeds, is still unknown. Viroids are often latent in wild plant species, whereas they are generally pathogenic in cultivated plants (Singh, 1973). In fact, the wild potato species Solanum phureja and S. stenotonum were found to harbour viroids, which are capable of inducing potato spindle tuber disease in cultivated potato. Also the seemingly healthy ornamental plant Columnea erythrophae appeared to be viroid-infected (Owens et al., 1978). Accordingly, wild plant species could serve as reservoir plants, from which viroids infect cultivated plant species, where they may become pathogenic and produce disease symptoms. Many observations lead to suspect that viroid diseases of cultivated plants are of recent origin and that their world-wide spreading is strongly favoured by the agricultural and horticultural activities of man expecially by the monoculture of crop plants and ornamentals.

Control Measures

The dissemination of viroid diseases may be controlled efficiently by prompt removal of all apparently diseased plants and by the selection of healthy plant material, which is then used for further propagation. In all cases, where the disease agent is readily transmitted mechanically by contaminated tools, their sterilisation with flaming or with chemicals like sodium hypochloride or sodium hydroxide plus formalin has proven successful in preventing further spreading of the disease. In addition, indexed viroid-free plant material to be used as propagating stock should be maintained in isolation and be treated with great caution to prevent recontamination. If the pertinent control measures are strictly followed the incidence and spreading of viroid diseases can be efficiently reduced.

STRUCTURE OF VIROIDS

For many years the molecular structure of viroids remained enigmatic, because physically recognisable entities were lacking even in highly infectious viroid preparations. Accordingly, the structural properties had to be inferred indirectly from different experimental approaches utilizing viroid infectivity as the sole parameter for its characterisation. All of this early work was carried out with PSTV and CEV and its interesting course has been reviewed in detail (Diener, 1972b; Diener, 1974; Diener and Hadidi, 1977). In summary, it turned out that in some experiments viroids behaved like double-stranded RNA, whereas others indicated that they were single-stranded molecules. These ambiguous properties led to assume that viroids may be a single-stranded molecule with some sort of hairpin structure and an extensive base-pairing or a double-stranded but incompletely base-paired molecule. From electron microscopy it was concluded that PSTV is a single-stranded RNA with some kind of a hairpin-like structure and with a molecular weight in the range of $8.0 - 9.0 \times 10^4$ daltons (Sogo et al., 1973). From runs in different gel systems the molecular weight estimates for PSTV and CEV ranged from 0.5 to 1.25×10^5 daltons.

Properties of Highly Purified Viroids

Considerable progress was made, when highly purified preparations became available in amounts sufficient to carry out direct biophysical and biochemical analysis. Then, it was unambiguously demonstrated with thermodynamic methods that viroids exist in both their native and denatured form as one single RNA strand with a molecular weight of about 120 000, which corresponds to a chain length of about 360 nucleotides (Sänger et al., 1976). From electron microscopic (Fig. 6) and hydrodynamic analysis it was concluded that the native viroid molecules are double-stranded, rod-like structures with an axial ratio of about 20 : 1. Electron

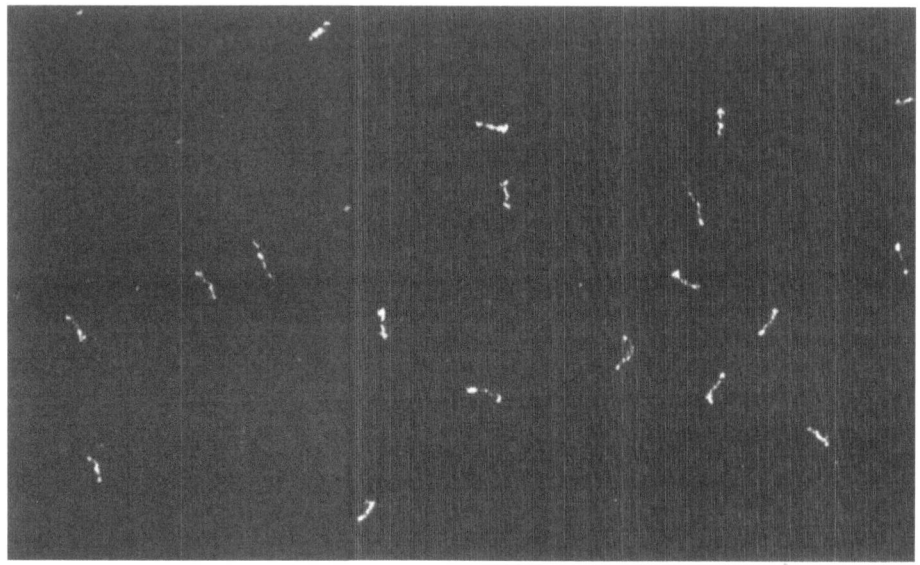

Fig. 6.   Dark-field electron micrograph of undenatured viroid
           (CPFV) molecules. The length of the native double-
           stranded rod-like structure is 37 ± 6 nm.

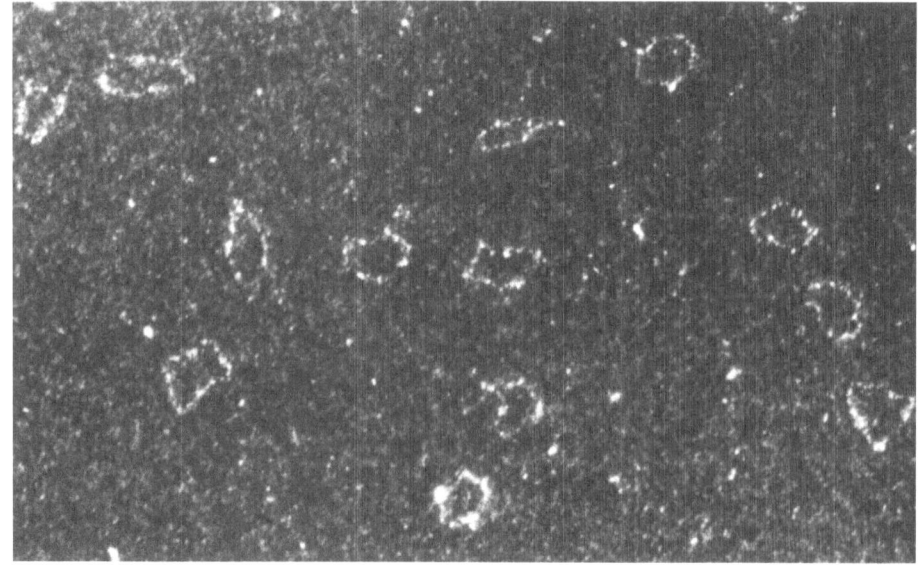

Fig. 7.   Dark-field electron micrograph of denatured PSTV. The
           contour length of the completely unfolded covalently
           closed circular viroid molecules is 100 ± 6 nm.

micrographs of denatured viroids finally demonstrated (Fig.7) that
they are actually single-stranded RNA circles, which due to intramolecular base-pairing exist in their native state as a double-
stranded, rod-like structure. A small fraction (usually less than
1 %) of the molecules were linears, which were considered to be
generated by nicking of circles during the isolation procedure.
The unexpected circularity of viroids was substantiated biochemically by attempted end-group labelling, which failed to show the
existence of 3'- and 5'-end groups in viroid molecules (Sänger et
al., 1976). The circularity of viroids was later confirmed by electron microscopy (McClements and Kaesberg, 1977), but the different
viroid preparations investigated contained variable amounts of linear molecules. Their possible origin and their properties are
still a matter of controversy. Viroids are the first example of a
single-stranded, circular RNA found in nature.

The Primary Structure of PSTV

     The unequivocal proof for the unique structural features of
viroids was finally provided, when the complete primary sequence
of PSTV became established (Gross et al., 1978). The sequencing
work was severely hampered by the fact that the rate of viroid replication is rather low, so that viroid molecules cannot be sufficiently labelled with ^{32}P in vivo for subsequent sequencing.
Therefore, purified viroid RNA had to be labelled in vitro after
controlled nuclease digestion using γ-^{32}P ATP and T4 phage-induced
5'-polynucleotide kinase from E. coli. The in vitro post-labelling
technique can only be applied, when viroid preparations free from
any contaminating cellular RNA are available and when the kinase
is free from any traces of nucleases. The suitability of this
approach was first proven by in vitro labelling of the complete
RNase A and T1 digests of different viroids (CEV, PSTV and CSV),
and by their comparison on the basis of their fingerprint patterns
(Gross, Domdey and Sänger, 1977), which established the existence
of different viroid species (Fig. 4). The high resolution achieved
in these fingerprints allowed the separation of the complete RNase
T1 and RNase A fragments of PSTV and their sequence analysis by a
combination of the conventional and the newly developed rapid gel-
sequencing techniques. For establishing the total sequence overlapping sequences are needed. They were obtained by controlled
digestion of PSTV with a set of different RNases by which longer
viroid-fragments could be generated (see legend of Fig. 8). These
overlapping fragments were also 5'-end-group labelled in vitro
and subjected to sequence analysis. The complete sequence of the
359 nucleotides of PSTV could be established with certainty by the
rather complex "puzzle" of arranging the large set of individual
sequences in the appropriate overlapping array (Fig. 8). The resulting complete nucleotide sequence clearly substantiated that
PSTV is a true single-stranded circular molecule consisting of
359 nucleotides.

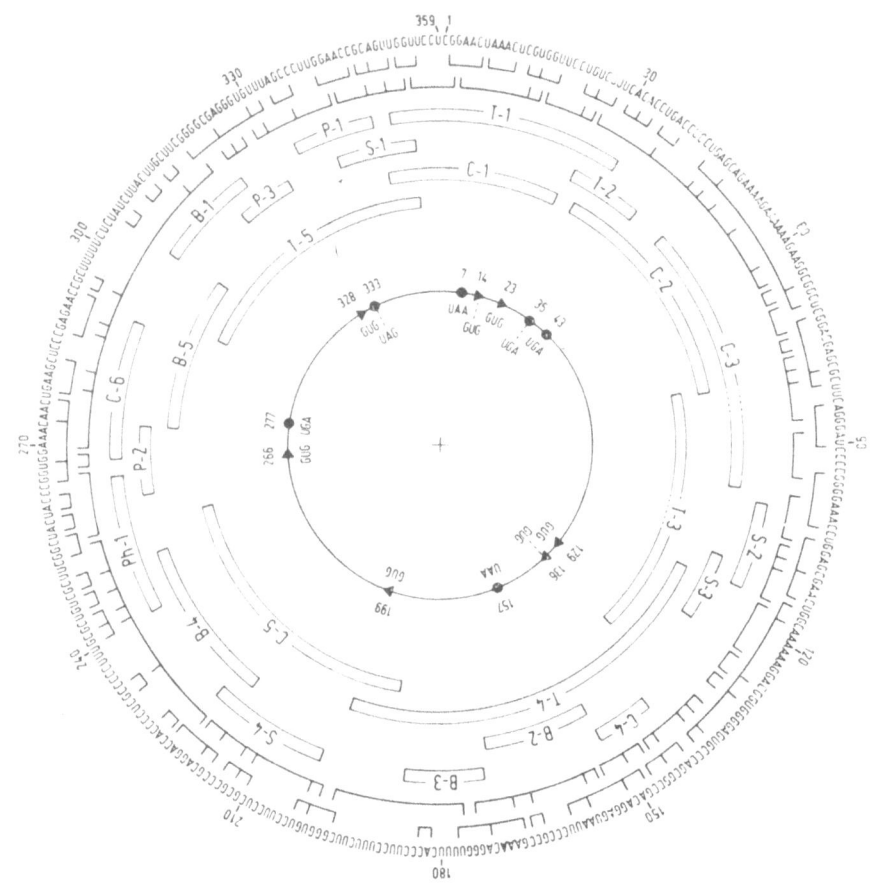

Fig. 8.   Primary structure of the PSTV molecule. The nucleotides
          have been numbered clockwise from 1 to 359. Fragments
          from complete pancreatic RNase and RNase T1 digestion are
          marked by brackets without numbers. A selection of over-
          lapping sequences needed to establish the total primary
          structure of PSTV is presented as labelled bars. B-1 to
          B-5: bacterial alkaline phosphatase containing an unknown
          nuclease. C-1 to C-6: CM-RNase (modified -carboxy-
          methyllysine-41 pancreatic RNase). T-1 to T-5: RNase T1.
          P-1 to P-3: pancreatic RNase. S-1 to S-4: nuclease from
          Staphylococcus aureus. Ph-1: RNase PhyI from Physarum
          polycephalum. The inserted solid circle indicates the
          location of possible initiator (►) and terminator (●)
          codons in PSTV. Numbers refer to the first nucleotide of
          the corresponding triplet.

The Secondary Structure

The secondary structure of the PSTV molecule was obtained when its primary nucleotide sequence was arranged in such a way that a maximum number of intramolecular base pairs occured. A largely double-stranded, rod-like structure was obtained (Fig. 9) which is characterised by the serial arrangement of 26 double-stranded stretches, which are separated by 25 segments of unpaired bases in single-stranded loops or bulge loops. In addition, there is one loop on each end of the rod-like molecule. The validity of this unique secondary structure was ascertained by considering the location of the enzymatic cleavage sites and the results from chemical modification of cytidines to uridines in single-stranded loops by bisulphite (Domdey et al., 1978). In detail, PSTV consists of 73 x AMP (= 20.3 %), 77 x UMP (= 21.4 %), 101 x GMP (= 28.1 %) and 108 x CMP (= 30.1 %). The two partially complementary halves of the PSTV circle have virtually the same length namely 179 and 180 nucleotides, respectively. There are 73 G : C (= 59.8 %), 38 A : U (= 31.2 %) and 11 G : U (= 9.0 %) pairs in the native secondary structure resulting in a total of 122 base pairs (= 67.7 %) out of the 179 base pairs potentially possible if complete intramolecular complementarity would exist. This rod-like secondary structure of PSTV agrees well with its thermal stability, the high cooperativity of the thermal denaturation process and the axial ratio of 20 as derived from hydrodynamic studies, and from electron microscopy (Sänger et al., 1976).

Exactly the same secondary structure model was obtained by an independent theoretical approach based on thermodynamics. With the established primary nucleotide sequence at hand the free energy of helix formation was systematically minimised for all 359 nucleotides of the PSTV molecule (Riesner et al., 1979). For this purpose the matrix procedure of Tinocco et al. (1971) was applied with a slight modification so that it became applicable to circular molecules. From the corresponding graphical representation of all possible double-stranded stretches the thermodynamically most stable secondary structure of PSTV under physiological conditions emerged, which is called the native structure. This structure is in complete agreement with the secondary structure model proposed earlier on the basis of maximal base-pairing and biochemical date and arguments (Gross et al., 1978). It should be noted here that all other secondary structure alternatives which can be constructed and which lead to a more globular shape of the molecule are not in accordance with the biochemical results and the thermodynamic data and calculations. Theoretical evaluations finally show, that the specific features of viroids such as the number of base pairs, the high cooperativity and the formation of very stable hair pins are improbable in random sequences (Riesner et al., 1979).

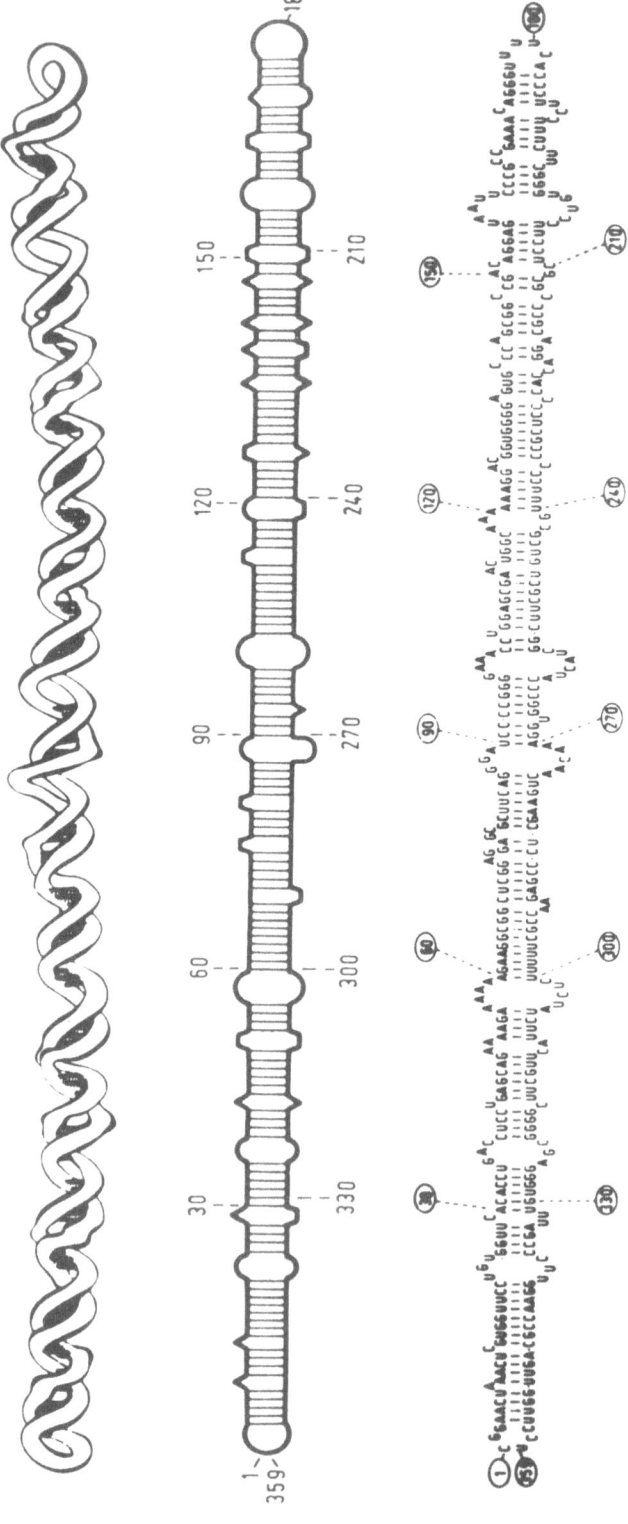

Fig. 9. Secondary structure of PSTV as shown in a two-dimensional base-pairing scheme (bottom), in the corresponding diagrammatic representation (middle) and in a three-dimensional manner (top). This structure is based on maximal intramolecular base-pairing taking into consideration the observed nuclease and bisulphite sensitivity of PSTV and the data obtained from detailed physicochemical investigations including binding studies with dyes and specific tRNAs.

Structure Formation and Multiple Forms

Nucleic acids may exert their biological functions through conformational changes. Such structural transformations can be studied in vitro by the analysis of their thermal denaturation process. The resulting curve representing the structural transitions is the "melting curve", which on the basis of thermodynamic theory allows certain conclusions on nucleic acid structure. In Fig.10 the differentiated melting curves for three structurally different nucleic acids are shown together with the corresponding structural changes occuring during the process of melting. The melting curve shows that the melting of viroid molecules is highly cooperative and that it occurs at a relatively low temperature as compared to the melting of a homogeneous double-stranded RNA. In fact, it combines the narrow range of transition of a RNA molecule consisting of two individual strands which are totally base-paired in a homogeneous way with the low temperature of transition of a branched single-stranded RNA species. It becomes evident that these characteristic features are due to the unique structure of the viroid molecule. The serial arrangement of double-stranded segments and single-stranded loops is responsible for the low temperature of transition, whereas the completely unbranched structure leads to the high cooperativity of the melting process. From this unique combination and the circularity of the molecules a general model for the secondary structure of viroids had already been proposed at a time before a single nucleotide sequence was known (Langowski et al., 1978; Klump, Riesner and Sänger, 1978). It turned out to be in perfect agreement with the detailed secondary structure model (Fig. 9), which was constructed after the first nucleotide sequence had been established (Gross et al., 1978; Riesner et al., 1979).

More recent investigations applying high resolution melting together with fast and low temperature jump techniques allowed a detailed analysis of the process of structure formation of PSTV on the basis of its established primary and secondary structure (Henco, Sänger and Riesner, 1979). They revealed (Fig. 11) that in addition to the highly cooperative main transition (T_m between 46.5 and 49 °C for different viroid species) all viroids show at higher temperatures an intermediate transition (T_m = 53 °C) and a high temperature transition (T_m = 68 °C). The amplitudes of both transitions comprise only about 1 % of that of the main transition, which represents the dissociation of 78 to 94 base pairs depending on the viroid species. During the intermediate transition two hairpins with 5 - 10 base pairs each and 10 - 20 nucleotides in the loops are dissociating. The high temperature transition corresponds to a hairpin of 9 G:C pairs and 1 A:U pair and more than 40 bases in the loop. It can be shown that these stable hairpins are not part of the native structure but are formed during the

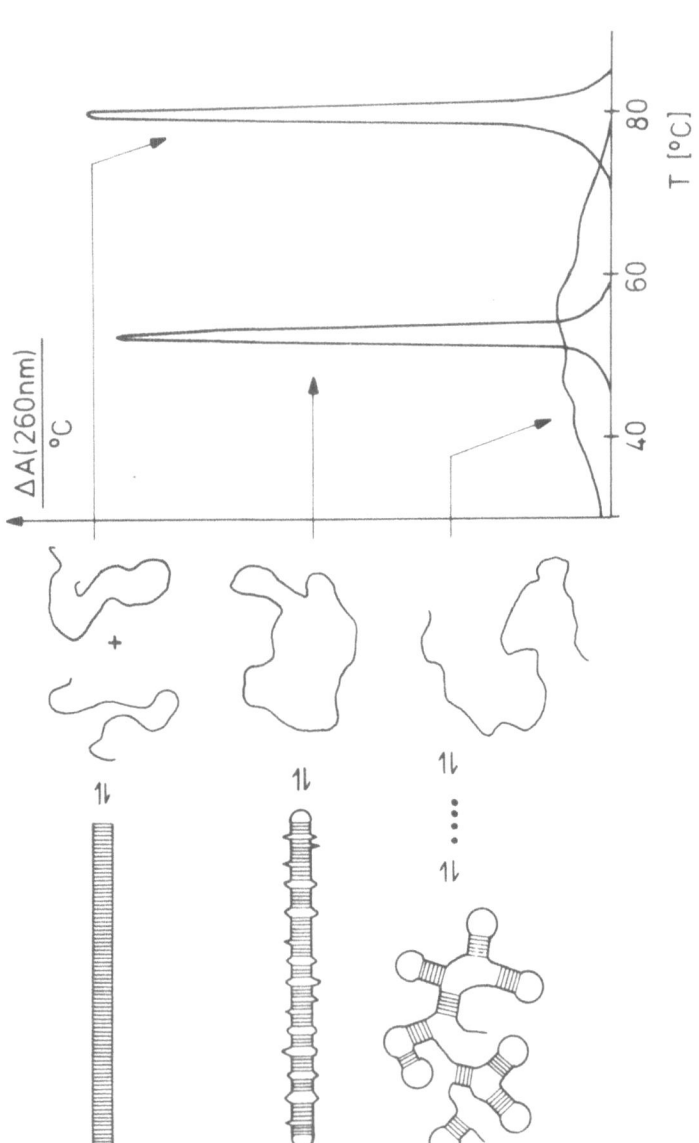

Fig. 10. Diagrammatic presentation of the differentiated melting curves as related to the secondary structure of different RNA molecules.

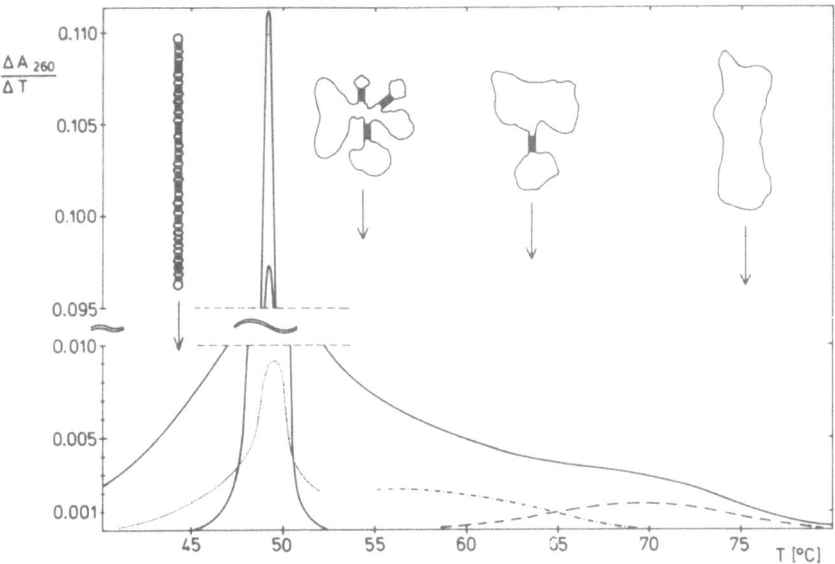

Fig. 11.   Composite melting curve of the structural
            transitions of PSTV.

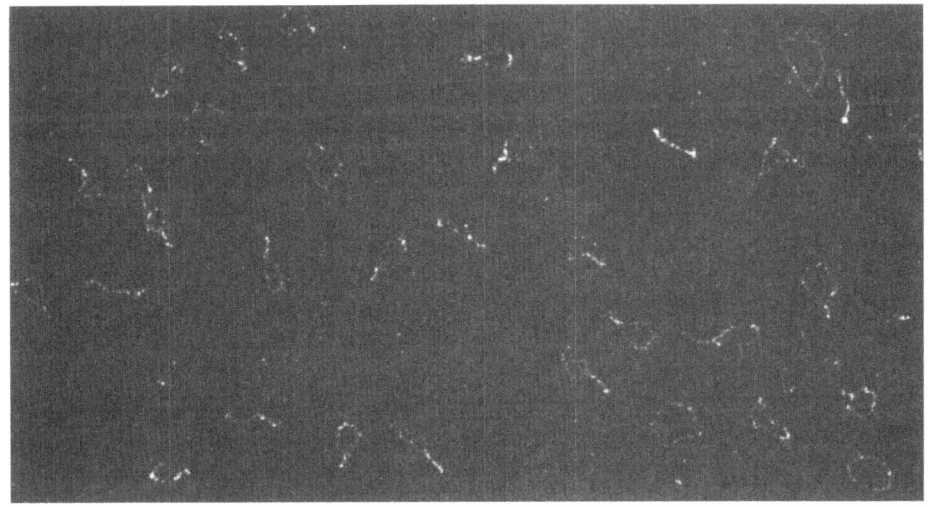

Fig. 12.   Dark-field electron micrograph of viroid molecules
            showing "tennis rackets" as structural intermediates
            trapped during the process of de- or renaturation.

main melting process. From these results the mechanism of structure
formation in PSTV can be described in detail on the basis of its
primary sequence as shown in Fig.13   (Henco, Sänger and Riesner,
1979).

According to this scheme after the opening of one of the
starting segments the left half of the PSTV molecule opens up at
about 50 ^{o}C with segments 1 - 13 in one cooperative step. Segments
14 and 15, however, are particularly stable and increase the T_m-
value of the dissociation of the right half of the molecule by
about 10 ^{o}C as compared to the left half. The dissociation of the
right half allows the formation of three new hairpins between
distant complementary regions (see curved arrows in Fig. 13).
Their high stability lowers the T_m-value of the complete right
half. This long range interaction prevents a biphasic melting and
allows the dissociation of the whole native structure in one cooperative transition with short-lived structural intermediates. The
high temperature transition with only one very stable base-paired
segment is of special interest, because it explains the nature and
origin of the "tennis rackets" and "ping pong paddles" in our
electron micrographs (Fig. 12) of denatured viroid molecules
(Sänger et al., 1976). Thus, the melting of viroids is not a mere
dissociation of base pairs but involves the concerted opening of
125 base pairs and the formation of 28 new base pairs. This process
may be described as a switching from the native rod-like structure
to a hairpin-containing circle. This process is unique as compared
to other nucleic acids and remarkably similar in all viroid species.
From the thermodynamic point of view the combination of high cooperativity and low denaturation temperature may be described as
an optimal combination between stability and flexibility. Although
viroids are the largest RNA molecules for which so many details of
structure and structure formation are known, the actual functional
implications of their structural properties remain to be elucidated.

Our finding of metastable viroid conformers would explain the
previous observation that viroids exist in multiple forms and that
viroid infectivity is found in well separated bands after chromatographic and electrophoretic separation (Singh et al., 1974;
Dickson et al., 1975; Randles, 1975; Randles et al., 1976). In fact,
purification of viroids involves several denaturation steps such
as phenolisation, treatment at elevated temperature and ethanol
precipitation. Electrophoretic analysis has shown that such structural intermediates may be obtained after redissolving ethanol
precipitates in aqueous solution and after rapid cooling of denatured viroid preparations. Therefore, the reported appearance of
viroid infectivity in well separated fractions does not imply molecular heterogeneity but is most probably due to conformational
heterogeneity.

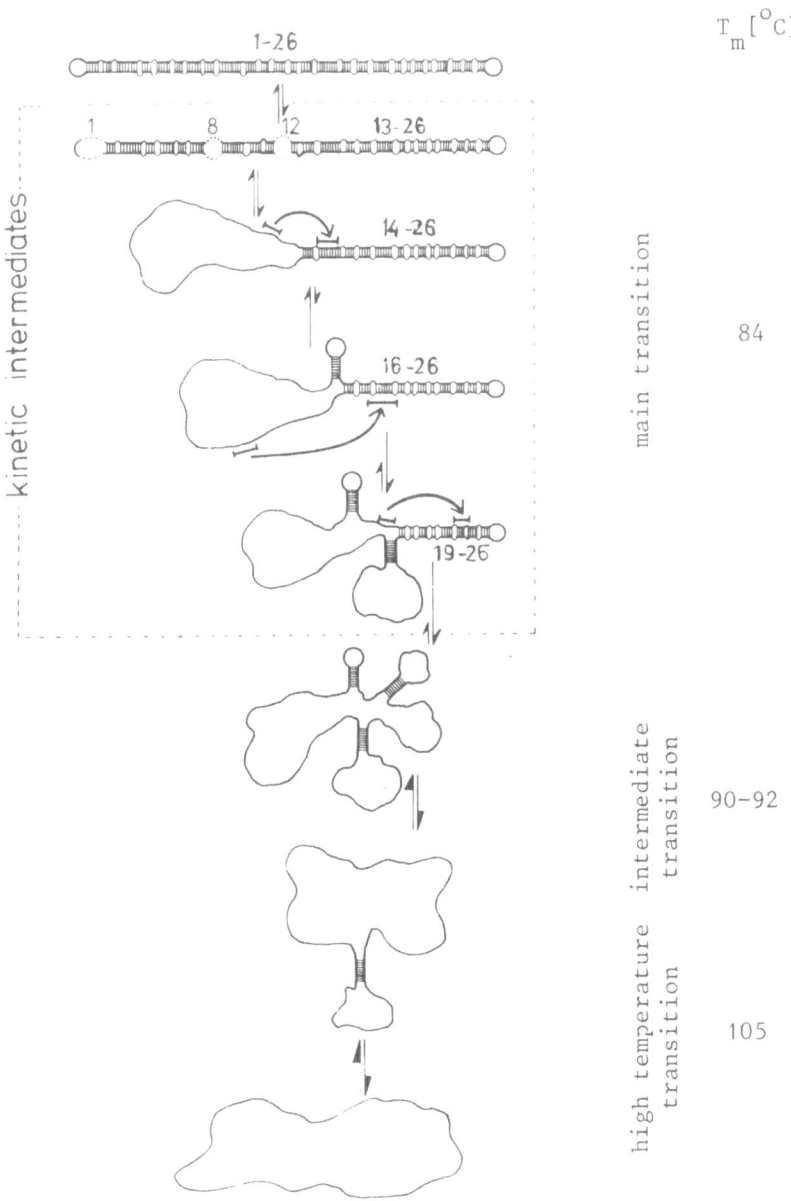

Fig. 13.   Mechanism of denaturation and renaturation of the
            PSTV molecule with different structural intermediates.
            The double-stranded segments of the molecule are
            numbered 1 - 26 from left to right. The T_m-values
            are given for 1 M NaCl.

Origin and Properties of Linear Viroid Molecules.

A controversy over circularity as the intrinsic structural feature of viroids arose when it was reported that PSTV preparations may contain up to 80 % linear viroid molecules only the rest being circles (Mc Clements and Kaesberg, 1977). The two forms were later separated by gel electrophoresis in the presence of high concentrations of formamide and urea, and the bioassay of the corresponding gel fractions indicated that both molecular species are infectious (Owens et al., 1977). From incorporation of ^{32}P into PSTV it was finally concluded that the observed linear forms are a major product of viroid replication and that they are not arising in vitro by nicking of circles with nucleases during isolation and purification (Hadidi and Diener, 1978).

A plausible explanation for the occurance of these unusually high concentrations of linear forms emerged when it was found that Mg^{++}-ions may produce linear molecules by nicking of viroid circles. Mg^{++}-ions have been used at slightly alkaline pH to cleave specifically individual tRNA species (Wintermeyer and Zachau, 1973) a reaction which generates 5'-hydroxyl endgroups and 2'- and 3'-phosphates. When we applied this Mg^{++}-catalysed phosphodiester cleavage to five different viroid species we found that a single nick occurred under a wide range of controlled conditions producing linear viroid molecules of the same size as the covalently closed viroid circles from which they arose (Sänger et al., 1979). The detailed analysis of the cleavage kinetics showed that 50 % of the viroid circles are converted into linear forms after about 25, 50, 150 and 360 min at 60, 50, 37 and 25 °C, respectively. All five viroid-species tested (PSTV, CPFV, CSV, CEV and CCMV) showed practically the same susceptibility to Mg^{++}-ions. Fig.14 shows the kinetics of the Mg^{++}-catalysed cleavage of PSTV circles on a fully denaturing 5 % polyacrylamide gel.

The detailed analysis of the migration of circular and linear molecules under non-denaturing conditions (i.e. in the presence of buffer of high ionic strength and at a gel temperature of 12-15 °C) showed that both forms comigrate in 5 % gels. Under these conditions the size and the secondary structure of both forms must be very similar if not identical. This finding is not surprising, because a single nick would certainly not change the rather stable rod-like secondary structure of the native viroid molecule. It should be pointed out that in our routine purification procedure all preparative electrophoretic runs are carried out under non-denaturing conditions in 5 % gels, where both intact and nicked circular viroid molecules comigrate. Therefore, we can exclude the possibility that we are selecting for viroid circles and thus inadvertantly loosing linear viroid molecules.

The end group analysis of the linearised molecules revealed that the ratio of 5'-terminal pA, pC, pU and pG was 38 : 17 : 25 : 20. This shows that in contrast to tRNA the Mg^{++}-catalysed cleavage of viroids is of rather low specificity. The nicking of viroids may also be detected in thermodynamic experiments. The midpoint-temperature and the cooperativity of the main denaturation process of the linear forms are significantly lower than those of the circular form. These results confirm our earlier theoretical prediction (Langowski et al., 1978) and support our interpretation of two melting peaks which appeared after repetitive heating in our previous calorimetric studies (Klump, Riesner and Sänger, 1978). The considerable amount of melting even at 48 °C indicates that a number of different transitions arising from heterogeneously nicked molecules are superimposed. This result is in good agreement with the end group analysis.

Mg^{++}-nicked and purified linear forms were found to be about one thousand to ten thousand times less infectious than previously reported for the isolated viroid circles (Sänger et al., 1976). The very low infectivity of our purified samples of linear molecules could be due to a contamination with viroid circles of less than 0.1 - 0.01 %. It is interesting to note that viroid circles were obtained again when inocula from individual symptom-bearing plants from the end-point dilution of the bioassay of isolated linear molecules were used for further propagation. Our finding that linear viroid molecules are non-infectious has been substantiated recently by Morris (1979) in the case of PSTV. The relatively high and varying percentage (30 - 80 %) of linear viroid molecules observed in certain preparations (Mc Clements and Kaesberg, 1977) can be attributed to Mg^{++}-catalysed cleavage of viroid circles, if the isolation and purification has been carried out in the presence of Mg^{++}-ions. In fact, according to the procedure reportedly used to purify PSTV (Diener, Hadidi and Owens, 1977; Hadidi and Diener, 1977; Diener and Hadidi, 1977) all the relevant samples must have been exposed for longer periods of time to 3 mM Mg^{++} in alkaline buffers of pH 9.0 at elevated temperatures. Most probably Mg^{++}-ions were added to stabilize viroid structure and the alkaline pH is obviously a precaution to inhibit the action of possible traces of nucleases. Unfortunately, however, it is inevitable that under such conditions Mg^{++} may exert the completely adverse effect of linearising viroid circles by the catalysis of the cleavage of phosphodiester bonds. When we followed this purification procedure as closely as possible we also obtained varying amounts (40 - 70 %) of linear forms and a corresponding lower yield of viroid circles. It should be emphasised here that all our present and earlier viroid preparations consisting of virtually a 100 % circular molecules always purified in the presence of an excess of EDTA to complex any polyvalent cations.

It is evident, therefore, that all traces of free polyvalent cations must be excluded during viroid purification if reliable conclusions concerning the occurance of linear viroid molecules in vivo are to be drawn. On the other hand, it is not possible at present to decide whether the few linear molecules in our completely Mg^{++}-free preparations are due to the action of nucleases or whether they are actual intermediates of viroid replication.

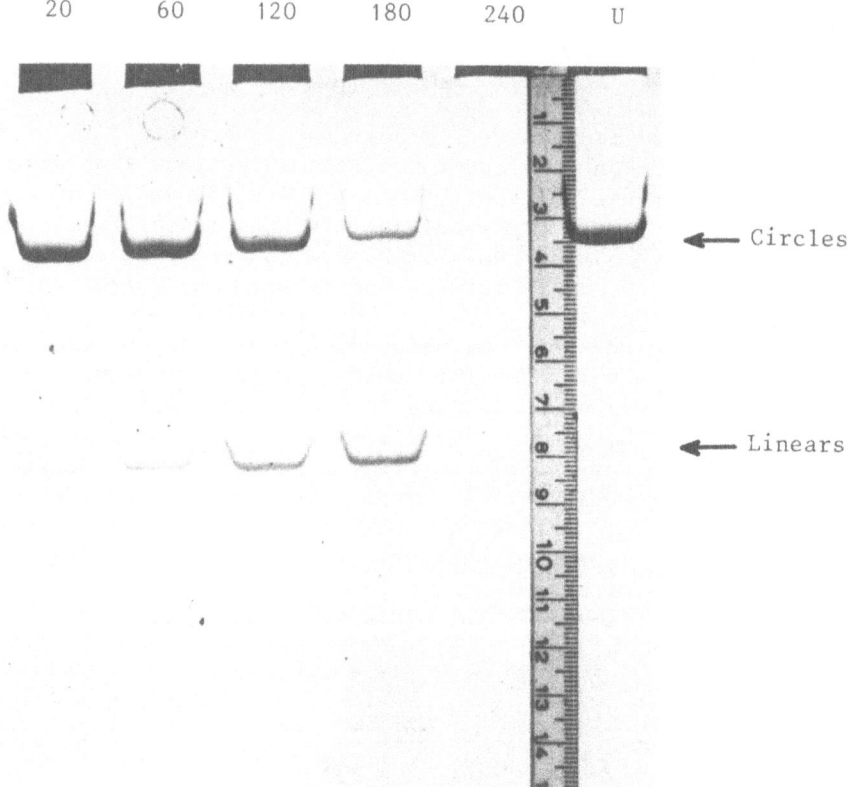

Fig. 14. Electrophoretic analysis of the kinetics of Mg^{++}-catalysed nicking of circular viroid (PSTV) molecules in a fully denaturing 5 % polyacrylamide gel. The number above the slots give the min of treatment with 5 mM Mg^{++} at 37 °C and pH 9.0. With increasing time of incubation linearisation of viroid circles increases. After 180 min about 50 % of the viroid circles are converted into linear forms. U = untreated control sample.

Absence of Tertiary Structure Folding

The base-pairing scheme of PSTV does not preclude interactions
between loops of distant parts of the molecule which would fold
the extended structure to a more globular shape. To probe for a
possible tertiary structure folding dye binding to the PSTV molecule was studied under a variety of ionic conditions. In case these
alteration of ionic strength would induce conformational changes
also a change in dye binding should be expected, because of the
impaired accessibility of the dye binding sites. Ethidium bromide,
which is well known as an intercalating dye without preference for
special base pairs of defined sequences was used to probe for
base-paired segments. The dye H 8208 2-(4'-aminophenyl)-5-(4'-
methylpiperazin-1"yl) benzimidazole, on the other hand, binds
specifically to A : U base pairs. It was found that these two
dyes are bound nearly quantitatively to the PSTV molecule up to
complete saturation under different ionic conditions. This shows
that changes of ionic strength do not induce any tertiary structure folding in the PSTV molecule. In fact, one is tempted to
correlate the 22 ethidium bromide molecules intercalating in the
native PSTV structure with its 26 double-stranded segments. On the
other hand, the 35 molecules of the dye H 8202 binding to the PSTV
molecule could be related to the 38 A : U base pairs actually
present in its native secondary structure (Riesner et al., 1979).

Further evidence for the absence of tertiary structure comes
from binding studies of specific tRNAs to PSTV (Wild et al., 1979).
Transfer RNA molecules are well suited for such studies, because
their anticodon loops are highly ordered by tertiary structure
folding. The specific anticodon region allows probing of certain
internal loops in the viroid molecule. It was found that five out
of the seven tRNA species tested are able to bind specifically to
those single-stranded PSTV loops which are characterised by sequence complementarity to the tRNA corresponding anticodon region
(Table 2). This finding is corroborated by the observation that
tRNA molecules for which an anticodon-complementary loop is lacking (tRNA fMet and tRNA Val1 from E. coli) do not bind to PSTV.
The free accessibility of the anticodon binding loops distributed
over the total viroid molecule clearly substantiates its completely extended structure and the absence of any tertiary structure
folding which would cover up these loops (Fig.15).This conclusion is
strengthened by the fact that tRNAs are rather voluminous conformational probes and that their binding would not only be hindered
sterically by tertiary structure folding in a limited area but
also by folding at more distant regions. From these results the
question arises if such complexes between viroids and cellular
RNAs might also exist in vivo. Although the concentration of
viroid molecules and tRNA might be sufficiently high in the cell
for complex formation, it is not possible at present to decide,
whether or not this feature is of any direct functional relevance.

tRNA	Origin	Anticodon		Molecules bound per PSTV molecule	K [M^{-1}]
Glu 2	E. coli	s^2U^+UC	(m^2A)	1	5.3×10^5
Asp	yeast	GUC	(m^1G)	1	not determined
Phe	yeast	G_m-AA	(Y)	2	7.5×10^5
Ser 1 + 2	yeast	IGA	(i^6A)	2	not determined
Lys	E. coli	s^2U^+UU	(t^6A)	3	7.0×10^4
fMet	E. coli	CAU	(A)	no binding!	–
Val 1	E. coli	VAC	(m^6A)	no binding!	–

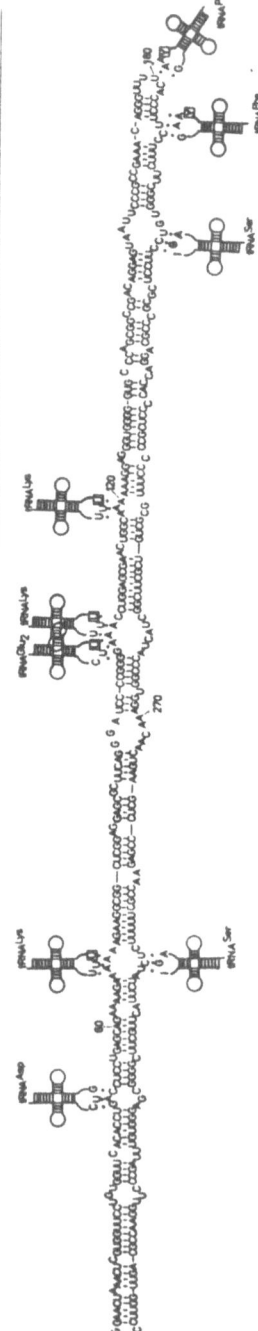

Fig. 15.   Binding of specific tRNAs to the PSTV molecule. The tRNAs are depicted smaller than compared to the size of PSTV.

The binding studies with specific dyes and tRNAs corroborate the previous evidence derived from electron microscopy, sequence analysis and thermodynamic studies that native viroid molecules are extended rod-like structures without any tertiary structure folding.

Properties of Hypothetical RNA Molecule Complementary to PSTV

The established sequence of PSTV allows to construct a hypothetical molecule of complementary sequence and it has been argued (Matthews, 1978) that viroids could exert their functions like negative strand viruses through such a complementary plus-strand RNA. But there is no conclusive evidence, as yet that such molecules occur as intermediates of viroid replication in vivo. Considering the experimental demonstration of their possible existence one should keep in mind, however, that they might be present in extremely low concentrations like the intermediate forms of replication in the case of conventional viruses. On the basis of the established sequence of PSTV the properties of the hypothetical circular RNA complement of PSTV were evaluated theoretically (Riesner et al., 1979). Despite the relatively high self-complementarity of the PSTV molecule itself the secondary structure of the complementary RNA cannot be the same, because only G : C and A : U pairs but not G : U pairs would be copied into the same type of base pairs. There are 16 G : C base pairs present in PSTV which will not lead to the same pairs in the complementary strand, whereas six new G : U pairs would be generated from six opposite A : C positions in the original molecule. Thermodynamic optimisation of the base-pairing scheme of the complementary strand leads to a PSTV-like secondary structure, which contains 117 base pairs instead of 122 in the actual PSTV molecule. Its melting temperature is about 12 $^{\circ}$C lower and the cooperativity of its melting is slightly decreased. From their rather similar structural features one can conclude that such complementary viroid molecules would most probably migrate together with the original PSTV. The vast excess of the latter would render the preparative separation of the complementary molecule as a rather difficult experimental problem. But, if successfully separated it can be clearly identified by RNase T 1 and RNase A fingerprints. On the basis of the complementary PSTV sequence and according to the previously established correlation between the brutto nucleotide composition of any RNase T 1 or RNase A oligonucleotide and its position in homochromatography fingerprints (Domdey et al., 1978) these fingerprints must be completely different from the PSTV fingerprints.

Another approach to demonstrate the existence of a PSTV RNA complement would be molecular hybridisation. Hadidi et al., (1976) failed to detect RNA sequences complementary to a major portion of PSTV in PSTV-infected or uninfected tomato leaves. Grill and Semancik (1978) on the other hand, claim that a different species

of complementary RNAs occur in CEV-infected Gynura and that the
corresponding hybrids exhibited the expected properties of RNA-
RNA duplex molecules. Interestingly, the same authors had previously reported that also CEV complementary DNA sequences exist in
Gynura (Semancik and Geelen, 1975). Unfortunately, however, they
have not tried, as yet, to clarify the relation between all these
different types of complementary molecules.

The unique structural features of viroids clearly indicate
that the existence of viroid complementary sequences can only be
conclusively demonstrated by molecular hybridisation when certain
experimental prerequisites have been fulfilled which take the circularity and the specific secondary structure of the probe into
account. Future investigations will show whether or not the
speculative concept of viroid replication and pathogenesis via a
complementary RNA molecule is correct or not.

Translation Properties of Viroids

The unique structure of the PSTV molecule raises many
questions related to viroid replication and function and the main
question is whether viroids code for peptides or proteins. Previous
attempts to translate PSTV and CEV into protein in several in vitro
systems have failed and viroid RNA does not interfere with the in
vitro translation of potent messenger RNA species in these systems
(Davies et al., 1974; Hall et al., 1974; Semancik et al., 1977).
Moreover, aminoacylation reactions carried out with CEV showed,
that amino acids are not accepted by this viroid (Hall et al.,
1974). This finding would preclude any tRNA-like functions for
viroids.

From the established primary sequence it appears that the
circular arrangement of the uneven number of 359 nucleotides makes
three rounds of translation potentially possible in PSTV. But the
common AUG initiator triplet and several other codons are missing.
There are seven possible GUG initiator and six possible terminator
triplets (see inner circle in Fig. 8). The shortest possible
translation product would be a tetrapeptide, whereas the longest
possible protein could results from more than two rounds of translation, provided that in this case the weak terminator codon UGA
would be translated. The lack of ribosome binding sites at the
5'-side of the possible GUG initiator codons, the absence in
viroids of the "cap" structure common to most eucaryotic messenger
RNAs and several other aspects of the primary and secondary structure of PSTV seem to preclude its translation into peptides or
proteins (Gross et al., 1978).

On the basis of these sequence data the hypothesis has been
put forward that viroids might excert their functions like negative
strand viruses via a complementary viroid molecule which would then

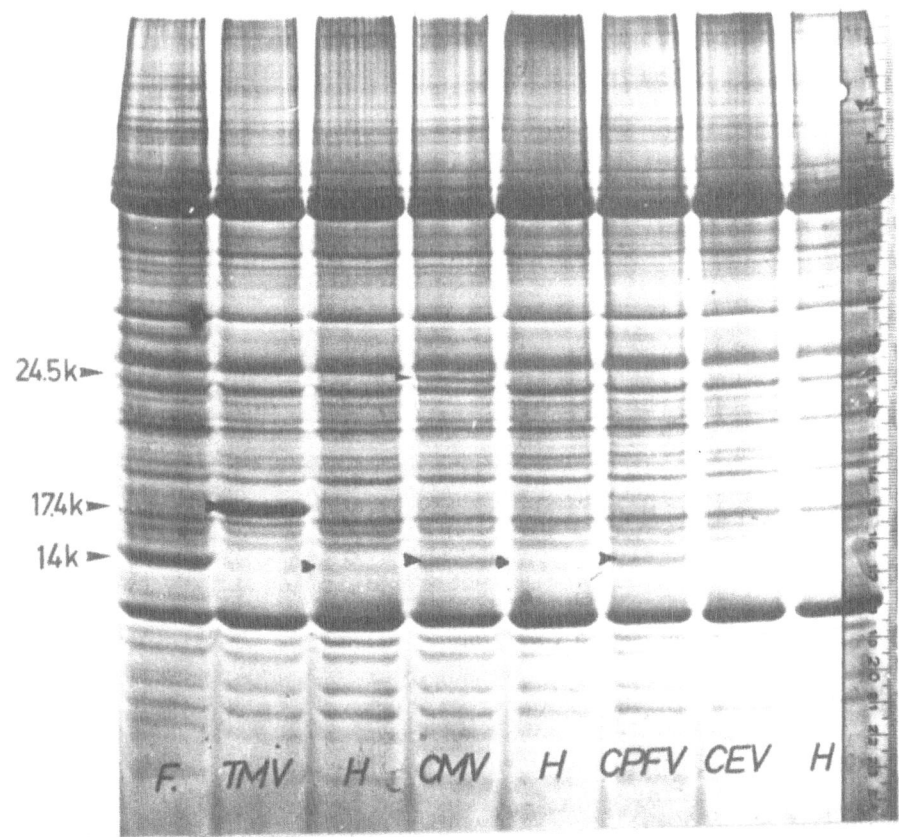

Fig. 18.   Protein pattern from tomato leaf tissue (cv. Rutgers) systemically infected with different pathogens.

      F = Cladosporium fulvum
    TMV = Tobacco mosaic virus
    CMV = Cucumber mosaic virus
   CPFV = Cucumber pale fruit viroid
    CEV = Citrus exocortis viroid
      H = Healthy control

Note the marked increase of the 14k protein band present in healthy tissue (H) after fungus (F), virus (CMV) and viroid (CPFV) infection which produce necrotic foliar symptoms. TMV and CEV infection only cause mild symptoms of leaf malformation. At 17.4k and 24.5k the viral coat proteins are visible.

be the positive coding strand (Matthews, 1978). In this case four
GUG initiator codons and nine possible terminators (4 x UAG, 2 x
UAA and 3 x UGA) could result in at least four proteins containing
28, 43, 79 or 108 amino acids. The two larger of these strictly
hypothetical viroid-specific polypeptides would have molecular
weights of about 8 500 and 11 300. Since the largest would contain 11 % arginine and 5 % lysine it should have histone-like
properties. This hypothesis would fit with the experimental data,
if the viroid-(CEV)-complementary RNA postulated by Grill and
Semancik (1978) on the basis of hybridisation data would exist and
if the two increased protein bands found in viroid-(CEV-)infected
Gynura tissue (Conejero and Semancik, 1977) would turn out to be
really viroid-specific. However, both these findings must be interpreted cautiously. First, the circularity and the unique secondary
structure of the viroid molecule require very specific techniques
to obtain reliable hybridisation data. Furthermore, the absence of
any trace of contaminating cellular RNA must be demonstrated rigorously. Second, the viroid-specificity of the increased protein
bands in CEV-infected Gynura is questionable, because these proteins are evidently present in lower concentration in healthy tissue
and accumulate upon viroid infection. This is also the case in
PSTV-infected tomato leaves, where a stimulation of two proteins
of high molecular weight (1.55 x 10^5 daltons) has been demonstrated by leucine-^{3}H incorporation (Zaitlin and Hariharasubramanian,
1972). More recent investigations clearly show that the considerable increase in the concentration of certain protein bands is
not viroid-specific, because it also occurs after infection with
conventional viruses and with a fungus. Interestingly, the intensity of these bands can be related to the severity of the necrotic
symptoms occuring after infection (Camacho-Henriquez and Sänger,
in preparation) (see Fig. 16).

     All these results indicate that the increased accumulation
of proteins in viroid-infected leaf tissue seems to be the result
of a general pathophysiological reaction of the host cell rather
than a viroid-specific response. Further investigations are needed
before the crucial question can be answered conclusively, whether
or not viroid-specific peptides or proteins are produced in vivo
and if they are possibly involved in viroid pathogenesis.

PROBLEMS OF VIROID REPLICATION

     Although the complete molecular structure of one viroid
(PSTV) has been established (Gross et al., 1978) the mechanism
of viroid replication and pathogenesis is still largely a matter
of conjecture. From different lines of evidence it can be concluded that viroids are replicating autonomously. The possibility
that their replication is dependent on a helper appeared rather
unlikely, because a viroid like PSTV replicates in a large number
of plant species (O'Brien and Raymer, 1964; Singh, 1973), which

then should all contain this helper. Moreover, especially designed experiments failed to demonstrate the existence of such helper viruses and their involvement in viroid replication (Diener, 1971 b; Diener et al., 1972). Whatever the mechanism of viroid replication will be there is little doubt that the progeny virus is a correct copy of the infecting molecule and not significantly altered by the corresponding host as shown by Dickson et al.(1978). They propagated PSTV and CEV, which have different fingerprints (Fig. 4) in their heterologous hosts Gynura and tomato and iso-lated, iodinated and fingerprinted the progeny viroid . The comparison of the RNase T$_1$ fingerprints clearly showed that both viroids do not undergo major sequence changes, when propagated in heterologous hosts.

Considering their chain length of about 360 nucleotides viroid molecules carry insufficient genetic information to code for a complete viroid-induced replicase even if three rounds of translation are assumed. This would still apply if viroids would excert their function like negative strand viruses, in which case the infecting molecule would be first copied into a complementary RNA. Because of this limited potential coding capacity one is forced to assume that viroid replication must largely depend on the enzyme machinery pre-existing in the host cell. But, despite the lacking translation properties of viroids in vitro the possibility still exists that those host enzymes involved in viroid replication are complemented or modified by small viroid-specific proteins or peptides, which would be coded for and synthesised in vivo.

In healthy plant tissue two types of known RNA-synthesising enzymes are presently known, namely the DNA-dependent RNA polymerase system (Duda, 1976) and the recently described RNA-dependent RNA replicase system (Ikegami and Fraenkel-Conrat, 1978; Romaine and Zaitlin, 1978). Theoretically viroids could therefore be replicated either through a complementary RNA like most RNA viruses or through a DNA template. This viroid-complementary DNA template could be newly synthesised upon viroid infection in a yet unknown way, because no reverse transcriptase-like enzymes have been found in plants, so far. On the other hand, a viroid-specific DNA could be already present in the host genome in a repressed state and its activation could be triggered by the infecting viroid molecule. In this case viroids would function as a regulatory RNA by depressing the viroid-complementary sequences in the host genome.

The DNA dependence of viroid RNA replication has previously been postulated from inhibition studies with actinomycin D which is known to inhibit DNA-dependent RNA-transcription. When this antibiotic was applied to leaf strips (Diener and Smith, 1975) and nuclei prepared from PSTV-infected tomato leaves (Takahashi

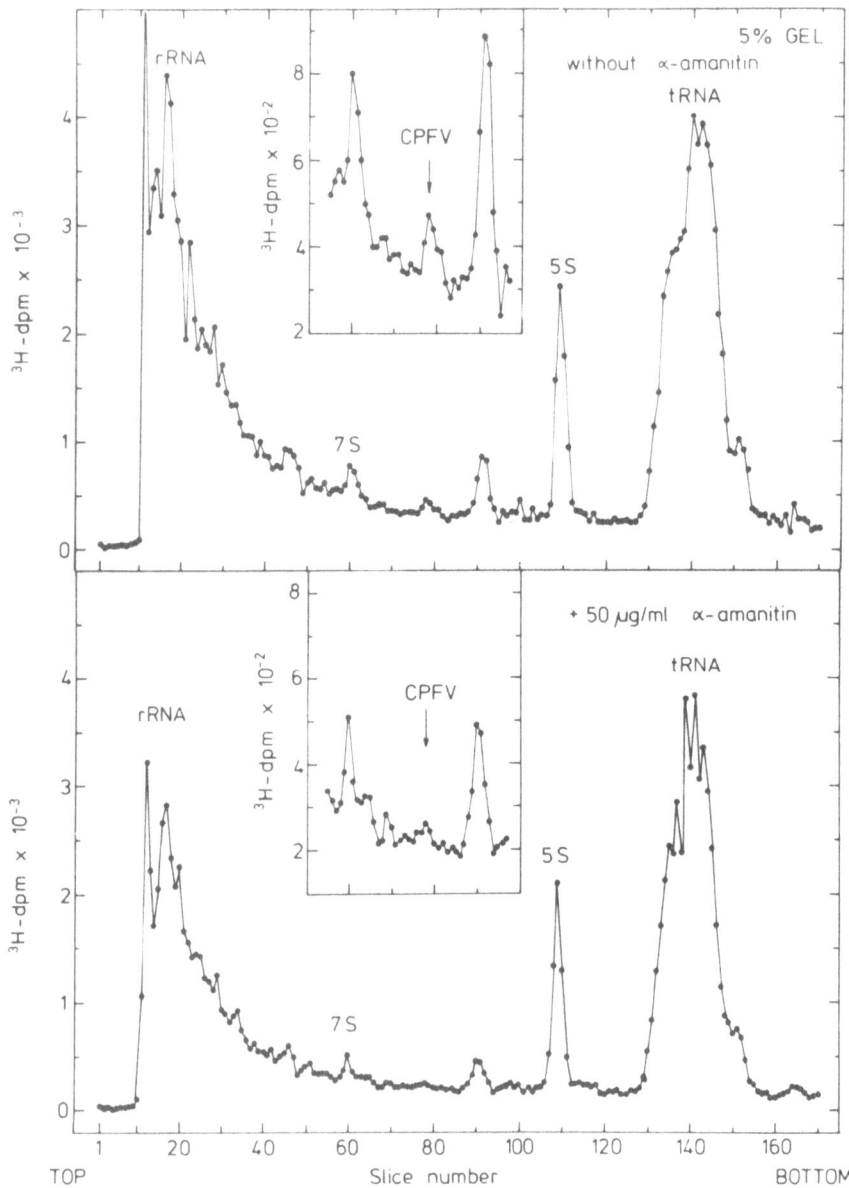

Fig. 17.   Analysis of 2M LiCl-soluble RNA from viroid (CPFV)-
            infected tomato protoplasts in 5 % gels. RNA pattern
            obtained in the absence (top) and in the presence of
            50 µg/ml α-amanitin (bottom). The inhibition of viroid
            replication by α-amanitin is about 75 %.

and Diener, 1975) viroid replication was inhibited. Additional evidence for the DNA dependence was inferred from the observed lack of any significant differences in the properties of DNA-dependent RNA polymerase fractions from healthy and from CEV-infected Gynura leaves. Therefore, it was concluded that probably no de novo synthesis of polymerase is required for CEV replication and that it is accomplished in vivo via a DNA template (Geelen et al., 1976).

There is less experimental evidence for RNA directed viroid replication, which would require the existence of viroid-complementary RNA in infected plants. With molecular hybridisation Hadidi et al. (1976) failed to detect such RNA sequences in PSTV – infected tomato leaves, whereas Grill and Semancik (1978) obtained evidence for the existence of a viroid-complementary RNA in CEV-infected Gynura. Although the published properties of the hybrid (resistance to RNase H, T_m-value of 90 $^\circ$C at low salt) indicated a RNA · RNA duplex, the presumed c-RNA itself behaved in such an unusual way that true viroid-complementarity becomes questionable.

Replication Studies in Protoplasts

Recent studies on viroid replication were carried out at the cellular level using protoplasts isolated from green leaves of the tomato cultivar "Hilda 72" (Mühlbach, Camacho-Henriquez and Sänger, 1977). The general metabolic capacity of these protoplasts is substantiated by their ability to regenerate new cell walls after 3 days, to divide within a week and to produce complete tobacco mosaic virus (TMV) particles after infection with TMV RNA in vitro (Mühlbach, Camacho-Henriquez and Sänger, 1977). Moreover, viroid replication can be followed in these protoplasts by incorporation of ^{3}H-uridine after infection in vitro (Mühlbach and Sänger, 1977). To discriminate unambiguously between the different DNA-dependent RNA polymerases possibly involved in viroid replication the mushroom toxin α-amanitin was applied. In eucaryotic cells α-amanitin is known to inhibit selectively the DNA-dependent RNA polymerases II and III at low and high concentrations, respectively, whereas RNA polymerase I is not inhibited at all (Wieland and Faulstich, 1978). It turned out that α-amanitin has no appreciable effect on the cellular and viroid RNA, if present at 10 µg per ml culture medium during the period of 48 – 72 h after inoculation, when viroid replication is detectably increased. At 50 µg per ml culture medium, however, α-amanitin inhibits viroid replication to about 75 %, whereas the biosynthesis of the prominent cellular RNA species tRNA, 5S RNA, 7S RNA and ribosomal RNA is not appreciably affected (Fig. 17). The determination of the intracellular α-amanitin concentration with the ^{3}H-labelled toxin under the experimental conditions showed that it is present in the protoplasts at 10^{-8} M, if 50 µg per ml culture medium are applied. At this concentration RNA polymerase II is known to be specifi-

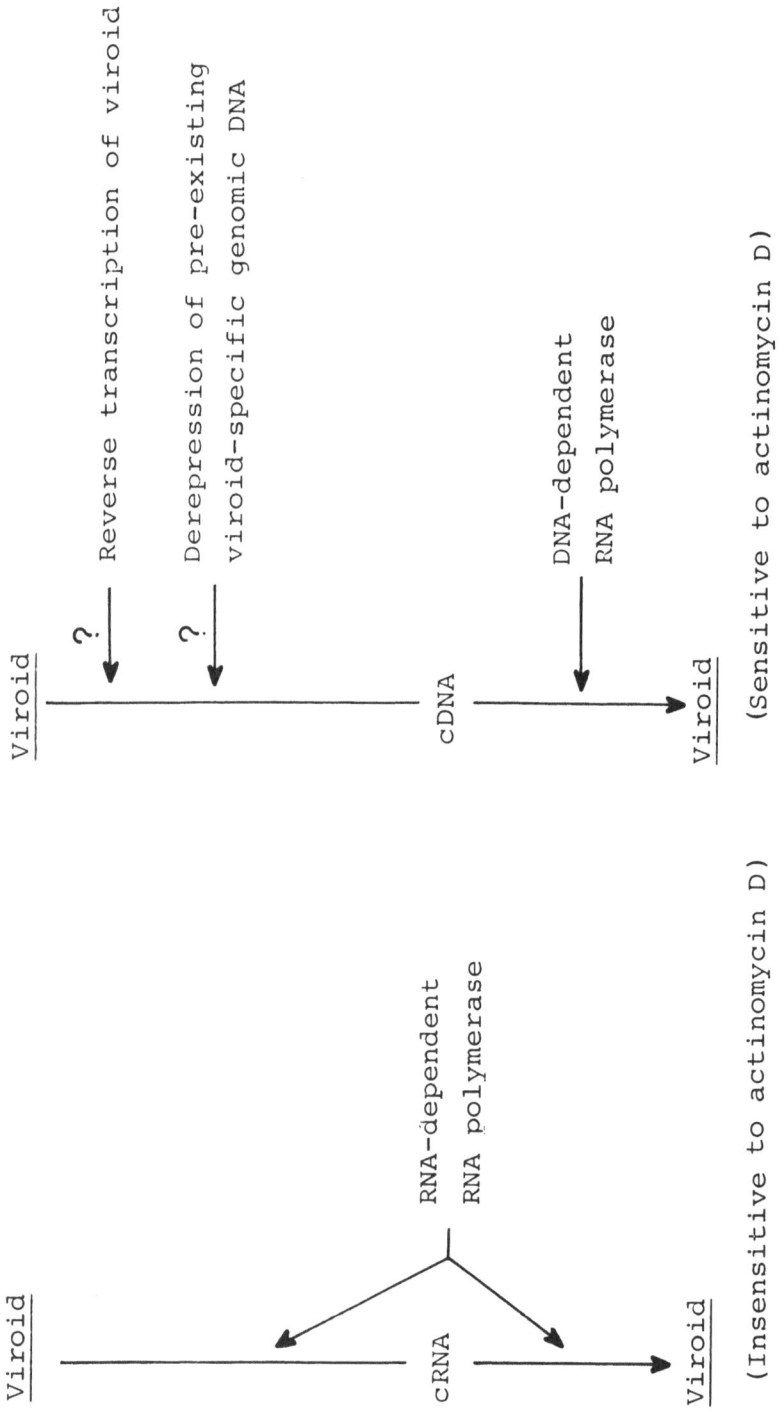

FIGURE 18. Possible Pathways of Viroid Replication

cally inhibited. Polymerase III requires a 1000-fold higher concentration and polymerase I is not affected at all by α-amanitin.

Control experiments with tomato protoplasts inoculated with tobacco mosaic virus (TMV) RNA showed that the replication and accumulation of TMV was not affected by α-amanitin. Therefore, the marked inhibition of viroid replication by α-amanitin is unlikely to be due to a secondary effect of this drug on cell metabolism in general (Mühlbach and Sänger, 1979). In protoplasts actinomycin D and even cycloheximide inhibited the biosynthesis of all RNA species including viroid RNA to about 85 % so that a non-specific secondary effect of these two inhibitors cannot be ruled out. In view of the non-specific side effects of actinomycin D the observed inhibition of viroid replication should be interpreted cautiously. It is only justified to exclude the involvement of a pre-existing RNA-dependent RNA replicase in viroid replication, because this pathway is known to be completely actinomycin D resistent. However, from the specific inhibition of viroid synthesis by 10^{-8} M α-amanitin the involvement of DNA-dependent RNA polymerase II in viroid replication can be safely deduced. This polymerase is located in the nucleoplasm and is responsible for the transcription of heterogeneous nuclear RNA species from the corresponding sections of the cellular DNA genome (Zylber and Penman, 1971; Weinman and Roeder, 1974). If viroid RNA is in fact a product of polymerase II the previous observation that the bulk of viroid infectivity is found within the nuclei and associated with the chromatin (Diener, 1971; Sänger, 1972) would find a plausible explanation.

It should be emphasised, however, that the observed α-amanitin inhibition cannot be considered as definitive proof of direct replication of viroid RNA by the nuclear DNA-dependent RNA polymerase II. There are several possible mechanisms, by which this enzyme system could be indirectly involved as a result of its central role in the formation of mRNAs in the host cell. There is also the remote possibility that viroid replication may proceed through a yet unknown pathway which is α-amanitin sensitive. If, however, DNA-dependent RNA polymerase II is directly involved it still remains to be determined whether viroid replication is based on derepresssion of a pre-existing viroid-specific DNA template or whether the infecting viroid molecule, because of its DNA-like secondary structure, is accepted directly as a template by DNA-dependent RNA polymerase II. The circularity and the unique secondary structure, which are characteristic for all viroids (Langowski, Henco, Riesner and Sänger, 1978) are undoubtedly of paramount functional importance. The possible pathways of viroid replication are shown in Fig.18 and further experiments are required before a definitive distinction between these theoretically possible modes can be made. However, from unusual structural

Fig. 19.   Healthy (left) and CEV-infected (right) plants of
            Gynura aurantiaca. Note the epinasty and malformation
            of the systemically infected leaves.

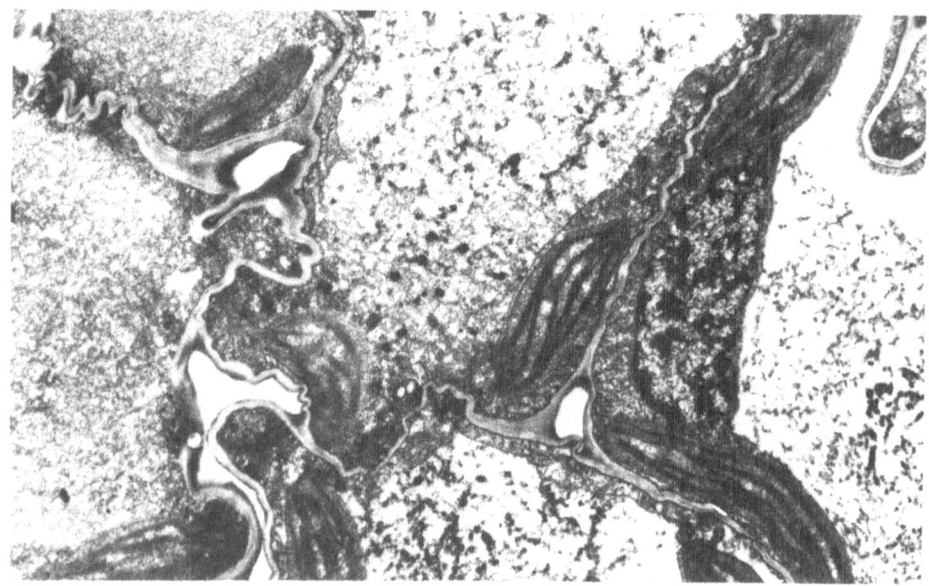

Fig. 20.   Electron micrograph of parenchyma cells from CEV-
            infected Gynura aurantiaca. Note the corrugated
            cell walls.

features an unusual mechanism of viroid replication and pathoge-
nicity is to be expected.

Cytopathic Effects of Viroid Infection

In susceptible plant species like in Gynura aurantiaca CEV-
infection produces conspicuous symptoms of mosaic, veinal necrosis, blistering and epinasty of leaves, which is generally combined with a severe stunting of the malformed whole plant (Fig.
19). Studies of the cytopathological changes in Gynura aurantiaca
leaf tissue systemically infected with CEV did not reveal any
significant changes in the fine structure of nuclei, mitochondria,
chloroplasts, ribosomes and the intraplasmatic membrane system.
The only cytopathic change observed in this viroid-host system
are malformations of the cell wall and of the cell wall associated so-called "paramural bodies" or "plasmalemmasomes". These
organelles are protrusions of the cellular membrane with variable
size, shape and internal structure and their origin and function
has not yet been clarified.

In individual healthy Gynura plants plasmalemmasomes with
either vesicular (Fig. 21) or with tubular (Fig. 22) internal
structures are found. In regions of cell wall thickening plasmalemmasomes in the process of spreading between cell wall and
plasmalemma are quite frequently observed. In CEV-infected
Gynura, however, both types of plasmalemmasomes exhibit pronounced malformations in their size, shape and internal structures (Fig. 23 and 24), and they were never found in the process of "spreading" between cell wall and plasmalemma, which is
characteristic for plasmalemmasomes in healthy tissue. Moreover,
malformed plasmalemmasomes are occuring exclusively in association with the irregular cell walls typical for CEV-infected and
symptom-bearing Gynura leaf tissue (Fig. 20). It should be emphasised that plasmalemmasomes are found at the frequency in
healthy as well as in CEV-infected Gynura (Wahn et al., 1979).
These results are at variance to the findings of Semancik and
Vanderwoude (1976) that plasmalemmasomes are only present in CEV-
infected tissue and that they are the primary cytopathic effect
of viroid infection. They were considered to be the sites, where
CEV RNA is replicated and accumulating. In contrast, plasmalemmasomes are normal constituents of many plant species including
Gynura and there are indications that they are involved in cell
wall formation. The finding of two types of plasmalemmasomes in
healthy tissue, their spreading in regions of cell wall thickening and the positive correlation between the pathological changes
of cell walls and of the internal structure of the plasmalemmasomes in viroid-infected tissue would support this contention.
One could assume that viroid infection may disturb this functional relationship in an yet unknown way, which could finally lead

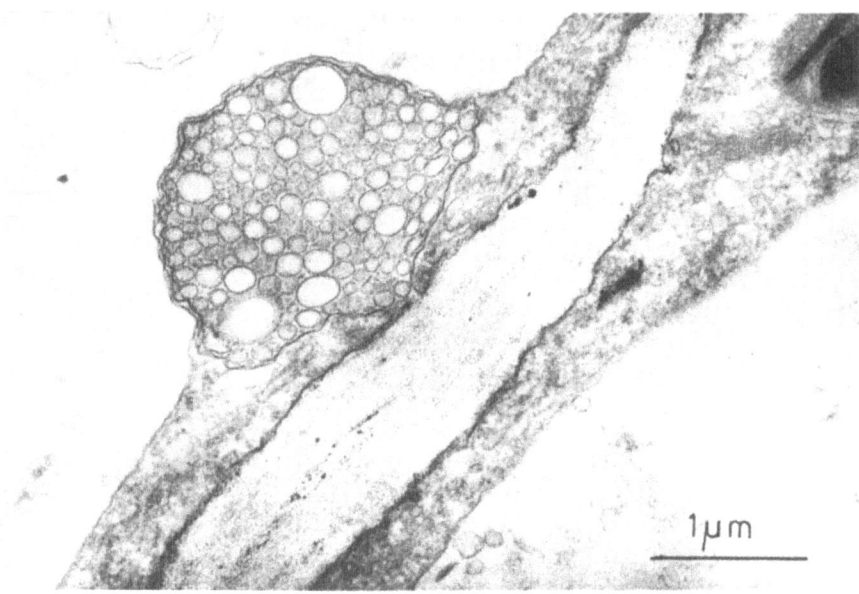

Fig. 21.   Vesicular plasmalemmasome in tissue of healthy
            Gynura aurantiaca.

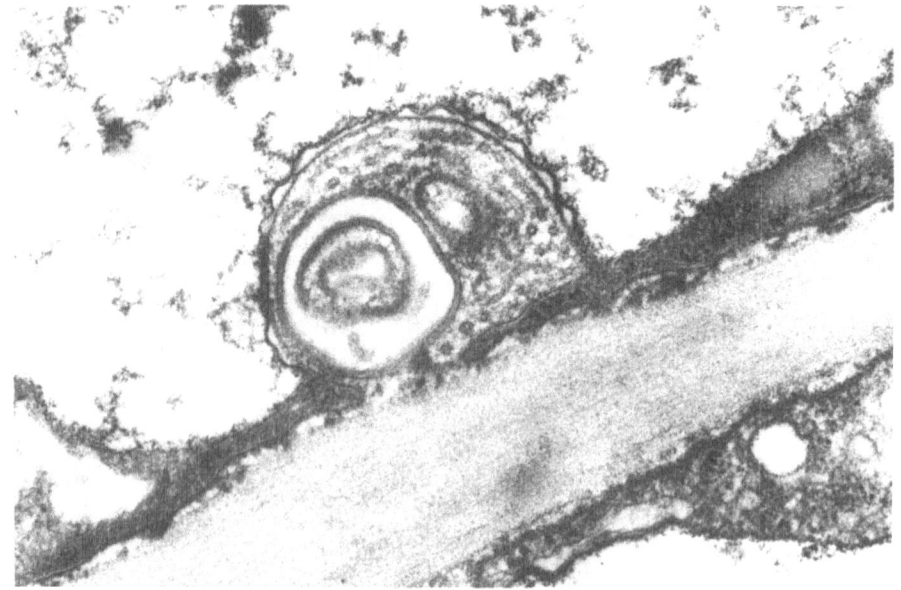

Fig. 22.   Tubular plasmalemmasome in tissue of healthy Gynura
            aurantiaca. Some of the tubules are sectioned
            longitudinally.

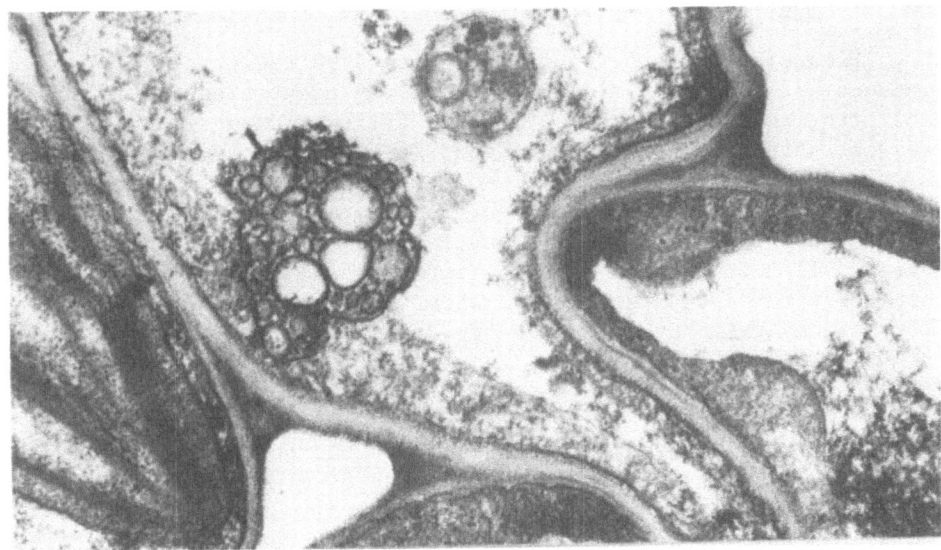

Fig. 23.   Vesicular plasmalemmasome in CEV-infected Gynura
           aurantiaca. Note its irregular internal structures
           and its association with malformed cell walls.

Fig. 24.   Tubular plasmalemmasome in CEV-infected Gynura
           aurantiaca. Note its heavily contrasted irregular
           internal structures and its association with a
           malformed cell wall.

to the expression of the macroscopic foliar symptoms.

Semancik and Vanderwoude (1976) related the presumed viroid-induced appearance of plasmalemmasomes in plant tissue etiologically to the membrane-bound vacuoles filled with vesicular and tubular structures in enlarged dendrites of scrapie-infected mamalian brain tissue, which leads to the neuronal degeneration. However, in view of the fundamental differences between plant cells and cells of the highly specialised mamalian brain tissue one should refrain from constructing any functional similarities between viroids and the scrapie agent on the basis of the pathological changes they induce on the cellular level.

VIROIDS IN ANIMAL AND MAN ?

With the discovery of plant viroids the question arose, whether viroids or viroid-like nucleic acids might be the causative agents of certain diseases of man and animal. Correspondingly, parallels have been drawn between some characteristics of viroids and the agent of the scrapie disease of sheep and goat (Diener, 1972, 1973). This disease belongs to a group of four known natural diseases of man and animal, the etiological agents of which exhibit, in fact, certain properties suggesting similarity with viroids. The other three diseases of this group are the transmissible encephalopathy of mink (Marsh and Hanson, 1969) and, in man, Kuru (Gajdusek, 1977) and Creutzfeldt-Jakob disease (Gibbs et al., 1968). These so-called "slow virus infections caused by unconventional viruses" (Gajdusek, 1977) exhibit long incubation periods and cause similar subacute spongiform encephalopathies, i. e. sponge-like vacuolation in the grey matter of the brain. These lesions result in progressive degenerative diseases of the central nervous system. All four diseases are characterised by the absence of virion-like structures,inclusion bodies or non-host proteins in infected cells and by their inability to elicit a specific immune response from their hosts.

The causative agents of these degenerative neuronal diseases exhibit rather unusual properties and their knowledge is mostly based on studies of the scrapie agent adapted to mice and hamsters. The agents are highly resistant to various physical and chemical treatment. But unlike in the case of viroids,extraction of nucleic acid with phenol destroys scrapie infectivity, which seemed to be based on the close association with membraneous structures. When a "target volume" of 1.5×10^5 was calculated from inactivation studies of the scrapie agent (Alper et al, 1966) the involvement of an infectious nucleic acid genome in the disease was discounted. It was argued that its potential coding capacity was far too low to explain the infectivity and replicability of the agent (Alper et al., 1966; Letarjet et al., 1970). Therefore, several uncon-

ventional hypotheses were put forward to explain the replication
of the scrapie agent without nucleic acid involvement.

   The discovery, isolation and characterisation of plant viroid
as infectious RNA molecules with a molecular weight of about
120 000 has certainly stimulated the study of the unconventional
agents of the slow neuronal diseases. The existence of a small infectious nucleic acid as their genome has now become a most reasonable possibility again after the small size of viroids has been un-
equivocally established. It should be remembered, however, that
in the meantime viroids have become biophysically and biochemically
well characterised entities, whereas the elusive agents of the subacute spongiform encephalopathies are still enigmatic in many ways.
Interestingly, recent studies have shown that the scrapie agent
sediments as an infectious particle smaller than any known animal
virus, that it possesses hydrophobic domains on its surface and
that the integrity of a presumed monomeric hydrophobic particle is
evidently required for infectivity (Prusiner et al., 1977; 1978 a,b).
These features may explain many of the unusual biological properties
associated with the scrapie agent, which could possibly be more
appropriately classified as a small hydrophobic virus.
   Recent inactivation studies of the agents of Kuru, Creutzfeldt-
Jakob disease and scrapie with ionising radiation have again shown
that a "nucleic acid target size" of no more than 100 000 daltons
may be calculated (Gibbs, Gajdusek and Letarjet, 1978) which is,
in fact, in the size range of viroids. Despite the suggestive
nature of the presumed similarities in early comparisons (Diener,
1972, 1973) it now appears that the small size of both, the viroids
and the postulated nucleic acid moiety of the unconventional agents
seems to be the only remaining well established fact. It has been
claimed recently that the agent responsible for scrapie infectivity
is an infectious low molecular DNA (Marsh et al., 1978; Malone et
al., 1979). However, the exact size and nature of this presumed
infectious nucleic acid and its relation to the infectious hydrophobic scrapie particle remain to be elucidated.

CONCLUDING REMARKS

   Viroids are the first plant pathogens the complete molecular
structure of which is known in detail. After the early years of
scepticism and even rejection their well established existence as
a rather unusual reality has removed a number of psychological
blocks in our thinking. Viroids have extended our concept of
disease agents into the region of small RNA molecules. Along that
way several unwritten dogmas have broken down. Infectious nucleic
acids were previously considered to only exist and survive in a
encapsidated form, the virus particle. Viroids showed, however,
that also "naked" infectious nucleic acid molecules can exist,
because they developed a unique structure, which confers the

necessary stability to the molecule. Furthermore, the minimal
size of infectious replicating nucleic acids is no longer restricted to molecules of 10^6 daltons which corresponds to a chain
length of about 3 600 nucleotides. Viroids are only one tenth of
that size and yet infectious and replicating. We are all precon-
ditioned by the common experience that proteins must be involved
whenever the replication of an infectious nucleic acid disturbs
cell functions. The failure to detect viroid-specific proteins
indicates that also this concept does possibly not apply for
viroid pathogenesis, which could be based on a regulatory interaction of the viroid molecule itself with the genome or/and
the metabolism of the host.

    Despite our detailed knowledge on viroid structure we know
little about viroid replication and pathogenesis. One can safely
assume, however, that both mechanisms must differ fundamentally
from the corresponding ones of conventional viruses, because the
potential genetic information encoded in the viroid molecule, if
expressed at all, is very limited. A number of technical problems
inherent in plant systems have to be overcome before the more
dynamic aspects of the mechanisms of viroid host plant interaction
can be solved. The study of viroid replication on the cellular
level has been greatly facilitated by suitable protoplast systems.
But one should remember that in higher plants certain pathogenic
responses are based on functions of the intact tissue or even the
entire plant.

    One of the most interesting questions centers around the
possible existence of viroids or viroid-like pathogens as causative agents of certain unconventional diseases of man and animal.
It must be emphasised that so far viroids have been found only in
higher plants and that there is no direct evidence yet that they
are also existing in other forms of life. Experience has shown,
however, that viruses, mycoplasms, bacteria and fungi are found
as pathogens in all organisms. It would be a rather unique situation, therefore, if viroids are confined to higher plants only.
Therefore, it is well justified to search for viroids or viroid-
like molecules in all those transmissible diseases, where the
involvement of conventional pathogens has been ruled out with the
classical methods of virology.

    In conclusion, viroids represent a completely new class of
molecular pathogens with unique structural features and a still
enigmatic mechanism of replication and pathogenesis. Viroids can
be considered as an optimal compromise between structural stability and functional flexibility, which combines maximal self-
protection with efficient replication. Because of their small size,
viroids are excellent objects to investigate the relationship
between molecular structure and biological function.

All these features render viroids   fascinating models for future
investigations by various scientific disciplines.

ACKNOWLEDGEMENTS:

    I wish to acknowledge the stimulation, continual collaboration
and excellent efforts of my many associates identified in the
respective publications. Their participation has made our viroid
project a most exciting and enjoyable enterprise. These studies
were supported by the Deutsche Forschungsgemeinschaft through
SFB 47 and by personal grants.

REFERENCES

Alper, T., Haig, D. A., and Clarke, M. C., 1966, The exceptionally
        small size of the scrapie agent, Biochem. Biophys. Res.
        Commun., 22:278-284.
Alper, T., Cramp, W. A., Haig, D. A., and Clarke, M. C., 1967,
        Does the agent of scrapie replicate without nucleic acid?,
        Nature, 214:764-766.
Calavan, E. C., Frolich, E. F., Carpenter, J. B., Roistacher, C. N.,
        and Christiansen, D. W., 1964, Rapid indexing for exocortis
        of citrus, Phytopathology, 54:1359-1362.
Conjero, V., and Semancik, J. S., 1977, Exocortis viroid: Altera-
        tion in the proteins of Gynura aurantiaca accompanying
        viroid infection, Virology, 77:221-232.
Davies, J. W., Kaesberg, P., and Diener, T. O., 1974, Potato
        spindle tuber viroid. XII. An investigation of viroid RNA
        as a messenger for protein synthesis, Virology, 61:281-286.
Dickson, E., Prensky, W., and Robertson, H.D., 1975, Comparative
        studies of two viroids: Analysis of potato spindle tuber
        and citrus exocortis viroids by RNA fingerprinting and
        polyacrylamide-gel electrophoresis, Virology, 68:309-316.
Dickson, E., Diener, T. O., and Robertson, H. D., 1978, Potato
        spindle tuber and citrus exocortis viroid undergo no major
        sequence changes during replication in two different hosts,
        Proc. Natl. Acad. Sci. U.S.A., 75:951-954.
Dickson, E., Robertson, H. D., Niblett, C. L., Horst, R. K., and
        Zaitlin, M., 1979, Minor differences between nucleotide
        sequences of mild and severe strains of potato spindle
        tuber viroid, Nature, 277:60-62.
Diener, T. O., 1971b, Potato spindle tuber "virus". IV. A repli-
        cating, low molecular weight RNA, Virology, 45:411-428.
Diener, T. O., 1972b, Viroids, Adv. Virus Res., 17:295-313.
Diener, T. O., 1972c, Is the scrapie agent a viroid?, Nature New
        Biology, 235:218-219.
Diener, T. O., 1973, Similarities between the scrapie agent and the
        agent of potato spindle tuber disease, Ann. Clin. Res.,
        5:268-278.

Diener, T. O., 1974, Viroids: The smallest known agents of infec-
    tious disease, Ann. Rev. Microbiol., 28:23-39.

Diener, T. O., Hadidi, A., and Owens, R. A., 1977, Methods for
    studying viroids, in: Methods in Virology, K. Maramosch
    and H. Koprowski, ed., Vol. 6:185-217, Academic Press,
    London, New York.

Diener, T. O., and Lawson, R. H., 1973, Chrysanthemum stunt: A
    viroid disease, Virology, 51:94-101.

Diener, T. O., and Smith, D. R., 1971, Potato spindle tuber viroid.
    VI. Monodisperse distribution after electrophoresis in
    20 % polyacrylamide gels, Virology, 46:498-499.

Diener, T. O., and Smith, D. R., 1975, Potato spindle tuber viroid.
    XIII. Inhibition of replication by actinomycin D, Virology,
    63:421-427.

Domdey, H., Jank, P., Sänger, H. L., and Gross, H. J., 1978,
    Studies on the primary and secondary structure of potato
    spindle tuber viroid: Products of digestion with ribonuc-
    lease A and ribonuclease T_1 and modification with bisul-
    fite, Nucleic Acids Res., 5:1221-1236.

Duda, C. T., 1976, Plant RNA Polymerases, Ann. Rev. Plant Physiol.,
    27:119-132.

Fernow, K. H., 1967, Tomato as a test plant for detecting mild
    strains of potato spindle tuber virus, Phytopathology,
    57:1347-1352.

Gajdusek,D. C., 1977, Unconventional viruses and the origin and
    disappearance of Kuru, Science, 197:943-960.

Geelen, J. L. M. C., Weathers, L. G., and Semancik, J. S, 1976,
    Properties of RNA polymerases of healthy and citrus
    exocortis viroid-infected Gynura aurantiaca DC, Virology,
    69:539-546.

Gibbs, C. J., Jr., Gajdusek, D. C., Asher, D. M., Alpers, M. P.,
    Beck, E., Daniel, P. M., and Matthews, W. B., 1968,
    Creutzfeldt-Jakob disease(spongiform encephalopathy):
    Transmission to the chimpanzee, Science, 161:388-389.

Gibbs, C. J., Jr., Gajdusek, D. C., and Letarjet, R., 1978,
    Unusual resistance to ionizing radiation of the viruses
    of Kuru, Creutzfeldt-Jakob disease, and scrapie, Proc.
    natn. Acad. Sci. U. S. A., 75:6268-6270.

Grill, L. K., and Semancik, J. S., 1978, RNA sequences complemen-
    tary to citrus exocortis viroid in nucleic acid prepar-
    ations from infected Gynura aurantiaca, Proc. Natl. Acad.
    Sci. U. S. A., 75:896-900.

Gross, H. J., Domdey, H., and Sänger, H. L., 1977, Comparative
    oligonucleotide fingerprints of three plant viroids,
    Nucleic Acids Res., 4:2021-2028.

Gross, H. J., Domdey, H., Lossow, C., Jank, P., Raba, M., Alberty,
    H., and Sänger, H. L., 1978, Nucleotide sequence and se-
    condary structure of potato spindle tuber viroid, Nature,
    273:203-208.

Hadidi, A., and Diener, T. O., 1977, De novo synthesis of potato
    spindle tuber viroid as measured by incorporation of ^{32}P,
    Virology, 78:99-107.
Hadidi, A., and Diener, T. O., 1978, In vivo synthesis of potato
    spindle tuber viroid: kinetic relationship between the
    circular and linear forms, Virology, 86:57-65.
Hadidi, A., Jones, D. M., Gillespie, D. H., Wong-Staal, F., and
    Diener, T. O., 1976, Hybridization of potato spindle tuber
    viroid to cellular DNA of normal plants, Proc. Natl. Acad.
    Sci. U. S. A., 73:2453-2457.
Hall, T. C., Wepprich, R. K., Davies, J. W., Weathers, L. G., and
    Semancik, J. S., 1974, Functional distinctions between the
    ribonucleic acids form citrus exocortis viroid and plant
    viruses: Cell-free translation and aminoacylation reactions,
    Virology, 61:486-492.
Henco, K., Riesner, D., and Sänger, H. L., 1977, Conformation of
    viroids, Nucleic Acids Res., 4:177-194.
Henco, K., Sänger, H. L., and Riesner, D., 1979, Fine structure
    melting of viroids as studied by kinetic methods, Nucleic
    Acids Res., 6:3041-3059.
Hollings, M., and Stone, O. M., 1973, Some properties of chrysan-
    themum stunt, a virus with the characteristics of an un-
    coated ribonucleic acid, Ann. Appl. Biol., 74:333-348.
Horst, R. K., 1975, Detection of a latent infectious agent that
    protects against infection by chrysanthemum chlorotic
    mottle viroid, Phytopathology, 65:1000-1003.
Horst, R. K., and Romaine, C. P., 1975, Chrysanthemum chlorotic
    mottle: A viroid disease, New York's Food and Life Sciences
    Quarterly, 8:11-14.
Hunter, G. D., Kimberlin, R. H., and Gibbons, R. A., 1968, Scrapie:
    A modified membrane hypothesis, J. Theoret. Biol.,
    79:101-108.
Ikegami, M., and Fraenkel-Conrat, H., 1978, RNA-dependent RNA
    polymerase of tobacco plants, Proc. Natl. Acad. Sci. U.S.A.,
    75:2122-2124.
Klump, H., Riesner, D., and Sänger, H. L., 1978, Calorimetric
    studies on viroids, Nucleic Acids Res., 5:1581-1587.
Langowski, J., Henco, K., Riesner, D., and Sänger, H. L., 1978,
    Common structural features of different viroids: Serial
    arrangement of double helical sections and internal loops,
    Nucleic Acids Res., 5:1589-1610.
Letarjet, R., Muel, B., Haig, D. A., Clarke, M. C., and Alper, T.,
    1970, Inactivation of the scrapie agent by near monochro-
    matic ultraviolet light, Nature, 227:1341-1343.
Malone, T. G., Marsh, R. F., Hanson, R. P., and Semancik, J. S.,
    1979, Evidence for the low molecular weight nature of
    scrapie agent, Nature, 278:575-576.

Marsh, R. F., and Hanson, R. P., 1969, Physical and chemical
    properties of the transmissible mink encephalopathy agent,
    J. Virol., 3:176-180.
Marsh, R. F., Malone, T. G., Semancik, J. S., Lancaster, W. D.,
    and Hanson, R. P., 1978, Evidence for an essential DNA com-
    ponent in the scrapie agent, Nature, 275:146-147.
Matthews, R. E. F., 1978, Are viroids negative-strand viruses?
    Nature, 276:850.
McClements, W. L., and Kaesberg, P., 1977, Size and secondary
    structure of potato spindle tuber viroid, Virology,
    76:477-484.
Morris, T. J., 1979, Evidence for a single infectious species of
    potato spindle tuber viroid, Intervirology, 11:89-96.
Mühlbach, H.-P., and Sänger, H. L., 1977, Multiplication of
    cucumber pale fruit viroid in inoculated tomato leaf proto-
    plasts, J. Gen. Virol., 35:377-386.
Mühlbach, H.-P., and Sänger, H. L., 1979, Viroid replication is
    inhibited by α-amanitin, Nature, 278:185-188.
Mühlbach, H.-P., Camacho-Henriquez, A., and Sänger, H. L., 1977,
    Isolation and properties of protoplasts from leaves of
    healthy and viroid-infected tomato plants, Plant Science
    Letters, 8:183-189.
Niblett, C. L., Dickson, E., Fernow, K. H., Horst, R. K., and
    Zaitlin, M., 1978, Cross protection among four viroids,
    Virology, 91:198-203.
O'Brien, M. J., and Raymer, W. B., 1964, Symptomless hosts of the
    potato spindle tuber virus, Phytopathology, 54:1045-1047.
Owens, R. A., Erbe, E., Hadidi, A., Steere, R. L., and Diener, T.O.
    1977, Separation and infectivity of circular and linear
    forms of potato spindle tuber viroid, Proc. Natl. Acad.
    Sci. U. S. A., 74:3859-3863.
Owens, R. A., Smith, D. R., and Diener, T. O., 1978, Measurement of
    viroid sequence homology by hybridization with complementary
    DNA prepared in vitro, Virology, 89:388-394.
Prusiner, S. B., Hadlow, W. J., Eklund, C. M., and Race, R. E.,
    1977, Sedimentation properties of the scrapie agent, Proc.
    Natl. Acad. Sci. U. S. A., 74:4656-4660.
Prusiner, S. B., Hadlow, W. J., Eklund, C. M., Race, R. E., and
    Cochran, P. S., 1978, Sedimentation characteristics of the
    scrapie agent from murine spleen and brain, Biochemistry,
    17:4987-4992.
Prusiner, S. B., Hadlow, W. J., Garfin, D. E., Cochran, P. S.,
    Baringer, J. R., Race, R. E., and Eklund, C. M., 1978,
    Partial purification and evidence for multiple molecular
    forms of the scrapie agent, Biochemistry, 17:4993-4999.
Randles, J. W., 1975, Association of two ribonucleic acid species
    with cadang-cadang disease of coconut palm, Phytopathology,
    65:163-167.

Randles, J. W., Rillo, E. P., and Diener, T. O., 1976, The viroid-
     like structure and cellular location of anomalous RNA
     associated with the cadang-cadang disease, Virology,
     74:128-139.
Riesner, D., Henco, K., Rokohl, U., Klotz, G., Kleinschmidt, A. K.,
     Gross, H. J., Domdey, H., and Sänger, H. L., 1979, Structure
     and structure formation of viroids, J. Mol. Biol., 133:
     (in the press).
Romaine, C. P., and Horst, R. K., 1975, Suggested viroid etiology
     for chrysanthemum chlorotic mottle disease, Virology,
     64:86-95.
Romaine, C. P., and Zaitlin, M., 1978, RNA-dependent RNA poly-
     merases in uninfected and tobacco mosaic virus-infected
     tobacco leaves: Viral-induced stimulation of a host poly-
     merase activity, Virology, 86:241-253.
Sänger, H. L., 1972, An infectious and replicating RNA of low
     molecular weight: The agent of the exocortis disease of
     citrus, Adv. Bioscience, 8:103-116.
Sänger, H. L., and Ramm, K., 1975, Radioactive labelling of
     viroid-RNA, in: Modification of the Information Content
     of Plant Cells, R. Markham, D. R. Davies, D. A. Hopwood,
     and R. W. Horne, eds., North-Holland/American Elsevier
     Publ. Co., Amsterdam: 229-252.
Sänger, H. L., Klotz, G., Riesner, D., Gross, H. J., and Klein-
     schmidt, A. K., 1976, Viroids are single-stranded covalent-
     ly closed circular RNA molecules existing as highly base-
     paired rod-like structures, Proc. Natl. Acad. Sci. U.S.A.,
     73: 3852-3856.
Sänger, H. L., Ramm, K., Domdey, H., Gross, H. J., Henco, K., and
     Riesner, D., 1979, Conversion of circular viroid molecules
     to linear strands, FEBS Letters, 99:117-122.
Sasaki, M., and Shikata, E., 1977a, Studies on the host range of
     hop stunt disease in Japan, Proc. Japan Acad., 53B:103-108.
Sasaki, M., and Shikata, E., 1977b, On some properties of hop
     stunt disease agent, a viroid, Proc. Japan Acad., 53B:
     109-112.
Semancik, J. S., and Geelen, J. L. M. C., 1975, Detection of DNA
     complementary to pathogenic viroid RNA in exocortis
     disease, Nature, 256:753-756.
Semancik, J. S., and Vanderwoude, W. J., 1976, Exocortis viroid:
     Cythopathic effects at the plasma membrane in association
     with pathogenic RNA, Virology, 69: 719-726.
Semancik, J. S., and Weathers, L. G., 1972b, Exocortis disease:
     Evidence for a new species of "infectious" low molecular
     weight RNA in plants, Nature New Biology, 237:242-244.
Semancik, J. S., and Weathers, L. G., 1972c, Pathogenic 10 S RNA
     from exocortis disease recovered from tomato bunchy-top
     plants similar to potato spindle tuber virus infection,
     Virology, 49:622-625.

Semancik,J. S., Magnuson, D. S., and Weathers, L. G., 1973a,
    Potato spindle tuber disease produced by pathogenic RNA
    from citrus exocortis disease: Evidence for the identity
    of the causal agent, Virology, 52:292-294.

Semancik, J. S., Conjero, V., and Gerhart, J., 1977, Citrus
    exocortis viroid: Survey of protein synthesis in Xenopus
    laevis oocytes following addition of viroid RNA, Virology,
    80:218-221.

Singh, R. P., 1973, Experimental host range of the potato spindle
    tuber "virus", Am.Potato J., 50:111-123.

Singh, R. P., and Clark, M. C., 1971b, Infectious low-molecular-
    weight ribonucleic acid, Biochem. Biophys. Res. Commun.,
    44:1077-1082.

Singh, R. P., and Clark, M. C., 1973b, Similarity of host response
    to both potato spindle tuber and citrus exocortis viruses,
    FAO Plant Protect. Bull., 21:121-125.

Singh, R. P., Michniewicz, J. J., and Narang, S. A., 1974, Multip-
    le forms of potato spindle-tuber metavirus ribonucleic acid,
    Can. J. Biochem., 52:809-812.

Sogo, J. M., Koller, T., and Diener, T. O., 1973, Potato spindle
    tuber viroid. X. Visualization and size determination by
    electron microscopy, Virology, 55:70-80.

Takahashi, T., and Diener, T. O., 1975, Potato spindle tuber
    viroid. XIV. Replication in nuclei isolated from infected
    leaves, Virology, 64:106-114.

Thomas, W., and Mohamed, N. A., 1979, Avocado sunblotch - A viroid
    disease?, Austr. Plant Path. Soc. Newsletter: 1-2.

Tinoco, I., Jr., Uhlenbeck, O. C., and Levine, M. D., 1971,
    Estimation of secondary structure in ribonucleic acids,
    Nature, 230:362-367.

Van Dorst, H. J. M., and Peters, D., 1974, Some biological obser-
    vations on pale fruit, a viroid-incited disease of cucumber,
    Neth. J. Pl. Path., 80:85-96.

Wahn, K., Rosenberg-de Gomez, F., and Sänger, H. L., 1979, Cyto-
    pathologie von viroid-infiziertem Pflanzengewebe. I. Ver-
    änderungen des Plasmalemmas und der Zellwand bei Gynura
    aurantiaca DC nach Infektion mit dem Viroid der Citrus
    Exocortis Krankheit (CEV), Phytopath. Z.,(in the press).

Weinman, R., and Roeder, R. G., 1974, Role of DNA-dependent RNA
    polymerase III in the transcription of the tRNA and 5S RNA
    genes, Proc. Natl. Acad. Sci. U. S. A., 71:1790-1794.

Wintermeyer, W., and Zachau, H. G., 1973, Mg^{2+}-katalysierte,
    spezifische Spaltung von tRNA, Biochim. Biophys. Acta,
    299:82-90.

Weathers, L. G., and Calavan, E. C., 1961, Additional indicator
    plants for exocortis and evidence for strain differences
    in the virus, Phytopathology, 51: 262-264.

Wild, U. Ramm, K., Sänger, H. L., and Riesner, D., 1979,
       Loops in viroids: Accessibility to tRNA anticodon
       binding, Eur. J. Biochem., (in the press).
Zaitlin, M., and Hariharasubramanian, V., 1972, A gel electro-
       phoretic analysis of proteins from plants infected with
       tobacco mosaic and potato spindle tuber viruses, Virology,
       47:296-305.
Zylber, E. A., and Penman, S., 1971, Products of RNA polymerases
       in HeLa cell nuclei, Proc. Natl. Acad. Sci. U. S. A.,
       68:2861-2865.